D1577311

PRINCIPLES OF CERAMICS PROCESSING

PRINCIPLES OF CERAMICS PROCESSING

Second Edition

JAMES S. REED
New York State College of Ceramics
Alfred University
Alfred, New York

A Wiley-Interscience Publication
JOHN WILEY & SONS, INC.
New York / Chichester / Brisbane / Toronto / Singapore

Library of Congress Cataloging in Publication Data:
Reed, James Stalford, 1938–
    Principles of ceramics processing / James S. Reed.—2nd ed.
        p.   cm.
    Rev. ed. of: Introduction to the principles of ceramic processing.
1988.
    "A Wiley-Interscience publication."
    Includes bibliographical references and index.
    ISBN 0-471-59721-X
    1. Ceramics. I. Title. II. Series: Reed, James Stalford, 1938–
Introduction to the principles of ceramic processing.
TP807.R36 1994
666—dc20   94-20838

17

Dedicated to Carol and Scott

CONTENTS

PREFACE TO THE SECOND EDITION

Ceramics has been identified as one of the prime fields where processing improvements and more advanced products can be anticipated. Such products will have an increased "technological knowledge content" and necessarily will have to be manufactured using processing technology that is both more advanced and better controlled. Advancements in ceramic processing technology will also be required for the commercialization of new products and to maintain the competitiveness of "standard products." Improvements in processing technology may also enable us to eliminate costly steps and improve productivity and product reliability.

The need for a periodically updated text which discusses the principles of ceramics processing is clear. In the first edition, each chapter presented scientific principles along with a discussion of processing technology and unit operations. The usefulness of this approach has been communicated to me by those working in the field and students alike. In preparing the second edition, attention was focused on updating the discussion of each topic without increasing the theoretical level and making the text more "user friendly."

Specifically, the second edition has been revised in the following ways: Parts V, VIII, and IX have been reorganized and expanded. New chapters discuss system consistency and batching, general ceramic forming principles, and machining and finishing processes. Chapters discussing binders, powder mechanics, and rheology, and all chapters discussing forming processes have been completely rewritten and expanded. Significant new material has been added to the chapters on raw materials, liquids and surfactants, vapor deposition, printing, and coating processes, and firing.

The second edition contains several new features. Processing flow diagrams are presented. Many new tables have been added to summarize important points.

More than 100 new figures are included. Descriptions of defects and their causes are either itemized in the text or summarized in a table. Each chapter contains a revised list of Suggested Reading, revised Problems, and a new feature, Examples. It is my firm belief that the questions posed and solutions presented in the Examples will serve as a tutorial for quick understanding and will also increase the quantitative appreciation of the subject matter.

In the preparation of the second edition, I have been most appreciative of the constant support and editing assistance of my wife, Carol Reed, and the many helpful comments and questions of students of ceramics processing working both in the field and in the classroom.

JAMES S. REED

Alfred, New York
December 1994

PREFACE TO THE FIRST EDITION

Ceramic processing has traditionally been discussed in terms of the material formulations and industrial arts used in the production of classes of products. This empirical approach was understandably necessary, because the material systems were complex and not well characterized. But as the characterization of materials has improved and the scientific principles underlying ceramic processing have been elucidated, especially in the past 20 years, ceramic processing can now be explained in the context of general principles and the engineering processes involved. Ceramic processing is an increasingly important aspect of ceramic engineering, and advances in understanding the principles of ceramic processing will be needed to produce more advanced ceramic products and components of integrated products. Information on the principles of ceramic processing is widely scattered in research papers and in conference proceedings. The need for a book that provides an overview and introduces the principles of ceramic processing has been impressed on me in my teaching experience and in conversations with industrial scientists and engineers. This book has been written to be used as a reference book and as a textbook.

Part I of this book describes applications and compositions of modern ceramics, presents a perspective of science in ceramic processing, and reviews fundamentals of surface chemistry. Part II describes starting materials for ceramics and emphasizes the chemical preparation of inorganic chemicals and more advanced materials. The characterization and specification of particulate materials and the structure and functions of processing additives are explained in Parts III and IV, respectively. Part V presents the fundamentals of particle packing and discusses the rheology of processing systems; this very complex topic is approached by considering the contributions of the component phases and insights of particle mechanics from the field of soil science. Principles involved in enhancing the physical and chemical character of the process system, called beneficiation processes, are explained along with industrial practices in Part VI. These four chapters include crushing and milling processes; batching and mixing; particle separation, concentration, and washing processes; and spray granulation and spray drying. Part VII presents the principles of forming in the categories of pressing, plastic forming, which includes extrusion, plastic molding, jiggering, and injection molding, casting processes, which include slip casting, pressure casting, solid casting, and tape casting, and molecular polymerization forming, which includes sol-gel processing and chemical vapor deposition. Causes and the prevention of defects in formed products are explained. Part VIII discusses drying fundamentals; surface processing which includes surface finishing, film printing, and glazing; and microstructural changes and development during sintering.

I want to express my appreciation to those who have helped me in many ways. My greatest thanks go to my wife, Carol Reed, for her constant help and encouragement. Many people have kindled and supported my interest in ceramic engineering. In particular I would like to acknowledge the late Professor George W. Brindley, who first introduced me to the field of ceramics; Dean John F. McMahon, who encouraged my interest in graduate study and ceramic processing; and Professor James E. Young, who supported my interest in developing courses and teaching ceramic processing. Last but not least, I would like to acknowledge the numerous students whose questions and research theses over the past 20 years have helped to clarify insights and provide new understanding.

In the preparation of the book I am indebted to Mr. Frank Cerra for his early encouragement, Mrs. Sandi Congelli for typing the manuscript, Ms. Hollis Findeisen for drafting the majority of the line drawings, and Mr. Ward Votava for assistance with the photomicrographs.

JAMES S. REED

Alfred, New York
December 1986

PRINCIPLES OF CERAMICS PROCESSING

PART I

INTRODUCTION

Ceramics processing is an ancient art but a young applied science. The history of ceramics processing indicates periods of rapid development interspersed within long periods of rather slow development; the second half of the twentieth century is a period of intense development. The transition of ceramics processing to an applied science is the natural result of an increasing ability to refine, develop, and characterize ceramic materials and systems of additives which aid in processing systems containing hard, brittle particles, improved equipment for processing these materials into products, and advances in understanding ceramic processing fundamentals. These topics are discussed in Chapter 1. Principles of physical chemistry that are a basis of ceramic processing science are reviewed in Chapter 2.

CHAPTER 1

CERAMICS PROCESSING AND CERAMIC PRODUCTS

This book is concerned primarily with understanding the scientific principles and technology involved in processing particulate ceramic materials into fabricated products. Our topic is commonly referred to as ceramics fabrication processes, ceramics processing technology, or simply ceramics processing. Ceramics processing technology is used to produce commercial products that are very diverse in size, shape, detail, complexity, material composition, structure, and cost. Several examples of both modern advanced ceramics and traditional ceramics are shown in Fig. 1.1. The applications of these products, as are indicated in Table 1.1, are also diverse and are dependent on the functions indicated in Table 1.2.

The functions of ceramic products are very dependent on their chemical composition and their atomic and microscale structure, which determines their properties. Compositions of ceramic products vary widely, and both oxide and nonoxide materials are used. Today the composition and structure of grains and grain boundary phases and the distribution and structure of pores is more carefully controlled to achieve greater product performance and reliability (Fig. 1.2). In the development and production of the more advanced ceramics, extraordinary control of the materials and processing operations is requisite to minimize microstructural defects. Recent successes in developing, producing, and applying advanced ceramics in high-tech applications have affected the consciousness of all engineers and heightened the interest in ceramics processing. An awareness that improvements in ceramic processing technology can improve manufacturing productivity and expand markets now pervades the industry.

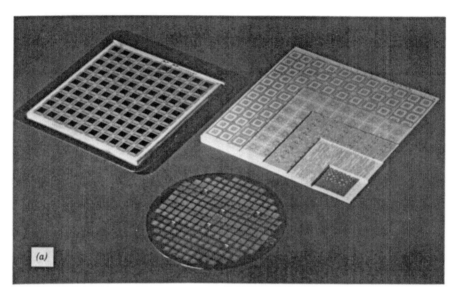

Fig. 1.1(a) Multilayer electronic packaging. The substrate used in the module is formed from sheets of unfired green ceramic. First, thousands of minute holes, or vias, are punched in each sheet. The wiring pattern that conducts the electrical signals is then formed by screening a metallic paste onto the sheet through a metal mask. The via holes are also filled with this paste to provide the electrical connections from one layer or sheet to another. The layers are then stacked and laminated together under heat and pressure, and the laminate is sintered in a process that shrinks it approximately 17%. (Photos courtesy of IBM, East Fishkill, NY.)

Fig. 1.1(b) Magnetic ceramic ferrites used in a wide variety of electrical power and electronic communications systems. (Photo courtesy of Magnetics Div., Sprang and Co., Butler, PA.)

Fig. 1.1(*c*) Refractory "honeycomb" cordierite catalyst support used in automobile exhaust system. (Photo courtesy of Corning Inc., Corning, NY.)

Fig. 1.1(*d*) Advanced alumina structural ceramics used in a wide variety of materials-processing technology applications (textile guides, valve seals, impellers, nozzle inserts, wear blocks, cutting tools, milling media, refractory insulation supports) and electrical applications (spark plug insulators, electronic substrates, electrical insulation). (Photo courtesy of Diamonite Products Div., Ferro Corporation, Shreve, OH.)

Fig. 1.1(e) Corrosion/erosion-resistant silicon carbide structural ceramic components used in the chemical processing industry. (Photo courtesy of Carborundum Co., Niagara Falls, NY.)

Fig. 1.1(f) Silicon nitride metal cutting tools (Photo courtesy of GTE Laboratories, Waltham, MA.)

Fig. 1.1(*g* **)** Traditional silicate ceramic products include electrical porcelain, household and institutional porcelain products, refractories, and ceramic wall and floor tile. (Photo courtesy of R.T. Vanderbilt Co., Norwalk, CT.)

TABLE 1.1 Products Produced by Ceramic Powder Processing

Electronics
    Substrates, chip carriers, electronic packaging
    Capacitors, inductors, resistors, electrical insulation
    Transducers, servisors, electrodes, igniters
    Motor magnets, spark plug insulators
Advanced structural materials
    Cutting tools, wear-resistant inserts
    Engine components
    Resistant coatings
    Dental and orthopedic prostheses
    High-efficiency lamps
Chemical processing components
    Ion exchange media
    Emission control components
    Catalyst supports
    Liquid and gas filters
Refractory structures
    Refractory lining in furnaces, thermal insulations, kiln furniture
    Recuperators, regenerators
    Crucibles, metal-processing materials, filters, molds
    Heating elements
Construction materials
    Tile, structural clay products
    Cement, concrete
Institutional and domestic products
    Cookware
    Hotel china and dinnerware
    Bathroom fixtures
    Decorative fixtures and household items

TABLE 1.2 Classification of Ceramics by Function

Function	Class	Nominal Composition[a]
Electrical	Insulation	α-Al_2O_3, MgO, porcelain
	Ferroelectrics	$BaTiO_3$, $SrTiO_3$
	Piezoelectric	$PbZr_{0.5}Ti_{0.5}O_3$
	Fast ion conduction	β-Al_2O_3, doped ZrO_2
	Superconductors	$Ba_2YCu_3O_{7-x}$
Magnetic	Soft ferrite	$Mn_{0.4}Zn_{0.6}Fe_2O_4$
	Hard ferrite	$BaFe_{12}O_{19}$, $SrFe_{12}O_{19}$
Nuclear	Fuel	UO_2, UO_2-PuO_2
	Cladding/shielding	SiC, B_4C
Optical	Transparent envelope	α-Al_2O_3, $MgAl_2O_4$
	Light memory	Doped $PbZr_{0.5}Ti_{0.5}O_3$
	Colors	Doped $ZrSiO_4$, doped ZrO_2, doped Al_2O_3
Mechanical	Structural refractory	α-Al_2O_3, MgO, SiC, Si_3N_4, $Al_6Si_2O_{13}$
	Wear resistance	α-Al_2O_3, ZrO_2, SiC, Si_3N_4, toughened Al_2O_3
	Cutting	α-Al_2O_3, ZrO_2, TiC, Si_3N_4, SIALON
	Abrasive	α-Al_2O_3, SiC, toughened Al_2O_3, SIALON
	Construction	Al_2O_3-SiO_2, CaO-Al_2O_3-SiO_2, porcelain
Thermal	Insulation	α-Al_2O_3, ZrO_2, $Al_6Si_2O_{13}$, SiO_2
	Radiator	ZrO_2, TiO_2
Chemical	Gas sensor	ZnO, ZrO_2, SnO_2, Fe_2O_3
	Catalyst carrier	$Mg_2Al_4Si_5O_{18}$, Al_2O_3
	Electrodes	TiO_2, TiB_2, SnO_2, ZnO
	Filters	SiO_2, α-Al_2O_3
	Coatings	NaO-CaO-Al_2O_3-SiO_2
Biological	Structural prostheses	α-Al_2O_3, porcelain
	Cements	$CaHPO_4 \cdot 2H_2O$
Aesthetic	Pottery, artware	Whiteware, porcelain
	Tile, concrete	Whiteware, CaO-SiO_2-H_2O

Source: Adapted from George B. Kenne and H. Kent Bowen, "High-Tech Ceramics in Japan: Current and Future Markets," *Am. Ceram. Soc. Bull.* **62**(5), 590–596 (1982).

"Whiteware is a family of porous-dense, fine-grained materials with a glassy matrix usually containing Al_2O_3, SiO_2, K_2O, and Na_2O. Porcelain is a type of whiteware that is nonporous, hard, and translucent. SIALON is a solid solution phase with the nominal composition $Si_4Al_2N_6O_2$. Toughened Al_2O_3 is a two-phase material containing a minor amount of doped or undoped ZrO_2.

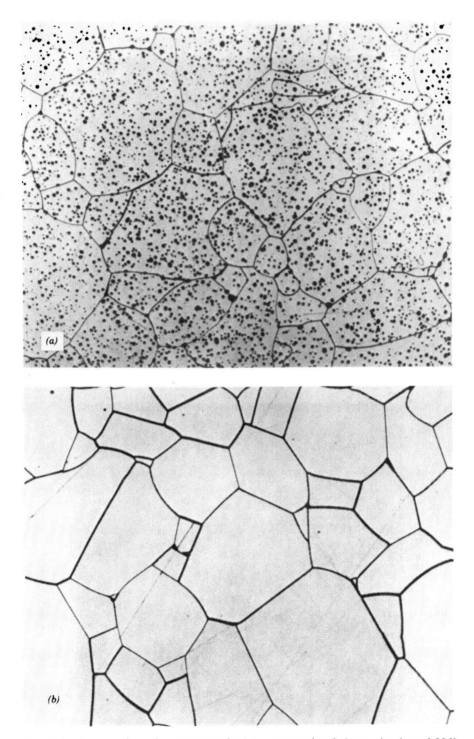

Fig. 1.2 Comparative microstructures in (*a*) a conventional dense alumina of 98% density, (*b*) an optically transparent alumina exceeding 99.9% density. (Photos *a* and *b* courtesy of General Electric Co., Richmond Heights, OH.)

9

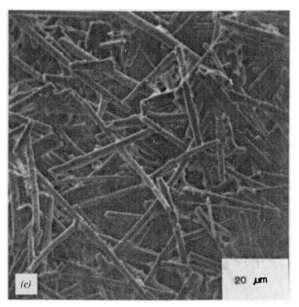

Fig. 1.2(c) Sintered, alumina fiber insulation of greater than 90% porosity. (Photo *c* courtesy of Zicar Products Inc., Florida, NY.)

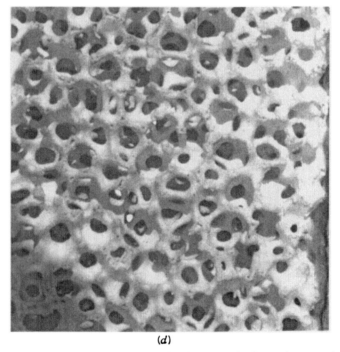

Fig. 1.2(d) Sintered, porous reticulated alumina ceramic for molten metal filtration. (Photo *d* courtesy of High Tech Ceramics Inc., Alfred, NY.)

1.1   A BRIEF HISTORY OF CERAMIC TECHNOLOGY

The history of ceramic processing technology is very interesting in that both simple processes developed in ancient times for natural materials, and recently developed, relatively sophisticated processes dependent on synthetic materials are used extensively near the end of the 20th century.

Hand mixing, hand building, and scratch and slip decorating of earthenware date back to before 5000 BC. The first forming machine was probably the potter's wheel, which was used earlier than 3500 BC for throwing a plastic earthenware body and later for turning a somewhat dried, leather, hard body. Shaping by pressing material in fired molds and firing in a closed kiln were subsequent developments.

The most notable achievement early in the Christian era was the development in China of pure white porcelain of high translucency. Duplication in the West was frustrated until 1708, when a young German alchemist, Fredrich Bottger, under the direction of the celebrated physicist Count von Tschirnhaus, discovered that fine porcelain could be produced on firing a body containing a fire-resistant clay with fusible materials. Other inventions in the eighteenth century included the use of a template for forming, slip casting in porous molds, auger extrusion, transfer decoration, and firing in a tunnel kiln.

The introduction of steam power in the nineteenth century led to the mechanization of mixing, filter pressing, dry pressing, and pebble mill grinding. Near the end of that century, separate phases of silica were distinguished using optical microscopy, and silicon carbide was synthesized in an electric furnace. Pyrometric cones were developed by Seager to control firing.

The first half of the twentieth century saw the rapid development of x-ray techniques for the analysis of the atomic structure of crystals and later electron microscopy for examining microstructure beyond the limit of the optical microscope. Material systems became more refined, and special compounds were developed, synthesized, and fabricated into products for refractory and electronic applications. Refined organic additives were purposefully introduced to improve the processing behavior. Industrial production became mechanized, and several stages of manufacturing were automated. Thermocouples were used routinely to monitor temperatures during firing.

The second half of the twentieth century has witnessed major advances in the synthesis, characterization, and fabrication of ceramic products. Scanning electron microscopy is now used for routine microstructural analysis for quality control in manufacturing. Several different instrumented techniques have been developed for bulk chemical analysis at a concentration of less than a fraction of one part in a million and surface concentrations a few atomic layers in thickness. The particle size distribution of a material can be determined to below 0.1 μm in a few minutes. The flow behavior during forming is developed and controlled using a multicomponent system of additives. Testing apparatus and processing machinery are much more advanced. Computers are now used throughout the industry to monitor and/or control raw-material handling and preparation, fabrication, and firing.

1.2 INDUSTRIAL CERAMICS PROCESSING

The realization of a product depends on material factors and nonmaterial factors such as the economics of the marketplace, consumer response, dimensional and surface finish tolerances, apparent quality, and manufacturing productivity. The manufacture of ceramics is a complex interaction of raw materials, tech-

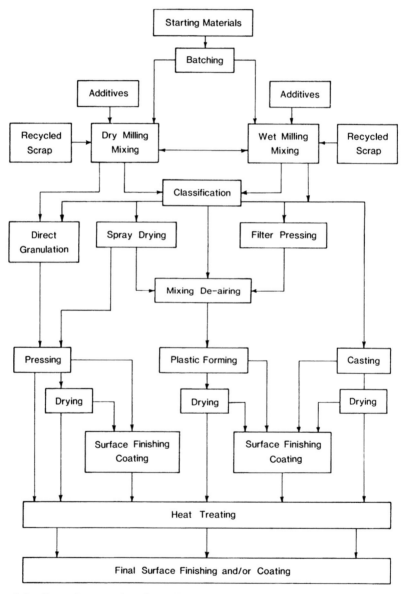

Fig. 1.3 General processing flow diagram illustrating different possible processing paths from starting materials to final product.

nological processes, people, and financial investment. Manufacturing managers are involved with all of these aspects. Production engineers, process design engineers, and scientists and engineers involved with ceramic materials and process research and development must be particularly knowledgeable of the applied science and industrial arts of ceramics processing.

Ceramics processing commonly begins with one or more ceramic materials, one or more liquids, and one or more special additives called processing aids. The starting materials or the batched system may be beneficiated chemically and physically using operations such as crushing, milling, washing, chemical dissolving, settling, flotation, magnetic separation, dispersion, mixing, classification, de-airing, filtration, and spray-drying. The forming technique used will depend on the consistency of the system (i.e., slurry, paste, plastic body, or a granular material) and will produce a particular unfired shape with a particular composition and microstructure. Drying removes some or all of the residual processing liquids. Additional operations may include green machining, surface grinding, surface smoothing and cleaning, and the application of surface coatings such as electronic materials or glaze. The finished material is then commonly heat-treated to produce a sintered microstructure. The sintered product may be a single component or a multicomponent composite structure. A general processing flow diagram indicating the sequences of operations used in forming a product is shown in Fig. 1.3.

1.3   SCIENCE IN CERAMICS PROCESSING

Up to about the first half of the twentieth century, processing engineers depended on empirical correlations and practical intuition for innovating new process designs and process controls. Most new products were seen as inventions rather than the planned outcome of research and development. Material systems were typically complex, and laboratory tests and analyses were tedious and time consuming. Wide margins existed for improving the performance and reliability of the products. Typically, a small amount of a new shipment of a raw material was processed through the plant to ascertain if it was usable or to suggest processing adjustments to minimize production losses.

Before the development of scientific insights of ceramics processing, the properties of the product were often correlated with changes in a processing operation to identify the more important superficial variables (Fig. 1.4). This empirical approach is still used and can be aided greatly using computers and statistical programs. However, empirical correlations such as these do not provide a scientific understanding of the fundamental causes of behavior during processing and forming. The probability that adjustments based on empirical correlations alone will produce significant advances is small, because the potential number of unsuccessful combinations of variables is always relatively large. Also, empirical correlations may be of little heuristic value when the processing engineer is faced with a lack of reproducibility in manufacturing,

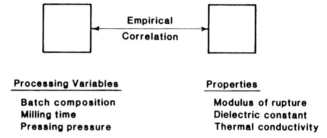

Processing Variables   **Properties**

Batch composition   Modulus of rupture
Milling time   Dielectric constant
Pressing pressure   Thermal conductivity

Fig. 1.4 Correlation of the final properties with processing variables may identify the more sensitive processing parameters but is empirical.

an insufficient reliability in the performance of nominally identical products, or when developing new products.

Viewed as a science, ceramics processing is the sequence of operations that purposefully and systematically changes the chemical and physical aspects of structure, which we call the characteristics of the system. The properties at each stage are a function of the characteristics of the system at each stage and the ambient pressure and temperature, as is shown in Fig. 1.5.

The objectives of the science of ceramics processing are to identify the important characteristics of the system and understand the effects of processing variables on the evolution of these characteristics. The objectives in process engineering should be to change these characteristics purposefully to improve product quality. Because of the key dependence on controlling characteristics, an understanding of techniques for characterizing the starting materials and the process system at each stage is an integral part of any discussion of ceramics processing.

Raw materials are now more beneficiated, more consistent, and often much simpler in composition. Modern instrumentation for analyzing ceramics materials and systems is more automated and precise, and with microcomputer accessories quantitative data are obtained quickly and presented in a convenient format. Computer-controlled, closed, raw material handling systems improve the precision, efficiency, health, and safety in batching and mixing particulate materials. The processing pressure and temperature are more precisely monitored and controlled. Wear-resistant materials are used in critical areas to min-

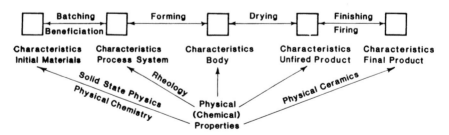

Fig. 1.5 Processing develops the characteristics of the system.

imize maintenance and contamination. In the factory, on-line monitoring and control of production processes is practiced in some stages of processing in some industries.

Principles of processing science can provide insights into fundamental causes of behavior, procedures for modifying and controlling materials and processes, and avenues for improving manufacturing productivity. The role of ceramics processing science in ceramic manufacturing will surely increase.

SUMMARY

Ceramic products are used in applications where the performance and reliability of the product must be predictable and assured and the product must be fabricated successfully in a productive manner. The manufacture of these products from a complex batch containing ceramic materials and processing additives into a finished product involves many operations. All of the materials and operations must be carefully controlled. Principles of science should be used in addition to empirical tests for understanding, improving, and controlling ceramics processing.

SUGGESTED READING

1. Frank Anderson, Christine Claypool, Mark Palmer, and George Phillips, "Ceramic Improvements for Market-Driven Quality," *Am. Ceram. Soc. Bull.*, **71**(12), 1787–1792 (1992).

2. Robert B. Heimann, "Technological Progress and Market Penetration of Advanced Ceramics in Canada," *Am. Ceram. Soc. Bull.*, **70**(7), 1120–1129 (1991).

3. James W. McCauley, "The Role of Characterization in Emerging High Performance Ceramic Materials," *Am. Ceram. Soc. Bull.*, **63**(2), 263–270 (1984).

4. George B. Kenney and H. Kent Bowen, "High-Tech Ceramics in Japan: Current and Future Markets," *Am. Ceram. Soc. Bull.*, **62**(5), 590–596 (1983).

5. W. D. Kingery, "Social Needs and Ceramic Technology," *Am. Ceram. Soc. Bull.*, **59**(6), 598–600 (1980); "Needs and Opportunities for Ceramic Science and Technology," *Proceedings of the 3rd CIMTEC Meeting on Modern Ceramics Technologies*, Elsevier, New York, 1980, pp. 11–66.

6. F. H. Norton, *Fine Ceramics*, Krieger, Malabar, FL, 1978.

7. Karl Schwartzwalder, "Processing Controls in Technical Ceramics," in *Ceramic Processing Before Firing*, George Y. Onoda Jr. and Larry L. Hench, Eds., Wiley-Interscience, New York, 1978, Chapter 2.

8. Robert J. Charleston, *World Ceramics*, Chartwell, Secaucus, NJ, 1977.

9. Haus Thurnauer, "Development and Use of Electronic Ceramics Prior to 1945," *Am. Ceram. Soc. Bull.*, **56**(2), 219–220, 224 (1977).

10. John F. McMahon, "Implications of Our Ceramic Heritage," *Am. Ceram. Soc. Bull.*, **56**(2), 221–224 (1977).

11. A. G. Pincus, "A Critical Compilation of Ceramic Forming Methods. 1. General Introduction," *Am. Ceram. Soc. Bull.*, **43**(11), 827–828 (1964).

PROBLEMS

1.1   Explain the difference between product uniformity and reproducibility of the product.

1.2   Distinguish clearly among the performance, reliability, and quality of a product.

1.3   Productivity can be defined as the ability to make a product sooner, faster, better, at lower cost, and meeting technical and legal requirements. Explain why each factor is important.

1.4   What are the driving forces for further improvements in processing ceramics? What are the restraining forces? Tabulate your answers for each question.

1.5   The key to expanding many markets for ceramics is a reduced variability of the product. What is the key to reducing variability?

1.6   Construct a process flow diagram for the production of tile by dry pressing. Consult Fig. 1.3.

1.7   How will the need for increased productivity and the need to retain manual skills be reconciled in future manufacturing?

1.8   Ceramics processing is said to be a "systems problem." Explain.

1.9   Do you agree with the statement, "Every processing scientist must also be a processing engineer?" Why or why not?

1.10   What was the beginning of "fine ceramics?"

1.11   The development of modern advanced ceramics has been called the "quiet revolution." What were key discoveries or developments?

1.12   Explain why the technology information content of pottery exceeded that of stone and bone. Give some examples of developments that were required for the production of pottery to be accomplished.

EXAMPLES

Example 1.1   Why is the correlation of the fracture strength of the fired ceramic with the ball milling time used in its processing an empirical correlation? What is an example of a parameter that could be scientifically correlated with fired strength?

Solution.   The correlation of fired properties with some processing variable may indicate important processing parameters, but is empirical because many changes in the characteristics occur after milling. An example of a scientific correlation would be the relationship between the strength and the porosity and pore size in the fired ceramic.

Example 1.2   Why are the capabilities of making a product sooner and better components of productivity?

Solution. A manufacturer must be able to adapt and respond to changes in the marketplace. Performance and dimensional tolerance specifications demanded by the customer will probably change. The ability to respond promptly to an upgrade in specifications without an expensive development program impacts productivity and business success.

Example 1.3   What is the meaning of "value added" in the manufacture of a product.

Solution. Consider these approximate financial data. For a barium titanate capacitor, the cost of the starting titanate is about $40/kg and the selling price of a finished capacitor is about $50/kg. The value added is then $10/kg. For a porcelain product, the raw material cost is about $0.12/kg and the selling price is about $3.32/kg. The value added is $3.20/kg. On a weight basis, the titanate has a greater added value of ($10 − $3.20)/$3.20 = 200%.

Example 1.4   What is the meaning of "technology information content" of a product?

Solution. Consider the diagram below which indicates the information contained in each new material developed. The development of pottery clearly required more knowledge than the development of implements of stone and bone. And the technological knowledge for iron making far exceeded that for pottery. Note that the information content of materials of technology increased nearly exponentially, except during the stagnant period of the Dark and Middle Ages. The information content of newer products continues to increase exponentially.

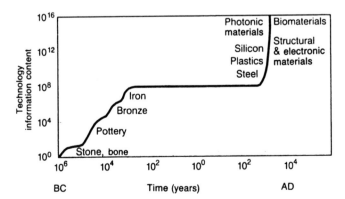

Example 1.4   Increase in technology information content during materials development over time. [From R. B. Heimann, *Am. Ceram. Soc. Bull.*, **70**(7), 1122 (1991); after L. L. Hench, *Adv. Ceram. Mater.*, **3**(3) 203–206 (1988).]

CHAPTER 2

SURFACE CHEMISTRY

Surface and interface phenomena play a particularly important role in ceramics processing because powder systems have a relatively high surface area/mass and the adsorption and distribution of additive phases on the surfaces may alter the microstructure and processing behavior quite markedly. In this chapter we review several general principles of physical chemistry which are especially important for understanding the processing behavior of particle systems.

2.1 THE ATOMIC STRUCTURE OF THE SURFACE DIFFERS FROM THAT IN THE INTERIOR OF THE PARTICLE

Surfaces of liquids and solids have special properties, because they terminate the phase. An atom at a free surface is bonded to fewer neighboring atoms than an atom within the particle. Since bonding reduces the potential energy, a surface atom has extra energy, called the surface energy, which can be partially reduced by slight adjustments in the composition, packing, and bonding between atoms in the surface. Nevertheless, surface atoms or ions are more active.

In terms of thermodynamics, the surface tension γ is defined as

$$\gamma = \left(\frac{\partial G}{\partial A}\right)_{P,T,N_i} \tag{2.1}$$

where G is the Gibb's free energy of the system. During the change in area A, the independent variables of pressure P, temperature T, and number of species

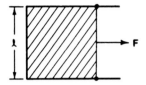

Fig. 2.1 A force is required to increase the surface area of a liquid film.

N_i in the system remain constant. Surface tension is a property of a surface and should not be confused with elastic tension.

Consider a liquid film between two slidewires, as shown in Fig. 2.1. The liquid can reduce its surface free energy when atoms move from the surface into the bulk, contracting the surface area. The work dW done by the force F on the slidewire of length l is

$$dW = F\,dx = 2\gamma l\,dx = \gamma\,dA \tag{2.2}$$

and

$$\gamma = \frac{F}{2l} \tag{2.3}$$

The factor 2 appears because the film has two surfaces. For liquids and low-viscosity suspensions free of external forces, the surface tension will cause spherical droplets to form, because this shape has the lowest surface area per unit volume.

TABLE 2.1   Representative Surface Tensions of Liquids in Air

Liquid	Temperature (°C)	Surface Tension (mN/m)
Water	0	76
	20	73
	25	72
	50	68
	80	63
Methanol	20	23
Ethanol	20	23
Acetone	20	24
Ethylene glycol	20	48
Oleic acid	20	33
Mercury	25	474
Water: 7.5% methanol	20	61
Water: 5% ethylene glycol	25	58
Water: 0.001% dimethyl silicone	25	39
Nitrogen	−196	8.8

Source: Handbook of Chemistry and Physics, R.C. Weast, Ed., CRC Press, Cleveland, 1983.

The surface tension of a crystal varies with the crystallographic direction of the faces. The more stable planes are usually those of lower surface free energy, which are usually more densely packed. Adsorbed atoms or ions can preferentially alter the surface tension of some planes and the crystal habit.

Surface tension values for ceramics are somewhat imprecise because of uncertainties concerning the purity of the surface and the accuracy of the technique. For solids, the surface tension is generally <100 mN/m for polymer materials, 100–2000 mN/m for oxides, and ≥ 1000 mN/m for metals and refractory carbides and nitrides. Representative values of the surface tension of several common liquids are listed in Table 2.1. The surface tension of most liquids decreases with temperature.

2.2 SURFACE TENSION CAUSES A PRESSURE DIFFERENCE ACROSS A CURVED SURFACE

Consider the bubble of radius r in a liquid, as shown in Fig. 2.2. Surface tension will tend to contract the surface area and the internal volume, increasing the internal pressure by an increment ΔP. At equilibrium, the work of contraction $\Delta P \, dV$ is equal to the decrease in surface free energy $\gamma \, dA$, and

$$\Delta P 4 \pi r^2 \, dr - \gamma 8 \pi r \, dr = 0 \qquad (2.4)$$

and

$$\Delta P = \frac{2\gamma}{r} \qquad (2.5)$$

Equation 2.4 is a special case of the Laplace equation. For a surface with principal radii of curvature r_1 and r_2 at a point A on the surface, the general Laplace equation is

$$\Delta P_A = \gamma \left(\frac{1}{r_1} + \frac{1}{r_2} \right) \qquad (2.6)$$

The radius r is negative when the curvature is concave.

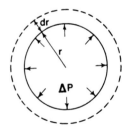

Fig. 2.2 Model for calculating the pressure in a bubble produced by surface tension and negative surface curvature.

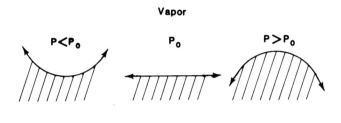

Fig. 2.3 The vapor pressure of a liquid varies with surface curvature.

This effect of surface curvature may cause the average chemical potential of atoms in microscopic particles to be greater than that in large particles. And atoms in a microscopic region of sharp positive curvature are of a higher chemical potential than atoms in a flat surface.

The surfaces we have considered are really interfaces between a solid or liquid and its vapor. A consequence of the surface curvature effect is that the equilibrium vapor pressure P is a function of the surface energy and surface curvature, given by the Kelvin equation

$$\ln \frac{P}{P_0} = \frac{2\gamma V_{\text{mol}}}{rRT} \tag{2.7}$$

where V_{mol} is the molar volume of the condensed phase at temperature T and P/P_0 is the vapor pressure over a curved surface relative to that over a planar surface (Fig. 2.3).

Processing consequences of the surface curvature effect are the more rapid evaporation or dissolving of finer particles in regions of sharp positive curvature (smoothing), preferential condensation in regions of sharp negative curvature, displacement of a curved phase boundary toward its center of curvature, and capillary phenomena.

2.3 WEAK VAN DER WAALS FORCES OR CHEMICAL BONDING CAUSES SOLID SURFACES TO ADSORB MOLECULES FROM GASES AND LIQUID SOLUTIONS

Primary chemical bonding between the surface and an adsorbed gas occurs in chemisorption, and the composition and structure of the surface are changed. Chemical adsorption is usually believed to provide monolayer molecular coverage. Chemisorption can occur above or below the critical temperature of the gas, and chemisorbed gas may be very difficult to remove. The oxidation of metals and the chemical hydration of oxides are examples of chemisorption.

General Van der Waals interactions cause a surface to physically adsorb a gas below its critical temperature. This physical adsorption may change the surface structure of a polymer solid but does not usually alter the structure of oxides and refractory metals. Physical adsorption is rapid and reversible, that is, it can be removed by lowering the pressure or increasing the temperature. Examples of physical adsorption in ceramic processing are the adsorption of polar molecules in solutions and gases such as CO_2, N_2, and water vapor on fine oxide particles. Physical adsorption may provide multilayer coverage.

Langmuir considered the adsorption of gas molecules to be a dynamic process. After partial adsorption giving a fractional surface coverage V_a/V_m, only those molecules striking a portion of the surface not already covered $(1 - V_a/V_m)$ may be expected to be adsorbed. Thermal agitation may cause the desorption of some molecules. When the rates of adsorption and desorption are equal, an adsorption equilibrium is established and

$$\frac{V_a/V_m}{1 - V_a/V_m} = bP \tag{2.8}$$

where V_a is the amount of gas adsorbed on the surface of the adsorbent at a pressure P, V_m is the volume of gas required for monolayer coverage, and b is a constant for particular system and temperature. Equation 2.8 may also be written in the linear form

$$\frac{P}{V_a} = \frac{1}{bV_m} + \frac{P}{V_m} \tag{2.9}$$

Brunauer, Emmett, and Teller expanded Langmuir's dynamic model for monolayer adsorption to include multilayer adsorption before complete monolayer coverage. Their analysis for isothermal adsorption is expressed by the BET equation*

$$\frac{P/P_s}{V_a(1 - P/P_s)} = \frac{1}{V_m C} + \frac{(P/P_s)(C - 1)}{V_m C} \tag{2.10}$$

where P_s is the saturation vapor pressure at the temperature of the adsorbate–absorbent system and C is a constant related to the energy of adsorption.

Of the classic adsorption isotherms illustrated in Fig. 2.4, Type I is characteristic of chemical adsorption on surfaces. Type II is typical of multilayer physical adsorption, and this, coupled with capillary adsorption in pores, produces type IV adsorption. Langmuir's analysis is best used for type I isotherms and the BET equation for types II and IV. The BET equation can be used to determine V_m for cases of multilayer adsorption using data for $0.05 < P/P_s < 0.35$ when C is a constant. The surface area is calculated by multiplying V_m by a monolayer packing constant for the adsorbed gas.

*Arthur W. Adamson, *Physical Chemistry of Surfaces*, Wiley-Interscience, New York, 1976.

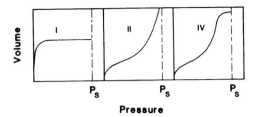

Fig. 2.4 Classical adsorption isotherms.

Above the critical temperature, adsorption is generally monomolecular unless adsorption occurs in pores a few multiples of the absorbate molecule size. In powder agglomerates or compacts, multilayer adsorption in pores is called capillary condensation. For a particular fractional pressure and temperature, Eq. 2.7 can be used to estimate the pore radius below which capillary condensation will occur.

In ceramics processing, adsorbed water vapor or other gases may promote the sticking of powders to surfaces and the agglomeration of ceramic powders, and soften binder phases. Heating above the boiling point is required to remove the adsorbed gas completely.

2.4 THE WETTING AND SPREADING OF A LIQUID ON A SOLID SURFACE DEPEND ON SHORT-RANGE MOLECULAR FORCES THAT CAN BE SIGNIFICANTLY MODIFIED BY A MONOLAYER COATING

Consider the liquid in contact with a planar surface, shown in Fig. 2.5. Spreading occurs when the contact angle θ measured through the liquid phase approaches zero. Wetting implies that the contact angle is less than 90°.

The interaction of the interfacial tensions at the liquid–vapor–solid juncture is described by the Young equation ($\theta > 0°$)

$$\cos \theta = \frac{\gamma_{SV} - \gamma_{SL}}{\gamma_{LV}} \qquad (2.11)$$

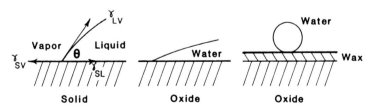

Fig. 2.5 Juncture of interfacial tensions for a liquid on a solid surface and cases of wetting and nonwetting.

where γ_{SV}, γ_{SL}, and γ_{LV} are the ffective interfacial tensions as indicated in Fig. 2.5.

A nonvolatile surfactant that is adsorbed by the liquid and lowers γ_{LV} will reduce the contact angle. Similarly a film of oil, wax, or polymer of lower γ_{SV} on an oxide particle may cause a greater apparent contact angle. Reaction between the liquid and solid, reducing γ_{SL}, may also improve the wetting.

In ceramics processing, wetting and spreading phenomena affect the coating of particles with liquid, the dispersion of agglomerates, and the stability of air bubbles in suspensions.

2.5 WETTING MAY CAUSE COMPRESSIVE FORCES BETWEEN PARTICLES AND LIQUID MIGRATION

Liquid that wets the surfaces of two particles will spread over the surface and concentrate in the contact region, forming a neck (Fig. 2.6). If the particles are finer than about 100 μm, buoyancy forces can be neglected. The pressure difference across the curved meniscus is given by Eq. 2.6. Wetting situations for spherical and angular particles are shown in Fig. 2.6; note that the principal radii are of opposite signs. When the smaller of these is negative, ΔP is negative, and a compressive stress occurs in the contact region. Detailed analyses show that for spheres in contact, the compressive stress is greatest when $\gamma_{LV} \cos \theta$ is large and the volume of liquid is small. For angular particles in point contact, calculations* indicate that the compressive pressure increases as the volume of liquid increases; torques and shear forces can cause rotation and sliding, causing flat sides to come together.

The pressure difference across a curved meniscus can also cause the migration of liquid between pores of different size or the migration of liquid from a saturated region to a less saturated region. As shown in Fig. 2.7, liquid will rise in a capillary if it wets the surface, but it will be depressed if nonwetting

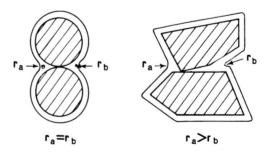

Fig. 2.6 Illustration of a liquid film of uniform thickness on two spherical particles and on two angular particles.

*J. W. Cahn and R. B. Heady, *J. Am. Ceram. Soc.*, **53**(7), 406–409 (1970).

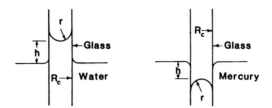

Fig. 2.7 Capillary rise and capillary depression.

occurs. For fine capillaries, the meniscus is approximately hemispherical, $R_C = r \cos \theta$, where R_C is the radius of the capillary, and from Eq. 2.6

$$\Delta P = \frac{2\gamma_{LV} \cos \theta}{R_C} \tag{2.12}$$

At equilibrium, this pressure difference will offset the hydrostatic pressure of a column of liquid of height H, which may rise above the meniscus external to the capillary, and

$$\Delta P = \frac{2\gamma_{LV} \cos \theta}{R_C} = D_L g H \tag{2.13}$$

or

$$H = \frac{2\gamma_{LV} \cos \theta}{D_L g R_C} \tag{2.14}$$

where D_L is the density of the liquid and g is the acceleration of gravity.

The average laminar flow velocity $\bar{v}$ of a column of liquid of length L and viscosity η in a horizontal cylindrical capillary is given by the Poiseuille equation:

$$\bar{v} = \frac{\Delta P R_C^2}{8\eta L} \tag{2.15}$$

Combining Eqs. 2.12 and 2.15

$$\bar{v} = \frac{\gamma_{LV} \cos \theta}{\eta} \frac{R_C}{4L} \tag{2.16}$$

These equations indicate that the rate of penetration of liquid into an agglomerate or through a porous medium will be greater for a liquid of lower viscosity and higher surface tension and for pores of larger radius. Finer pores produce

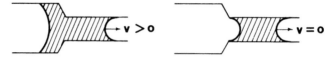

Fig. 2.8 Unbalanced capillary forces may cause liquid migration.

a greater suction. An increase in temperature may reduce η/γ_{LV} and improve penetration. Surfactants that increase cos θ also tend to reduce γ_{LV}.

Pore capillaries in packed particles will vary in radius along their length. As shown in Fig. 2.8, a wetting liquid will migrate into the smaller-diameter segment. Repellency occurs when the wetting angle is greater than 90°.

In ceramics processing, the capillary force provides a mechanism for the cohesion of wetted agglomerates, the migration of liquid in pores, and the rearrangement of particles during mixing. Capillary suction produces a driving force for the migration of liquid in ordinary slip casting and drying.

2.6 WEAK INTERMOLECULAR FORCES CAN CAUSE THE ADHERENCE OF MOLECULES ON A SURFACE AND THE AGGLOMERATION OF SMALL PARTICLES

The attractive Van der Waals forces have their origin in the attractive interaction between fluctuating electrical dipoles in one molecule and the induced dipole in a neighboring molecule and, if present, the attractive interaction between permanent dipoles of particular chemical species. The former are called London dispersion forces and are important in hydrogen bonding.

These forces, which can be summed for a particular geometry to obtain an estimate of the energy of interaction, are of the form of the product of an attraction constant and a geometrical term.

For two small spherical particles of diameter a, the potential energy of attraction U_A is

$$U_A = \frac{-Aa}{24h} \tag{2.17}$$

The force of attraction F_A is

$$F_A = \frac{+Aa}{24h^2} \tag{2.18}$$

when $h \ll a$, where h is the separation between the surfaces of the two particles, and A is the Hamaker attraction constant. The attraction constant is a function of the composition and molecular structure of the particles and the medium between them. When the particles are dispersed in a liquid, the Hamaker constant is

$$A = (\sqrt{A_2} - \sqrt{A_1})^2 \qquad (2.19)$$

where A_2 and A_1 are the Hamaker constants for the particles and dispersion medium, respectively. The attraction constant A is smaller when the particles and medium are more similar in composition.

In addition to the adsorption of molecules, the attractive Van der Waals forces are significant in ceramics processing, because they promote the agglomeration of dry powders and the coagulation of particles in liquid suspension. Mechanisms to counter these effects are discussed in Chapters 9 and 10.

SUMMARY

Atoms and molecules at surfaces may have unsatisfied chemical bonds that produce a surface tension that modifies the behavior of a material. The surface behavior is altered by surface curvature and material adsorbed on the surface. Surface wetting produces capillary phenomena in porous ceramic systems. Weak Van der Waals forces may produce agglomerates in powder systems when repulsion forces are insufficient to inhibit their formation.

SUGGESTED READING

1. Arthur W. Adamson, *Physical Chemistry of Surfaces*, Wiley-Interscience, New York, 1976.
2. W. D. Kingery, H. K. Bowen, and D. R. Uhlmann, *Introduction to Ceramics*, 2nd ed., Wiley-Interscience, New York, 1976.
3. Gordon M. Barrow, *Physical Chemistry*, 3rd ed., McGraw-Hill, New York, 1973.
4. S. J. Gregg and K. S. W. Sing, *Adsorption, Surface Area, and Porosity*, Academic Press, New York, 1967.

PROBLEMS

2.1 Calculate the radius of curvature for which the relative vapor pressure $P/P_0 = 0.9$ for water at 20°C. What is the implication for evaporation of water from flat surfaces versus necks at particle contacts.

2.2 The wetting angle of a drop of water on a glass surface is 16° when the temperature is 20°C. Estimate the interfacial tension γ_{SL} when $\gamma_{SV} = 300$ mN/m at 20°C.

2.3 Water does not completely wet glass at 20°C. Should wetting be improved on heating to 60°C? What is assumed in your calculation?

2.4 If the surface in Problem 2.2 is first coated with a wax film for which $\gamma_{SV} = 30$ mN/m and $\gamma_{SL} = 30$ mN/m, is the wetting improved?

2.5 Calculate the capillary suction of ethyl alcohol in a cylindrical capillary ($R_C = 1$ μm) at 20°C. Assume $\cos \theta = 1$.

2.6 Calculate the relative migration velocity for water and ethyl alcohol in identical capillaries at 20°C.

2.7 What is the pressure required for mercury intrusion into a pore with a radius of 0.1 μm? Assume 25°C and a wetting angle of 140°.

2.8 Calculate and compare the capillary suction of a mold for pressing with $R_C = 2$ μm and a mold for jiggering with $R_C = 1.3$ μm. Assume complete wetting and 25°C.

2.9 Calculate the time for liquid to penetrate to the center of a 1-cm agglomerate ($R_C = 0.5$ μm) when the liquid viscosity is 5000 mPa s. Assume $\cos \theta = 1$, $\gamma_{LV} = 68$ mN/m.

2.10 Explain why the liquid shown in Fig. 2.8 will migrate to the right.

2.11 Does the surface roughness of dry particles influence the maximum Van der Waals force between them?

EXAMPLES

Example 2.1 The following wetting angles were obtained for drops of alcohol–water solutions of increasing water content when placed on a substrate. Estimate the critical liquid–vapor surface tension γ_c for which the wetting angle $\theta = 0$.

γ (mN/m)	θ (deg)
20	17.5
22	30.0
24	36.0
26	42.0

Solution. The wetting equation is $\gamma_{LV} = (\gamma_{SV} - \gamma_{SL})/\cos \theta$. Assuming ($\gamma_{SV} - \gamma_{SL}$) remains essentially invariant with alcohol–water composition, the data can be plotted as γ_{LV} vs. $1/\cos \theta$. The intercept of γ_{LV} when $1/\cos \theta = 1$ is an estimate of the maximum value γ_c for which complete wetting can occur. From the graph shown below, the value of γ_c is 19 mN/m.

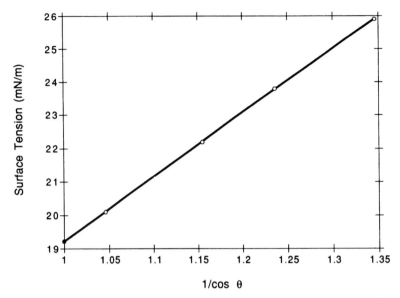

Example 2.1 Surface tension versus 1/cosine of wetting angle and critical surface tension.

Example 2.2 Estimate the capillary suction produced by a gypsum mold used for slip casting at 25°C. Assume that the wetting angle is zero degrees.

Solution. As indicated in Fig. 25.15, the effective pore diameter in a casting mold is approximately 2.0 μm. The capillary suction is estimated as

$$\Delta P = 2\gamma_{LV} \cos \theta / R_C = 2(72 \times 10^{-3} \text{ N/m})(1)/(1 \times 10^{-6} \text{ m})$$

$$\Delta P = 144 \text{ kPa}$$

Note: This value is close to measured values.

Example 2.3 Estimate the height of liquid elevation produced by capillary action in an extruded ceramic with a pore radius of 1.0 μm and water temperature of 80°C. Assume that $\cos \theta = 1$.

Solution. Liquid elevation is calculated from the equation

$$H = \frac{2\gamma_{LV} \cos \theta}{D_L g R_C}$$

$$H = \frac{2(63 \times 10^{-3} \text{ N/m})(1)}{(0.972 \times 10^{3} \text{ kg/m}^3)(9.8 \text{ m/s}^2)(1 \times 10^{-6} \text{ m})}$$

$$H = 13 \text{ m}$$

Note: Water elevation to the surface is not commonly a problem when drying ceramics.

Example 2.4   Estimate the effect of heating on the velocity of water migration in identical capillaries. Contrast the mean velocity at 20 and 50°C.

Solution. The mean velocity is calculated from the equation

$$\bar{v} = \frac{\gamma_{LV} \cos \theta \, R_C}{4\eta L}$$

and for identical capillaries,

$$\bar{v}_{50}/\bar{v}_{20} = (\gamma_{50}/\gamma_{20})(\eta_{20}/\eta_{50})$$

$$\bar{v}_{50}/\bar{v}_{20} = [(68 \text{ mN/m})/73 \text{ mN/m}] (1.00 \text{ mPa} \cdot \text{s}/0.55 \text{ mPa} \cdot \text{s})$$

$$\bar{v}_{50} = 1.7\bar{v}_{20}$$

Note: Temperature has a larger effect on viscosity than surface tension, and heating increases the velocity.

Example 2.5   An agglomerated powder is charged into a mixing tank containing water at 20°C. Assuming $\cos \theta = 1$ and air displaced by water penetrating agglomerates does not impede penetration, estimate the depth of penetration in 1 min into agglomerates with $R_C = 0.5 \ \mu m$.

Solution. The velocity of penetration depends on the liquid depth L, that is

$$\bar{v} = \frac{\gamma_{LV} \cos \theta \, R_C}{4\eta_L L}$$

The depth as a function of time is

$$\frac{dL}{dt} = \frac{\gamma_{LV} \cos \theta \, R_C}{4\eta_L L}$$

which on integrating gives

$$L = [\gamma_{LV} \cos \theta \, R_C \, t/2 \, \eta_L]^{0.5}$$

$$L = \frac{(73 \times 10^{-3} \text{ N/m})(0.5 \times 10^{-6} \text{ m})(60 \text{ s})}{2(1.01 \times 10^{-3} \text{ N} \cdot \text{s/m}^2)}$$

$$L = 32 \text{ mm} = 3.2 \text{ cm}$$

Note: Penetration into agglomerates is quite rapid for a low-viscosity liquid in contrast to a high-viscosity liquid. For a liquid viscosity of 10,000 mPa · s, the time to penetrate 3.2 cm is 10,000 min or 170 hr.

Example 2.6 Why do fine powders spontaneously agglomerate, whereas coarser materials flow freely?

Solution. We can understand these observations by examining the ratio of the Van der Waals attraction F_A between two spherical particles and the weight of one particle F_W tending to cause their separation.

$$\frac{F_A}{F_W} = \frac{(Aa/24\ h^2)}{(\pi a^3/6)\ D_p g}$$

or, $F_A/F_W \sim$ (constant/a^2). Below some particle size a $F_A > F_W$ and the particles spontaneously agglomerate. For ceramic powders agglomeration occurs when the particles are finer than about 40 μm.

Example 2.7 Calculate the particle diameter for which the capillary adhesion force is equal to the weight of the particle. Assume alumina particles of $D_p =$ 3.98 Mg/m^3, water at 20°C, and cos $\theta = 1$. Refer to the figure below.

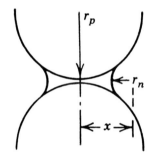

Example 2.7 Model used for calculation.

Solution. From the Laplace equation, $\Delta P = \gamma(1/r_1 + 1/r_2)$, or $\Delta P = \gamma(1/(x - r_n) + 1/r_n)$. When $r_n \ll r_p$ and $r_n \ll x$, $\Delta P = \gamma (1/x + 1/r_n)$ and $\Delta P \cong + \gamma (1/r_n)$. From geometry, $r_p^2 + x^2 = (r_p + r_n)^2$, and $r_n = x^2/2r_p$.

$$\frac{F_{capillary\ adhesion}}{F_{gravity}} = \frac{\Delta P(\pi x^2)}{(4/3)\pi r_p^3 D_p g}$$

Substituting for ΔP and x^2,

$$\frac{F_{capillary\ adhesion}}{F_{gravity}} = \frac{(3/2)\gamma}{r_p^2 D_p g}$$

For the case $F_{capillary\ adhesion}/F_{gravity} = 1$,

$$r_p^2 = \frac{3\gamma}{2D_p g} = \frac{3(73 \times 10^{-3}\ \text{N/m})}{2(4.0 \times 10^3\ \text{kg/m}^3)9.8\ \text{m/s}^2}$$

$$r_p = 1.7\ \text{mm}$$

PART II

CERAMIC RAW MATERIALS

In studying ceramics processing it is necessary to be familiar with the types of raw materials available. Clay minerals, which provide plasticity when mixed with water; feldspar, which acts as a nonplastic filler on forming and a fluxing liquid on firing; and silica, which is a filler that resists fusion, have been the backbone of the traditional ceramic porcelains. Other silicate minerals are used in whitewares such as ceramic tile, thermal shock-resistant cordierite products, and steatite electrical porcelains.

Silica, aluminosilicates, tabular aluminium oxide, magnesium oxide, calcium oxide, and mixtures of these minerals have long been used for structural refractories. Alumina, magnesia, and aluminosilicates are now used in some advanced structural ceramics. Silicon carbide and silicon nitride are used for refractory, abrasive, electrical, and structural ceramics. Finely ground alumina, titanates, and ferrites are the backbone of the electronic ceramics industry. Stabilized zirconias are used for advanced structural and electrical products and zircon, zirconia, and other oxides doped with transition and rare-earth metal oxides are widely used as ceramic pigments. These materials are commonly prepared by calcining particle mixtures, but some are now produced using special chemical techniques.

In Chapter 3, the more common ceramic materials produced in large tonnage and widely used in ceramics are considered. Special materials of exceptional purity and homogeneity which are being developed for research and some very advanced products are discussed in Chapter 4.

CHAPTER 3

COMMON RAW MATERIALS

In this chapter we briefly consider the nature of the starting materials, traditionally called raw materials, that can be purchased from a vendor and received at a manufacturing site. These materials can vary widely in nominal chemical and mineral composition, purity, physical and chemical structure, particle size, and price. Categories of raw materials include (1) nonuniform crude material from natural deposits, (2) refined industrial minerals that have been beneficiated to remove mineral impurities to significantly increase the mineral purity and physical consistency, and (3) high-tonnage industrial inorganic chemicals that have undergone extensive chemical processing and refinement to significantly upgrade the chemical purity and improve the physical characteristics.

The choice of a raw material for a particular product will depend on material cost, market factors, vendor services, technical processing considerations, and the ultimate performance requirements and market price of the finished product. For products in which processing adds considerable dollar value, the cost of the starting material is a relatively small component of the production costs. Accordingly, a higher-quality and more expensive material may be acceptable for microelectronics, coatings, fibers, and some high-performance products. But the average cost of raw materials for building materials and traditional ceramics such as tile and porcelain must be relatively low. Cost–benefit considerations may suggest substitutions of materials of lower cost that do not impair the quality, or alternatively, a more expensive material, which may be more economically processed and/or which will increase the quality and performance of the product.

3.1 CRUDE MATERIALS

Many early ceramics industries were based near a natural deposit containing a combination of crude minerals that could be conveniently processed into usable products. Construction materials such as brick and tile and some pottery items are historical examples, and many are still identified by the regional name.

Some crude materials are of sufficient purity to be used in heavy refractories: Crude bauxite, a nonplastic ore containing hydrous alumina minerals, clay minerals, and mineral impurities such as quartz and ferric oxides, is used in producing some refractories. Today, however, most ceramics are produced from more refined minerals.

3.2 INDUSTRIAL MINERALS

Industrial minerals are used in large tonnages for producing construction materials, refractories, whitewares, and some electrical ceramics. They are used extensively as additives in glazes, glass, and raw materials for industrial chemicals. Common examples are listed in Table 3.1.

Clays are produced by the weathering of aluminosilicate rocks and sedimentation. Clay minerals are layer-type hydrous aluminosilicates which can be

TABLE 3.1 Starting Materials for Ceramics

Category	Purity (%)	Materials
Crude materials	Variable	Shales, stoneware clay, tile clay, crude bauxite, crude kyanite, natural ball clay, bentonite
Industrial minerals	85–98 (99 quartz)	Ball clay, kaolin, refined bentonite, pyrophyllite, talc, feldspar, nepheline syenite, wollastonite, spodumene, glass sand, potter's flint (quartz), kyanite, bauxite, zircon, rutile, chrome ore, calcined kaolin, dolomite
Industrial inorganic chemicals	98–99.9	Calcined alumina (Bayer process), calcined magnesia (from brines, seawater), fused alumina, fused magnesia, aluminum nitride, silicon carbide, silicon nitride, barium carbonate, titania, calcined titanates, iron oxide, calcined ferrites, zirconia, stabilized zirconia, calcined zirconates
Special inorganic chemicals	>99.9	Various materials (see Chapter 4)

dispersed into fine particles (Fig. 3.1). Kaolin is a relatively pure, white firing clay composed principally of the mineral kaolinite $Al_2Si_2O_5(OH)_4$ but containing other clay minerals, as indicated in Table 3.2, and a minor amount of impurity minerals such as quartz SiO_2, ilmenite $FeTiO_3$, rutile TiO_2, and hematite Fe_2O_3. Ball clay is a sedimentary clay of fine particle size containing

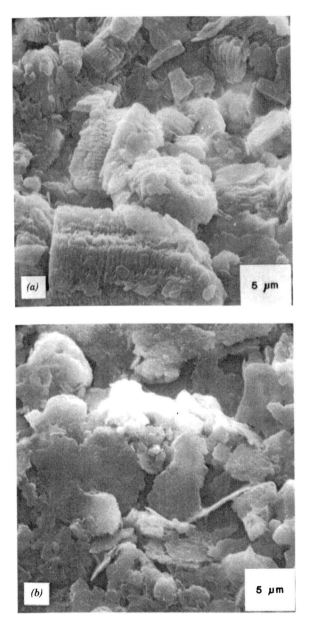

Fig. 3.1   Kaolin that is (a) aggregated and (b) dispersed into platelets.

TABLE 3.2   Clay Minerals

Mineral	Ideal Chemical Formula
Kaolinite	$Al_2(Si_2O_5)(OH)_4$
Halloysite	$Al_2(Si_2O_5)(OH)_4 \cdot 2H_2O$
Pyrophyllite	$Al_2(Si_2O_5)_2(OH)_2$
Montmorillonite	$(Al_{1.67}Na_{0.33}Mg_{0.33})(Si_2O_5)_2(OH)_2$
Mica	$Al_2K(Si_{1.5}Al_{0.5}O_5)_2(OH)_2$
Illite	$Al_{2-x}Mg_xK_{1-x-y}(Si_{1.5-y}Al_{0.5+y}O_5)_2(OH)_2$

Source: After W. D. Kingery, *Introduction to Ceramics*, 1st ed., Wiley-Interscience, New York, 1960.

complex organic matter ranging down to a submicron size. Bentonite is a complex clay containing a relatively high proportion of the clay mineral montmorillonite. Clays are used in whiteware formulations and aluminosilicate refractories to produce plasticity in forming and resistance to deformation when partial fusion occurs during firing.

Other layer-type hydrous silicates are talc $Mg_3Si_4O_{10}(OH)_2$ and pyrophyllite $Al_2Si_4O_{10}(OH)_2$, which are used extensively in compositions for ceramic tile, cordierite, and steatite porcelain. Commercial grades contain impurities such as calcite $CaCO_3$ or dolomite $(Mg,Ca)CO_3$ and other mineral impurities that depend on the source.

Crushed and milled quartz SiO_2 derived from relatively pure deposits of sandstone is a granular silicate mineral used extensively in whitewares, refractories, and glaze compositions (Fig. 3.2). Feldspars composed of the minerals albite $NaAlSi_3O_8$ and microcline or orthoclase $KAlSi_3O_8$ and nepheline syenite containing albite, microcline, and nephelite $K_{0.5}Na_{1.5}(Al,Si)_2O_8$ are the principal fluxes used in whitewares and silicate glazes. Wollastonite $CaSiO_3$ is used in some tile compositions and glazes. Petalite $LiAlSi_4O_{10}$ and spodumene $LiAlSi_2O_6$ are used as a secondary flux and to reduce the thermal expansion of the fired material.

Chrome ores composed principally of a complex solid solution of spinels $(Mg,Fe)(Al,Cr,Fe)_2O_4$ and impurities such as dolomite and magnesium silicates are used in combination with calcined magnesia MgO in basic refractories. Lime CaO produced by calcining limestone $CaCO_3$ and calcined dolomite $(Ca, Mg)O$ is bonded with tar and used for lining basic oxygen steel furnaces. Beneficiated kyanite Al_2SiO_5, bauxite, and zircon $SiZrO_4$ are also used in refractory compositions. Milled zircon is also used as an opacifier in glazes and in producing zircon pigments and is a precursor for zirconia ZrO_2. Calcined kaolin (Fig. 3.2) is used as a nonplastic filler in refractory mixes and mortars.

The beneficiation of industrial minerals begins with crushing and grinding to a small enough size to liberate undesired mineral phases. Further beneficiation may include settling and flotation to segregate minerals by density or size, the separation of magnetic minerals using powerful electromagnets, blending of different processing runs for consistency, and perhaps particle size classifi-

Fig. 3.2 Crushed (*a*) 140-mesh quartz and (*b*) calcined kaolin particles have rough surfaces.

cation. Solids may be concentrated by filtration or centrifugation, and a portion of the soluble impurities are eliminated with the liquid. A typical flow diagram for refining kaolin is shown in Fig. 3.3. Materials, especially clays, refined and ultimately used in suspension form, may be purchased as an aged slurry of controlled viscosity that is shipped in railroad tank cars.

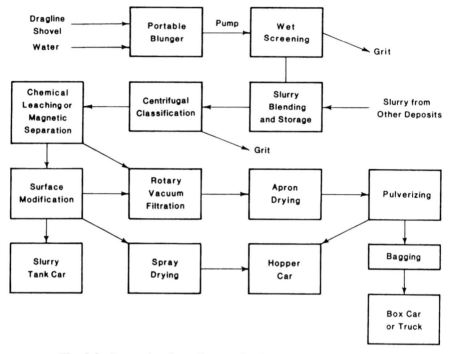

Fig. 3.3   Processing flow diagram for the beneficiation of kaolin.

Concentrated solids are usually dried using a rotary or belt dryer or by spray drying. Some materials are calcined, and a hard aggregate is formed. Dried cake or calcined materials may be pulverized or ground and then sized or air-elutriated before bagging or loading in hopper cars. Many fine materials are loaded and unloaded using pneumatic fluidization and are stored at the plant site in large silos.

3.3   INDUSTRIAL INORGANIC CHEMICALS

Important industrial ceramic chemicals include tabular and calcined aluminas, magnesium oxide, silicon carbide, silicon nitride, alkaline earth titanates, soft and hard ferrites, stabilized zirconia, and inorganic pigments. Extensive chemical beneficiation reduces the content of accessary minerals and may increase the chemical purity up to about 99.5%. For many materials, the scale of operation is extremely large, which aids in lowering the unit processing costs and selling price.

Alumina Al_2O_3 is the most widely used inorganic chemical for ceramics (Table 1.2) and is produced worldwide in tonnage quantities for the aluminum and ceramics industries using the Bayer process. The principal operations in the Bayer process are the physical beneficiation of the bauxite, digestion (in the presence of caustic soda NaOH at an elevated temperature and pressure), clarification, precipitation, and calcination, followed by crushing, milling, and sizing (see Fig. 3.4). During the digestion, most of the hydrated alumina goes into solution as sodium aluminate:

$$\text{Impurity}_{(solid)} + \text{Al(OH)}_{3(solid)} + \text{NaOH}_{(soln)} \rightleftharpoons \tag{3.1}$$
$$\text{Na}^+{}_{(soln)} + \text{Al(OH)}_4^-{}_{(soln)} + \text{Impurity}_{(solid)}$$

and insoluble compounds of iron, silicon, and titanium are removed by settling and filtration. After cooling, the filtered sodium aluminate solution is seeded with very fine gibbsite Al(OH)$_3$, and at the lower temperature the aluminum hydroxide reforms as the stable phase. The agitation time and temperature are carefully controlled to obtain a consistent gibbsite precipitate. The gibbsite is continuously classified, washed to reduce the sodium content, and then calcined. Material calcined at 1100–1200°C is crushed and ground to obtain a range of sizes (Fig. 3.5). Tabular aluminas are obtained by calcining to a higher temperature, about 1650°C.

Magnesium oxide MgO of greater than 98% purity is prepared by precipitating magnesium hydroxide in a basic mixture of treated dolomite and natural brines or seawater containing MgCl$_2$ and MgSO$_4$, followed by washing, filtration, drying, and calcination.

Zirconia ZrO$_2$ of 99% purity is obtained by the caustic fusion of zircon ZrSiO$_4$:

$$\text{ZrSiO}_4 + 4\text{NaOH} \rightarrow \text{Na}_2\text{ZrO}_3 + \text{Na}_2\text{SiO}_3 + 2\text{H}_2\text{O} \tag{3.2}$$

Chemical dissolving of the silicate in water simultaneously hydrolyzes the sodium zirconate to hydrated zirconia. Zirconia is also produced by hot chlorination of zircon in the presence of carbon, and the hydrolysis of the zirconium tetrachloride product to form ZrOCl$_2$. The ZrOCl$_2$ can be calcined directly or reacted with a base in water to form hydrous zirconia. Zircon may also be dissociated to ZrO$_2$ + SiO$_2$ by heating above 1750°C and the zirconium separated by leaching with sulfuric acid:

$$\text{ZrO}_2 + \text{SiO}_2 + 2\text{H}_2\text{SO}_4 \rightarrow \text{Zr(SO}_4)_2 + \text{SiO}_2 + 2\text{H}_2\text{O} \tag{3.3}$$

Silicon carbide SiC is produced in large tonnages using the Acheson process by reacting a batch consisting principally of high-purity sand and low-sulfur coke at 2200–2500°C in an electric arc furnace.

$$\text{SiO}_2 + 3\text{C} \rightarrow \text{SiC} + 2\text{CO}_{(gas)} \tag{3.4}$$

The crystalline product is crushed, washed in acid and alkali, and then dried after iron has been removed magnetically. Granular material is used in refractories and bonded abrasives. Milled material chemically treated to remove

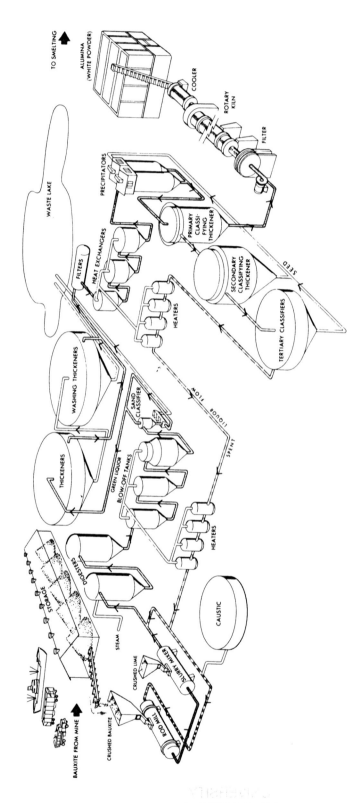

Fig. 3.4 The Bayer process for chemically refining bauxite into alumina. (Courtesy of Alcoa Inc., Pittsburgh, PA.)

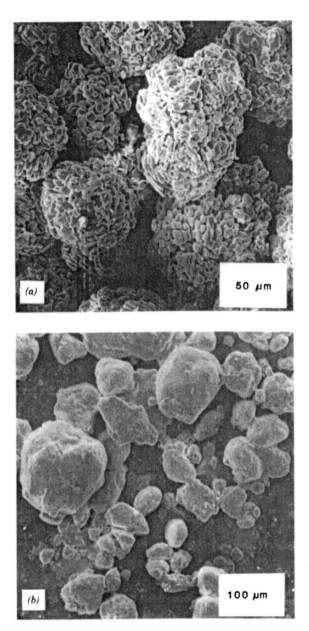

Fig. 3.5   Scanning electron micrographs of particles in calcined Bayer process alumina: (a) aggregates of grains before milling and (b) particle agglomerates of irregular shape in milled powder.

impurities introduced in milling is used industrially for structural ceramics. Silicon nitride Si_3N_4 is prepared by reacting silicon metal powder with nitrogen or a mixture of silica and carbon powders with nitrogen at a high temperature:

$$3SiO_2 + 6C + 2N_{2(gas)} \rightarrow Si_3N_4 + 6CO_{(gas)} \tag{3.5}$$

Silicon nitride and aluminum nitride powders may be produced by the carbothermal process:

$$3SiO_2 + 6C + 2N_{2(gas)} \rightarrow Si_3N_4 + 6CO_{(gas)} \tag{3.6}$$

$$Al_2O_3 + 3C + N_{2(gas)} \rightarrow 2AlN + 3CO_{(gas)} \tag{3.7}$$

Aluminum nitride is also formed by direct nitridation:

$$2Al_{(solid)} + N_{2(gas)} \rightarrow 2AlN \tag{3.8}$$

The oxynitride SIALON is produced by the reaction of mixed powders of silicon nitride, aluminum nitride, and alumina at a high temperature; the reaction is

$$Si_3N_4 + AlN + Al_2O_3 \rightarrow Si_3Al_3O_3N_5 \tag{3.9}$$

The production of mixed metal oxides for electronic ceramics such as barium titanate $BaTiO_3$, ferrites such as $Mn_{0.5}Zn_{0.5}Fe_2O_4$ and $BaFe_{12}O_{19}$, mixed metal oxide resistors, and ceramic colors such as doped zirconia involves the batching and reaction of industrial inorganic chemicals, as is shown for the ferrite in Fig. 3.6. The concentration of chemical dopants is carefully controlled. Soluble material is sometimes removed by filtering before drying. Precursor industrial chemicals for these compounds are commonly powders finer than a few microns in size. Barium carbonate $BaCO_3$ and titania TiO_2 are commonly used for preparing the titanates, and manganese carbonate $MnCO_3$, zinc oxide ZnO, hematite Fe_2O_3, and barium carbonate for the ferrites.

Titania TiO_2 is produced by the sulfate or chloride process. In the sulfate process, ilmenite $FeTiO_3$ is treated with sulfuric acid at 150-180°C to form the soluble titanyl sulfate $TiOSO_4$:

$$FeTiO_3 + 2H_2SO_4 + 5H_2O \rightarrow FeSO_4 \cdot 7H_2O + TiOSO_4 \tag{3.10}$$

After removing undissolved solids and then the iron sulfate precipitate, the titanyl sulfate is hydrolyzed at 90°C to precipitate the hydroxide $TiO(OH)_2$:

$$TiOSO_4 + 2H_2O \rightarrow TiO(OH)_2 + H_2SO_4 \tag{3.11}$$

The titanyl hydroxide is calcined at about 1000°C to produce titania TiO_2. In the chloride process, a high-grade titania ore is chlorinated in the presence of

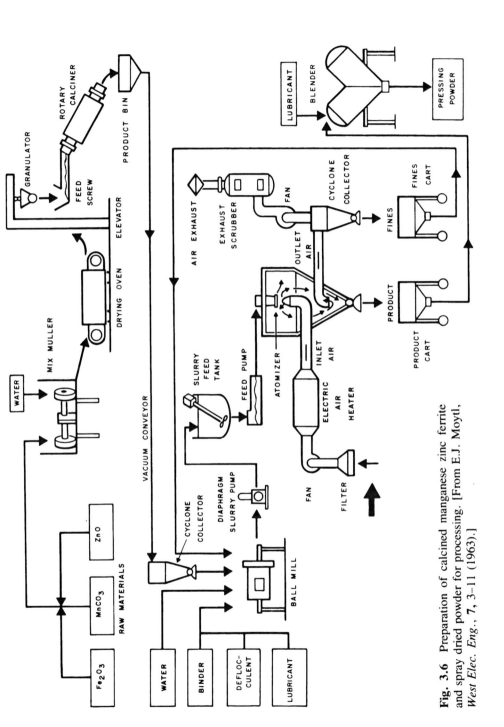

Fig. 3.6 Preparation of calcined manganese zinc ferrite and spray dried powder for processing. [From E.J. Moytl, *West Elec. Eng.*, **7**, 3–11 (1963).]

carbon at 900-1000°C and the chloride $TiCl_4$ formed is subsequently oxidized to TiO_2.

Barium carbonate $BaCO_3$ is the primary source of barium oxide BaO for ceramics. Barite ore, nominally $BaSO_4$, is reduced at a high temperature to barium sulfide BaS which is water soluble. The reaction of an aqueous sulfide solution with sodium carbonate Na_2CO_3 or carbon dioxide CO_2 produces a barium carbonate precipitate which is then washed, dried, and ground.

The commercial iron oxide hematite α-Fe_2O_3 used for preparing ferrites is produced from the thermal decomposition of hydrated ferrous sulfate $FeSO_4 \cdot 7H_2O$ or by the precipitation of hematite and goethite α-$Fe_2O_3 \cdot H_2O$ from an oxygenated sulfate solution containing dispersed iron metal. The size and shape of the ultimate crystals of Fe_2O_3 are very dependent on the pH, temperature, time, and impurities during precipitation. Zinc oxide ZnO is produced by roasting a concentrate of the mineral sphalerite ZnS in air. Manganese carbonate $MnCO_3$ is derived from manganese sulfate $MnSO_4$.

When thermally reacting titanates and ferrites, the temperature, time, and atmosphere must be adequate to permit decomposition of the carbonate and promote interdiffusion of the reactants through the reaction product that may be several microns in thickness (Fig. 3.7). The time dependence of the relative amount x of reactant A of radius r_A transformed into reaction product is given by the Carter equation:

$$[1 + (z - 1)x]^{2/3} + (z - 1)(1 - x)^{2/3} = z + 2(1 - z)\frac{Kt}{r_A^2} \quad (3.12)$$

where K is the apparent rate constant and z is the volume of product formed from a unit volume of reactant A.* The effect of temperature T is commonly

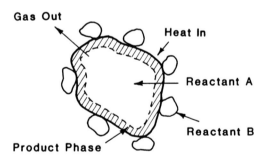

Mixed Powder Reaction

Fig. 3.7 Model for mixed powder reaction when particles are well dispersed, indicating maximum diffusion length for reaction is controlled by radius of reactant particle.

*H. Schmalzried, *Solid State Reactions*, Academic Press, New York, 1974.

expressed by the Arrhenius relation and

$$K = K_0 \exp \frac{-Q}{RT} \tag{3.13}$$

where K_0 is the limiting rate constant that depends on the diffusion path length and Q is the apparent activation energy for diffusion. The reaction is a function of both time and temperature, but temperature has a greater influence on the rate. The time for total reaction eliminating the reactants varies directly with the maximum size of agglomerates or segregated material in the batch. When reacting micron-size oxide powders, thermal processing at a temperature in excess of 1200°C is commonly requisite, as is shown in Fig. 3.8 for the formation of spinel $MgAl_2O_4$.

Other important variables affecting solid-state reactions during calcining are the particle size distributions of the reactants, the mixedness of the reactants, the composition and flow of gases, the depth and turnover of material, and endothermic and exothermic effects. Sintering during calcination produces par-

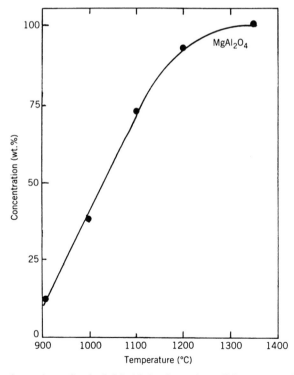

Fig. 3.8 The formation of spinel $MgAl_2O_4$ from the solid-state reaction of micron-size MgO and α-Al_2O_3 powder as a function of the reaction temperature for a constant reaction time of 8 h.

ticle aggregates (Fig. 3.9), and considerable grinding of the hard aggregates is commonly required to produce a fine powder.

As indicated by the nominal solid-state reaction for the formation of barium titanate at a temperature above 1250°C

$$BaCO_3 + TiO_2 \rightleftharpoons BaTiO_3 + CO_{2(gas)} \tag{3.14}$$

the partial pressure of CO_2 in the pores of the product influences the reaction kinetics. Also the nonequilibrium phase Ba_2TiO_4 initially forms between $BaTiO_3$ and unreacted $BaCO_3$ and is undesirable in the calcined product; this phase is minimized by dispersing agglomerates of titania and mixing thoroughly to maximize the particle contacts and reduce the diffusion path between $BaCO_3$ and TiO_2. Calcination in a furnace, which provides a more uniform temperature in the material and mixing of material with air, as shown in Fig. 3.10, may produce a more uniform product. In calcining ferrites and pigments, the oxygen pressure of the air must be controlled to obtain the requisite oxidation states of the transition metal ions. The calcining temperatures and atmosphere must also be controlled to prevent the loss of nonrefractory chemical dopants.

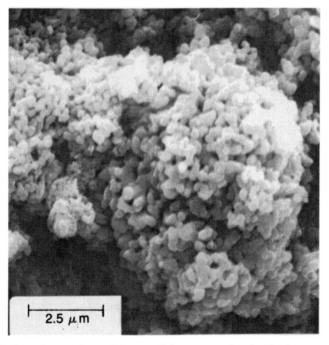

Fig. 3.9 Barium titanate formed by a solid-state reaction is highly aggregated and milling is required to produce a fine powder.

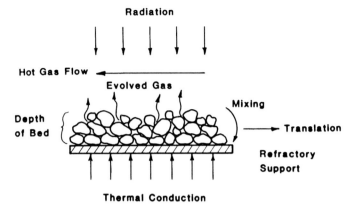

Fig. 3.10 Thermal transport, gas flow, and particle movement during rotary calcination and solid-state reaction.

SUMMARY

Few ceramics are produced today using crude raw materials. Industrial minerals are refined physically to reduce the concentration of undesirable mineral impurities and to produce a particular particle size distribution; water-soluble impurities are removed by washing. Industrial inorganic chemicals used to produce the majority of technical ceramics are chemically processed on a large scale to improve both the chemical and the mineral purity; the calcined product containing hard aggregates is commonly milled to disperse the aggregates and obtain a product of controlled size distribution. Mixed-oxide industrial chemicals are commonly produced by calcining a mixture of these industrial chemicals. The completeness of the reaction and uniformity of the product depend on the particle size and mixedness of the reactants and the time, temperature, and atmosphere and their uniformity during calcination. Different lots of processed materials are blended to maintain a higher level of uniformity.

SUGGESTED READING

1. Magnus Ekelund and Bertil Forslund, "Carbothermal Preparation of Silicon Nitride: Influence of Starting Material and Synthesis Parameters," *J. Am. Ceram. Soc.*, **75**(3), 532–539 (1992).

2. Julie M. Schoenung, "Analysis of the Economics of Silicon Nitride Powder Production," *Am. Ceram. Soc. Bull.*, **70**(1), 112–116 (1991).

3. Martin R. Houchin, David H. Jenkins, and Hari N. Sinha, "Production of High-Purity Zirconia from Zircon," *Am. Ceram. Soc. Bull.*, **69**(10), 1706–1710 (1990).

4. L. D. Hart, *Alumina Chemicals*, The American Ceramic Society, Westerville, OH, 1990.

5. H. Okada, H. Kawakami, M. Hashiba, E. Miura, Y. Nurishi, and T. Hibino, "Effect of Physical Nature of Powders and Firing Atmosphere on $ZnAl_2O_4$ Formation," *J. Am. Ceram. Soc.*, **68**(2), 58–63 (1985).

6. *Materials of Advanced Ceramics and Traditional Ceramics*, Ceramic Industry Magazine, Corcoran Publishers, Solon, OH, 1985.

7. *Process Mineralogy of Ceramic Materials*, Wolfgang Baumgart et al., Eds., Elsevier, New York, 1984.

8. *Kirk–Othmer, Encyclopedia of Chemical Technology*, Wiley-Interscience, New York, 1983.

9. Betty L. Milliken, "Color Control in a Pigment Manufacturing Plant," *Am. Ceram. Soc. Bull.*, **62**(12), 1338–1340 (1983).

10. T. Nomura and T. Yamaguchi, "TiO_2 Aggregation and Sintering of $BaTiO_3$ Ceramics," *Am. Ceram. Soc. Bull.*, **59**(4), 453–455, 458 (1980).

11. F. H. Norton, *Fine Ceramics*, Krieger, Malabar, FL, 1978.

12. W. D. Kingery, H. K. Bowen, and D. R. Uhlmann, *Introduction to Ceramics*, 2nd ed, Wiley-Interscience, New York, 1976.

13. W. E. Worrall, *Clays and Ceramic Raw Materials*, Halsted Press Div., Wiley-Interscience, New York, 1975.

14. Rex W. Grimshaw, *The Chemistry and Physics of Clays and Other Ceramic Materials*, Wiley-Interscience, New York, 1971.

15. *Annual Ceramic Industry Data Book*, Cahners, Boston, MA.

16. *Ceramic Source*, American Ceramic Society, Columbus, OH.

PROBLEMS

3.1 Categorize the materials in the following pairs of starting materials as crude material, industrial mineral, or industrial inorganic chemical and explain your assignment: bauxite–Bayer process alumina, kaolin–calcined kaolin, calcined alumina–calcined kaolin, silicon carbide–silicon nitride, and titania–barium titanate.

3.2 When forming a compound by a solid-state reaction, does the completion of the reaction depend on the average or the maximum particle size? Explain. Does the aggregation of the starting materials influence the reaction?

3.3 Write the chemical reaction for the hydrolysis of sodium zirconate which follows the reaction in Eq. (3.2).

3.4 State several reasons why a pigment calcined in an open crucible in a gas-fired kiln may differ in color from the same pigment batch fired in a closed sagger in an electric furnace.

3.5 Construct a processing flow diagram for the formation of $BaTiO_3$ powder.

3.6 Why are hard powder aggregates commonly formed in a solid-state re-acted batch of material?

3.7 Calculate the amount of $BaCO_3$ and TiO_2 required to produce 1 kg of $BaTiO_3$.

3.8 Calculate the amounts of SiO_2, C, and N_2 required to form 1 kg of Si_3N_4.

3.9 Write the nominal reaction equation for the formation of doped barium titanate $Ba_{1-x}M_xTiO_3$ on heating a mixture of $BaCO_3$, MSO_4, and TiO_2 powders. Illustrate the diffusion paths.

3.10 Substitute the Arrhenius relation into the Carter equation and comment on the dependence of amount reacted on increasing time and increasing temperature.

3.11 Make a sketch similar to Fig. 3.7 for the formation of $ZnFe_2O_4$ from the reaction of Fe_2O_3 and ZnO.

EXAMPLES

Example 3.1   Contrast the mineralogical structure of the kaolin and mont-morillonite families of clay minerals.

Solution. The kaolin minerals include kaolinite, nacrite, dickite, and halloy-site, and kaolinite is the most abundant and important. The kaolin minerals are two-layer minerals. The disilicate layer with the composition Si_2O_5 has Si tetrahedrally coordinated with $-O$ bonds. The second layer has the composition $Al_2(OH)_6$ and is called the gibbsite layer. The Al atoms are octahedrally co-ordinated with $-O$ bonds common to both layers and $-OH$ within the layer. Partial substitution of Al^{3+} for Si^{4+} in the octhedral layer and Mg^{2+} and Fe^{2+} in the tetrahedral layer commonly occurs when formed and the basal plane is negatively charged. Chemisorbed alkalis and Ca^{2+} occur for charge neutrality. The resultant octahedral layer is distorted and this weakens the bonding between the structural units. Slight differences of the stacking of the units produces the particular types of clay minerals.

The montmorillonites are three-layer minerals having a gibbsite sheet sand-wiched between two disilicate layers. The parent mineral is pyrophyllite. Iso-morphous cation substitution produces related minerals. Partial substitution of Mg^{2+} for Al^{3+} in the octahedral layer with surface adsorbed alkali for charge neutrality produces montmorillonite. One-fourth substitution of Al^{3+} for Si^{4+} in the tetrahedral layer causes the surface alkali to be strongly bonded, and the material mica is produced. Significant substitution in both layers, as is indicated in Table 3.2, produces the mineral illite. When Mg^{2+} completely replaces Al^{3+} in the gibbsite layer, the mineral is talc.

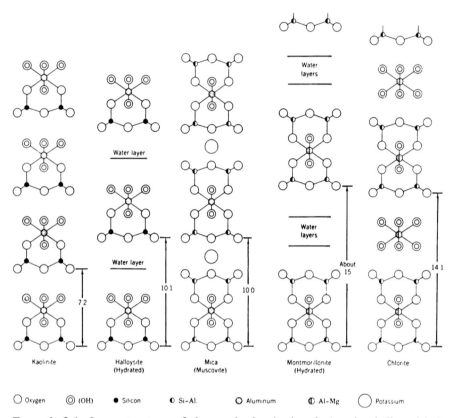

Kaolinite   Halloysite   Mica   Montmorillonite   Chlorite
(Hydrated)   (Muscovite)   (Hydrated)

○ Oxygen   ◎ (OH)   ● Silicon   ◑ Si–Al   ○ Aluminum   ◉ Al–Mg   ○ Potassium

Example 3.1 Layer structures of clays and related minerals (spacing indicated is in angstroms). [From W. D. Kingery, H. K. Bowen, and D. R. Uhlmann, *Introduction to Ceramics*, 2nd ed., Wiley-Interscience, New York, 1976; after R. E. Newnham and G. W. Brindley, *Acta Cryst.*, **9**, 759–764 (1956); **10**, 88 (1957).]

Example 3.2 Compare and contrast the processing of ceramic grade silica and Bayer process alumina.

Solution. The operations of crushing, grinding, and classification are involved in the production of both materials. The big difference in the processing of the alumina is major chemical refining: the chemical dissolving of the aluminum constituents of the bauxite, the chemical precipitation of gibbsite, the washing of the precipitate, and the calcination of the precipitate in the Bayer process.

Example 3.3 Explain why the dispersion of aggregates and agglomerates in reactant materials and the uniform mixing of reactants are essential to obtain a uniform product from a solid-state reaction technique.

Solution. As is indicated by the Carter equation and Fig. 3.7, the time to produce a particular amount of reaction product depends directly on r_A^2/K.

Dispersion of aggregates will reduce the maximum reactant size r_A which is especially important because the dependence is on the square of the size. Increasing temperature increases the rate constant K.

A higher temperature or time will be required to complete the reaction in poorly mixed microscopic regions where the effective r_A is large. Aggregate in precursors may not be completely reacted and may have a different grindability than the majority of the product.

CHAPTER 4

SPECIAL INORGANIC CHEMICALS

New and potential uses of ceramics in what are called high-performance or high-technology applications have stimulated much interest in novel techniques for preparing special ceramic powders with special characteristics. Characteristics sought include a purity in excess of 99.9%, a precisely controlled, reproducible chemical composition including dopants, chemical homogeneity on an atomic scale, and a precisely controlled and consistent submicron particle size. In some applications, a special particle shape may be a goal. The variety of compounds prepared in the laboratory is extensive. Although ceramics have been produced for years from these special powders on a laboratory scale, relatively few of these special powders have been used in industrial processing. However, the successful commercial applications of special materials for products such as optical fibers and thick film electronic ceramics, and large potential markets for more advanced ceramics have increased the interest in and the evaluation of these techniques for industrial fabrication. In this chapter we consider general technical aspects of different techniques and examples of materials that have been prepared.

4.1 POWDERS FROM CHEMICAL SOLUTION TECHNIQUES

Chemical solution techniques provide a relatively convenient means for achieving powders of high purity and fine size. First a suitable liquid solution containing the cations of interest is prepared and analyzed. A solid particulate phase may be formed by precipitation, solvent evaporation, or solvent extraction. Segregation is minimized by combining the ions in a precipitate or gel phase or by extracting the solvent in a few milliseconds from a microscopic

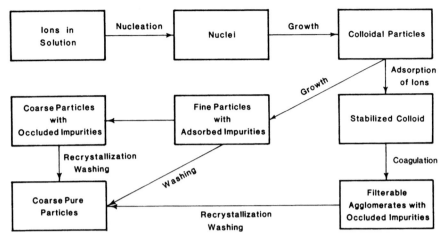

Fig. 4.1   Processing flow diagram for the preparation of a pure precipitate.

drop. The solid phase is usually a salt that can be decomposed without melting by calcination at a relatively low temperature. A porous, friable calcine is ground relatively easily to a submicron size.

Precipitation Techniques

Chemical precipitation techniques of the type used for classical wet quantitative analyses can be used to prepare a wide variety of inorganic salts (Fig. 4.1). The addition of a chemical precipitant to the solution or a change in temperature or pressure may decrease the solubility limit and cause precipitation. Precipitation occurs by nucleation and growth. Impurity ions in solution that are adsorbed on particular surfaces of the particles may alter their growth rates. Relatively slow growth rates along particular crystallographic directions will cause the precipitate particles to have an anisometric shape. A higher degree of supersaturation may increase the nucleation rate and produce a smaller particle size, but if precipitation is extremely rapid, foreign ions tend to be occluded in the particle. The mixing rate and temperature must be controlled to obtain a controlled precipitate. When the cations in solution are of about the same size and chemically similar, the precipitation of a salt containing the cations in solid solution may occur; this is called coprecipitation. In heterogeneous precipitation, the concentration of an ion in the salt differs from that in the solution, and the composition of the coprecipitate may change as precipitation progresses. Less soluble isomorphs tend to concentrate in the salt.

The coprecipitation of ferrous oxalates for the preparation of soft ferrites has been studied in detail by Gallagher et al.* A nickel–iron solid solution

*P. K. Gallagher, H. M. O'Bryan, F. Schrey, and F. R. Monteforte, *Am. Ceram. Soc. Bull.,* **48**(11), 1053–1059 (1969).

oxalate is precipitated on admixing hot ammonium oxalate in a heated aqueous sulfate solution:

$$NiSO_4 + 4FeSO_4 + 5(NH_4)_2C_2O_4 \cdot H_2O \xrightarrow{H_2O(20°C)}$$

$$5Ni_{0.2}Fe_{0.8}C_2O_4 \cdot 2H_2O + 10NH_4^+ + 5SO_4^{2-} \tag{4.1}$$

The thermal decomposition of the oxalate salt in air at a temperature below 500°C produces a well-ordered nickel ferrite compound and is described by the reaction

$$3Ni_{0.2}Fe_{0.8}C_2O_4 \cdot 2H_2O_{(s)} + \frac{4+x}{2} O_2(g)$$

$$\rightarrow Ni_{0.6}Fe_{2.4}O_{4(s)} + xCO_{(g)} + (6-x)CO_{2(g)} + 6H_2O_{(g)} \tag{4.2}$$

The atomic scale mixedness in the salt crystal enables the direct formation of the ferrite of fine particle size (see Example 4.3).

In some cases, it may be possible to precipitate a specific compound containing two cations, and the composition of the precipitate will be uniform regardless of the concentration of ions in solution. An example of the latter is $Ba(TiO(C_2O_4)_2) \cdot 4H_2O$, formed by the reaction*

$$BaCl_2 + TiCl_4 + 2H_2C_2O_4$$

$$\xrightarrow{H_2O(20°C)} BaTiO(C_2O_4)_2 \cdot 4H_2O + 6H_3O^+ + 6Cl^- \tag{4.3}$$

The thermal decomposition of the barium titanyl oxalate salt produces barium titanate indirectly, but at a relatively low temperature (700°C), because of the fine particle size and mixedness:

$$BaTiO(C_2O_4)_2 \cdot 4H_2O_{(s)}$$

$$\xrightarrow{(25-225°C)} BaTiO(C_2O_4)_{2(s)} + 4H_2O_{(g)} \tag{4.4}$$

$$BaTiO(C_2O_4)_{2(s)} + \tfrac{1}{2}O_{2(g)}$$

$$\xrightarrow{(225-456°C)} BaCO_{3(s)} + TiO_{2(s)} + CO_{(g)} + 2CO_{2(g)} \tag{4.5}$$

$$BaCO_{3(s)} + TiO_{2(s)} \xrightarrow{(465-700°C)} BaTiO_{3(s)} + CO_{2(g)} \tag{4.6}$$

The hydrolysis of mixed alkoxides has been used to produce a variety of oxides with particle sizes finer than 20 nm. When water is admixed into alcohol

---

*M. D. Rigterink, *J. Can. Ceram. Soc.*, 37, LVI-LX (1968).

solutions of the alkoxides, barium isopropoxide $Ba(OC_3H_7)_2$ and titanium amyl-oxide $Ti(OC_5H_{11})_4$, the simultaneous hydrolytic decomposition produces barium titanate according to the reaction*

$$Ba(OC_3H_7)_{2(soln)} + Ti(OC_5H_{11})_{4(soln)}$$

$$\xrightarrow{H_2O(20°C)} BaTiO_{3(powder)} + 2C_3H_7OH_{(soln)} + 4C_5H_{11}OH_{(soln)} \qquad (4.7)$$

An oxide containing several different but chemically similar ions may be prepared by admixing a solution of the ion in an excess of the precipitant. The very high degree of supersaturation causes the rapid precipitation of all ions. The precipitation system and conditions must be well controlled to obtain a reproducible precipitate; important variables include solution concentrations, pH, mixing and stirring rates, and temperature. Precipitates may be purified by digestion, washing, and, in some cases, reprecipitation prior to filtration (Fig. 4.1). Digestion is growth of the larger precipitate particles at the expense of the finer particles while the precipitate is in the solution; surface-adsorbed impurities decrease as the specific area decreases. Washing will improve the purity if surface-adsorbed impurities are removed without precipitating other ions in solution films on particles. Dissolving and reprecipitation in a fresh solution may reduce the concentration of minor impurities. Alum $NH_4Al(SO_4)_2 \cdot 12H_2O$ dissolves in a hot aqueous solution. The reprecipitated alum formed on cooling has a lower concentration of alkali and transition metal impurities:

$$NH_4Al(SO_4)_2 \cdot 12H_2O_{(impure)} \xrightarrow{H_2O \ (heat)} solution$$

$$solution \xrightarrow{(cool)} NH_4Al(SO_4)_2 \cdot 12H_2O_{(purified)} + H_2O_{(impure)} \qquad (4.8)$$

This technique is used commercially to produce alumina with a purity exceeding 99.995% (Fig. 4.2).

Precipitation techniques have been widely investigated for preparing sub-micron-size, high-purity oxide powders. Particle sizes as small as 2 nm have been produced for some systems, and these have been used as commercial catalysts. Precipitation techniques are currently being considered for the industrial production of more advanced ferrites.

Solvent Evaporation and Extraction Techniques

An alternative procedure used to prepare special powders is to disperse the solution containing the ions of interest into microscopic volumes and then remove the solvent as a vapor, forming a salt. Maintenance of atomic-scale homogeneity will be possible for multicomponent systems only when the components are of about equal solubility or when the salt forms extremely rapidly.

*K. S. Mazdiyasni, R. T. Dolloff, and J. S. Smith, *J. Am. Ceram. Soc.*, **52**(10), 523–526 (1969).

Fig. 4.2   Scanning electron micrograph of an aggregate of α-Al_2O_3 produced on calcination of alum.

Spray drying has been used to produce and dry salt particles 10–20 μm in diameter. The fast drying that occurs in a few milliseconds reduces segregation. This technique has been reported to give good results for the preparation of ferrites from mixed sulfates and Mg-stabilized beta alumina from mixed nitrates.* Porous calcined agglomerates are reported to be easily comminuted into the component particles. A variation of this general procedure is the atomization of the solutions in a hot furnace to combine the drying and calcination in one step; because the heating rate is several hundred degrees per second, complete decomposition occurs only if the salt decomposes at a relatively low temperature. Unaggregated submicron particles of magnesia were produced from the spray pyrolysis of an acetate solution:[†]

$$Mg(C_2H_3O_2)_2 \cdot 4H_2O_{(s)} \xrightarrow{H_2O(20°C)} solution$$

$$\xrightarrow{air(500°C)} MgO_{(s)} + CO_{2(g)} + CO_{(g)} + H_2O_{(g)} \tag{4.9}$$

*J. G. M. de Lau, *Am. Ceram. Soc. Bull.*, **49**(6), 570–574 (1970).
†T. Gardner and G. Messing, *Am. Ceram. Soc. Bull.*, **63**(12), 1498–1501 (1984).

Another approach used has been to absorb the chemical solution into a microporous organic material such as cellulose. The loaded fiber is first pyrolyzed and then calcined in a controlled atmosphere. The porous agglomerates are easily comminuted. This technique has been used to produce colloidal-sized particles or fibers of a wide variety of oxides, carbides, and metals (Fig. 4.3).

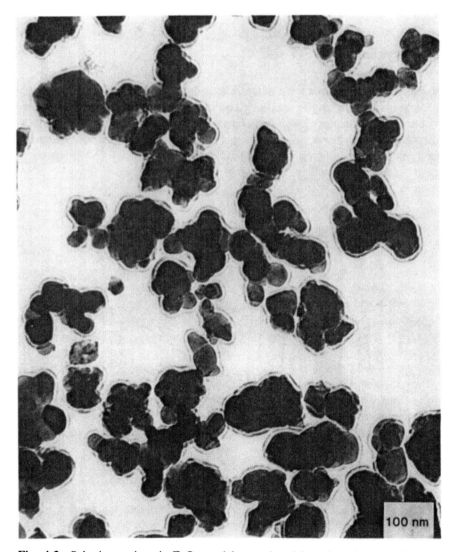

Fig. 4.3 Submicron zirconia ZrO_2 particles produced by adsorption, pyrolysis, and calcination in air. (Note ammonium polyacrylate hull from adsorbed deflocculant; transmission electron micrograph.)

Other approaches for extracting the solvent have included spraying of the solution into an immiscible desiccating liquid (liquid/liquid drying) or spraying into a hot immiscible liquid to produce vaporization.

In freeze drying, a salt solution is sprayed into a cold liquid such as hexane ($-60°C$) and quickly frozen into a salt and ice. The solvent is sublimated at a pressure below the triple point by slowly raising the pressure and temperature, as indicated in Fig. 4.4. Calcination of the homogeneous salt yields porous aggregates which are milled to a micron-size powder with good sinterability.*

Sol-Gel Techniques

Sol-gel processing has attracted much interest both for the preparation of special powders and for forming thin coatings and cast or extruded shapes. In sol-gel processing, colloidal particles or molecules in a suspension, a sol, are mixed with a liquid, which causes them to join together into a continuous network, called a gel. Polymerization greatly restricts chemical diffusion and segregation. The gel is dried, calcined, and milled to form a powder.

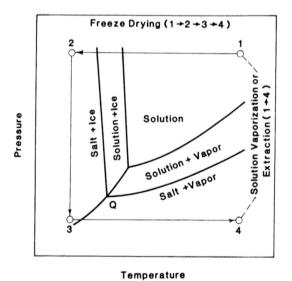

Fig. 4.4 A pressure–temperature phase diagram for an aqueous salt solution and the processing path for freeze drying. [After M. Rigterink, *Am. Ceram. Soc. Bull.*, **51**(2), 161 (1972).]

*M. D. Rigterink, *Am. Ceram. Soc. Bull.*, **51**(2), 158–161 (1972).

Silica gel is produced by the hydrolysis and polymerization of the alkoxide tetraethyl orthosilicate in a solution, as indicated by the following reactions:

$$(OC_2H_5)_4Si + H_2O \xrightarrow{\text{hydrolysis}} (OC_2H_5)_3SiOH + C_2H_5OH \tag{4.10}$$

$$(OC_2H_5)_3SiOH + OH^- \xrightarrow{\text{dehydration}} (OC_2H_5)_3SiO^- + H_2O \tag{4.11}$$

$$(OC_2H_5)_3SiO^- + (OC_2H_5)_3SiOH \xrightarrow{\text{gelling}} (OC_2H_5)_3SiOSi(H_5C_2O)_3 + OH^- \tag{4.12}$$

Polymerized gels may also be formed by the hydrolysis of other simple or mixed alkoxides such as titanium and zirconium esters. Gels may also be formed when an aqueous sol of a hydrated aluminum or iron oxide is mixed with a dehydrating agent; a hydrogen-bonded alumina gel is produced by mixing an aqueous boehmite AlO(OH) sol with acetone. This technique has also been used to produce titania and zirconium oxide powders. In the citrate process, citric acid is added to a prepared salt solution; after partial alcohol dehydration to a viscous liquid, the solution is further dehydrated and decomposed by spraying into a furnace.

4.2 POWDERS FROM VAPOR PHASE REACTIONS

Vapor phase reactions have been used to produce special oxide and nonoxide powders (Fig. 4.5). Oxides can be formed by reacting a metal chloride with water vapor at a high temperature, as is indicated for the formation of titania by the reaction

$$TiCl_{4(g)} + 2H_2O_{(g)} \rightarrow TiO_{2(s)} + 4HCl_{(g)} \tag{4.13}$$

Salts dissolved in alcohol have been pyrolyzed in an atomizing burner to produce compound oxide powders. Fine oxide powders have also been produced by oxidizing evaporated metals.

Silicon nitride powder may be formed by reacting silicon tetrachloride and ammonia in a plasma below 1000°C:

$$3SiCl_{4(g)} + 4NH_{3(g)} \rightarrow Si_3N_{4(s)} + 12HCl(g) \tag{4.14}$$

The thermal decomposition of $(CH_3)_2SiCl_2$ and CH_3SiH_5 vapor has also been used to produce high-purity silicon carbide SiC.

Vapor phase techniques produce submicron-size, well-dispersed particles

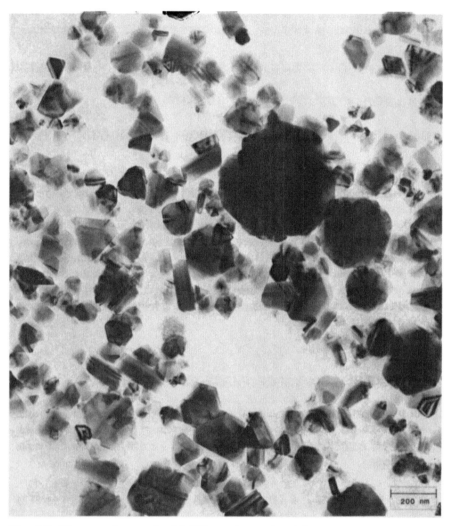

Fig. 4.5 Electron micrograph of β-SiC powder produced from a vapor phase reaction.

entrained in large volumes of gas. Large, complex collection systems are required to remove the powder.

4.3 OTHER TECHNIQUES

Techniques somewhat intermediate between the solid-state reaction of mixed-oxide particles and the special chemical techniques may offer a compromise between added expense and improved homogeneity. Mixtures of oxide powders or gels may form a crystalline mixed hydroxide compound when heated to 180–

700°C in the presence of steam having a pressure of 1–100 MPa. A mixed-oxide compound of fine grain size is formed on calcining to a relatively low temperature. This technique, called hydrothermal synthesis, has been used to produce a variety of mixed oxide compounds.*

Decomposition of a salt commonly produces a relatively fine, reactive product. Mixtures of salts that decompose at a similar temperature, such as an alkaline earth carbonate and an aluminum or iron hydroxide, may produce a mixed-oxide compound of fine particle size at a relatively low temperature. Silicon nitride and silicon aluminum oxynitride have been produced by heating discrete oxide particles in a nitrogen atmosphere at a high temperature.

A special chemical technique may be used to dope a fine powder or gel. After dispersion of the powder in a chemical solution, a precipitation or gelation reaction may precipitate or immobilize the dopant on the surface. The doped powder produced on drying and then heating to a relatively low temperature may be finer and require less comminution to disperse the aggregates.

SUMMARY

Powders and colloidal materials that have special or very precisely controlled characteristics have been developed for research studies and for the industrial production of some advanced ceramics. In preparing the special materials, relatively pure precursors are first mixed on an atomic scale in a liquid or gas. A salt or gel containing the ions of interest is formed by a reaction producing a precipitate or by very rapidly removing the solvent. Chemical segregation which reduces the homogeneity must be prevented during the preparation process. Decomposition should occur at a temperature below that causing sintering of the product into a hard aggregate. Powders prepared using special chemical techniques may range widely in chemical and physical form. Special chemical techniques continue to be developed and evaluated for the production of powders for advanced ceramics. To be beneficial, they must enable production of a better product.

SUGGESTED READING

1. Tetsuo Yamada, "Preparation and Evaluation of Sinterable Silicon Nitride Powder by Imide Decomposition Method," *Am. Ceram. Soc. Bull.*, **72**(5), 99–106 (1993).
2. Edward S. Martin and Mark L. Weaver, "Synthesis and Properties of High-Purity Alumina," *Am. Ceram. Soc. Bull.*, **72**(7), 71–77 (1993).
3. N. Balagopal, H. K. Varma, K. G. K. Warrier, and A. D. Damodaran, "Citrate Precursor Derived Alumina–Ceria Composite Powders," *Ceram. Int.*, **18**, 107–111 (1992).

*S. Komarneni et al., *Adv. Ceram. Mater.*, **1**(1), 87–92 (1986).

4. Shigeyuki Somiya and Yoshihiro Hirata, "Mullite Powder Technology and Applications in Japan," *Am. Ceram. Soc. Bull.*, **70**(10), 1624–1640 (1991).

5. Wolfgang F. Kladnig and Wilhelm Karner, "Pyrohydrolysis for the Production of Ceramic Raw Materials," *Am. Ceram. Soc. Bull.*, **69**(5), 814–821 (1990).

6. William H. Rhodes and Samuel Natansohn, "Powders for Advanced Structural Ceramics," *Am. Ceram. Soc. Bull.*, **68**(10), 1804–1812 (1989).

7. William J. Dawson, "Hydrothermal Synthesis of Advanced Ceramic Powders," *Am. Ceram. Soc. Bull.*, **67**(10), 1673–1678 (1988).

8. Shin-Ichi Hirano, "Hydrothermal Processing of Ceramics," *Am. Ceram. Soc. Bull.*, **66**(9), 1342–1344 (1987).

9. Timothy J. Gardner and Gary L. Messing, "Preparation of MgO Powder by Evaporative Decomposition of Solutions," *Am. Ceram. Soc. Bull.*, **63**(12), 1498–1502 (1984).

10. K. S. Mazdiyasni, "Fine Particle Perovskite Processing," *Am. Ceram. Soc. Bull.*, **63**(4), 591–594 (1984).

11. "Better Ceramics through Chemistry," in *Materials Research Society Proceedings*, Vol. 32, C. J. Brinker, D. E. Clark, and D. R. Ulrich, Eds., North Holland, New York, 1984.

12. D. W. Johnson Jr., "Powder Preparation-Ceramics," in *Advances in Powder Technology*, Gilbert Y. Chin, Ed., American Society of Metals, Metals Park, OH, 1982, Chapter 2.

13. David W. Johnson Jr., "Nonconventional Powder Preparation Techniques," *Am. Ceram. Soc. Bull.*, **60**(2), 221–224, 243 (1981).

14. P. E. D. Morgan, "Chemical Processing for Ceramics," in *Processing of Crystalline Ceramics, Materials Science Research*, Vol. 11, H. Palmour, R. E. Davies, and T. M. Hare, Eds., Plenum, New York, 1978, pp. 67–77.

15. D. W. Johnson Jr. and P. K. Gallagher, "Reactive Powders from Solution," in *Ceramic Processing Before Firing*, George Y. Onoda Jr. and Larry L. Hench, Eds., Wiley-Interscience, New York, 1978, Chapter 12.

16. P. K. Gallagher, H. M. O'Bryan, F. Schrey, and F. R. Monforte, "Preparation of a Nickel Ferrite from Coprecipitated $Ni_{0.2}Fe_{0.8}C_2O_4 \cdot 2H_2O$," *Am. Ceram. Soc. Bull.*, **48**(11), 1053–1059 (1969).

PROBLEMS

4.1 Compare the reaction temperatures for the formation of $BaTiO_3$ by solid state reaction versus the titanyl oxalate process.

4.2 Design two different chemical processes for making MgO powder doped with NiO.

4.3 Contrast spray pyrolysis and vapor phase reaction techniques.

4.4 What is the difference between the growth and the recrystallization of a precipitate?

4.5 Explain why the selection of the salt phase is important both at the precipitation stage and the calcination stage when preparing a mixed ion compound.

4.6 Describe the preparation of a doped alumina using the freeze-drying process.

4.7 Describe the citrate process which may be considered a sol-gel process.

4.8 Design a specific process for producing an iron oxide powder doped with cobalt only in the near-surface region.

4.9 Calculate the volume of HCl gas when producing 1 kg of Si_3N_4 by a vapor phase reaction.

4.10 For a heterogeneous coprecipitation, the distribution of dopant I and host H at some moment in the surface layer of precipitate is

$$dI_{ppt}/dH_{ppt} = \lambda \, (I_0 - I_{ppt})/(H_0 - H_{ppt})$$

How does the amount of dopant I vary with the amount of host precipitated when the partition coefficient $\lambda = 1$, and when $\lambda = 0.4$?

EXAMPLES

Example 4.1 Using the precipitation of hydroxides of Ba^{2+}, Al^{3+}, and Cr^{3+} from aqueous solution of the chlorides as an example, explain the basic difference between simultaneous precipitation and coprecipitation in a solution.

Solution. In each case a liquid solution is the precursor phase in the process. However, the much larger cation size and lower valence of Ba^{2+} will cause it to precipitate as a separate hydroxide phase $Ba(OH)_2$. The Al^{3+} and Cr^{3+} coprecipitate in a single solid-solution hydroxide phase $Al_xCr_{1-x}(OH)_3$.

Example 4.2 How may a special chemical technique be used to dope the surface of an oxide or hydroxide particle with another cation?

Solution. Submicron particles of the primary oxide or hydroxide can be dispersed in a liquid solution containing the dopant ion of interest. A change in pH, temperature, or pressure, or the addition of a precipitant or polymerizing agent which causes a slow precipitation, gelling, or adsorption of the dopant on the particle surface can effectively produce a surface-film phase containing the dopant. Subsequent low temperature calcination may produce a surface-modified particulate product.

Example 4.3 Why are the reaction temperatures for the preparation of a powder from the decomposition of a salt commonly much lower than the temperature for the solid-state reaction between mixed powders?

Solution. In salt decomposition the cations are well mixed on an angstrom scale and the diffusion length is on the nanometer scale. But for the solid-state reaction of mixed powders the diffusion length is of micron dimensions (see Fig. 3.7). Also, the nucleation and growth process during salt decomposition produces particles smaller than the size of the parent particle, as is shown in the figure below.

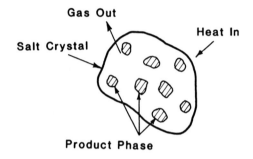

Salt Decomposition

Example 4.3 Product grain size formed on nucleation and growth during decomposition of a salt.

PART III

MATERIALS CHARACTERIZATION

The properties of a fired product are very dependent on the characteristics of the starting material and their subsequent modification during processing. In order to select and control these materials, it is necessary to have a knowledge of their more important characteristics. Some of this information is supplied in the material specifications provided by the raw-material supplier. However, the material system will have to be characterized at different processing stages, and the processing engineer must be familiar with the techniques used and the information obtained. The general characteristics of a material system and common specifications of several commercial ceramic materials are presented in Chapter 5. Chemical and microstructure analyses are discussed in Chapter 6. Chapters 7 and 8 describe principles and techniques for characterizing the size, shape, density, surface area, and porosity of ceramic particle systems.

CHAPTER 5

CHARACTERISTICS AND SPECIFICATIONS OF CERAMIC MATERIALS

The characteristics of a material are those parameters that specify the chemical and physical aspects of its composition and structure. "Composition" denotes the proportions of chemically and physically different constituents. "Structure" refers to the spatial distribution, orientation, and association of these constituents.

The properties of a material are its responses to changes in the physical or chemical environment. Every particle system will have particular properties— for example, a particular thermal conductivity, elastic modulus, and dielectric constant. Flow and deformation properties are commonly referred to as rheological properties. Responses to the chemical environment such as adsorption or dissolution are chemical properties. Porous particle systems may have special properties such as capillarity, permeability, and electroosmotic flow. Dispersed systems have special properties such as settling rate, electrophoretic mobility, and optical scattering. A system is said to be anisotropic with respect to a particular property if the property varies with direction in the material.

This chapter will consider the general characteristics of particle systems and specifications provided for ceramic raw materials. The student should keep in mind the distinct difference between the properties and characteristics of a material system.

5.1 PARTICLES, POWDERS, COLLOIDS, AND AGGLOMERATES

The size range of particles used in ceramics processing covers a wide range of sizes, as is indicated in Fig. 5.1. A particle is a discrete, solid unit of material and may be single or multiphase in composition. Groups of particles that are

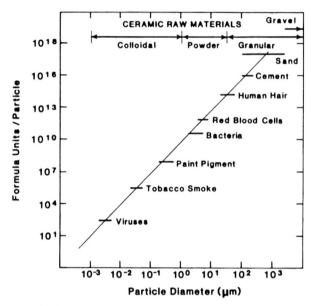

Fig. 5.1 Increase in formula units per Al_2O_3 particle with particle size and particle size range of granular materials, powders, and colloidal materials and some common substances.

weakly bonded together may behave as a fragile, larger pseudoparticle called an agglomerate (see Fig. 3.5). If strongly bonded together, the larger particle is not easily dispersed and is referred to as an aggregate or a hard agglomerate. Bonds in hard agglomerates are generally primary chemical bonds formed by a chemical reaction or sintering. In soft agglomerates, the relatively weak bonds may be of electrostatic, magnetic, Van der Waals, or capillary adhesion type.

The magnitude of the inertial force of a particle relative to surface forces has a major effect on particle behavior. A particle system is said to be granular if the gravitational force is predominant (the material is free-flowing), a powder if the surface force is of the same order as the gravitational force (naturally agglomerates), and colloidal if the particles are so fine that the inertial force of a particle is insignificant and the surface forces dominate the behavior. The surface forces are dependent on the environment of the particles. But for practical purposes, particles larger than 44 μm (opening in a 325-mesh sieve) can generally be considered to be granular and particles smaller than 1 μm as colloidal. Colloids dispersed in a low-viscosity liquid typically exhibit Brownian motion at 20°C. The behavior of powders and colloids can be markedly altered by adsorbed surfactants that modify the surface forces.

5.2   RAW-MATERIAL SPECIFICATIONS

The general characteristics of one particle and a system of particles are listed in Table 5.1. Complete characterization is an impossible task, and for each

TABLE 5.1 Characteristics of a Particle System

Single Particle	Particle System
1. Primary chemical composition	1. Distribution of chemical composition
2. Impurity composition, distribution, and partitioning	2. Distribution of impurities
	3. Distribution of phase composition
3. Phase composition	4. Distribution of crystal defects
4. Point and line defects, domains, etc.	5. Porosity and pore structure
5. Structure of phases, boundaries	6. Particle structure distribution
6. Porosity and pore structure	7. Particle size distribution
7. Size	8. Particle shape distribution
8. Shape	9. Particle density distribution
9. Density	10. Bulk density
10. Specific surface area	11. Specific surface area

TABLE 5.2 Specification of Special High-Purity Aluminas

Characteristic	Calcined[a]	Calcined[a]	Calcined[b]
Crystal phase	>90% gamma	85% alpha	alpha
Purity (%)	99.99	99.99	99.99
Impurity analysis of ceramic grade (ppm)			
Na	20	20	
Pb	4	4	
Si	18	18	
Cr	4	4	
Fe	10	10	
Ga	15	15	
Ca	10	10	
Mg	5	5	
Zn	4	4	
Ti	5	5	
Mn	3	3	
V	3	3	
Cu	2	2	
SiO_2			<50
Fe_2O_3			<20
CaO			<10
Na_2O			<10
Ga_2O_3			<10
Others			<10
Ultimate particle size (μm)	0.01	0.15	<0.5
Specific surface area (m²/g)	115	10	5–50
Agglomerate size (μm) mean	2	0.6	0.5
Crystal density (Mg/m³)	3.67	3.98	3.98
Apparent bulk density (Mg/m³)	0.12	0.51	

[a]Products of Baikowski International Corp., Charlotte, NC.
[b]Product of Aluminum Company of America, Pittsburgh, PA.

material and application we must consider what characterization is necessary and sufficient. Some of this information may be supplied by the raw-material vendor on a specification sheet for the material. Table 5.2 lists typical specifications for special very high purity, very fine alumina materials. The impurity composition is quite complete, and the crystalline phase is identified. The specific surface area and nominal information about agglomeration and crystal size are presented. Since these are relatively expensive materials, the customer will certainly determine additional characteristics of each lot and perhaps process and fabricate a small amount of the material in the laboratory or factory to verify that it is satisfactory.

Typical specifications of three calcined Bayer process aluminas are listed in Table 5.3. The parameters specified are similar to those for the purer alumina, but we can readily see from the specifications that these aluminas are quite different in chemical purity and particle size (see Figs. 3.5 and 4.2). A plant

TABLE 5.3 Specifications of Three Bayer Process Aluminas

Characteristic	Calcined Intermediate Soda	Reactive Low Soda	Tabular (−325 Mesh)
Chemical analysis (wt%)			
Al_2O_3	99.4	99.7	99
SiO_2	0.02	0.02	0.2
Na_2O	0.25	0.08	0.10
Fe_2O_3	0.04	0.01	0.3
CaO	0.04	0.01	0.07
LOI (1100°C)	0.2		
Total water[a]	0.3		
α Alumina phase (%)	90+	~ 100	~ 100
Ultimate crystal size (μm)	<5.0	>0.5	
Particle size distribution			>95
Sieve analysis (wt %)			
+100 mesh			
+200 mesh			
+325 mesh			
−325 mesh			> 95
Sedimentation analysis[b] (μm)			
90% <	40	1.5	
50% <	12	0.5	
10% <	3	0.2	
Specific surface area[c] (m^2/g)	1.0	3–6	
Specific gravity	3.8	3.98	>3.4
Bulk density (Mg/m^3)	1.0		

Source: Products of Aluminum Company of America, Pittsburgh, PA.
[a]1100°C ignition loss after adsorption at 44% relative humidity.
[b]Gravity settling.
[c]Nitrogen adsorption.

engineer examines the preshipment specifications supplied by the vendor for the particular lot of material. Depending on past experience and the particular application, a set of characteristics may be determined for a small sample using standardized test procedures, in the purchaser's laboratory, before authorizing or rejecting a shipment.

Specifications for three different barium titanate powders used for electronic ceramics are listed in Table 5.4. These materials are prepared with different chemical and particle size characteristics, as indicated in the specifications. The concentration of CO_2 and SO_3 indicates the incomplete decomposition of reactants during calcination. Differences in the electrical properties of a fired body reflect the variations in the chemical stoichiometry and physical characteristics of the raw materials and microstructure developed during firing.

Specifications for the commercial kaolins listed in Table 5.5 include the basic chemical and particle size characteristics. The composition of clay mineral types and mineral impurities such as free quartz are not listed. The MBI index is a relative indication of the specific surface area determined by adsorption of methylene blue dye. Clay bodies are usually processed as suspensions, and the pH index of a suspension may suggest the compatibility or

TABLE 5.4 Typical Specifications of Calcined Barium Titanates

Character	Capacitor	MLC	Piezoelectric
Chemical analysis (wt%)			
SiO_2	0.10	0.12	0.15
Al_2O_3	0.10	0.14	0.16
TiO_2	34.64	33.95	33.38
SrO	0.90	0.78	0.91
BaO	63.59	64.28	64.18
Na_2O	0.10	0.17	0.15
SO_3	0.15	0.14	0.18
CO_2	0.09	0.15	0.43
LOI	0.17	0.32	0.57
Size analysis $(\mu m)^a$			
90% <	4.5	5.5	5.4
50% <	1.6	2.3	2.0
10% <	0.8	1.0	0.8
+325 Mesh (%)	0.02	0.02	0.02
Bulk density (Mg/m^3)	1.8	2.4	2.0
Electrical Property Analyses (body contains 10% calcium zirconate and 1% magnesium zirconate):			
Dielectric constant (25°C)	5250	4000	4400
Dissipation factor (% at 25°C)	1.18	0.83	0.67
Δ Dielectric constant (100°C)	−52.9	−48.9	−54.0
Δ Dielectric constant (−10°C)	−33.7	−2.4	−4.9
Fired density (Mg/m^3)	5.30	5.54	5.60

^{a}Products of TAM Ceramics Inc., Niagara Falls, NY.

TABLE 5.5 Typical Specifications of Ceramic-Grade Kaolins

Characteristic	NC[a]	GA-P[b]	GA-C[b]
Chemical analysis (%)			
SiO_2	47.72	45.36	45.74
Al_2O_3	37.53	38.26	38.25
Fe_2O_3	1.16	0.36	0.41
TiO_2	0.08	1.52	1.55
CaO		0.47	0.06
MgO		0.04	0.12
K_2O	1.17	0.21	0.06
Na_2O	0.15	0.11	0.14
LOI	14.04	13.47	13.66
Total	99.85	99.80	99.99
Particle size analysis			
(cumulative mass percent finer)			
20 (μm)	97.5	98.0	97.0
10	89.0	93.5	88.0
5	75.0	83.0	74.5
2	53.0	65.0	54.0
1	35.0	48.5	38.0
0.5		32.0	21.5
0.2		15.0	11.0
MBI (meq/100 g)	—[c]	7.8	2.0
pH	5	7.2	4.3
PCE	33–34	34–35	34–35
Dry MOR (MPa)	1.2	3.4	0.9

[a]North Carolina kaolin, Harris Mining Co. Inc., Spruce Pine, NC.
[b]Georgia kaolin, Cyprus Industrial Minerals, Inc., Sandersville, GA. (GA-P for plastic forming, GA-C for casting.)
[c]Contains halloysite with tubular particle shape.

change in pH if one clay is substituted for another. The pyrometric cone equivalent (PCE) indicates the relative resistance of a material to vitrification and creep on heating. The MOR is the flexural strength of dried bars formed by extrusion. The pH, PCE, and MOR are not characteristics; rather, they are indices that indicate something about effects of soluble chemical impurities, impurity phases, and the particle size distribution on the chemical, thermal, and mechanical behavior, respectively.

SUMMARY

The characteristics of a material are the parameters necessary for its identification or description. Specifications provided by suppliers of materials provide some of these characteristics. More complete specifications or tighter specifications of the lot-to-lot reproducibility of a material commonly increase the material cost. Processing alters the characteristics of the particle system. Materials processors should determine the materials characteristics that are requisite for control of the processing and the properties of their products.

SUGGESTED READING

1. F. H. Norton, *Fine Ceramics*, Krieger, Malabar, FL, 1978.
2. W. M. Flock, "Characterization and Process Interactions," in *Ceramic Processing before Firing*, George Y. Onoda Jr. and Larry L. Hench, Eds., Wiley-Interscience, New York, 1978, Chapter 4.
3. G. Y. Onoda Jr. and L. L. Hench, "Physical Characterization Terminology," in *Ceramic Processing before Firing*, George Y. Onoda Jr. and Larry L. Hench, Eds., Wiley-Interscience, New York, 1978, Chapter 5.
4. Y. S. Kim, "Effects of Powder Characteristics," in *Treatise on Materials Science and Technology*, Vol. 9, Franklin F. Y. Wang, Ed., Academic Press, New York, 1976, pp. 51–67.

PROBLEMS

5.1 What is the relationship between parts per million and percent when reporting the impurity in a solid substance?

5.2 Calculate and compare the impurity analysis of the five major impurities in the two alpha aluminas in Table 5.2. Use an elementary oxide basis.

5.3 Calculate and compare the sharpness index = (size 90% < − size 10% <)/ size 50% < for the two Bayer process aluminas in Table 5.3.

5.4 Calculate the bulk density as a percent for the two aluminas in Table 5.2 and compare your results to the value for the calcined alumina in Table 5.3.

5.5 Calculate the impurity level of CO_2 and SO_3 in parts per million for the barium titanate powders in Table 5.4. What is the source of these impurities?

5.6 The chemical analyses (wt%) of three production lots of calcined capacitor grade barium titanate are as follows:

SrO	0.95	0.82	0.84
BaO	63.71	64.00	63.94
TiO_2	34.67	34.61	34.59

What is the reproducibility of the molar ratio $(BaO + SrO)/TiO_2$?

5.7 Which parameters in Table 5.5 are characteristics and which are properties? Explain your reasoning.

5.8 Compare the content of colloidal sizes for the two Georgia kaolins in Table 5.5.

5.9 Explain why the tubular particles in NC kaolin may cause a misinterpretation of dye adsorption index MBI (see Table 5.5).

EXAMPLES

Example 5.1   What are differences in agglomerate characteristics for the special calcined aluminas in Table 5.2.

Solution. The mostly gamma phase alumina has the largest mean agglomerate size of 2 μm and the smallest crystallite size (ultimate particle size) of 0.01 μm; the size ratio is 2/0.01 or 200. For the other aluminas calcined at a higher temperature, the agglomerate size is smaller and the crystallite size is larger; therefore, the agglomerate size and the number of particles in a mean agglomerate is lower. The greater agglomeration of the gamma alumina is reflected in its lower apparent bulk density.

Example 5.2   What is the nominal purity of the barium titanates in Table 5.4?

Solution. The elementary oxides forming the perovskite structure are BaO, SrO, and TiO_2. Summing the weights of these gives 99.1% purity for the capacitor, 99.0% for the MLC, and 98.5% for the piezoelectric type of material. The nominal purity is 99%.

Example 5.3   Compare the stoichiometry for the three types of barium titanates in Table 5.4.

Solution. Ideally, stoichiometric barium titanate contains 1 mole of alkaline earth oxides (BaO + SrO) and 1 mole of TiO_2. The moles of the oxide components are found by dividing each by its molecular weight. For the capacitor grade material,

$$\text{moles BaO} = 63.54 \text{ g}/153.34 \text{ g/mol} = 0.415$$

$$\text{moles SrO} = 0.90 \text{ g}/103.62 \text{ g/mol} = 0.009$$

$$\text{moles TiO}_2 = 34.64 \text{ g}/79.88 \text{ g/mol} = 0.434$$

$$(\text{BaO} + \text{SrO})/\text{TiO}_2 \text{ mole ratio} = (0.415 + 0.009)/0.434 = 0.976$$

Calculating in a similar manner, the mole ratio for the other types is

| MLC Type | 1.005 |
| Piezoelectric | 1.02 |

It is observed, after sintering, that a much larger grain size is obtained when the mole ratio > 1; the grain size strongly influences the electrical properties.

Example 5.4   What are the important differences between granular materials, powders, and colloids?

Solution. The differences in particle characteristics and behavior may be seen in the table below:

Parameter	Granular Material	Powder	Colloid
Size (μm)	> 44	$44 - 1$	< 1
F_A vs. F_W	$F_A \ll F_W$	$F_A = F_W$	$F_A \gg F_W$
Flowability	Good (free flowing)	Poor	Very poor
Agglomeration	Minimal	Spontaneous	Spontaneous
Vol. Ads./Vol. Par.	Insignificant	Significant	Very significant

Note: F_A is Van der Waals attractive force. F_W is particle weight. Vol. Ads./Vol. Par. is the volume of adsorbed processing additive relative to the volume of the particle.

CHAPTER 6

CHEMICAL AND PHASE COMPOSITION

Wet chemical techniques have been used routinely for the analyses of major elements in ceramic materials. However, the analysis of impurities and many major elements is now commonly performed using instrumented techniques, which are faster and more accurate. Microscale chemical analyses are determined using electron beam techniques. Surface analysis techniques coupled with ion beam machining may provide an analysis of surface material. Structural techniques such as x-ray diffraction, infrared spectroscopy, and light and electron microscopy are used to determine the identity of the phases, their structure, and the microstructure. Thermal analysis techniques are used to infer a change in the composition and structure of the system from the effects of a change in temperature or atmosphere on the chemical and physical properties of the material system.

6.1 BULK CHEMICAL ANALYSIS

The conventional qualitative elemental analysis of a ceramic material does not usually pose problems. Most industrial ceramic minerals contain at least 30 detectable elements, but fewer than 10 are commonly present at a level greater than 0.01–0.05%. Care must be taken to provide a representative sample.

Small samples are removed systematically from the bulk to obtain a representative sample for analysis. The samples may be analyzed individually or pooled and split down using a riffler or by cone and quartering to obtain a statistically random sample for analyses. When individual samples are analyzed, the data may indicate variations in the bulk and, when pooled, the mean analysis. Airborne samples may be collected in a filter using a suction device. The sample for analysis is often less than 1 g.

In classical wet chemical techniques, the material is fused by heating in the presence of a flux, dissolved in acid, and then analyzed using a standard precipitation, titration, or colorimetric technique. Wet chemical techniques are still widely used for the routine analyses of major elements in a wide variety of materials, and the sensitivity may sometimes exceed 1 ppm using colorimetric analysis.

Instrumented spectroscopic techniques listed in Table 6.1 are used for qualitative survey analyses, quantitative impurity analyses, and for the quantitative analyses of some major elements in systems that are not easily or accurately analyzed using wet chemistry techniques. The energy of free atoms, ions, and molecules is quantized, and each species has a set of characteristic energy levels. The absorption of radiant energy may cause a transition from a lower to a higher energy level; the return to a lower energy state causes the emission of radiation that is characteristic of the particular species. Electron transitions between higher energy levels produce radiation ranging from the near infrared to the ultraviolet. Transitions of electrons nearer the nucleus produce x-rays. In solids, liquids, and gases, transitions can occur between energy levels associated with the vibration and rotation of molecules and cause adsorption in the infrared.

In emission spectrographic analysis, the sample of compacted powder is excited in an electric arc or by a laser flash. Chemical information is provided by the wavelength and intensities of the characteristic line spectra emitted in the visible to ultraviolet region of the spectrum (Fig. 6.1). Inductively coupled plasma emission spectrometry (ICP) may be used for the analysis of more than 70 elements to below 1 ppm.

TABLE 6.1 Spectroscopy Techniques

Method	Principle	Detect
Emission spectroscopy (powder)	Thermal stimulation of electrons in atom	Emitted line spectrum (visible-UV)
Flame emission spectroscopy (liquid solution)	Thermal stimulation of electrons in atom	Emitted line spectrum (visible-UV)
Atomic absorption spectroscopy (liquid solution)	Thermal stimulation of electrons in atom	Emitted line spectrum (visible-UV)
X-ray fluorescence (powder)	X-ray stimulation of electrons in atom	Emitted line spectrum (x-rays)
Mass spectroscopy (gas from solid, liquid, gas)	Ions deflected in a magnetic field	Mass/charge of ion
Infrared spectroscopy (solid, liquid, gas)	Molecular vibrations with a change in dipole moment absorb IR radiation	Absorption spectrum (infrared)

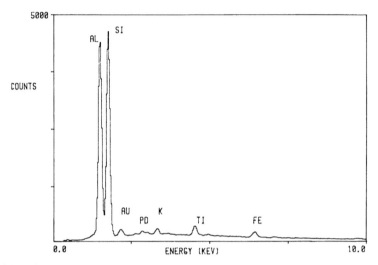

Fig. 6.1 Line spectrum for a mullite refractory material determined using energy-dispersive spectroscopy (EDS). (Au and Pd from specimen preparation; courtesy of W. Votava, Alfred University.)

Alkalies emit in the visible range; samples containing alkali are commonly dissolved in a liquid, and the alkali is analyzed using flame emission spectrometry. Atomic absorption spectrometry may be used to analyze as many as 30–40 elements present at a concentration of <0.1%. It has become the standard technique for the industrial analyses of impurities and is now used for analyzing major elements in some systems. The sample in solution form is sprayed into a flame to dissociate it into its elements. Radiation from a cathode lamp containing the element of interest is also passed through the flame. Dissociated atoms in the flame absorb the spectrum emitted by the lamp, which reduces the transmitted intensity. Concentrations down to the parts-per-million level are determined by comparing the absorbence of standards and the sample solution.

X-ray fluorescence in which x-rays are used to stimulate secondary characteristic x-radiation may be used for the qualitative analysis of major and minor elements with an atomic number greater than that of sodium. Characteristic x-rays are diffracted by an analyzing crystal, and the diffraction angle and intensity provide information about the element and its concentration. This technique is accurate and fast for routine analyses.

The components of a gas in pores, or a gas produced by heating or sputtering, may be determined using gas chromatography or mass spectroscopy. The par-

Fig. 6.2 Diffuse Fourier transform infrared spectrum (DRIFT) for (top) as-received Bayer process alumina powder and (bottom) hydrated alumina phases. The adsorption bands between 3000 and 4000 cm^{-1} for the alumina indicate chemisorbed water and the presence of a Bayerite–Gibbsite surface phase. (Courtesy of Cynthia Incorvati, Alfred University.)

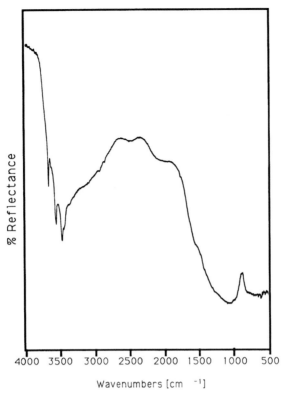

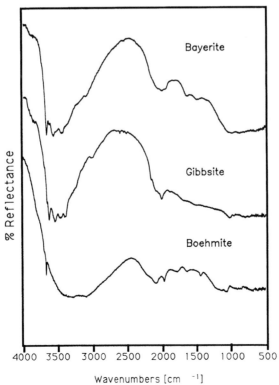

TABLE 6.2 Instrumental Analysis Techniques

Bulk Techniques	Comments
Emission spectroscopy (ES)	Elemental analyses to the ppm level, frequently used for qualitative survey analyses, 5 mg powder sample
Flame emission spectroscopy (FES)	Quantitative analyses of alkali and Ba to the ppm level, ppb detectability for some elements, solution sample
Atomic absorption spectroscopy (AAS)	Industry standard for quantitative elemental impurity analyses; detectability to ppm level, solution sample
X-ray fluorescence (XRF)	Elemental analyses, detectability to 10 ppm, $Z > 11$, solid/liquid samples
Gas chromatography/mass spectrometry (GC/MS)	Identification of compounds and analysis of vapors and gases
Infrared spectroscopy (IRS)	Identification and structure of organic and inorganic compounds, mg dispersed power in transparent liquid or solid or thin-film sample
X-ray diffraction (XRD)	Identification and structure of crystalline phases, quantitative analysis to 1%, mg powder sample
Nuclear magnetic resonance (NMR)	Identification and structure of organic (NMR) compounds, sample to 5 mg for H and 50 mg for C

titioning of the unknown gas between the carrier gas and a sorbent as it flows through an absorption column may provide a characteristic adsorption spectrum, enabling identification of the compounds in the gas phase. Mass spectrometry may be used to determine the identity and concentration of inorganic or organic ions present at concentrations of less than 10 ppm in a gas produced by heating or sputtering, because their deflection in a magnetic field is proportional to the charge divided by the mass of the ion.

Infrared spectrometry determines the adsorption of infrared radiation due to characteristic vibrations and rotations of atoms in molecules and solid compounds. It is used to determine molecular structure and the presence of trace molecular anion impurities in calcined materials (Fig. 6.2).

Neutron activation analysis is used to determine quantitatively the concentration of radioactive isotopes in a material. A neutron source may be used to convert susceptible elements to a radioactive isotope. Capabilities of instrumented analysis techniques are summarized in Table 6.2.

6.2 PHASE ANALYSIS

Crystalline phases diffract x-rays according to the Bragg law,

$$n\lambda = 2d \sin \theta \qquad (6.1)$$

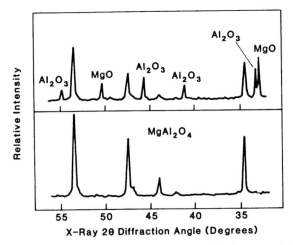

Fig. 6.3 X-ray diffraction pattern for (top) an incompletely reacted 1:1 mixture of periclase MgO and corundum Al_2O_3 containing spinel $MgAl_2O_4$ and (bottom) a completely reacted mixture.

where θ is the diffraction angle for a lattice spacing d, λ is the wavelength of the x-rays, and n is an integer. Powder or polished polycrystalline specimens are used, and the diffraction 2θ angles are recorded. The identification of a phase is accomplished by comparing the d spacings and relative intensities of the sample material with reference data for known materials (Fig. 6.3). Quantitative phase analysis to about 1% is possible when phases are randomly oriented and diffraction lines of different phases are clearly distinguished.

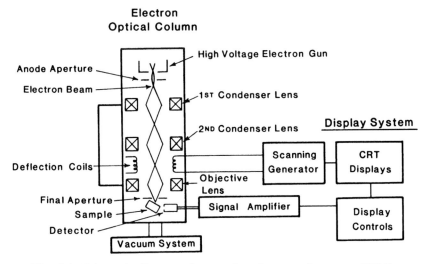

Fig. 6.4 Schematic diagram of a scanning electron microscope (SEM).

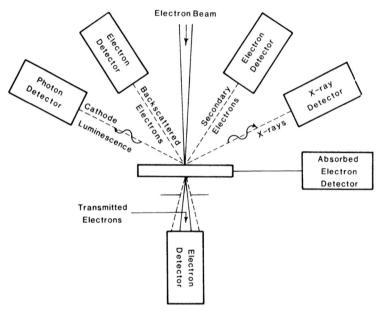

Fig. 6.5 Signals and detection modes for electron beam microanalysis.

Optical microscopy has long been used to identify phases in thin sections of polycrystalline and partially vitrified systems. Optical microscopy is also used routinely to examine surface topography and the microstructure of polished and etched specimens down to about 0.2 μm. Scanning electron microscopy (Fig. 6.4), having a greater resolution, is now widely used for microstructure analyses, because it is convenient and versatile and because micrographs with

TABLE 6.3 Microscopic Characterization Techniques

Technique	Comments
Light microscopy (LM)	Microstructure of etched, polished, or thin sections; surface topography; phase analyses; resolution to 0.2 μm
Scanning electron microscopy (SEM) with dispersive x-ray spectroscopy (EDS)	Microstructure of fracture, polished, or etched surface; resolution 10 nm; qualitative and semiquantitative analyses with 2 μm resolution using EDS; 0.1% detectability Z > 11, microstructure of dispersed organic in backscatter mode
Transmission electron microscopy (TEM)	Microstructure of thin sections 20–200 nm thick, resolution 1 nm, identification of crystal structure by electron diffraction, qualitative and semiquantitative analyses using scanning TEM with EDS with 30- to 50-nm resolution

a large depth of focus can be obtained quickly to document the microstructure. Detection modes provided by the interaction of the electron beam with the sample are shown in Fig. 6.5. With an energy-dispersive spectroscopy attachment, microscale qualitative chemical analyses can be obtained rather easily, and this chemical information can aid greatly in the interpretation of microstructures. Although less convenient to use, TEM can provide analysis down to a resolution of about 1 nm; the structure of defect phases and grain boundaries can be determined using the electron diffraction or scanning modes with EDS analysis (Table 6.3).

6.3 SURFACE ANALYSIS

The analysis of surface and near-surface material, which may vary from that of the bulk, has been aided greatly by recent improvements in electron and ion beam instrumentation (Table 6.4). In Auger electron spectroscopy, a scanning beam of electrons excites the surface of the specimen, and the energy of emitted "Auger electrons" provides information about the atomic numbers of the elements present. Bombarding the surface with ions, called ion milling, can remove atomic layers of material for depth profiling. Subsequent analyses provide information about near-surface concentration gradients. In electron microprobe analyses, the characteristic x-rays emitted when electrons scan a microscopic region of the surface are detected and used to identify and quantify chemical elements present.

If the surface is excited using monochromatic x-rays, the photoelectrons emitted from the surface contain information about the type of atoms and their oxidation state and structure in the surface. This technique is called x-ray photoelectron spectroscopy and is used for chemical analyses. Bombarding the

TABLE 6.4 Surface Analysis Techniques

Surface Technique	Comments
Auger electron spectroscopy (AES)	Elemental analyses to a resolution of 3 μm lateral and 5 nm thick, detection to 0.1%, depth profile by ion milling
Electron microprobe analyzer (EMA)	Qualitative analyses $Z > 5$, quantitative analyses of polished sections to $<0.1\%$, lateral and depth resolution of 2 μm
Secondary ion mass spectrometry (SIMS)	Identification of elements of all Z by mass spectrometry of sputtered ions, depth profiles by ion milling, depth resolution to 3 nm, detectability to ppm level
X-ray photoelectron spectroscopy (XPS)	Elemental analyses of layers 3 nm thick, lateral resolution 1 nm, depth profiles by ion milling, chemical bonding information
Fourier transform infrared spectroscopy (FTIR)	Identification of structure of adsorbed molecules and coating

surface with monoenergetic low-energy ions will remove surface ions by sputtering, and these can be analyzed using a mass spectrograph. This technique is called secondary ion mass spectroscopy. Because of the importance of surfaces in ceramic processing, these surface techniques are becoming more available, and their use is increasing in the development of more advanced ceramics.

6.4   THERMOCHEMICAL AND THERMOPHYSICAL ANALYSES

Thermochemical techniques are used to determine thermodynamic changes in individual materials and reactions between materials in a batch or between a

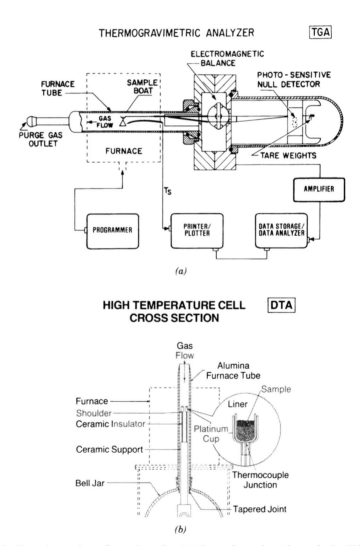

Fig. 6.6   Experimental configurations for (*a*) thermal gravimetric analysis (TGA) and (*b*) differential thermal analysis (DTA).

material and the atmosphere, usually when the temperature is changed. For thermogravimetric analysis (TGA), material is suspended from a balance, and the weight is monitored during controlled heating or cooling or under isothermal conditions (Fig. 6.6). In differential thermal analysis, DTA, thermocouples in contact with a powder specimen and a reference powder, respectively, indicate the test temperature and any differential temperature due to an endothermic or exothermic transition or reaction in the sample as a function of temperature or time (Fig. 6.7). When the enthalpy change is determined, the technique is called differential scanning calorimetry (DSC). Important parameters that must

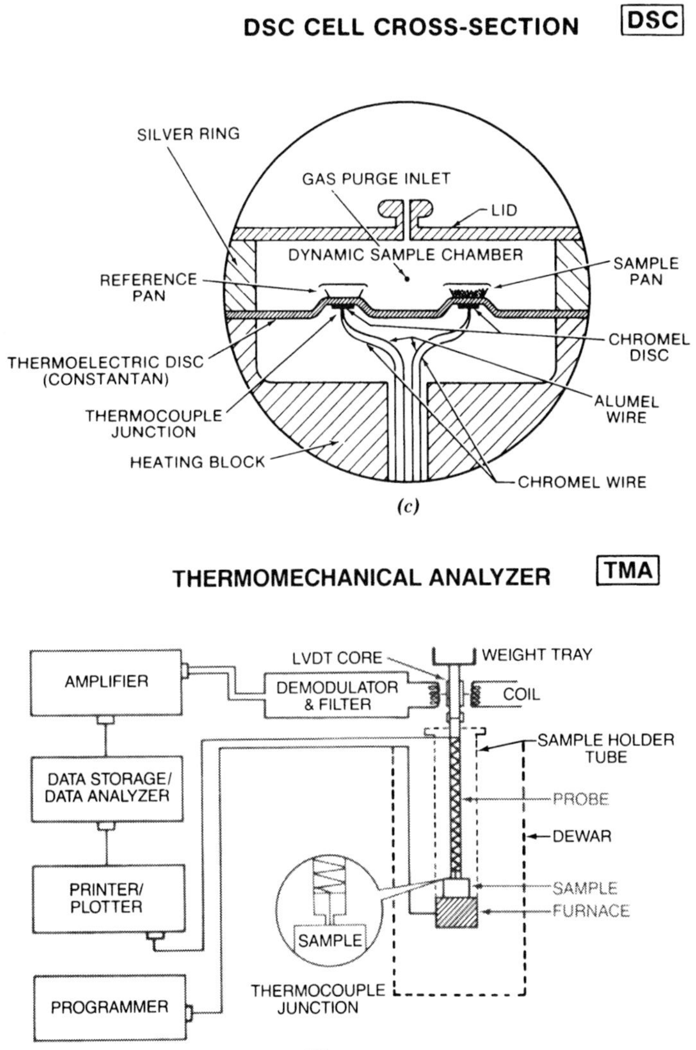

DSC CELL CROSS-SECTION DSC

(c)

THERMOMECHANICAL ANALYZER TMA

(d)

Fig. 6.6 (c) Differential scanning calorimetry (DSC) and (d) thermal mechanical analysis (TMA). (Figures courtesy of Du Pont Inc., Wilmington, DE.)

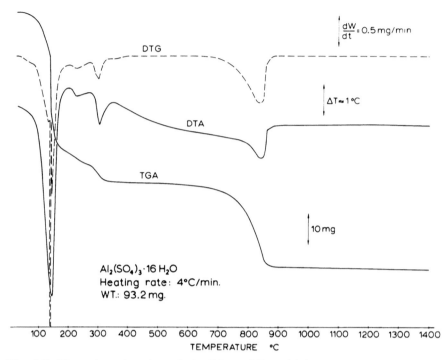

Fig. 6.7 Thermal gravimetric analysis (TGA), differential thermal gravimetric analysis (DTG), and differential thermal analysis (DTA) for the decomposition of alum $Al_2(SO_4)_3$ $16H_2O$. Decomposition yields water up to 300°C and SO_3 between 600 and 875°C. (Courtesy of Mettler Instrument Corp., Hightstown, NJ.)

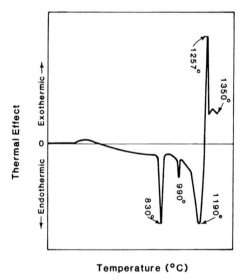

Temperature (°C)

Fig. 6.8 Differential thermal analysis for a mixture of $BaCO_3$ and TiO_2 heated to 1350°C in air. Endothermic peaks at 830 and 990°C correspond to structural transitions in $BaCO_3$. The peaks at 1190 and 1257°C correspond to the decomposition of $BaCO_3$ and the formation of the transient phase Ba_2TiO_4 and the desired phase $BaTiO_3$. [After L. Templeton and J. Pask, *J. Am. Ceram. Soc.*, **42**(5), 213 (1959).]

be controlled include the rate of heating or cooling, the thermal conductivity of the container and packed powder sample and reference, the particle size of the sample, and the composition and flow of the atmosphere. Thermochemical information supplemented with chemical, phase, and microstructural analyses of heated material is used to identify the changes, such as the elimination of liquids, the oxidation and vaporization of organic additives, transitions in materials, and reactions between materials (Fig. 6.8) or between materials and the gaseous environment, vitrification, and recrystallization.

Thermophysical analyses include the monitoring of expansion or shrinkage during heating or cooling and the resistance to mechanical penetration or transmission of mechanical vibrations during heating (TMA). Thermophysical analyses are used to identify phase changes and sintering of inorganic materials during calcination and changes in the properties of organic binders.

SUMMARY

Ceramic materials must be judiciously characterized to control and reproduce the material. The analyses and techniques selected should be planned wisely so that those characteristics that are determined supply the essential information. Nominal chemical characterization includes the major chemical composition and the impurity analysis. Phase analysis provides information about the type and amount of crystalline and amorphous constituents. Microanalytical techniques provide information about the surface composition and chemical and phase segregation on a microscopic scale. Changes in the material and its reaction with the environment during a thermal change are determined using thermal analysis techniques. Basic materials characterization can be tedious, expensive, and time-consuming, but, wisely practiced, it can provide information essential to developing or producing a higher-quality ceramic. Numerous contract service laboratories provide standard and special analyses.

SUGGESTED READING

1. J. E. Enrique, E. Ochandio, and M. F. Gazulla, "Chemical Analysis," in *Engineered Materials Handbook*, Vol. 4, ASM International, Materials Park, OH, 1991.

2. P. K. Gallagher, "Characterization of Materials by Thermoanalytical Techniques," *MRS Bull.*, (7), 23–27 (1988).

3. Sharon L. Smith, "Small Samples Yield to FTIR Microanalysis," *Res. Dev.*, (9), 113–118 (1986).

4. Franklin F. Y. Wang, "Powder Characterization," in *Advances in Powder Technology*, Gilbert Y. Chin, Ed., American Society of Metals, Metals Park, OH, 1982, Chapter 3.

5. Carlo G. Pantano, "Surface and In-Depth Analysis of Glass and Ceramics," *Am. Ceram. Soc. Bull.*, **60**(11), 1154–1163, 1167 (1981).

6. Byron Kratochuil and John K. Taylor, "Sampling for Chemical Analysis," *Anal. Chem.*, **53**(8), 924A-938A (1981).

7. R. Nathan Katz, "Characterization of Ceramic Powders," in *Treatise on Materials Science and Technology*, Vol. 9, Academic Press, New York, 1976.

8. H. Bennett, "Trends in Ceramic Analysis II," *Ceram. Age.*, **5**, 9-10 (1974).

9. L. L. Hench and R. W. Gould, *Characterization of Ceramics*, Marcel Dekker, New York, 1971.

10. Robert A. Condrate, "Infra-red Spectroscopy," in *Analytical Methods for Material Investigation*, G. A. Kirkendale, Ed., Gordon and Breach, New York, 1971, Chapter 6.

11. P. D. Garn, *Thermoanalytical Methods of Investigation*, Academic Press, New York, 1965.

PROBLEMS

6.1 The impurity analysis (ppm) of a material is reported as follows: SiO_2, 50–100; Fe_2O_3, 50–100; MgO, < 10; CaO, < 10; TiO_2, < 10; others, < 10. What is the maximum purity of the material in weight percent?

6.2 Two specimens of zinc aluminate spinel nominally $ZnO \cdot Al_2O_3$ are prepared by reacting zinc oxide and aluminum oxide. The weights of zinc oxide and alumina are 80 and 100 g in batch 1 and 81.38 and 101.94 g in batch 2. Calculate the stoichiometry of each spinel assuming the components react completely.

6.3 What technique would you use for a survey of impurities in zinc manganese ferrite?

6.4 What technique would you use for a quantitative analysis of lithium, sodium, and strontium impurity in barium titanate?

6.5 How are the concentrations of Fe^{2+} and Fe^{3+} in a ferrite determined?

6.6 How would you determine quantitatively the reaction temperature and amount of unreacted zinc oxide for the reaction in problem 6.2?

6.7 How would you determine if zinc oxide volatilizes during the reaction in problem 6.2?

6.8 How would you analyze for the presence of carbonate impurity in barium titanate made by reacting barium carbonate and titania? How would you determine the amount present?

6.9 What technique would you use to analyze the concentration of impurities in grain boundaries of sintered alumina of 10-μm grain size?

6.10 How would you determine the amount of monoclinic and tetragonal zirconia on the surface of a zirconia wear plate?

6.11 How would you analyze for surface oxide and adsorbed hydrocarbon on dense polycrystalline SiC?

6.12 A clay has the following chemical analysis: SiO_2 (45.5), Al_2O_3 (38.3), Fe_2O_3 (0.64), TiO_2 (1.70), CaO (0.07), MgO (0.25), K_2O (0.08), Na_2O (0.12), and ignition (13.34 wt%). Estimate the composition of kaolinite, montmorillonite, hematite, and rutile. Assume that the MgO is present in the montmorillonite.

EXAMPLES

Example 6.1   What information about a particle sample might be obtained using microscopy, EDS, and XPS?

Solution. Optical microscopy will show any dispersed phase, crystallinity, and other physical structure. Scanning electron microscopy will show morphology over a broad range of magnifications, is relatively easy to use, and is quick. EDS is used to identify the chemical composition of elements with $Z > 11$, and XPS may provide information about the chemical bonding.

Example 6.2   The thermal gravimetric analysis for the decomposition of $Al_2(SO_4)_3 \cdot 16H_2O$ in Fig. 6.7 indicates a loss of 27 mg on heating from 600 to 1400°C. The sample weight was 93.2 mg. Can we conclude that all of the SO_3 has been eliminated from the material?

Solution. The formula weight of $Al_2(SO_4)_3 \cdot 16H_2O$ is calculated to be 630.384 g and the weight of 3 moles of SO_3 is 240.183 g. For complete decomposition, the weight loss (ΔW) would be

$$\Delta W_{(SO_3)}/W_{total} = 240.183 \text{ g}/630.384 \text{ g} = 0.381$$

The theoretical weight loss is 0.381 (93.2 mg) = 35.5 mg. The experimental weight loss is about 24% less than that expected and it can be concluded that some SO_3 remains in the material after calcining to 1400°C. This conclusion is corroborated by the negative slope of the TGA curve at 1400°C. Determination of SO_3 in the calcined material by chemical analysis is needed to verify this conclusion based on the observed weight loss.

CHAPTER 7

PARTICLE SIZE AND SHAPE

The particle sizes of most conventional materials fall in the range of 50 nm to 1.0 cm. However, sizes as large as 10 cm are used in some refractory castables and concrete and particles as small as 5 nm are observed in some chemically prepared materials. This range of sizes extends from about the size of coarse gravel to the size of viruses, as shown in Fig. 5.1.

Two or more size analysis techniques may be required to cover the size range of interest. Fortunately, recent advances in instrumentation have greatly improved the precision and reduced the time and tedium involved in obtaining an analysis, and we can now use particle size results in a more timely and effective way. The accuracy of the particle size data depends somewhat on the sample preparation, the particle shape, and the technique used for the analysis. Because the particle size distribution is one of the most important characteristics of a particle system, it is important that we understand principles involved in instrumental techniques used and factors that can alter the data and their interpretation.

7.1 ANALYSIS TECHNIQUES

Just as for chemical analyses, it is important to obtain a representative sample of the appropriate amount. In particle size analysis, we must recognize that sampling and handling may change the physical state of the sample. Dispersing the material in a liquid may reduce the concentration and sizes of agglomerates. Conversely, agglomeration may occur in a well-dispersed suspension if a slow analysis technique is used. We should not forget the intended use of the size information when selecting an instrument and specifying the sample preparation procedure. Techniques for reducing the size of a representative sample and causes of error in size analysis are indicated in Fig. 7.1.

Microscopy Techniques

Microscopy can quickly show the nominal size and shape of the particles in a sample, and obtaining a representative micrograph of the sample is often the

Sampling of Powders
Laboratory

1. Grab sample — Select small amount from larger sample

2. Cone and quarter — Split sample into 4 sections

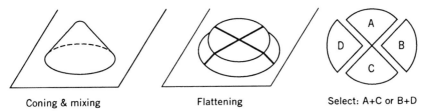

Coning & mixing   Flattening   Select: A+C or B+D

3. Riffling:

Riffle splitter   Rotary riffle

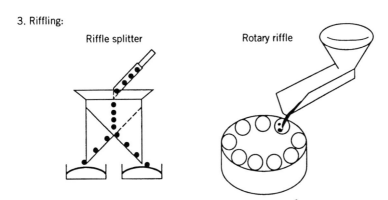

Fig. 7.1 Laboratory sampling involves cone and quartering or riffling to obtain a smaller representative sample for analysis. The major source of error in size analysis depends on the size range of the sample (see page 94).

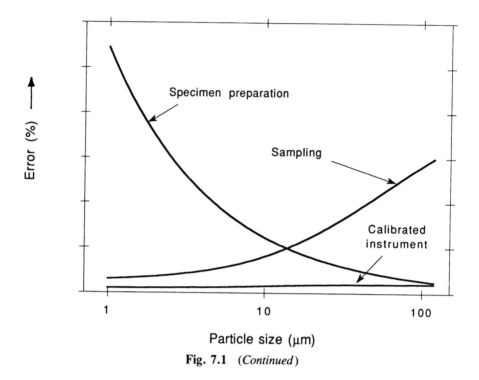

Fig. 7.1 (*Continued*)

first step (Fig. 4.5). Additional micrographs are often taken at a higher magnification to observe atypical particles and finer details of the shape and surface characteristics. The full range of sizes can be examined using optical and electron microscopes. Often a scanning electron microscope will cover the size range of interest. Dry powder can be mounted on the sample stub by inserting a stub coated with an adhesive into the powder. Alternatively, a dilute suspension of particles can be placed on the stub, which will form a deposit of particles on evaporation of the liquid. Extreme dilution, chemical dispersion, fast drying, or freeze-drying may be used to minimize agglomeration.

Quantitative image analysis requires a photographic enlargement or an electronic display. For statistical accuracy, it is necessary to measure and count at least 700 particles lying in a plane. It will be difficult to distinguish and measure fine particles in a micrograph when the particles are poorly dispersed and when coarse particles are more than 1 order of magnitude larger than the fines are present (Fig. 7.2). Agglomerates in powders will be apparent, if present, and with care it will be possible to determine if these are hard or soft agglomerates (Fig. 7.3).

Particles may vary widely in shape (Figs. 3.1, 3.2, 3.5). Crushed minerals and calcined aggregates are generally angular and have rough surfaces. Milled materials are commonly more rounded, but aggregates may have a rough surface. Coarser kaolin particles may be in the form of lineated or fairly isometric

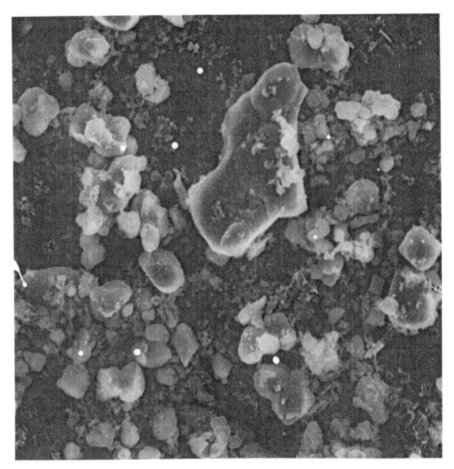

Fig. 7.2 Microscopic analysis may not distinguish all particles when the size range is very large. (Size range is 0.1–20 μm and small particles are indistinct.)

stacks of platelike particles; fine kaolin contains both small stacks and platelike particles.

The characteristic size of particles that are spherical or cubical may be defined in terms of the diameter or edge length. For most real particles, the definition of the characteristic size is less precise. The maximum chord length in a particular reference direction between two tangents on opposite sides of the image is often used as the characteristic size of the particle, as shown by the parameter a in Fig. 7.4. Another characteristic size can be defined as the length of the chord that bisects the area of the particle. The diameter of the circle with an equal projected area is sometimes used as the size. Particle dimensions can be measured manually on a photographic enlargement using vernier calipers or a piezoelectric matrix pad. Images from light and electron microscopes can also be displayed on the screen of an image-analyzing com-

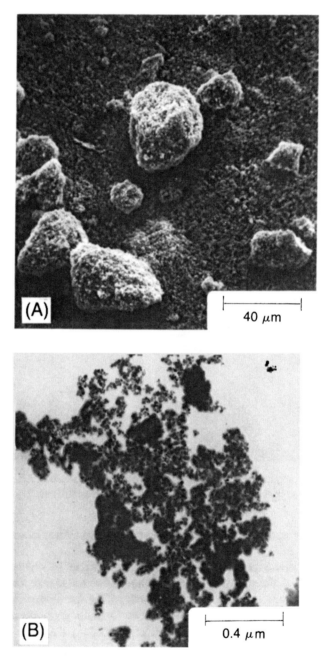

Fig. 7.3   As-received ultrafine zirconia powder contains (A) soft agglomerates that are easily dispersed into (B) small aggregates of very fine crystallites.

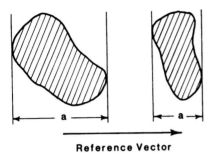

Reference Vector

Fig. 7.4 The distance between two normals tangent to the image of the particle may be used as the "characteristic length" of the particle.

puter which can rapidly and automatically scan the image and determine a preselected characteristic dimension of each region of differing optical density. A light-pen accessory enables the operator to select or omit particular features (such as agglomerates) in the analysis.

The aspect ratio of a particle is defined as the longest dimension divided by the shortest dimension, and a particle with an aspect ratio greater than 1 is said to be anisometric. The aspect ratio of the particles may be estimated by measuring the major and minor axes of particles suitably oriented in a photomicrograph. The constant of proportionality between the particle size and its area A or volume V can be considered a numerical shape factor. Examples are given in Fig. 7.5. The ratio ψ_A/ψ_V relative to that for a sphere is sometimes used as an index of angularity.

Error in microscopy techniques can occur owing to an inaccurate magnification, a nonrandom arrangement of particles, and insufficient sampling. The magnification of the microscope must be calibrated. Anisometric particles will tend to rest with their longer dimension parallel to the place of the image, and this introduces a bias in the data. Extreme fines are often incompletely analyzed because of human bias or agglomeration in the specimen.

Sieving Techniques

Sieving is the classification of particles in terms of their ability or inability to pass through an aperture of controlled size. Particles are introduced onto a stack of sieves with successively finer apertures below, and the particles are agitated to induce translation until blocked by an aperture smaller than the particle size. Sieves with openings greater than 37 μm are constructed using

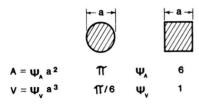

$$A = \psi_A a^2 \qquad \pi \qquad \psi_A \qquad 6$$
$$V = \psi_V a^3 \qquad \pi/6 \qquad \psi_V \qquad 1$$

Fig. 7.5 Figure showing the definition of the area shape factor ψ_A and the volume shape factor ψ_V. The size of the particle is a and the area and volume are A and V, respectively.

wire mesh and are identified in terms of mesh size and the aperture size; the relationship between mesh size and aperture size for U.S. Standard sieves is listed in Appendix A. The aperture size can be calibrated using an optical micrometer. Monolithic electrodeposited metal sieves are now commercially available with uniform apertures as small as 2 μm.

Sieving is the most widely used technique for sizes down to 44 μm. A set of sieves often follows a $\sqrt{2}$ progression of sizes. Agglomeration becomes a problem below about 44 μm and can introduce error in the analysis. In some cases, dry sieving analyses can be extended to 20 μm by admixing a desiccant of known size into a very dry powder sample. Alternatively, the sample can be dispersed in a liquid and sieved wet.

For an equal weight of sample, the number of particles increases as (size)$^{-3}$ and apertures of the finer sieves become blocked ("blinded") if the sample size is too large and the sieving mechanism is inefficient. Older sieving apparatus depended on the low-frequency mechanical shaking for particle motivation. Modern apparatus that imparts a relatively high frequency air pulse produces a more efficient sieving of the small sizes. The combination of a low-frequency mechanical pulse to translate coarse particles and the higher-frequency air pulse can provide precise dry sieving analyses for a wide range of sizes if the particles are regular or only slightly anisometric in shape. An elongated particle does not have a single "characteristic size," and the precision of sieve analyses depends somewhat on the aspect ratio of the particle.

Wet sieving with "micromesh" sieves finer than 37 μm requires suction to pull liquid and fine entrained particles through the apertures. Particle motivation can be assisted by supporting the sieve in an ultrasonic tank or using a mechanical vibrator. Wet sieving on a 44- or 37-μm sieve is often used when analyzing the "subsieve" portion of a clay. Wet sieving to 2 μm for routine control analysis has been reported.

Sedimentation Techniques

A spherical particle of density D_p and diameter a released in a viscous fluid of viscosity η_L of lower-density D_L will momentarily accelerate and then fall at a constant terminal velocity v. Laminar liquid flow occurs when the particle Reynolds number Re is less than 0.2. The Reynolds number is calculated from the equation Re $= (vaD_L)/\eta_L$. For ceramic particles, the upper size limit is about 50 μm. The terminal velocity is related to the particle diameter according to the Stokes equation (Fig. 7.6):

$$v = a^2(D_p - D_L)g/18\eta_L \qquad (7.1)$$

where g is the acceleration due to gravity or centrifuging. The time t for settling a height H is

$$t = \frac{18H\eta_L}{a^2(D_p - D_L)g} \qquad (7.2)$$

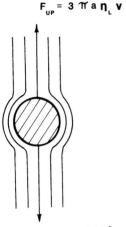

$$F_{UP} = 3 \pi a \eta_L v$$

$$F_{DOWN} = \frac{\pi a^3}{6}(D_p - D_L)g$$

Fig. 7.6 Equilibrium of forces during the settling of a particle in a Newtonian fluid with laminar flow.

For particles of alumina in water, the time for gravity settling 1 cm is about 1 min for a 10-μm particle but about 2 h for a 1-μm particle.

Commercial apparatus utilizes either a layer technique or a homogeneous technique. In the layer technique, a very dilute suspension is carefully placed on a column of settling liquid, and the fraction of particles that have settled to a depth H is measured with time. Settling times are reduced by centrifuging, for which the Stokes equation is

$$t = \frac{18\eta_L \ln (r_t/r_0)}{a^2(D_p - D_L)\omega^2} \tag{7.3}$$

where ω is the angular velocity of the centrifuge and r_0 and r_t are the radial positions of the particles before and after centrifuging.

In the homogeneous technique, the fraction of particles in a plane at some depth H in a dilute (<4 vol %) homogeneous suspension is determined. Particles of all sizes may settle across the plane, but only particles of a particular size and smaller, in accordance with Stokes' law, will remain at or above the plane after some particular settling time. The analysis time for sizes finer than 1 μm is reduced considerably by centrifuging. The concentration in suspension is normally determined from the relative intensity I/I_0 of transmitted light or x-rays:

$$-\ln (I/I_0) = k\Sigma N_i a_i^2 \tag{7.4}$$

where k is a constant and N_i is the number of particles of size a_i and

$$\Sigma \text{ particle volume } = -a_i \ln (I/I_0) \tag{7.5}$$

The analysis time varies with the particle density and fineness, but analysis times are typically in excess of 10 min and may range up to several hours for a gravitational settling to 0.02 μm using commercial instrumentation.

Possible sources of error in sedimentation size analyses include hindered settling due to particle interactions, the tendency of fine particles to be pulled along behind larger ones, and agglomeration caused by Brownian motion. An adsorbed hull of liquid molecules and/or deflocculant can increase the size but reduce the average density or submicron particles (Fig. 4.3). Incomplete dispersion and agglomeration during settling may also affect the size data. Ultrasonic dispersion and an admixed deflocculant are commonly used for an ultimate size analysis.

Settling techniques cannot be used for samples containing particles of different density. For nonspherical particles, the diameter of a sphere that settles at the same rate, the equivalent spherical diameter, is obtained from the analyses. Theoretical models indicate that the Stokes' drag force is insignificantly affected by small distortions from sphericity.

Electrical Sensing Techniques

The resistance of an electrolyte current path through a narrow orifice between two electrodes increases when a ceramic particle passes through the orifice. The pulse in resistance is proportional to the volume of electrolyte displaced. In electrical sensing techniques, resistance pulses for a stream of dispersed particles passing through the orifice are converted into voltage pulses, amplified, scaled, and counted electronically (Fig. 7.7).

Monosize samples of known size are used for calibration. Using one orifice tube, the analysis time for a predispersed sample is less than 1 min with very high statistical confidence. The working range is typically 2–40% of the ap-

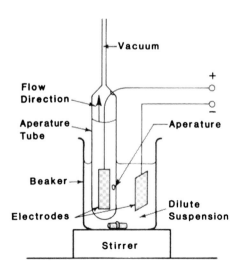

Fig. 7.7 Schematic of the electrical sensing zone technique.

erture size, or 0.3–400 μm using several aperture tubes. This technique can be used for samples containing particles of different density and an aspect ratio up to about 10. Extreme dilution to less than a few parts per million is required to minimize coincident counting. Samples must be dispersed in special electrolytes.

Laser Diffraction Techniques

Dispersed particles momentarily traversing a collimated light beam will cause Fraunhofer diffraction of light outside of the cross section of the beam when the particles are larger than the wavelength of the light. The intensity of the forward-diffracted light is proportional to the particle size squared, but the diffraction angle varies inversely with particle size. A He-Ne laser is commonly used for the light source. The combination of an optical filter, lens, and photodetector or a lens and multielement detector coupled with a microcomputer enables computation of the particle size distribution from the diffraction data (Fig. 7.8). The sample may be either a liquid or gas suspension of particles or droplets of about 0.1 vol % concentration. A typical analysis time is about 2 min, which permits on-line analysis. The possible size range capability of this technique is about 1–1800 μm. However, the working size range of a particular instrument varies with the model, and some selection is possible. This technique is independent of the density of the particulates and is factory calibrated. Particles finer than the detection limit are not included in the data.

Submicron sizing analysis may be obtained from dynamic light scattering. In dynamic light scattering the Doppler frequency shift related to the particle velocity produced by Brownian motion is measured. Wide-angle Mie scattering is also used for submicron size analysis. The popularity of light diffraction instruments for size analysis has grown rapidly because of their precision, speed of analysis, ease of operation, versatility, and ease of maintenance.

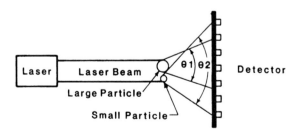

Diffraction Angle θ1 For Large Particles.
Diffraction Angle θ2 For Small Particles.

Fig. 7.8 Particles larger than the wavelength of the laser beam produce Fraunhofer forward diffraction.

Light Intensity Fluctuation Techniques

Dispersed particles finer than 5 μm constantly diffuse randomly through the liquid and can scatter light. The frequency of the fluctuations may be detected using a photomultiplier, and the particle size is calculated from the frequency using the Stokes–Einstein equation

$$a = (k_B T)/(3\pi\eta_L D_T) \qquad (7.6)$$

where D_T is the translational diffusion constant at a temperature T. An average size and polydispersity index is rapidly obtained using a digital correlation computer. The working size range is from about 5 to 0.005 μm.

Other Techniques

X-ray line-broadening techniques may be used to determine the average crystallite size in a powder finer than about 0.5 μm and small-angle x-ray scattering below 0.1 μm. Calibration is required, and particles should not be polycrystalline.

Air permeability techniques have also been used to determine an index of average particle size in the range 50–0.2 μm. Air permeability is a function of the number and geometry of capillaries in a bed of powder. The geometry of the capillaries depends on particle packing characteristics and is not simply a function of some nominal particle size. However, the air permeability technique is quick and is sometimes used for quality control purpose.

An average particle size may also be calculated from the specific surface area of the powder. This technique is discussed in Chapter 8.

7.2   SELECTING A TECHNIQUE

The specification of an instrument and sample preparation procedures should be based on experience as well as principles. Instrument cost, flexibility, size range capability, specimen preparation, and analysis time are all important factors. Ease of operation and maintenance and the direct readout of results in printed or graphical form is preferable. But most important, the result should be useful. Capabilities of the techniques discussed above are summarized in Table 7.1.

Comparative size analyses for representative materials using different size analysis techniques and optimum procedures are quite limited. Zwicker* has reported very good agreement in the size analysis for finely ground alumina using micromesh sieving, centrifugal sedimentation, and electronic counting. Kahn† showed that control of agglomeration was critical to obtain a precise

---

*J. D. Zwicker, *Am. Ceram. Soc. Bull.*, **46**(3), 303–306 (1967).
†M. Kahn, *Am. Ceram. Soc. Bull.*, **57**(4), 448–451 (1978).

TABLE 7.1   Summary of Particle Size Analysis Techniques

Method	Medium	Size Capability (μm)	Sample Size (g)	Analysis Time[a]
Microscopy				
Optical	Liquid/gas	400–0.2	<1	S-L
Electron	Vacuum	20–0.002	<1	S-L
Sieving				
	Air	8000–37	50	M
	Air	5000–37	5–20	M
	Liquid	5000–5	5	L
	Inert gas	5000–20	5	M
Sedimentation				
(Gravity)	Liquid	100–0.2	<5	M-L
(Centrifuge)	Liquid	100–0.02	<1	M
Electrical sensing zone				
	Liquid	400–0.3	<1	S-M
Light scattering				
Fraunhofer	Liquid/gas	1800–1	<5	S
Mie	Liquid	1–0.1	<5	S
Doppler	Liquid	6–0.003	<1	S
Intensity fluctuation				
	Liquid	5–0.005	<1	S

[a] S = short (<20 min); M = moderate (20–60 min); L = long (>60 min).

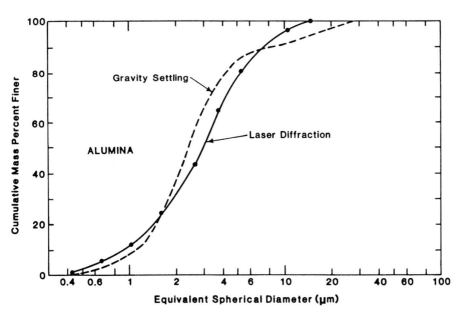

Fig. 7.9 Comparative size data for a calcined alumina obtained by gravity settling and laser diffraction techniques. (After H. Frock and P. Plantz, *Processing Consequences of Raw Material Variables*, Alfred University Press, Alfred, NY, 1985.)

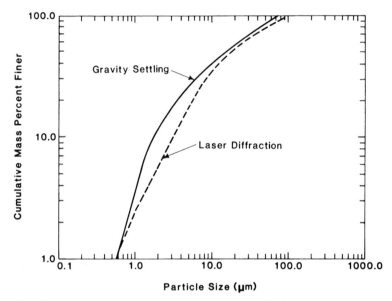

Fig. 7.10 Comparative size data for well-dispersed, milled wollastonite CaSiO₃ particles having an aspect ratio of about 4.

analysis for barium titanate finer than a few microns. Croft* reported that sedimentation analyses were more precise and more sensitive to the submicron size distribution of clay suspensions than an electrical sensing zone technique. Frock and Plantz† reported reproducible differences in the size analyses of a common alumina powder when using different techniques, as indicated in Fig. 7.9. For a sample of columnar-shaped particles of wollastonite, the size distribution determined by gravity settling was reproducibly finer than that obtained by light diffraction, as shown in Fig. 7.10.

ASTM studies for centrifugal sedimentation and electrical sensing zone analyses have indicated that a precision of ±1% (15% confidence level) at each size and ±3% between laboratories can be expected for each method using very careful procedures.

7.3 PRESENTATION OF PARTICLE SIZE DATA

Particle size data are usually presented in tabular or graphical form for analyzing and comparing the mean size and distribution of sizes. A description of the data in terms of a particular mathematical distribution function is also desirable but not always possible.

*T. Croft, in *Processing Consequences of Raw Material Variables*, J. Varner et al., Eds., Alfred University Press, Alfred, NY, 1985, Chapter 1.
†H. Frock and P. Plantz, *ibid*.

Tabulated size data were presented in the specifications of materials in Chapter 5. A histogram may be constructed from data classified according to the number, mass, or volume fraction for particular intervals of size (Fig. 7.11). Data determined on a volume basis are equal to data on a mass basis when the density of the particles does not vary with size. The cumulative size distribution is obtained by summing the fractions of particles that are finer (CNPF or CMPF) or alternatively larger (CNPL or CMPL) than specific sizes between the smallest and largest sizes of the distribution (Table 7.2).

When the number fraction ΔN of particles in each size interval Δa is plotted as a function of size, the fractional distribution function $f_N(a)$ is obtained in the limit as Δa approaches 0:

$$f_N(a) = \lim_{\Delta a \to 0} \frac{\Delta N}{\Delta a} = \frac{dN}{da} \tag{7.7A}$$

Similarly, the fractional distribution by mass $f_M(a)$ is

$$f_M(a) = \lim_{\Delta a \to 0} \frac{\Delta M}{\Delta a} = \frac{dM}{da} \tag{7.7B}$$

The fractional distribution functions are related to the cumulative size distribution $F(a)$ by the equations

$$f_N(a) = \frac{dF_N(a)}{da} \tag{7.8A}$$

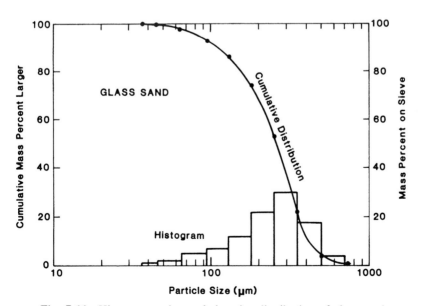

Fig. 7.11 Histogram and cumulative size distribution of glass sand.

TABLE 7.2 Particle Size Distribution of Ball Milled Petalite LiAlSi$_4$O$_{10}$

Sieve Size (μm)	Mass on Sieve		CMPF (%)	CMPL (%)
	(g)	(%)		
1000	0	0.0	100.0	0.0
500	0.18	0.9	99.1	0.9
355	0.50	2.5	96.6	3.4
250	1.45	7.3	89.3	10.7
180	3.00	15.0	74.3	25.7
125	3.96	19.8	54.5	45.5
90	3.26	16.3	38.2	61.8
63	3.18	15.9	22.3	77.7
44	1.94	9.7	12.6	87.4
0	2.52	12.6	0.0	100.0
Total	20.00	100.0		

and

$$f_M(a) = \frac{dF_M(a)}{da} \tag{7.8B}$$

where $F(a)$ is the cumulative fraction finer than size a and the subscripts N and M indicate number and mass bases, respectively. For distributions extending over a wide range of size, the fractional distribution with the logarithm of size is used:

$$f_N(\ln a) = \frac{dF_N(a)}{d(\ln a)} = \frac{a[dF_N(a)]}{da} \tag{7.9}$$

Particle sizes corresponding to $F_M(a)$ of 90, 50, and 10% are now often listed in the specifications of a material, as is indicated in Tables 5.3 and 5.4. These values provide an approximate basis for comparing the apparent mean size and range of sizes.

It is common practice to fit a mathematical function to the cumulative data. The fractional distribution is then obtained by taking the derivative of the cumulative function. Mathematical functions used to describe common particle size data are discussed in Section 7.5.

7.4 CALCULATING A MEAN PARTICLE SIZE

Different mean particle sizes may be defined for a material in terms of the fractional distribution function $f_N(a)$ and an appropriate shape factor. The total number of particles N is

$$N = \int_{a_{min}}^{a_{max}} f_N(a) \, da \tag{7.10}$$

The sum of the lengths L of all of the particles is

$$L = \int_{a_{min}}^{a_{max}} a f_N(a) \, da \tag{7.11}$$

and the mean length $\bar{a}_L$ is

$$\bar{a}_L = L/N = \frac{\int_{a_{min}}^{a_{max}} a f_N(a) \, da}{N} \tag{7.12}$$

The total area A of the particles is

$$A = \int_{a_{min}}^{a_{max}} \psi_A a^2 f_N(a) \, da \tag{7.13}$$

If $\bar{a}_A$ is the size of the particle having the average surface area,

$$A = N \psi_A \bar{a}_A^2 \tag{7.14}$$

When ψ_A is a constant

$$\bar{a}_A = \sqrt{\int_{a_{min}}^{d_{max}} \frac{a^2 f_N(a) da}{N}} \tag{7.15}$$

Definitions of other mean particle sizes are given in Table 7.3. When the mathematical distribution function is not known, the average size can be calculated using a finite summation (Table 7.3). Using finite summation, the number of particles N is

$$N = \sum_{i=1}^{i=N_c} f_N(\bar{a}_i) \tag{7.16}$$

where $f_N(\bar{a}_i)$ is the fraction of particles in an interval Δa_i having a mean size $\bar{a}_i$, and N_c is the number of size classes.

TABLE 7.3   Definitions of Mean Particle Size

Mean	Equation	Finite Summation
Length ($\bar{a}_L$)	$\bar{a}_L = \dfrac{\displaystyle\int_{a_{\min}}^{a_{\max}} a f_N(a)\, da}{N}$	$\bar{a}_L = \dfrac{\displaystyle\sum_{i=i}^{i=N_c} \bar{a}_i f_N(\bar{a}_i)}{N}$
Surface ($\bar{a}_A$)	$\bar{a}_A = \sqrt{\dfrac{\displaystyle\int_{a_{\min}}^{a_{\max}} a^2 f_N(a)\, da}{N}}$	$\bar{a}_A = \sqrt{\dfrac{\displaystyle\sum_{i=1}^{i=N_c} \bar{a}_i^2 f_N(\bar{a}_i)}{N}}$
Volume ($\bar{a}_V$)	$\bar{a}_V = \sqrt[3]{\dfrac{\displaystyle\int_{a_{\min}}^{a_{\max}} a^3 f_N(a)\, da}{N}}$	$\bar{a}_V = \sqrt[3]{\dfrac{\displaystyle\sum_{i=1}^{i=N_c} \bar{a}_i^3 f_N(\bar{a}_i)}{N}}$
Volume/surface ($\bar{a}_{V/A}$)	$\bar{a}_{V/A} = \dfrac{\displaystyle\int_{a_{\min}}^{a_{\max}} a^3 f_N(a)\, da}{\displaystyle\int_{a_{\min}}^{a_{\max}} a^2 f_N(a)\, da}$	$\bar{a}_{V/A} = \dfrac{\displaystyle\sum_{i=1}^{i=N_c} \bar{a}_i^3 f_N(\bar{a}_i)}{\displaystyle\sum_{i=1}^{i=N_c} \bar{a}_i^2 f_N(\bar{a}_i)}$

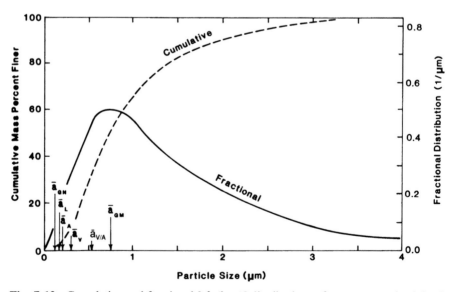

Fig. 7.12   Cumulative and fractional [$f_M(\ln a)$] distributions of a superground calcined alumina powder. The distribution is log normal; calculated mean sizes are indicated on the size scale.

If the size data are on a mass basis, $f_N(a)$ may be calculated from $f_M(a)$ for substitution into the equations in Table 7.3, that is,

$$f_M(a) = (D_p \psi_v a^3) f_N(a) \tag{7.17A}$$

and

$$f_M(\bar{a}_i) = (D_p \psi_v \bar{a}_i^3) f_N(\bar{a}_i) \tag{7.17B}$$

Because of the different dependences on particle size, the rank of the different mean sizes is $\bar{a}_L < \bar{a}_A < \bar{a}_V$, as is shown in Fig. 7.12. The mode for the distribution by number will appear at a smaller size than the mode corresponding to the mass distribution because of the greater number of particles in a unit mass of finer particles.

7.5 PARTICLE SIZE DISTRIBUTION FUNCTIONS

When a particular mathematical function can be used to satisfactorily approximate the size data for a material, the characteristic parameters of the distribution and the mathematical form may be used for purposes of describing and comparing materials. The characteristic parameters are a characteristic mean size or maximum size and a distribution modulus indicating the range of sizes.

The size distribution of many granulated powders produced by spray-drying and many powders produced by milling fined-grained minerals and calcined aggregate is often approximated by the log-normal LN distribution function

$$f(\ln a) = \frac{1}{\sqrt{2\pi} \ln \sigma_g} \exp - \frac{(\ln a_i - \ln \bar{a}_g)^2}{2(\ln \sigma_g)^2} \tag{7.18}$$

where the frequency function $f(\ln a)$ may be either a number, volume, or mass basis. The parameter $\bar{a}_g$ is the geometric mean size, and σ_g is the geometric standard deviation indicating the range of sizes. The log-normal distribution is characteristically skewed as shown in Fig. 7.12. Using logarithmic probability paper, data for $F_N(a)$ or $F_M(a)$ will plot as a straight line if the distribution is log-normal. The geometric standard deviation σ_g is obtained from the ratio of sizes for which $F_N(a)$ or $F_M(a)$ is equal to 84.13% and 50.00%. The geometric mean size ($\bar{a}_{gN}$ or $\bar{a}_{gM}$) is the size corresponding to 50.00%. The geometric mean size $\bar{a}_{gM}$ for mass data is larger than the geometric mean $\bar{a}_{gN}$ for number data for the same material, as is indicated in Fig. 7.13. One geometric mean can be calculated from the other using the equation

$$\ln \bar{a}_{gM} = \ln \bar{a}_{gN} + 3(\ln \sigma_g)^2 \tag{7.19}$$

Linear, area, and volume mean sizes may be calculated for a LN size distribution using the equations in Table 7.4. Size data for an industrial spray-dried

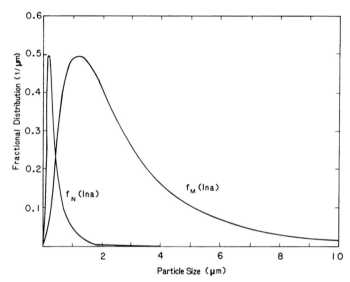

Fig. 7.13 Fractional number and mass size distributions for a milled zircon powder having a log-normal size distribution ($\bar{a}_{gM} = 1.2\ \mu$m, $\sigma_g = 2.25$).

ferrite powder, obtained using two different techniques, and the characteristic log normal parameters are listed in Table 7.5. And size data for ball-milled zircon and barium titanate powders that are log-normal in distribution are listed in Table 7.6.

Another empirical distribution function that has been observed to describe the particle size data of some milled coarse-grained materials is the Rosin–Rammler (RR) equation:

$$F_M(a) = 1 - \exp - (a/a_{\text{RR}})^n \qquad (7.20)$$

where a_{RR} is a characteristic size parameter for which $F_M(a) = 0.63$ and n^{-1} is an index indicating the dispersion of size. For the RR distribution, a plot of $-\ln [1 - F_M(a)]$ versus size on log-log paper is linear; n corresponds to the slope, and a_{RR} is the size for which $[-\ln (1 - F_M(a)] = 1$. The mean size

TABLE 7.4 Equations for the Log-Normal Distribution

$$\ln \bar{a}_L = \ln \bar{a}_{gN} + 0.5(\ln \sigma_g)^2$$
$$\ln \bar{a}_A = \ln \bar{a}_{gN} + 1(\ln \sigma_g)^2$$
$$\ln \bar{a}_V = \ln \bar{a}_{gN} + 1.5(\ln \sigma_g)^2$$
$$\ln \bar{a}_{V/A} = \ln \bar{a}_{gN} + 2.5(\ln \sigma_g)^2$$

TABLE 7.5   Size Data and Log-Normal Parameters for an Industrial Spray-Dried Manganese Zinc Ferrite Powder

Size (μm)	CNPF (%)[a]	Sieve Size (μm)	CMPF (%)[b]
120	100.0	177	100
90	94.2	125	91.4
70	82.0	88	60.2
60	65.2	62	28.0
50	50.9	44	4.0
40	22.6		
30	11.0		
		LN Parameters	
$\bar{a}_{g_N}$ (μm)	50.5		—
$\bar{a}_{g_M}$ (μm)	—		79.0
σ_g (μm)	1.43		1.39

[a]SEM analysis.
[b]Seive analysis.

TABLE 7.6   Size Data and Log-Normal Parameters for Finely Milled Zircon and Barium Titanate Powders

	Milled Barium Titanate		Milled Zircon
Size (μm)	CNPF (%)[a]		CMPF (%)[b]
10.0			100
6.0			97.5
4.0	100		93.0
3.0	99[c]		86.5
2.0	98		71.3
1.5	92		60.0
1.0	80		40.5
0.6	58		19.4
0.4	36		9.0
0.2	10		1.4
0.1	2		
	LN Parameters		
$\bar{a}_{g_N}$ (μm)	0.52		—
$\bar{a}_{g_M}$ (μm)	—		1.2
σ_g (μm)	2.12		2.25

[a]Scanning electron microscopy.
[b]Stokes sedimentation.
[c]Aggregates.

TABLE 7.7 Size Data and Characteristic Parameters for Two Milled Materials

Calcined Alumina		−200-Mesh Quartz	
Size (μm)	CMPF (%)	Size (μm)	CMPF (%)
50	79.5	50	92.0
40	67.0	40	81.6
30	53.5	30	69.0
20	35.0	20	50.0
15	24.5	15	39.2
10	14.0	10	27.5
6	6.0	6	17.5
4	2.5	4	12.3
RR parameters		*GS parameters*	
a_{RR} (μm)	36	a_{max} (μm)	47
n	1.5	n	0.85

$\bar{a}_{V/A} = a_{RR}/\Gamma(1 - 1/n)$ where Γ is the gamma function. The RR distribution is generally more skewed than the log-normal distribution. Size data and the calculated RR parameters are listed in Table 7.7 for a ball-milled alumina initially of coarse grain size.

The size distribution of rather coarse industrial minerals produced by crushing and grinding can sometimes be approximated by the Gaudin–Schuhmann

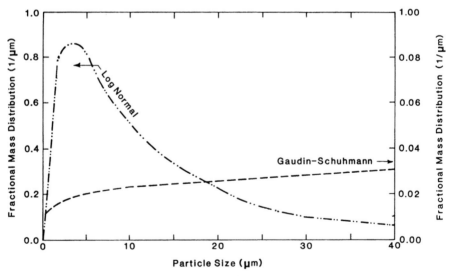

Fig. 7.14 Comparative fractional mass size distributions for materials <40 μm having a log-normal ($\bar{a}_{gM} = 3.4$ μm; $\sigma_g = 2.9$) and a Gaudin–Schuhmann particle size distribution ($n = 1.2$).

(GS) equation:

$$F_M(a) = (a/a_{max})^n \tag{7.21}$$

where a_{max} is the apparent maximum particle size obtained by extrapolation and n is the constant indicating the range of sizes. The size distribution of milled quartz in Table 7.7 is approximated by the GS distribution and is nearly linear when plotted on log-log graph paper. Fig. 7.14 shows the GS distribution is very different than the log-normal distribution.

The size data of real-particle systems depend on the type of fracture, milling, blending of sizes, and other processing. Particles size data are illustrated and discussed in several chapters presenting information on particle packing, milling, spray-drying, forming, and sintering processes.

SUMMARY

The particle size distribution of a particle system may be analyzed using several different techniques. Microscopy techniques provide information about both the size and shape of the particles and the presence of particle agglomerates; the potential range of sizes that may be analyzed is extremely large, and the analysis time is greatly reduced by using a computerized image analyzer. Sieving is convenient and widely used for the size analysis of particles larger than 44 μm. Laser diffraction techniques are very fast and convenient and have become very popular for size analysis from about 100 to <1 μm. Sedimentation techniques are versatile and enable analyses from about 44 to less than 0.1 μm in size in one analysis. Electrical sensing techniques are fast and convenient when the range of sizes is moderate. Light intensity fluctuation analysis permits size analysis to below 0.01 μm. The physical principle of each technique is the basis on which the size is defined, and precise size data for the same powder may vary somewhat using different techniques.

Particle size data are commonly presented in tabular form or graphically for purposes of analysis and comparison. Several different mean particle sizes and the range of sizes may be calculated to summarize the size data. The size distribution of powders produced by milling calcined aggregates and spray-dried powders may often be approximated by a regular mathematical equation such as the log-normal distribution function. When the distribution function is known, the size data are summarized in terms of the type and characteristic parameters of the distribution.

SUGGESTED READING

1. *Modern Methods of Particle Size Analysis*, H. Barth, Ed., Wiley-Interscience, New York, 1985.

2. Bruce Weiner, "Introduction to Particle Sizing Using Photon Correlation Spectroscopy," in *Processing Consequences of Raw Material Variables*, J. Varner, J. Funk, P. Johnson, and J. Reed, Eds., Alfred University Press, Alfred, NY, 1985.

3. Harold N. Frock and P. E. Plantz, "Rapid Analyses of Fine Particles Using Light Scattering," ibid.

4. Terrance Allen, *Particle Size Measurement*, Wiley-Interscience, New York, 1981.

5. Harold N. Frock and Richard Walton, "Particle Size Instrumentation for the Ceramic Industry," *Am. Ceram. Soc. Bull.*, **59**(6), 650–651 (1980).

6. T. Allen and M. G. Baudet, "The Limits of Gravitational Sedimentation," *Powder Tech.*, **18**, 131–138 (1977).

7. J. P. Olivier, G. K. Hickin, and Clyde Orr, Jr., "Rapid Automatic Particle Size Analyses in the Subsieve Range," *Powder Tech.*, **4**, 257–263 (1970/71).

8. J. D. Zwicker, "Comparison of Particle Size Analysis Methods," *Am. Ceram. Soc. Bull.*, **4**(3), 303–306 (1967).

9. John Elvans Smith and Myra Lee Jordan, "Mathematical and Graphical Interpretation of the Log-Normal Law for Particle Size Distribution Analyses," *J. Colloid Sci.*, **19**, 549–559 (1964).

PROBLEMS

7.1 How is the characteristic size parameter specified for each of the techniques in Table 7.1?

7.2 For the coins penny, nickel, and quarter, which has the largest and which has the smallest aspect ratio?

7.3 Compare the shape factors calculated for a nickel in Example 7.1 with the values for a sphere.

7.4 Compare the ratio ψ_A/ψ_V for a cube and a sphere.

7.5 Calculate the maximum size of a spherical alpha alumina particle settling in water at 25°C for which Re < 0.2.

7.6 Plot the data in Table 7.2 in the form of CMPF and CMPL.

7.7 Plot the data in Table 7.2 in the form of a histogram.

7.8 The diameters of 12 spherical particles are 2 μm (one), 4 μm (two), 6 μm (three), 8 μm (one), 10 μm (two), and 12 μm (three). Calculate each of the mean sizes indicated in Table 7.3.

7.9 Using the CNPF data in Table 7.5, calculate the mean sizes $\bar{a}_L$, $\bar{a}_A$, $\bar{a}_V$, $\bar{a}_{V/A}$ by finite summation and using the equations in Table 7.4.

7.10 From the CMPF data in Table 7.5, calculate $\bar{a}_{gN}$ and $\bar{a}_{V/A}$.

7.11 When are the CMPF data and the CVPF (cumulative volume percent finer) data the same?

7.12 Calculate $\bar{a}_{gN}$ for the distribution shown in Fig. 7.13 and correlate your value to the distribution shown.

7.13 Plot the CMPF data in Table 7.6 indicating an uncertainty of ± 2 wt% and determine the geometric mean and the geometric standard deviation.

7.14 Show that the size data in Table 7.2 are approximated by the RR distribution equation. Calculate a_{RR} and n.

7.15 Using the method of finite summation, calculate $\bar{a}_{gM}$, $\bar{a}_{gN}$, and $\bar{a}_{V/A}$ from the data in Table 7.7 for milled quartz. Assume $a_{max} = 60$ μm and reasonable shape factor.

7.16 Contrast the settling times for alumina particles of 0.1 μm diameter for a settling height of 1 cm under gravitational conditions and in a 3600-rpm centrifuge with an initial radial position of 10 cm (water 20°C).

7.17 Estimate the fluctuation time constant t_c (μsec) and D_T for the Brownian motion of an alumina particle of 0.1 μm diameter in water at 20°C, given $(t_c)^{-1} = 2D_T K^2$. Here, $K = (4\pi n/\lambda) \sin(\theta/2)$ where $n = 1.33$ is the index of refraction of the liquid, $\lambda = 633$ nm for the He–Ne laser, and $\theta = 90°$.

7.18 Calculate the mass and number of formula units in MgO particles having a diameter of 1 nm, 1 μm, and 1 mm.

7.19 The LN parameters for a milled alumina powder are $\bar{a}_{gM} = 1.3$ μm and $\sigma_g = 2.4$. Graph $f_M(\ln a)$ and $f_M(a)$ using a linear size scale.

7.20 What is the value of $f(\ln a)$ when $a = \bar{a}_g$?

7.21 Derive ψ_A/ψ_V as a function of the aspect ratio AR for a flat cylindrical shape (such as a coin).

EXAMPLES

Example 7.1 A nickel coin is 1.95 mm thick H and 21.3 mm in diameter Dia. What is the aspect ratio AR and what are its shape factor parameters ψ_A and ψ_V? Assume that the characteristic size is 21.3 mm as would be determined by sieving.

Solution. $AR = \text{Dia}/H = 21.3$ mm/1.95 mm $= 11/1$.

$$\psi_A = A/a^2 = [2(\pi a^2/4) + \pi aH]/a^2 = \pi/2 + \pi(1.95 \text{ mm}/21.3 \text{ mm}) = 1.86$$

$$\psi_V = V/a^3 = (\pi a^2/4)H/a^3 = \pi H/4a = \pi 1.95 \text{ mm}/4(21.3 \text{ mm}) = 0.072$$

The shape factor ratio $\psi_A/\psi_V = 1.86/0.072 = 25.9$, which is much larger than $\psi_A/\psi_V = 6$ for a sphere or a cube.

Example 7.2 For the gravity settling of alumina particles of 10, 1.0 and 0.1 μm diameter in water at 20°C, what are the differences in settling time?

Solution. The Stokes equation is $t = 18 H\eta_L/a^2(D_p - D_L)g$. For $a = 10$ μm,

$$t = \frac{18(0.01 \text{ m})(1.005 \times 10^{-3} \text{ Pa s})}{(10^{-5} \text{ m})^2(3.98 - 0.9982)\text{Mg/m}^3(9.8 \text{ m/s}^2)}$$

and $t = 0.0175$ h $= 1.1$ min. Note that time varies inversely with a^2. For $a = 1.0$ μm, $t = 1.75$ h, and for $a = 0.1$ μm, $t = 175$ h $= 7.3$ days.

Particles larger than 10 μm settle quickly in a low-viscosity liquid during handling. But colloidal particles remain in suspension for several days unless centrifuged.

Example 7.3 Compare the mean sizes of six particles with size $a = 2, 4, 8, 16, 32, 64$ μm. Definitions of the mean sizes are in Table 7.3.

Solution. $\bar{a}_L = (2 + 4 + 8 + 16 + 32 + 64)/6 = 21$ μm

$$\bar{a}_A = [(2^2 + 4^2 + 8^2 + 16^2 + 32^2 + 64^2)/6]^{1/2} = 30 \text{ μm}$$

$$\bar{a}_V = [(2^3 + 4^3 + 8^3 + 16^3 + 32^3 + 64^3)/6]^{1/3} = 37 \text{ μm}$$

$$\bar{a}_{V/A} = \bar{a}_V^3/\bar{a}_A^2 = 37^3/30^2 = 56 \text{ μm}$$

Note the strong effect of the extreme coarse particle on $\bar{a}_V$ and $\bar{a}_{V/A}$.

Example 7.4 Using the size data for barium titanate powder in Table 7.6, calculate the geometric mean size $\bar{a}_{gN}$ and the geometric mean σ_g by the method of finite summation.

Solution. Parameters may be calculated for each size interval as indicated in the table below:

f_N	a_i (nm)	$\ln(a_i)$	$f_N \ln(a_i)$	$(\ln a_i - \ln \bar{a}_{gN})^2$	$f_N(\ln a_i - \ln \bar{a}_{gN})^2$
1	3500	8.16	8.16	3.61	3.61
1	2500	7.82	7.82	2.43	2.43
6	1750	7.47	44.82	1.46	8.76
12	1250	7.13	85.56	0.76	9.12
22	800	6.68	146.96	0.18	3.96
22	500	6.21	136.62	0.00	0.00
26	300	5.70	148.20	0.31	8.06
8	150	5.01	40.08	1.56	12.48
2	50	3.91	7.82	5.52	11.05
100			626.04		59.47

$$\ln \bar{a}_{gN} = \Sigma f_N(a_i) \ln a_i / N = 626.04/100 = 6.26; \qquad \bar{a}_{gN} = 520 \text{ nm}$$

$$\bar{a}_{gN} = 0.52 \ \mu m$$

$$\ln \sigma_g = [\Sigma f_N(a_i)(\ln a_i - \ln \bar{a}_{gN})^2/N]^{0.5} = (59.47/100)^{0.5}$$

$$= 0.771; \ \sigma_g = 2.16$$

The results calculated agree closely with results obtained graphically, which are indicated in Table 7.6.

Example 7.5 The specific surface area S_m of a powder may be calculated using the equation

$$S_m = \frac{\psi_A/\psi_V}{\bar{a}_{V/A} D_a}.$$

Estimate S_m for the milled zircon powder in Table 7.6. Assume that $\psi_A/\Psi_V = 6$ and $D_a = 4.7 \ \text{Mg/m}^3$. For the milled zircon, $\bar{a}_{gM} = 1.2 \ \mu m$ and $\sigma_g = 2.25 \ \mu m$.

Solution. From Table 7.4, $\bar{a}_{V/A} = \ln \bar{a}_{gN} + 2.5(\ln \sigma_g)^2$ and $\ln \bar{a}_{gN} = \ln \bar{a}_{gM} - 3(\ln \sigma_g)^2$. Combining these equations, $\ln \bar{a}_{V/A} = \ln \bar{a}_{gM} - 0.5(\ln \sigma_g)^2$.

$$\ln \bar{a}_{V/A} = \ln 1.2 - 0.5(\ln 2.25)^2 = -0.147; \qquad \bar{a}_{V/A} = 0.86 \ \mu m$$

$$S_m = 6/(0.86 \times 10^{-6} \text{ m})(4.7 \times 10^6 \text{ g/m}^3) = 1.5 \text{ m}^2/\text{g}$$

CHAPTER 8

DENSITY, PORE STRUCTURE, AND SPECIFIC SURFACE AREA

Knowledge of the densities of ceramic phases, particles, processing additives and bulk systems is important in ceramic processing for several reasons. The raw-material storage capacities are directly dependent on the bulk density of the material. We need to know the particle density of raw materials and the phase density of liquids and additives to calculate volumetric proportions from the gravimetric batch composition and to ascertain settling tendencies. A change in a very precisely determined particle density may indicate a change in the phase structure, chemical composition, or porosity of the material.

Pore size analysis is used to obtain information about the sizes and structure of pores in calcined aggregates, soft agglomerates and granules, and unfired and fired materials.

The specific surface area of a powder is very dependent on the particle size and shape and the content of submicron size particles and finer pores or fissures in the surfaces of particles. Specific surface area analysis may detect the presence of colloidal material that is not detected in some size analysis techniques. The specific surface area of a powder containing nonporous particles may be used as a relative index of absorptivity for approximating the amount of surface-modifying processing aid required in a particular processing operation.

In this chapter we will examine principles involved in determining the density, specific surface area, and pore structure of different ceramic materials.

8.1 DENSITY

The bulk density of a particle system is the mass per unit volume of the particles and interstices. The density of a particle is the mass/volume ratio of the particle.

118

In a system of particles, the particle density refers to the mean density of all the different size particles. When the particles are nonporous, the particle density is referred to as the ultimate particle density and is the mean density of the solid phases constituting the particles. The density calculated from the chemical formula weight and the volume of the unit cell is termed the x-ray density. Often the ultimate density and x-ray density are nearly identical, but heavy elemental impurities may cause a measurable difference if these impurities are neglected in calculating the x-ray density.

For porous particles, the particle density must be defined more particularly. Particles dispersed in a fluid may adsorb molecules of the fluid into surface-accessible pores. The accessible porosity may be all or only a fraction of the porosity of the particles. Penetration of the fluid depends on the size of the molecules in the fluid relative to the pore size, wetting, and absence of another fluid in the pores. The ultimate volume is the volume of the solid phases. For a porous particle dispersed in a fluid, the apparent volume is the volume of the solid and the inaccessible pores (volume of fluid displaced). The particle bulk volume is the volume of solid and all porosity within the particles. Each of these particular volumes may be used to define the ultimate (D_u), apparent (D_a), and bulk (D_b) particle density (Fig. 8.1). The specific gravity is the weight of the particle relative to the weight of an equal volume of water.

Liquid pycnometry, as shown in Fig. 8.2, is commonly used for determining the mean apparent particle density of coarser powders to a precision of about ± 0.005 Mg/m^3. A calibrated pycnometer bottle containing powder is weighed, liquid of known density is added, and the bottle is reweighed. Accuracy depends on wetting the particles completely and eliminating air bubbles. A drop of wetting agent (antifoaming agent), which reduces the surface tension of water aids in wetting and minimizes the floating of micron-size particles. Non-aqueous liquids of low vapor pressure such as kerosene are also used. A calibrated pycnometer bottle may be used to determine the density of another liquid.

A gas pycnometer may be used for measuring the density of powders and is more convenient when a significant fraction is finer than 10 μm in size. The powder density determined using gas pycnometry may be slightly higher than

Porous Fragment

19.65 g
10.0 cm^3 Total
0.5 cm^3 Closed Porosity
2.0 cm^3 Open Porosity
D_u= 2.62 Mg/m^3
D_a= 2.46
D_b= 1.96

Fig. 8.1 Illustration of a porous fragment and bulk, apparent, and ultimate density values.

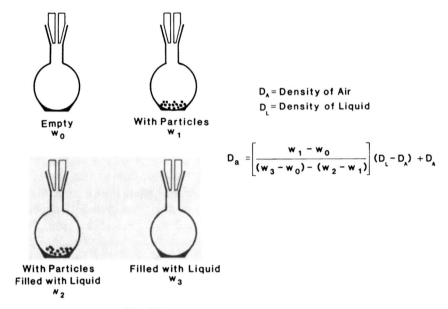

Fig. 8.2 Liquid pycnometer technique.

that obtained using liquid pycnometry when a small gas molecule such as He is able to penetrate into very small pores. In some materials, adsorption and condensation of gas molecules at room temperature in pores of a nanometer size can cause an erroneously high density, as indicated in Table 8.1.

The crushing and grinding of porous particles increases the relative fraction of surface pores (Fig. 8.3), and the apparent density of the finer particles may be higher, as indicated in Fig. 8.4 for tabular alumina. In many calcined powders that have been milled, the extremely coarse particles may be porous aggregates of slightly lower density, although most of the particles in the distribution are nonporous. The apparent particle density is equal to the ultimate particle density when all pores in particles are penetrated by the fluid and the volume of fluid displaced is equal to the volume of solid phases.

TABLE 8.1 Apparent Particle Densities Determined by Liquid and Gas Pycnometry at 20°C

Material	Nominal Size (μm)	Fluid	Density (Mg/m^3)
Tabular alumina	590–2000	Water	3.65
(Al_2O_3)		Nitrogen	3.65
Rutile	44–250	Water	4.25
(TiO_2)		Nitrogen	4.35
Bentonite[a]	0.3–20	Water	2.22
(clay minerals/		Kerosene	2.53
montmorillonite)		Nitrogen	4.86[a]

[a]Significant error due to adsorption and condensation in pores.

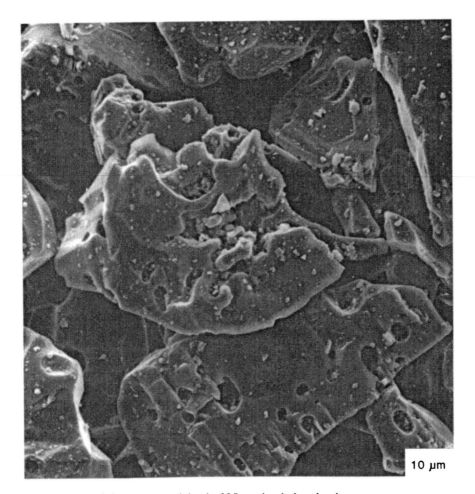

10 µm

Fig. 8.3 Large particles in 325 mesh tabular alumina are porous.

The density of a particle larger than about 1 mm can be determined from its buoyancy in a calibrated heavy liquid in the range of 1–5 Mg/m³. Small calibrated divers may be used to calibrate the density of a uniform liquid or the density gradient in a liquid solution to an accuracy of 0.001 Mg/m³ (Fig. 8.5). The liquid should completely wet the particle.

An Archimedes balanced may be used to determine the density of a particle larger than about 1 cm³ to a precision of about ±0.0002 Mg/m³. Liquid wetting the filament suspending the particle produces a force F_γ and viscous drag,

$$F_\gamma = 2\pi r_f \gamma_{LV} \cos \theta \qquad (8.1)$$

where r_f is the radius of the filament.

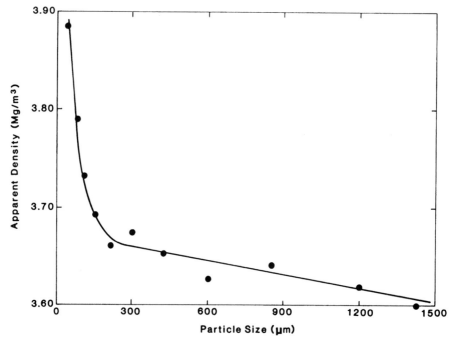

Fig. 8.4 Apparent density of sized tabular alumina particles.

The drag force is reduced by using an extremely thin filament and a liquid system that minimizes $\gamma_{LV} \cos \theta$. When the filament is very thin or the meniscus of the liquid intersects the filament at the same position during each weighing, the buoyancy of the wire does not need to be considered, and the particle volume V_p and mass M_p are

$$V_p = \frac{F_\gamma/g + M_A - M_S}{D_L - D_A} \tag{8.2}$$

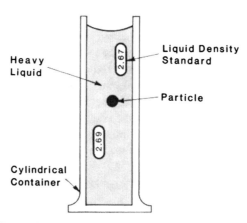

Fig. 8.5 Sink–float behavior of a particle and density standards in a heavy liquid.

and

$$M_p = M_A + D_A V_p \qquad (8.3)$$

where M_A is the mass in air, and M_S is the mass suspended in the liquid. The apparent density D_a of the particle is

$$D_a = M_p/V_p \qquad (8.4)$$

8.2 POROSITY AND PORE STRUCTURE

Many natural mineral particles, calcined aggregates, and soft agglomerates contain pore interstices, microfissures, and pores. These pore defects range widely in size and structure and may or may not be surface accessible. General details of surface and internal pores can be observed by microscopically examining external surfaces and fracture or polished surfaces of internal structure. The approximate range of pore sizes is readily obtained by image analysis. However, the quantitative, volumetric description of the pore size distribution and structure from two-dimensional images is a very complex problem.

Pore size information determined microscopically may be augmented using mercury intrusion porosimetry (Fig. 8.6). Mercury does not wet most ceramic materials, and pressure must be applied to force it into an evacuated surface pore. A special sample holder is partially filled with dried sample, evacuated, and then filled with mercury. Pressure applied to the mercury causes penetration of pores larger than a particular size. The intruded radius R at any applied pressure P is calculated from the Washburn equation for a cylindrical pore:

$$R = \frac{-2\gamma_{LV} \cos \theta}{P} \qquad (8.5)$$

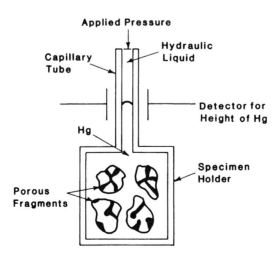

Fig. 8.6 Schematic of mercury penetration porosimetry technique.

The contact angle of mercury on most oxides is about 130–140° and in the range of 110–140°. Pressures up to 200 MPa are commonly used, and instruments with a 420-MPa capability are available. This range of pressures enables penetration of pores ranging from about 200 μm to 2 nm, as indicated in Table 8.2.

Mercury intrusion results are commonly presented in the form of cumulative volume penetrated per mass of specimen as a function of pore radius, as shown in Fig. 8.7. A rise in the curve indicates penetration of pores of a particular entry size. For unfired bodies containing packed particles or a bulk sample of particles or granules, the initial penetration normally corresponds to penetration into interstices among particles. Penetration into smaller pores in agglomerates or aggregates occurs at a higher pressure. The total penetration volume at the highest pressure is used to calculate the apparent volume. When two regions of penetration are clearly resolved, the relative fraction of pores in each size range may be estimated, and both the bulk and apparent particle density can be determined from a porosimetry analysis.

The fractional pore size distribution $f(R) = dV/dR$, where V is the volume of penetration, may be derived directly from the cumulative pore size distribution or from the cumulative penetration with pressure. Differentiating Eq. 8.5,

$$P\ dR + R\ dP = 0 \qquad (8.6)$$

and

$$f(R) = \frac{P\ dV}{R\ dP} \qquad (8.7)$$

The relation between the cumulative and fractional pore size distribution curves is shown in Fig. 8.8. The fractional curve more clearly illustrates small changes in pore size.

The entry radius for pores that are constricted at the surface will be smaller than the effective radius of the cavity being filled with mercury. Pores of this

TABLE 8.2   Dependence of Intruded Radius on Pressure
($\gamma_{H_g} = 474$ mN/m)

Radius (μm)	$\theta = 130°$ Pressure (MPa)	$\theta = 140°$ Pressure (MPa)
100	0.006	0.007
10	0.06	0.07
1	0.6	0.7
0.1	6.0	7.0
0.01	60	70
0.001	600	700

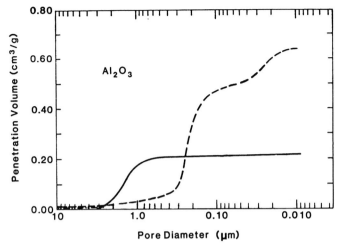

Fig. 8.7 Cumulative mercury penetration porosimetry into nodules of compacts of (lower curve) nonporous calcined alumina particles and (upper curve) calcined alum-derived powder containing porous aggregates.

shape produce a hysteresis in the penetration volume, and the penetration is dependent on the size of the specimen. Hysteresis and residual mercury in the pores after decompression indicate the presence of these "ink well" pores. This effect is minimal in highly porous agglomerates, but can be quite pronounced in relatively dense, sintered aggregates. The migration of binders and segregation of fine particles at the surface can also reduce the size of surface pores.

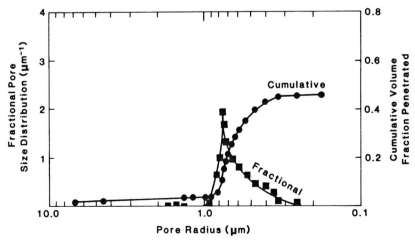

Fig. 8.8 Cumulative and fractional pore size distributions in a compact of calcined alumina powder.

8.3 SPECIFIC SURFACE AREA

The specific surface area is the surface area of the particles per unit mass or volume of material. It is commonly determined by the physical adsorption of a gas or by the chemical adsorption of a dye such as methylene blue. For a porous material, the surface area determined experimentally depends on the size of the adsorbed molecule relative to the size of the pores. Small gas molecules may penetrate pores smaller than 2 nm, whereas a bulky molecule is excluded.

Physical adsorption of a gas at a cryogenic temperature may be used to determine the specific surface area of a ceramic powder. For type II or type IV adsorption isotherms (Fig. 2.4), the volume of gas providing monolayer coverage on adsorption V_m may be calculated using Eq. 2.10. The specific surface area per unit mass S_M is given by the equation

$$S_M = \frac{N_A V_m A_m}{V_{mol} M_s} \tag{8.8}$$

where N_A is Avogadro's number, A_m is the area occupied by one adsorbate molecule (16.2×10^{-20} m^2 for N_2 and 19.5×10^{-20} m^2 for Kr), V_{mol} is the volume of 1 mole of gas at the standard temperature and pressure of V_m, and M_s is the mass of the sample.

Prior to adsorption, the surface of the powder sample is degassed by vacuum drying or by flushing an inert gas through the mass of heated powder. The adsorption of N_2 at the boiling point of liquid nitrogen (77.4°K) occurs quickly and is used routinely when S_M of the powder exceeds 1 m^2/g. Desorption on heating is commonly used for the analysis (Fig. 8.9).

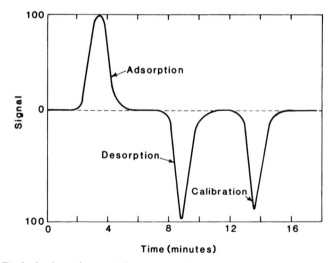

Fig. 8.9 Typical adsorption and desorption of N_2 gas during cooling to 77.4 K and heating to room temperature. (Courtesy of Quantachrome Corp., Greenvale, NY.)

The specific surface area of a powder consisting of nonporous particles can be calculated from the mean particle size $\bar{a}_{V/A}$ using the equation

$$S_M = \frac{\psi_A/\psi_V}{\bar{a}_{V/A} D_a} \tag{8.9}$$

A shape factor ratio ψ_A/ψ_V of 6 is usually assumed when the specific shape factors of the particles are unknown. The specific surface area is sensitive to variations in the shape and size of particles finer than 1 μm (Figs. 8.10, 8.11) or a change in the concentration of a minor phase of much higher surface area. It is not normally sensitive to agglomeration if chemical additives have not been introduced.

Type II isotherms (Fig. 2.4) are characterized by an initial portion with a diminishing slope, an intermediate portion (often linear), and a final upturn at a P/P_s of about 0.7–0.9. Behavior in these three regions is interpreted as monolayer adsorption, multilayer adsorption, and capillary condensation in pores between particles respectively. If the particles contain pores a few multiples of the size of the adsorbate gas, the capillary condensation may occur at a much lower relative pressure.

The chemical adsorption of bulky molecules from a solution in contact with the powder may be used to determine a quantitative index of the external surface area of clays, bentonites, and powders containing particles that do not adsorb the dye into internal pores. Bulky molecular dyestuffs such as methylene blue

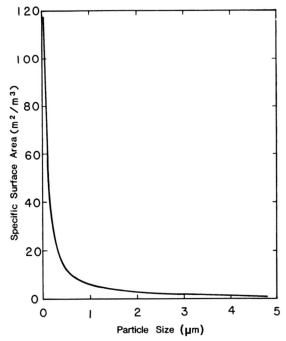

Fig. 8.10 Variation of specific surface area with mean particle size $\bar{a}_{V/A}$.

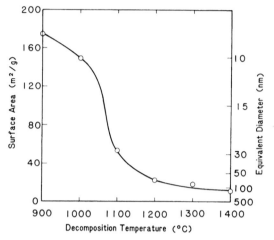

Fig. 8.11 Dependence of specific surface area of an alumina powder derived from $Al_2(SO_4)_3$ on the calcining temperature.

(organic anion with formula weight of 320) are chemisorbed on the surface of clay particles owing to an exchange mechanism. The clay suspension is first acidified to free montmorillonite particles and replace surface-adsorbed ions with H^+. A standardized methylene blue solution is admixed incrementally in a dilute suspension of the clay. The end point of the titration is reached when

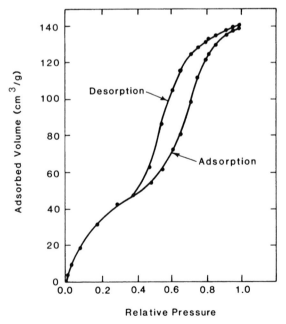

Fig. 8.12 Cryogenic adsorption and desorption of N_2 on a porous alumina powder. (Courtesy of Quantachrome Corp., Greenvale, NY.)

TABLE 8.3   Values of Kelvin Radius and Absorbed
Thickness nX for Nitrogen Desorption

P/Ps	Kelvin Radius (nm)	Film Thickness (nm)
1.00		
0.98	47.3	5.0
0.97	31.3	4.2
0.96	23.4	3.6
0.95	18.6	3.1
0.90	9.1	2.0
0.85	5.9	1.6
0.80	4.3	1.2
0.70	2.7	0.9
0.50	1.4	0.6
0.30	0.8	0.5

Source: F. M. Nelsen and F. T. Eggertsen, *Anal. Chem.*, **30**, 1387
(1958).

blue tint initially migrates with the liquid medium, as when a drop of suspension
is placed on a piece of filter paper. The methylene blue index (MBI), defined
as the milliequivalent of methylene blue adsorbed per 100 g of dry sample, is
commonly reported for clay materials (Table 5.5).

For pores larger than about 1 nm, the pore size distribution in particles may
be estimated from the hysteresis in the physical adsorption and desorption
behavior of a gas at a crygenic temperature (Fig. 8.12). The pore radius is
estimated using the Kelvin equation (Eq. 2.7) and relative pressure values in
the desorption loop, since the Kelvin equation describes liquid behavior. For
a cylindrical pore shape, the radius of the pore R_p is $R_p = R_K + nX$, where
R_K is the radius of the liquid in the pore, calculated from the Kelvin equation,
containing n statistical layers of thickness X in the adsorbed film. Values of
the Kelvin radius and adsorbed thickness are given in Table 8.3 for liquid N_2
at 77.4°K. The shape of the adsorption-desorption hysteresis may provide
qualitative information about the pore shape.

SUMMARY

The density of a ceramic material depends on the proportion and density of
component phases. Internal voids in the material reduce the mass, and the
apparent density is less than the ultimate density of the solid phase. The particle
size of the sample and accuracy desired are factors in selecting a technique for
determining the density. Microscopy, penetration analysis, and desorption behavior may be used to obtain the size and amount of open porosity in particles.
Specific surface area analysis complements the particle size analysis and may
provide information about the presence of a minor amount of extremely fine
particles not detected in the particle size analysis.

SUGGESTED READING

1. S. Lowell and Joan E. Shields, *Powder Surface Area and Porosity*, Chapman and Hall, New York, 1984.
2. C. Orr Jr., "Physical Characterization Technique for Particles," in *Ceramic Processing Before Firing*, George Y. Onoda Jr. and Larry L. Hench, Eds., Wiley-Interscience, New York, 1978, Chapter 6.
3. E. S. Paleck, "Specific Surface Area Measurements on Ceramic Powders," *J. Powder Tech.*, **18**, 45–48 (1977).
4. Arthur W. Adamson, *Physical Chemistry of Surfaces*, Wiley-Interscience, New York, 1976.
5. S. Karp, S. Lowell, and A. Mustacciuolo, "Continuous Flow Measurement of Desorption Isotherms," *Anal. Chem.*, **44**(12), 2395–2397 (1972).
6. Clyde Orr Jr., "Application of Mercury Penetration to Materials Analyses," *J. Powder Tech.*, **3**, 117–123 (1969–1970).
7. H. Gobrecht and A. Richter, "Density Measurement and Disorder in Lead Selenide," *J. Phys. Chem. Solids*, **26**, 1889–1893 (1965).
8. S. Brunauer, P. H. Emmett, and E. Teller, *J. Am. Chem. Soc.*, **60**, 309 (1938).

PROBLEMS

8.1   Calculate the porosity values from the data in Fig. 8.1.

8.2   The data obtained from a liquid pycnometry test are as follows: pycnometer mass = 32.383 g; pycnometer + powder = 41.923 g; pycnometer + water = 84.229 g; pycnometer + powder + water = 91.170 g. The water and ambient temperature were 24 °C. Calculate the apparent density of the powder.

8.3   If the ultimate density of the powder in Problem 8.2 is 3.98 Mg/m^3, what is the closed porosity in the particles?

8.4   The unit cell parameters of the hexagonal ferrite $BaFe_{12}O_{19}$ are c = 23.194 Å and a = 5.893 Å. The unit cell contains two formula units. What is the "x-ray density"?

8.5   Calculate the mean density of a powder blend consisting of 20 wt% zirconia (6.03 Mg/m^3) and 80 wt% alumina (3.98 Mg/m^3).

8.6   The specific surface areas for a conventional and a chemically prepared powder are 1.2 and 10.3 m^3/g, respectively. Compare the mean size of each. Assume apparent density is 6.01 Mg/m^3.

8.7   Estimate S_M for the two powders in Table 7.6.

8.8   Estimate $a_{V/A}$ for the alumina powders in Table 5.2 and compare your answers to the ultimate particle size and agglomerate size for each. Comment.

8.9 Verify the equivalent diameter calculated from the S_M on calcination at 1100°C in Fig. 8.11.

8.10 Estimate the specific surface area of kaolinite and montmorillonite particles assuming the particles are square platelets. The dimensions of the kaolinite are $1 \times 1 \times 0.1$ μm and those for the montmorillonite are $100 \times 100 \times 1$ nm. Assume a particle density of 2.60 Mg/m^3 for each. How many grams of each have an area equal to that of a 100×50 m soccer field?

8.11 Nitrogen gas is adsorbed on an alpha alumina powder at 77.4 K. The area occupied by an adsorbate molecule is 16.2×10^{-20} m^2. At STP, the volume of a mol of gas is 2.24×10^4 mL, and the volume of gas required to form a monolayer on 1 g of powder is 2.52 mL. Calculate the specific surface area.

8.12 For the data in Problem 8.10, calculate the water content of each powder if each adsorbs a water film of 0.5-nm thickness (about 2 molecules thick).

8.13 When the adsorptive capacity of other clay minerals is insignificant compared to that of the bentonite component, the MBI is based on 75 meq/100 g dry bentonite. Estimate the bentonite content of the clays in Table 5.5. Why is this test not accurate for a clay containing a high content of halloysite?

8.14 Estimate the range of pore size for the results shown in Fig. 8.12. Assume that the molar volume of adsorbed N_2 is 34.69 cm^3/mol.

EXAMPLES

Example 8.1 Calculate the bulk D_b, apparent D_a, and ultimate density D_u of the porous material in Fig. 8.1.

Solution. Density is defined as mass/volume. The total volume, which includes solids and both surface-accessible and closed porosity, is used for calculating D_b. The volume of solids and closed porosity is used when calculating D_a. Only the volume of solid material is used when calculating the ultimate density D_u. Using the data given in Fig. 8.1,

$$D_b = 19.65 \text{ g}/10.0 \text{ cm}^3 = 1.96 \text{ g/cm}^3 = 1.96 \text{ Mg/m}^3$$

$$D_a = 19.65 \text{ g}/(10.0 - 2.0)\text{cm}^3 = 2.46 \text{ g/cm}^3 = 2.46 \text{ Mg/m}^3$$

$$D_u = 19.65 \text{ g}/(10.0 - 2.0 - 0.5)\text{cm}^3 = 2.62 \text{ g/cm}^3 = 2.62 \text{ Mg/m}^3$$

Example 8.2   A piece of dry sintered material is weighed in air (M_A = 15.384 g). The specimen was evacuated in a desiccator before water was added to infiltrate open pores. The balance with a fine wire hanger was tared to zero and the suspended weight of the specimen was obtained (M_S = 11.325 g). After removal, the surface was carefully blotted and the weight in air was measured (M_W = 15.436 g). The temperature of the air and water was 24°C. What are the apparent density D_a and bulk density D_b of the sample calculated to three significant figures?

Solution.   The volume of the material may be calculated using Eq. 8.2 and the correction for the wetting of the filament (F_γ/g) may be neglected. The density of air and water at 24°C are found in Appendix 3. From Eq. 8.2,

$$V_p = (M_A - M_S)/(D_L - D_A)$$

$$V_p = (15.384 - 11.325)g/(0.997296 - 0.001188)g/cm^3 = 4.059 \text{ cm}^3$$

$$M_p = M_A + D_A V_p = 15.384 \text{ g} + (0.001188 \text{ g/cm}^3)4.059 \text{ cm}^3 = 15.389 \text{ g}$$

$$D_a = M_p/V_p = 15.389 \text{ g}/4.059 \text{ cm}^3 \qquad D_a = 3.79 \text{ Mg/m}^3$$

The bulk volume is the sum of the apparent volume of the particle and the volume of the open pores in the particle:

$$V_b = V_p + V_{(open\,pore)}$$

$$V_b = (M_A - M_S)/(D_L - D_A) + (M_W - M_A)/(D_L - D_A)$$

$$V_b = (M_W - M_S)/(D_L - D_A)$$

$$V_b = (15.436 - 11.325)g/0.996 \text{ g/cm}^3 \qquad V_b = 4.127 \text{ cm}^3$$

$$D_b = M_p/V_b = 15.389 \text{ g}/4.127 \text{ cm}^3 \qquad D_b = 3.73 \text{ Mg/m}^3$$

Example 8.3   Mercury penetration porosimetry results for compacts of two alumina powders are shown in Fig. 8.7. The particles corresponding to the upper curve are porous aggregates of corundum (D_u = 3.98 g/cm³). Estimate the volume fraction of pores between aggregates and the fraction within aggregates in the sample. What is the mean apparent porosity of the aggregates?

Solution.   From the figure, it is estimated that the penetration into the interstices between the aggregates is 0.50 cm³/g and the total penetration which includes the penetration into the aggregates is 0.65 cm³/g. For 1 g of solid material the volume is (1/3.98 g/cm³) or 0.25 cm³/g. From the definitions for each type of porosity:

> Interparticle porosity = interparticle volume/total volume
> Interparticle porosity = 0.50 cm³/g/(0.65 + 0.25)cm³/g = 0.556
> Sample interparticle porosity = 55.6%

Intraaggregate porosity = volume within aggregates/total volume
Intraaggregate porosity = $(0.65 - 0.50)$ cm³/g $100/(0.65 + 0.25)$ cm³/g
Sample intraaggregate porosity = 16.7%

Mean porosity of aggregates = volume of penetration into aggregates/bulk
volume of the aggregates
Porosity = $(0.65 - 0.50)$ cm³/g $100/[(0.65 - 0.50) + 0.25]$ cm³/g
Mean porosity of aggregates = 37.5%

Example 8.4 What is the apparent porosity of the sintered material in Example 8.2?

Solution. The apparent porosity is the volume of open pores $V_{(open\ pores)}$ divided by the bulk volume of the material V_b.

$$\text{Apparent Porosity } (\%) = V_{(open\ pores)}\ 100/V_b$$

From Example 8.2, $V_{(open\ pores)} = (M_W - M_A)/(D_L - D_A)$, and $V_b = (M_W - M_S)/(D_L - D_A)$, and

Apparent Porosity $(\%) = (M_W - M_A)\ 100/(M_W - M_S)$
Apparent Porosity $(\%) = (15.436 - 15.384)(100)/(15.436 - 11.325)$
Apparent Porosity $(\%) = 0.052(100)/4.111 = 1.26\%$

PART IV

PROCESSING ADDITIVES

When processing ceramics, several different additives must be introduced into the batch to produce the particle dispersion and flow behavior requisite for forming. These additives may be categorized as follows:

1. Liquid/solvent
2. Surfactant (wetting agent)
3. Deflocculant
4. Coagulant
5. Binder/flocculant
6. Plasticizer
7. Foaming agent
8. Antifoam
9. Lubricant
10. Bactericide/fungicide

Other than the liquid/solvent, these additives are added in a small amount and most are eliminated in a later stage of processing and do not appear in the final product. But from a processing perspective, these additives are essential materials. The wise selection and control of these additives is often the key to successful processing with high yields or for the development of an improved process or product.

An elementary understanding of organic chemistry is needed to comprehend the nature and structure of organic additives. The selection of these additives is now based much more on recognizing their chemical characteristics and functions rather than simply empirical experimentation. The types, characteristics, and functions of processing additives are described and discussed in Chapters 9–12.

CHAPTER 9

LIQUIDS AND WETTING AGENTS

Liquids are used in ceramic processing to wet the ceramic particles and provide a viscous medium between them and to dissolve salts, compounds, and polymeric substances in the system. The admixed liquid changes the state of dispersion of the particles and alters the mechanical consistency.

A surfactant is a substance added to reduce the surface tension of the liquid or the interfacial tension between the surface of the particle and the liquid to improve wetting and dispersion. In this chapter we consider the composition and properties of water, nonaqueous liquids, and surfactants used in ceramic processing.

9.1 WATER

The principal liquid used in ceramic processing is water, and until recently its availability, low cost, and consistency have been taken for granted. Process water must be of satisfactory quality, and water used in industrial processing is now monitored routinely. At some manufacturing sites, the available water must be treated to make it usable or consistent, and process water is sometimes treated and recycled in the plant. Water is now viewed by some plant engineers as a raw material that must be refined to improve processing and productivity.

Pure water contains polar H_2O molecules and the ions H_3O^+ and OH^-; it has a pH 7 and a specific conductivity of 0.055 μmho/cm at 20°C. Water exposed to air combines with CO_2 to form a weak acid:

$$CO_{2(gas)} + H_2O \rightleftharpoons H_2CO_3 \tag{9.1}$$

and

$$H_2CO_3 + H_2O \rightleftharpoons H_3O^+ + HCO_3^- \tag{9.2}$$

The additional H_3O^+ decreases the pH, and the greater concentration of ions increases the specific conductivity of the water.

Well water and municipal water may contain suspended organic and inorganic substances and dissolved mineral salts, as indicated in Table 9.1. The hardness of water is generally due to the dissolved calcium, magnesium, and ferrous salts. Calcium carbonate is relatively insoluble, but calcium bicarbonate $Ca(HCO_3)_2$ is significantly soluble. Heating reduces the hardness according to the reaction

$$Ca^{2+} + 2HCO_3^- \rightarrow CaCO_{3(solid)} + H_2O + CO_{2(gas)} \tag{9.3}$$

However, heating does not reduce hardness due to dissolved sulfates and chlorides.

The total dissolved solids TDS is estimated from the specific conductivity C using the equation

$$TDS \ (ppm) = \frac{C(\mu mho/cm) - 0.055}{2.5} \tag{9.4}$$

The constant 2.5 is a weighted average of the specific conductivities of common ionized solids in a water sample.

TABLE 9.1   Composition of Water (ppm)

Component	Well	City	Treated
Ca	300	78	Trace
Mg	172	42	Trace
Na	8	360	Trace
Bicarbonate	350	350	0
Carbonate	0	0	0
Sulfate	100	125	Trace
Chloride	30	5	Trace
Nitrate	0	0	0
pH	7.8	7.9	4.5–5.5
Sp conductivity			
(μmho/cm)	670	650	10–40
TDS (ppm)	268	260	4–16

Source: R. Thomas, "*Water—Its Processing Consequences*" in *Processing Consequences of Raw Material Variables*, J. Varner et al., Eds., Alfred University Press, Alfred, NY, 1985, p. 53.

TABLE 9.2   Representative Properties of Liquids (20°C)

Liquid	Formula	Dielectric Constant	Surface Tension (mN/m)	Viscosity (mPa · s)	Boiling Pt. (°C)	Flash[a] Pt. (°C)
Water	H_2O	80	73	1.0	100	None
Methyl alcohol	CH_3OH	33	23	0.6	65	18
Ethyl alcohol	C_2H_5OH	24	23	1.2	79	8
n-Propyl alcohol	C_3H_7OH	20	24	2.3		
Isopropyl alcohol	C_3H_7OH	18	22	2.4	49	21
n-Butyl alcohol	C_4H_9OH	18	25	2.9	100	38
n-Octyl alcohol	$C_8H_{17}OH$	10	28	10.6	171	
Ethylene glycol	$C_2H_6O_2$	37	48	20	>197	>116
Glycerol	$C_3H_8O_2$	43	48	20	290	
Trichloroethylene	C_2HCl_3	3		5.5	87	
Methylethyleneketone	C_4H_8O	18	25	0.4	80	2

Source: Handbook of Chemistry and Physics, R. C. Weast, Ed., CRC Press, Boca Raton, FL, 1958.
[a]More recent values provided by suppliers are somewhat lower.

In-plant treatment of city or well water using a commercial filtration–deionizing system can reduce the dissolved salts to trace concentrations (Table 9.1). Chlorine and suspended matter may be removed in a carbon filter. Cation and anion impurities are removed using rechargeable exchange resins.

Reagent-grade water should have a specific conductance below 5.0 μmho/cm, and treated water of this purity can be produced by filtration and ion exchange. Higher-quality reagent-grade water with a conductivity of about 1 μmho/cm is produced by distillation and deionization.

Water is a polar liquid and has a high dielectric constant and surface tension relative to other liquids, as indicated in Table 9.2. It is a good solvent for polar and ionic compounds. Water can form a hydrogen bond with substances containing an −OH or −COOH group, and substances such as polymerized alcohols and carbohydrates dissolve in water if the molecule is not too large. The viscosity of water decreases significantly on heating from 20 to above 50°C, as shown in Table 9.3.

TABLE 9.3   Variation of Viscosity (mPa·s) of Water and Alcohols with Temperature

	Temperature (°C)				
Liquid	0	20	30	50	70
H_2O	1.8	1.0	0.8	0.55	0.4
CH_3OH	0.8	0.6	0.5	0.4	
C_2H_5OH	1.8	1.2	1.0	0.7	0.5
C_3H_7OH	3.9	2.3	1.7	1.1	0.8
C_4H_9OH	5.2	3.0	2.3	1.4	0.9

In the production of ceramic tile, it has been reported that recycling of all waste water can reduce the external water requirement by about one-half. Recycled water is sometimes conveniently used for the wet grinding of raw materials, to add moisture to the dry body, and for washing equipment and the work areas.

9.2   ORGANIC LIQUIDS

Nonaqueous liquids such as trichloroethylene, alcohols, ketones, and refined petroleum oil or liquid wax are used as the liquid and solvent medium in suspensions of materials that react with water or when dispersion or drying is a particular problem such as in casting electronic substrates and for printing resistive or conductive films on surfaces of formed ceramics. Organic liquids may be flammable and toxic, and special precautions and monitoring are required where they are used and disposed. The flash point is the lowest temperature in air at which ignitable or explosive vapors are given off and must be high enough to provide a safe range of working temperatures. Because of their higher cost and the environmental concerns, recycling is often cost effective.

Nonpolar molecules tend to be soluble in nonpolar solvents. The dissolving power of the liquid used should be sufficient to dissolve processing additives but should not prevent the adsorption of these additives on the particles. Organic liquids used in ceramic processing have a lower surface tension and dielectric constant than water, as indicated in Table 9.2. The lower surface tension γ_{LV} aids in the wetting of solids if γ_{SL} is not too high. A solution of solvents may often provide the best compromise of low dielectric constant and surface tension for dispersion and dissolving additives and a satisfactory viscosity, boiling point, and flash point for ease in handling and drying. Nonaqueous solutions that have been used in preparing a slurry for casting include ethyl alcohol and trichloroethylene or methylethyl ketone. Mixtures of alcohols may offer a satisfactory compromise of chemical and physical properties and be more acceptable for health and safety reasons. These solutions have a low boiling point and dissolve any water impurity that might cause hydration of the solids. Terpineol and butyl carbitol are liquids of low volatility used in paste printing compositions.

Petroleum oils are composed of hydrocarbon molecules that may vary in molecular weight. Longer molecules have a higher viscosity and boiling range. Waxes are relatively high molecular weight hydrocarbons that are solid at room temperature but that soften to form a low-viscosity liquid (see Chapter 11).

The viscosity η_s of ideal solvent solutions may be calculated using the equation

$$\ln \eta_s = \sum f_i^w \ln \eta_i \qquad (9.5)$$

where f_i^w is the weight fraction and η_i the viscosity of each liquid component. For solutions of interacting solvents such as the alcohols in hydrocarbon solvents, the solution viscosity is lower than that calculated using Eq. 9.5, and an effective viscosity of the minor component must be used.

9.3 POLAR LIQUIDS NEAR OXIDE SURFACES

A polar liquid may be both physically and chemically adsorbed on the surface of dispersed oxide particles. For oxides such as SiO_2, Al_2O_3, and TiO_2 in water, evidence of chemical surface hydration is provided by infrared adsorption studies, heats of immersion, and the thermal behavior of the adsorption–desorption kinetics. Reactions of the type

$$\text{physical adsorption} \quad MO_{(surface)} + H_2O \rightarrow MO{-}H_2O_{(surface)} \quad (9.6)$$

and

$$\text{chemical adsorption } M + MO_{(surface)} + H_2O \rightarrow 2MOH_{(surface)} \quad (9.7)$$

have been suggested (see Fig. 9.1). The polar hydroxyl ($-OH$) groups may cause the surface to attract and physically adsorb one to several additional

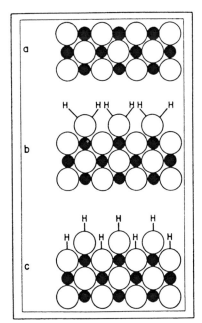

Fig. 9.1 Cross section of atomic structure of an oxide showing (*a*) dry surface, (*b*) surface with physically adsorbed water, and (*c*) surface with chemisorbed water (chemically hydrated surface). [Courtesy of W. H. Morrison, *J. Coatings Tech.*, **57**(721), 56 (1985)].

layers of polar water molecules. This physically bonded water is immobilized and is of a different structure from bulk water. Alcohols and liquids containing the carboxyl ($-$COOH) group also are chemisorbed on oxide surfaces. The absorption behavior depends significantly on the thermal history and prior hydration of the surface.

9.4 SUFACTANTS

Methyl ($-$CH$_3$) and ethyl ($-$C$_2$H$_5$) molecular groups are nonpolar. But hydroxyl ($-$OH), carboxyl ($-$COOH), sulfonate ($-$SO$_3^-$), sulfate ($-$OSO$_3^-$), ammonium ($-$NH$_4^+$), amino ($-$NH$_2$), and polyoxyethylene ($-$CH$_2$CH$_2$O$-$) groups are polar in nature. A polar group attracts polar liquid molecules and is called a lyophilic group. A nonpolar group such as a hydrocarbon chain ($-$C$_x$H$_y$) is a lyophobic group. In aqueous systems, these groups are often referred to as hydrophilic and hydrophobic groups, respectively.

Surfactants are molecules of a particular design which have one end that is polar and the other end that is nonpolar. Examples of nonionic and anionic surfacts are shown in Fig. 9.2. Nonionic surfactants do not ionize when dis-

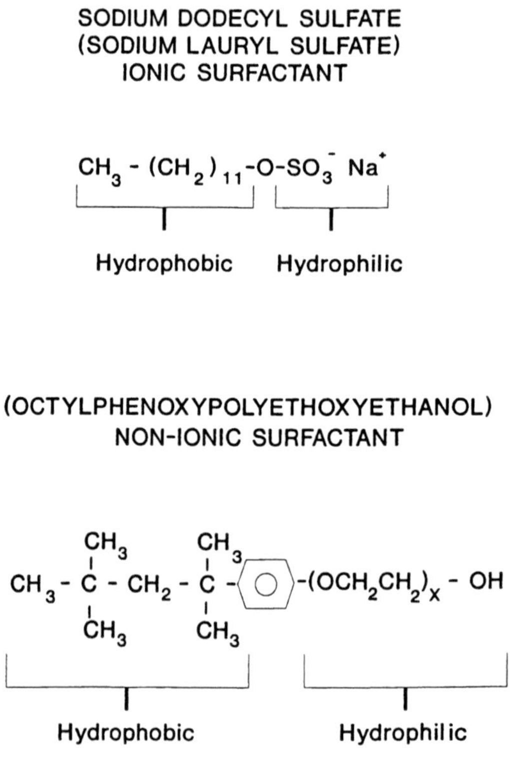

SODIUM DODECYL SULFATE
(SODIUM LAURYL SULFATE)
IONIC SURFACTANT

$$\text{CH}_3 - (\text{CH}_2)_{11} - \text{O-SO}_3^- \ \text{Na}^+$$

Hydrophobic   Hydrophilic

(OCTYLPHENOXYPOLYETHOXYETHANOL)
NON-IONIC SURFACTANT

$$\text{CH}_3 - \underset{\underset{\text{CH}_3}{|}}{\overset{\overset{\text{CH}_3}{|}}{\text{C}}} - \text{CH}_2 - \underset{\underset{\text{CH}_3}{|}}{\overset{\overset{\text{CH}_3}{|}}{\text{C}}} - \bigcirc - (\text{OCH}_2\text{CH}_2)_x - \text{OH}$$

Hydrophobic   Hydrophilic

Fig. 9.2 Molecular structure of a nonionic surfactant and an anionic surfactant indicating hydrophobic and hydrophilic portions of molecule.

TABLE 9.4   Examples of Surfactants

Type	Generic Name	Composition
Nonionic	Ethoxylated nonylphenol	$C_9H_{19}(C_6H_4)O(CH_2CH_2O)_{10}H$
	Ethoxylated tridecyl alcohol	$C_{13}H_{27}O(CH_2CH_2O)_{12}H$
Anionic	Sodium stearate	$C_{17}H_{35}COO^-Na^+$
	Sodium disopropylnaphtalene sulfonate	$(C_3H_7)_2C_{10}H_5SO_3^-Na^+$
Cationic	Dodecyltrimethylammonium chloride	$[C_{12}H_{25}N(CH_3)_3]^+Cl^-$

solved in the liquid. Anionic surfactants have a relatively large lyophobic group that is commonly a long-chain hydrocarbon and a negatively charged lyophilic group that is the surface-active portion of the molecule. Anionic surfactants are widely used in industry. Family types include the alkali sulfonates, sulfates, lignosulfonates, carboxylates, and phosphates. Alkali and ammonium anionic surfactants are more soluble in water than in organic liquids. Cationic surfactants have a positively charged lyophilic group. They are commonly toxic and are not widely used in ceramic processing. Examples of ionic and nonionic surfactants are listed in Table 9.4.

The driving force for the adsorption of a surfactant at a surface or interface is the reduction of the Gibbs free energy ΔG on its adsorption. When added to a polar liquid of high surface tension such as water, surfactant molecules will concentrate at the surface with the lyophilic end adsorbed in the polar liquid (see Fig. 9.3). An apparent concentration of only 0.01–0.2% may reduce the surface tension γ_{LV} of the liquid very significantly (see Table 2.1) and greatly improve the wetting of suspended solids by reducing the wetting angle. Accordingly, surfactants are often called wetting agents.

Surfactant molecules may also improve the compatibility of the solid with the liquid medium when they are adsorbed at the interface and reduce γ_{SL}. When added along with an oxide powder into a nonpolar liquid, the surfactant will be adsorbed with the lyophilic end on the particle and the lyophobic end

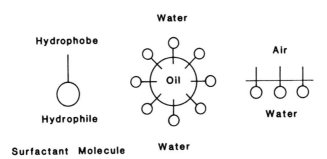

Fig. 9.3 Structural symbol for surfactant molecule, oriented adsorption of surfactant adsorbed at oil–water interface in oil–water emulsion, and oriented adsorption of surfactant at water–air interface.

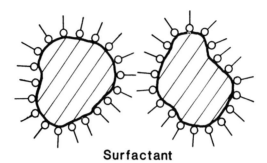

Surfactant

Fig. 9.4 Polar end of surfactant molecule is adsorbed on the surface of an oxide particle dispersed in oil.

in the liquid, as is shown in Fig. 9.4. When the adsorption preference is not strong, such as on the addition of stearic acid $CH_3(CH_2)_{16}COOH$ to a suspension of clay in water, the lyophilic end will project into the water, and the lyophobic end will be attached to the clay surface.* Anionic surfactants tend to be adsorbed on neutral and positive particles. Cationic surfactants are strongly adsorbed on the negative surfaces of minerals. The adsorption of surfactants is discussed further in Chapter 10.

Polymer molecules with repeating ionizable groups, called polyelectrolytes, which promote dispersion without necessarily reducing the surface tension or interfacial tension, are sometimes included in the category of surfactants. These additives are discussed under the topic of deflocculants in Chapter 10. In this text, polyelectrolytes used for dispersion will be called deflocculants.

9.5 EMULSIONS

After being mixed mechanically, water and oil separate spontaneously because the interfacial tension is quite high. The addition of a surfactant, which lowers the interfacial tension, may produce a stable dispersion of fine droplets of one liquid in the other (Fig. 9.3). In this role, the surfactant is called an emulsifier. The Bancroft rule says that the continuous phase is the one in which the emulsifier is more soluble. Stability is also enhanced by the emulsifier forming a physical barrier to the coalescence of droplets, during stirring or mixing.

One method for rating the relative strengths of the hydrophilic and hydrophobic groups of a surfactant molecule is the HLB method. Using an empirical scale ranging from 0 to 20, a high value is assigned to a strongly hydrophilic group and a low value to a strongly hydrophobic group. The HLB number is an index of the balance between these groups.

The HLB numbers are reported for commercial surfactants and application

*W. E. Worral, *Clays and Ceramic Raw Materials*, Halsted Press, New York, 1975.

TABLE 9.5   HLB Application Guidelines

Range	Application
3–6	Water–oil emulsifier
7–9	Wetting agent
8–18	Oil–water emulsifier
13–15	Detergent
15–18	Solubilizer

Source: Charles Martens, *Emulsion and Water-Soluble Paints and Coatings*, Reinhold Publ. Co., 1964. HLB = 20(1-S/A); S = saponification number and A = acid number of separated acid.

guidelines are indicated in Table 9.5. The stability of the dispersion should be higher when the HLB number of the surfactant approximates the weighted mean for the dispersed phase and liquid. Nominal HLB values of some common materials are listed in Table 9.6. As indicated in Table 9.5, for dispersion in water, the HLB of the surfactant should exceed 8. For water dispersed in oil, HLB is less than 6. Representative HLB values for commercial emulsifiers (surfactants) are listed in Table 9.7. When using a mixture of two surfactants, an arithmetic mean is used. To determine the appropriate HLB value for emulsification, compatible surfactants with quite different HLB values may be selected and blends of the two are evaluated for effectiveness to obtain the approximate HLB value required. A surfactant family with a HLB value near the target can then be found from reference information.

Surfactants are effective in very small concentrations < 0.1 mol/L, and when added in excess of that required for adsorption, microscopic clusters of oriented surfactant molecules called micelles form in the solution. Above the critical concentration at which clusters form, called the critical micelle concentration CMC, the surface tension is essentially constant. Other properties

TABLE 9.6   HLB Values for Common Materials

Material	HLB Value
Glyceryl trioleate	1.0
Cottonseed oil	7.5
Paraffin wax	9
Microcrystalline wax	9.5
Mineral oil	10
Silicone oil	10.5
Kerosene	12.5
Carnuba wax	14.5
Dimethyl phthalate	15
Stearic acid	17

TABLE 9.7   HLB Values for Some Emulsifiers

Emulsifier	HLB Value
Propylene glycerol monostearate	3.4
Glycerol monooleate	4.3
Diethylene glycol monooleate	4.7
Diethylene glycol monolaurate	6.1
Tetraethylene glycol monooleate	7.7
Polyoxyethylene monolaurate	12.8
Polyoxyethylene monostearate	17.9

of the solution, such as the electrical conductivity and refractive index, exhibit a different dependence on surfactant concentration above the CMC.

SUMMARY

Liquids used in ceramic processing provide a viscous medium for dissolving chemical additives and dispersing particles. A surfactant is an additive that is preferentially adsorbed at a surface or interface and promotes wetting and dispersion by reducing the surface or interfacial tension.

Water is the major liquid used in ceramic processing, and it is often refined to improve its purity or consistency. It is polar and has a high surface tension. Nonaqueous liquid systems are less polar in nature and are good solvents for relatively nonpolar substances; their solutions provide greater ranges in properties. The liquid system used must dissolve additives yet permit their adsorption at interfaces to assist in dispersion. Precautions must be exercised in the handling and disposing of nonaqueous liquids.

SUGGESTED READING

1. Ronald J. Thomas, "Water—Its Processing Consequences," in *Processing Consequences of Raw Materials Variables*, James R. Varner et al., Eds., Alfred University Press, Alfred, NY, 1985, pp. 47–54.

2. M. V. Parish, R. R. Garcia, and H. K. Brown, "Dispersions of Oxide Powders in Organic Liquids," *J. Mater. Sci.*, **20**, 996–1008 (1985).

3. ASTM D1193-77, Part 31, Water, in *1980 Annual Book of ASTM Standards*, American Society for Testing Materials, Philadelphia, PA, 1980.

4. Temple C. Patton, *Paint Flow and Pigment Dispersion*, Wiley-Interscience, New York, 1979.

5. Milton J. Rosen, *Surfactants and Interfacial Phenomena*, Wiley-Interscience, New York, 1978.

6. J. C. Williams, "Doctor Blade Process," in *Treatise on Materials Science and Technology*, Vol. 9, Franklin F. Y. Wang, Ed., Academic Press, New York, 1976, pp. 173-198.

7. *Merck Index*, Merck and Company, Inc., Rahway, NJ, 1976.

8. *Detergents and Emulsifiers*, McCutcheon's Division, Allured Publishing Corp., Ridgewood, NJ, 1974.

9. Ludwig F. Audrieth and Jacob Kleinberg, *Non-Aqueous Solvents*, Wiley, New York, 1953.

PROBLEMS

9.1 What is the significance of the flash point for a nonaqueous liquid?

9.2 A sample of water has a specific conductivity of 560 μmho/cm. Calculate the TDS. Is this distilled water?

9.3 Explain the difference between the impurity chemical analysis results and the TDS in Table 9.1.

9.4 Explain the difference in pH values for the treated and untreated water samples in Table 9.1.

9.5 What is the relative difference in viscosity of water and ethyl alcohol at 20 and 50°C.

9.6 Draw the structure of isopropyl alcohol and comment on the structure relative to that of a surfactant molecule.

9.7 Sketch the structure of a micelle of nonionic surfactant molecules in water.

9.8 For water at 25°C on a substrate, $\gamma_{SV} = 300$ mN/m and the wetting angle is 18°. Calculate the wetting angle when a surfactant is adsorbed only on the surface of the liquid and reduces γ_{LV} to 65 mN/m.

9.9 The HLB numbers of three different surfactants are 3.7, 6.7, and 13.5. Which would you select for dispersion of an oxide in water and which for dispersion in oil?

9.10 Write the reactions for the physical adsorption and the chemisorption of water on the surface of SiO_2. What is the name of the chemisorbed specie?

EXAMPLES

Example 9.1 What are the interactions between the liquid, particles, and additives in a processing system?

Solution. Key processes in developing a satisfactory processing system are wetting, dissolving, and adsorption, as indicated on the figure. The liquid must commonly dissolve molecular and ionic processing additives so that they become homogeneously distributed. Wetting of the particles by the liquid is essential for particle dispersion and to develop a uniform coating of additives

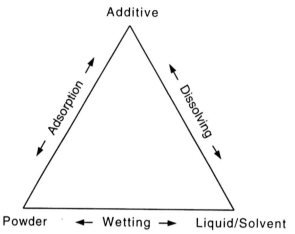

Example 9.1 Powder, liquid/solvent, and additive components of system and wetting, dissolving, and adsorption functions.

on the particles. The adsorption of additives on particles must be controlled and the solvent power of the liquid should be adequate for dissolving, but not too strong to prevent the adsorption of additives.

Example 9.2 A sample of water used to wash a powder in a development trial is determined to have a specific conductivity of 15.4 μmho/cm (0.15 S/m). The aqueous supernatant after ball milling the powder is determined to be 489 μmho/cm. What can you say about dissolved ions in the water before and after milling?

Solution. The specific conductivity of 15.4 is significantly higher than the value of 5.0 for reagent grade water. The TDS is calculated using Eq. 9.4 to be 6 ppm for the water used, which on comparison to the results in Table 9.1 would indicate that treated water is being used. The TDS = 196 ppm in the water after milling indicates a significantly higher concentration of ions dissolved from the powder during milling.

Example 9.3 What is the effect of heating on the viscosity of water?

Solution. On heating water above its freezing point, its molar volume (molecular weight/density) increases and its viscosity decreases. Relative to the viscosity at 20°C, the viscosity decreases 45% on heating to 50°C and 60% on heating to 70°C. This significant decrease on heating has the effect of increasing its velocity of migration due to capillary forces.

Example 9.4 What is the estimated viscosity of a solution of 70 wt% trichloroethylene and 30 wt% ethyl alcohol? The effective viscosity of ethyl alcohol often used as a cosolvent is 1.1 mPa·s.

Solution. The viscosity of trichloroethylene is 5.50 mPa·s from Table 9.2. Equation 9.5 may be used to estimate the viscosity of the solution, but note that an effective viscosity must be used for the alcohol in the nonideal solution.

$$\log \eta_{soln} = 0.70 \log 5.50 + 0.3 \log 1.10 \qquad \eta_{soln} = 3.4 \text{ mPa·s}$$

Example 9.5 The surfactant and lubricant oleic acid $C_{18}H_{34}O_2$ may aid in wetting oxide particles with oil and polyethylene in a powder injection molding body, which enables the proportion of powder/liquids (powder loading) to be increased. What is the effective surfactant mechanism of the oleic acid in this batch?

Solution. The surface tension of the nonaqueous liquids is low. But adsorption of the oleic acid at the particle–liquid interface may reduce the interfacial tension γ_{SL} and improve wetting by increasing $\cos \Theta$:

$$\cos \Theta = (\gamma_{SV} - \gamma_{SL})/\gamma_{LV}$$

The polar $-COOH$ would adsorb on the oxide particle and the $-C_xH_y$ would protrude into the nonpolar liquid.

Example 9.6 What emulsifier composition would you select for emulsifying 20 wt% paraffin wax in 80 wt% oil having an HLB of 13? Why?

Solution. Emulsification depends on both the composition (HLB) and the concentration of the two phases; both play a role in determining the emulsification power and the continuous phase in the emulsion. As a first approximation, the HLB of the emulsifier should be equal to the weighted average of the ingredients to be emulsified. The HLB of the paraffin is 9.0. The average HLB of the emulsifier should be

$$\text{Average HLB} = 0.2(9.0) + 0.8(13) = 12.2$$

For an emulsifier composed of polyoxyethylenated laural alcohol (HLB = 16.9) and polyoxyethylenated cetyl alcohol (HLB = 5.3), the proportion x of polyoxyethylenated laural alcohol would be

$$12.2 = 16.9\,(x) + 5.3\,(1 - x) \qquad x = 6.9/11.6 = 0.59$$

A mixture of 59 wt% polyoxyethylenated laural alcohol and 41 wt% polyoxyethylenated cetyl alcohol should be evaluated. Other mixtures of other emulsifying agents having the same weighted HLB number should also be evaluated experimentally to determine which structural type is most efficient.

CHAPTER 10

DEFLOCCULANTS AND COAGULANTS

Additives called deflocculants and coagulants are very important in ceramic processing. Dispersion stabilized by an additive adsorbed on the particles which increases the repulsive forces by electrical charging and/or by sterically hindering the close approach of particles is called deflocculation, and the additive is called a deflocculant. A coagulant is a simple electrolyte that promotes particle agglomeration by reducing the particle repulsion forces or steric hindrance. Deflocculation and coagulation processes must be carefully controlled to produce a satisfactory process system.

Particle agglomeration may also be produced by the bridging action of adsorbed polymer molecules and coagulated colloidal particles. This type of agglomeration is called flocculation and is discussed in Chapter 11.

10.1 PARTICLE CHARGING IN LIQUID SUSPENSIONS

Ceramic powders are of high surface area and relatively low solubility, and surface chemistry tends to control their charging behavior. The surface of a particle may become charged by (1) the desorption of ions at the surface of the material, (2) a chemical reaction between the surface and the liquid medium changing the composition of the surface, and (3) the preferential adsorption of specific additive or impurity ions from the chemical solution adjacent to the particle.

Desorption and Dissolving

A typical example of charging by desorption is the liberation of alkali from the surface of a clay mineral such as kaolinite. When the kaolinite mineral is formed, lattice substitution of the type

$$Al^{3+}_{(lattice)} + K^+_{(surface)} = Si^{4+}_{(lattice)} \tag{10.1}$$

and

$$Mg^{2+}_{(lattice)} + K^+_{(surface)} = Al^{3+}_{(lattice)} \tag{10.2}$$

occurs to some extent. Surface-adsorbed alkali and large alkaline earth ions required for charge neutrality are weakly bonded to the crystallite and are exchangeable ions. The cation exchange capacity CEC is the maximum amount of exchangeable cations per 100 g of clay material. In clay minerals, the exchangeable cations are adsorbed at interfaces between crystallites in the stacked clay aggregates. The dispersion of the clay aggregates in water will liberate alkali into the aqueous medium, leaving the surface of the particle negatively charged, as shown in Fig. 10.1.

Chemical Reaction with an Aqueous Medium

For oxides with a hydrated surface, the surface chemistry in water is dominated by the chemical reactions

$$MOH^+_{2(surface)} \overset{K_1}{\rightleftharpoons} MOH_{(surface)} + H^+_{(solution)} \tag{10.3}$$

and

$$MOH_{(surface)} \overset{K_2}{\rightleftharpoons} MO^-_{(surface)} + H^+_{(solution)} \tag{10.4}$$

where M represents a metal ion at the surface, such as Ba^{2+}, Al^{3+}, Si^{4+}, Ti^{4+}, Zr^{4+}, and so on. Each M is also bonded to bulk oxygen ions to satisfy its

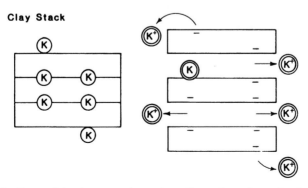

Fig. 10.1 Kaolin particles become charged on dispersion when adsorbed alkali ions are liberated.

bonding needs. The point of zero charge PZC of the surface may be defined in terms of the pK's of reactions (10.3) and (10.4):

$$PZC = \frac{pK_1 + pK_2}{2} \tag{10.5}$$

and indicates the average acid–base character of the surface. When bonded to a tetravalent cation, the surface hydroxyls are more acidic, and the PZC is lower. Approximate values of the PZC of several ceramic materials are listed in Table 10.1.

TABLE 10.1 Isoelectric Points of Oxides

Material	Nominal Composition	IEP
Muscovite	$KAl_3Si_3O_{11} \cdot H_2O$	1
Quartz	SiO_2	2
Delta manganese oxide	MnO_2	2
Soda lime silica glass	$1.00Na_2O \cdot 0.58CaO \cdot 3.70SiO_2$	2–3
Albite	$Na_2O \cdot Al_2O_3 \cdot 6SiO_2$	2
Orthoclase	$K_2O \cdot Al_2O_3 \cdot 6SiO_2$	3–5
Silica (amorphous)	SiO_2	3–4
Zirconia	ZrO_2	4–5
Rutile	TiO_2	4–5
Tin Oxide	SnO_2	4–7
Apatite	$10CaO \cdot 6PO_2 \cdot 2H_2O$	4–6
Zircon	$SiO_2 \cdot ZrO_2$	5–6
Anatase	TiO_2	6
Magnetite	Fe_3O_4	6–7
Hematite	αFe_2O_3	6–9
Goethite	$FeOOH$	6–7
Gamma iron oxide	γFe_2O_3	6–7
Kaolin (edges)	$Al_2O_3 \cdot SiO_2 \cdot 2H_2O$	6–7
Chromium oxide	αCr_2O_3	6–7
Mullite	$3Al_2O_3 \cdot 2SiO_2$	7–8
Gamma alumina	γAl_2O_3	7–9
Alpha alumina	αAl_2O_3	9–9.5
Alumina (Bayer process)	Al_2O_3	7–9.5
Zinc oxide	ZnO	9
Copper oxide	CuO	9
Barium carbonate	$BaCO_3$	10–11
Yttria	Y_2O_3	11
Lanthanum oxide	La_2O_3	10–12
Silver oxide	Ag_2O	11–12
Magnesium Oxide	MgO	12–13

Source: Temple C. Patton, *Paint Flow and Pigment Dispersion*, Wiley-Interscience, New York, 1979; E. G. Kelly and D. J. Spottiswood, *Introduction to Mineral Processing*, Wiley-Interscience, New York, 1982; J. M. Cases, *Silic. Ind.*, **36**, 145 (1971); R. H. Toon, T. Salman, and G. Donnay, *J. Colloid Interface Sci.*, **70**, 483 (1979).

The charge on a hydrated surface of a pure oxide particle dispersed in water is determined by its reaction with H_3O^+ or OH^- ions, as is shown for alumina in Fig. 10.2. The addition of H_3O^+ ions will reduce the pH and cause the uncharged surface to become protonated and positively charged. The addition of OH^- ions will remove hydrogen from the surface and produce a negative surface charge when the pH is greater than the PZC of the surface. Exchange at the surface is reversible but time-dependent, and the relative concentration of OH^- or H_3O^+ ions is potential determining. Generally, the rate of dissolving of a particle is a minimum at the PZC.

In layer minerals such as kaolinite, the exchange process occurs predominantly at the edges of the crystals. The faces are not surfaces with broken bonds (as for the surfaces of quartz or gibbsite), and exchange on the faces is limited. The faces of the kaolinite crystal dispersed in water are usually considered to have a negative charge. On titrating of a clay with an acid or base, ion exchange with little change in pH occurs near the PZC of the edges of the particles (Fig. 10.3). In the acidified suspension, negative gold sol particles are adsorbed preferentially on the edges of kaolinite crystals, as is shown in Fig. 10.4.

Adsorption of Specific Ions

Ions in solution that are not a component of the pure liquid medium may also interact with the surface. Simple ions are adsorbed on oppositely charged surfaces. Complete adsorption of the simple ions M^+ or A^- can neutralize the surface charge as indicated by the reactions

$$MO^-_{(surface)} + M^+_{(solution)} \rightarrow MOM_{(surface)} \tag{10.6}$$

and

$$MOH^+_{2(surface)} + A^-_{(solution)} \rightarrow MOH_2A_{(surface)} \tag{10.7}$$

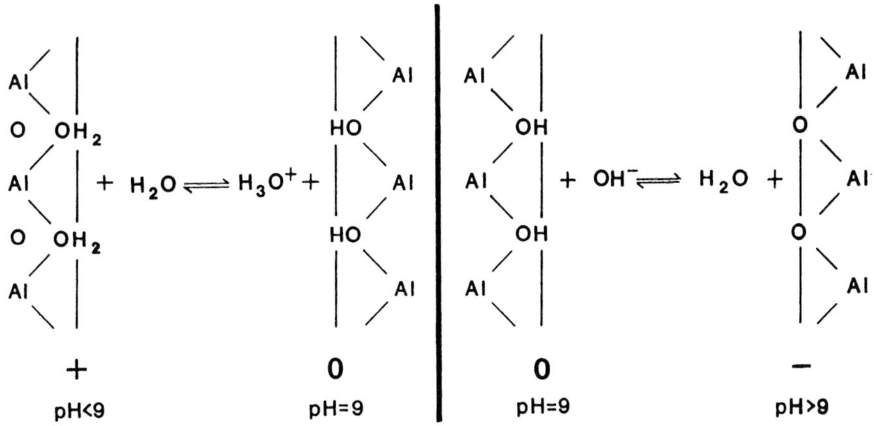

Fig. 10.2 Reaction of the hydrated surface of alumina with (left) an acid and (right) a base.

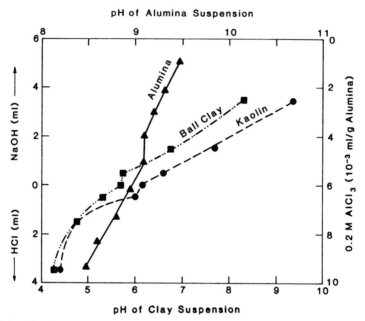

Fig. 10.3 Titration curves for alumina, kaolin, and ball clay indicate inflection near the PZC of the alumina and the PZC of the edges of clay particles.

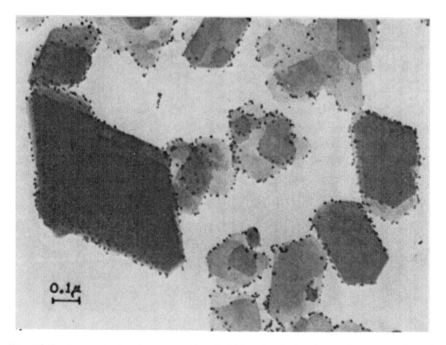

Fig. 10.4 A negatively charged gold colloid is adsorbed only on the edges of kaolin in an acidified suspension (pH < IEP). (From H. van Olphen, *An Introduction to Clay Colloid Chemistry*, Wiley-Interscience, New York, 1977.)

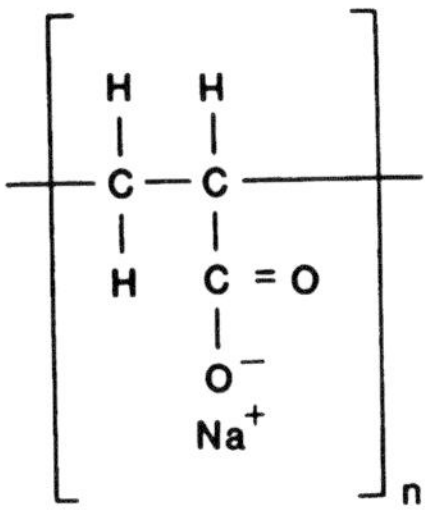

Sodium Polyacrylate 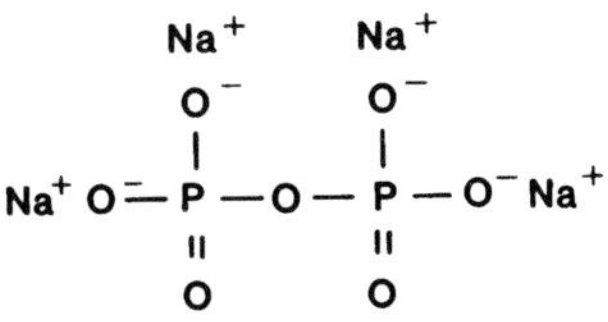   **Sodium Pyrophosphate**

Fig. 10.5 Structure of sodium polyacrylate and sodium pyrophosphate deflocculants; the dissociation of Na^+ increases with pH.

Complex ions such as sodium pyrophosphate (Fig. 10.5) may adsorb on neutral or charged surfaces. The adsorption of a multivalent ion may reverse the surface charge;

$$MO^-_{(surface)} + M^{n+}_{(solution)} \rightarrow MOM^{(n-1)+}_{(surface)} \qquad (10.8a)$$

$$MOH^+_{2(surface)} + A^{n-}_{(solution)} \rightarrow MOH_2A^{(n-1)-}_{(surface)} \qquad (10.8b)$$

Studies of the adsorption of simple hydrated ions on hydrated oxide surfaces indicate that the specific ion adsorption from the aqueous solution occurs rather abruptly at a particular pH.

Ions such as Al^{3+} and Fe^{3+} may hydrolyze in aqueous solutions, and the adsorption is more complicated. The hydrolysis of aluminum salts is represented by the following equations:

$$Al(H_2O)_6^{3+} \rightleftharpoons Al(OH)(H_2O)_5^{2+} + H_3O^+ \qquad (10.9a)$$

$$Al(OH)(H_2O)_5^{2+} \rightleftharpoons Al(OH)_2(H_2O)_4^+ + H_3O^+ \qquad (10.9b)$$

$$Al(OH)_2(H_2O)_4^+ \rightleftharpoons Al(OH)_3(H_2O)_3 + H_3O^+ \qquad (10.9c)$$

$$Al(OH)_3(H_2O)_3 \rightleftharpoons Al(OH)_3 + 3H_2O \qquad (10.9d)$$

The ions $Al(OH)(H_2O)_5^{2+}$ and $Al(OH)_2(H_2O)_4^+$ occur at pH < 7. In the process of hydrolysis, hydrated aluminum ions lose protons, and the final neutral complex loses water to form the hydroxide precipitate. The hydroxide $Al(OH)_3$ is amphoteric, and at pH > 8.5 the aluminate ion $Al(OH)_4^-$ is formed.

Using Auger spectroscopy for surface analysis, Horn and Onoda* observed that aluminum is rapidly adsorbed from an aqueous $AlCl_3$ solution onto a silica glass surface in the pH range from about 5 to 9; particles were observed to be positively charged for pH below 9 and negatively charged for pH greater than 9. Their results indicated that a rather stable aluminum hydroxide coating was formed between pH 5 and 9. Similarly, adsorption onto alumina occurs at a

*J. M. Horn and G. Y. Onoda Jr., *J. Am. Ceram. Soc.*, **61**(11), 523–527 (1978).

TABLE 10.2 Common Deflocculants Used in Water

Inorganic	Organic
Sodium carbonate	Sodium polymethacrylate
Sodium silicate	Ammonium polyacrylate
Sodium borate	Sodium citrate
Tetrasodium pyrophosphate	Sodium succinate
	Sodium tartrate
	Sodium polysulfonate
	Ammonium citrate

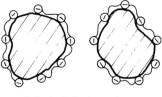

Anionic Polyelectrolyte

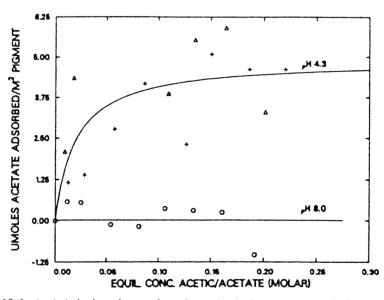

Fig. 10.6 (top) Anionic polymer electrolyte adsorbed on neutral or slightly positive particle creates a negative skin on particle. (bottom) Negative acetate molecules are adsorbed on titania particles for pH < IEP but are not adsorbed for pH > IEP of titania. [From W. H. Morrison, *J. Coatings Technol.* **57** (721), 57 (1985).]

pH of about 9, as indicated in Fig. 10.3. James and Healy* also showed that the adsorption of hydrolyzable ions can cause a reversal in polarity at (1) the PZC of the substrate surface, (2) the pH of the nucleated metal hydroxide, and (3) the PZC of the adsorbed metal hydroxide coating.

An ionic polymer has regular ionizable side groups, as shown in Fig. 10.5; ionization in solution produces charged side groups and oppositely charged counterions. The adsorption of a polymer electrolyte may reverse the polarity and increase the charge on the particle, as indicated by Eqs. 10.8 and 10.9. Molecules are adsorbed at multiple points on the surface of the particle (Fig. 10.6). Polymer electrolytes of low molecular weight may be powerful deflocculants in systems containing polar liquids. Relatively short polymer molecules are used where milling or high-shear mixing may degrade the polymer molecules.

Compositions of common polyelectrolytes used for deflocculation are listed in Table 10.2. Polyelectrolytes containing the sulfonate ($-SO_3^-$) group have an extreme affinity for water, and alkali polysulfonates are powerful surfactants and deflocculants.

A dilute solution of sodium silicate (water glass) is an inorganic polyelectrolyte that is widely used for deflocculating slurries of clays and other industrial minerals. The sodium silicate acts as a buffer, emulsifies oil, and precipitates highly charged cations from the solution. Grades are available with a SiO_2/Na_2O weight ratio of about 1.6–3.3; the chain length increases as the weight ratio increases. Sodium phosphates may also be used as deflocculants, but their disposal in waste water is now restricted for environmental reasons.

10.2   DEVELOPMENT OF AN ELECTRICAL DOUBLE LAYER

Ions and polar molecules in the solution surrounding a particle will respond to the charged surface. Coulombic forces will repel like-charged ions but attract polar liquid molecules and oppositely charged ions into a region near the surface, increasing their concentration relative to that in the bulk. Although overall charge neutrality must obtain, a difference in electrical potential between the surface and the bulk solution can be effected, as shown in Fig. 10.7. The potential gradient will not be sharp, because thermal vibration of molecules of the liquid medium causes the diffusion of some counterions. Our model is a charged particle with a relatively immobile adsorbed layer of counterions, called the Stern layer, and a concentration gradient of counterions and polar liquid molecules in a diffuse layer which attains some steady-state configuration. This model is now commonly called the diffuse electrical double-layer model.

Gouy† and Chapman‡ independently derived a model for the potential gra-

---

*R. O. James and T. W. Healy, *J. Colloid Interface Sci.*, **40**, 52–64 (1972).
†G. Gouy, *Ann. Phys.*, **1**(9), 129 (1917).
‡D. L. Chapman, *Phil. Mag.*, **25**(6), 475 (1913).

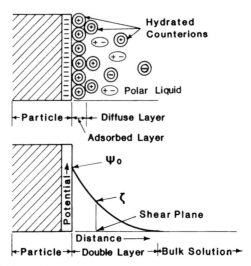

Fig. 10.7 Electrical double layer model for particle charging in a polar liquid and Gouy–Chapman profile of the electrical potential.

dient in the diffuse layer based on a uniformly charged surface, an adjacent solution with uniform dielectric constant ϵ_r, and point charges. Assuming the distribution of charge is described by the Boltzman equation, the concentration of counterions N_i in the diffuse layer relative to the concentration in the bulk solution N_{io} is

$$N_i/N_{io} = \exp\left(-U_i/k_B T\right). \tag{10.10}$$

The potential energy of the ion U_i is a function of the valence of the ion Z_i and the electrical potential ψ at that position:

$$U_i = Z_i(e\psi) \tag{10.11}$$

For the case of a surface potential $\psi_0 \leq 100$ mV,

$$\psi = \psi_0 \exp - (x/\kappa^{-1}) \tag{10.12}$$

where κ^{-1} is the distance from the charged surface to the plane where $\psi = \psi_0/2.718$ and is called the thickness of the double layer. The double-layer thickness is calculated from the equation

$$\kappa^{-1} = \left(\frac{\epsilon_r \epsilon_0 N_A k_B T}{F^2 \Sigma N_i Z_i^2}\right)^{1/2} \tag{10.13}$$

where N_i is the concentration of each ion type in the solution phase and F is the Faraday constant. The variation of the potential with position predicted by

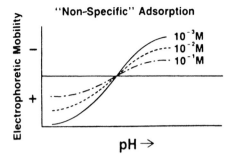

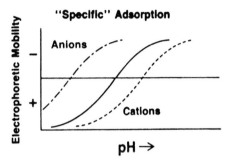

Fig. 10.8 Specific adsorption of counterions in the Stern layer changes the IEP in contrast to nonspecific adsorption.

the Gouy–Chapman model is shown in Fig. 10.7. For a 0.01 M aqueous solution of 1:1 electrolyte at 20°C, the effective double-layer thickness κ^{-1} is calculated to be about 3 nm. Increasing the concentration or valence of the counterions compresses the diffuse layer and increases the potential gradient. The use of a liquid of a lower dielectric constant or temperature also compresses the diffuse layer.

We will see in the next section that the stability of a suspension deflocculated with a simple electrolyte depends importantly on $d\psi/dx$. Although the Gouy–Chapman model does not consider the specific structure of adsorbed ions of finite size, it is very useful for qualitatively understanding and predicting deflocculation behavior.

10.3 ELECTROKINETIC PROPERTIES

Charged particles in a suspension will respond to an imposed potential difference, and the particle velocity is called the electrophoretic velocity. In a similar manner, the ionic solution adjacent to the wall of a capillary with an electrical double layer will be motivated to flow if a potential difference is imposed; this is called electroosmotic flow. During flow, a hydrodynamic plane of slippage must occur somewhere in the double layer. The location of the slippage plane X_s is a matter of some disagreement, but is certainly beyond the first adsorbed layer of simple or polymer counterions (called the Stern layer) and perhaps additional layers of adsorbed polar liquid molecules and ions. The potential at the slippage plane is called the zeta potential ζ and can be calculated from an electrokinetic property. For the electrophoretic transport of nonconducting particles,

$$\zeta = \frac{f_H \eta \nu_e}{\epsilon_r \epsilon_0 E} \tag{10.14}$$

where η is the viscosity of the electrolyte and ν_e is the electrophoretic velocity for an imposed electrical field E. The ratio ν_e/E is the electrophoretic mobility. The Henry constant f_H is equal to 1 when the product of the particle diameter a and κ is greater than 100 and 3/2 when $a\kappa$ is less than 1.

In systems with simple counterions, the zeta potential is an indication of the gradient in electrical potential when the surface potential remains constant. The

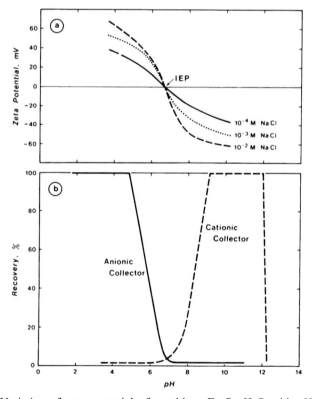

Fig. 10.9 Variation of zeta potential of goethite $\alpha Fe_2O_3 \cdot H_2O$ with pH and ion attraction. (From E. G. Kelly and D. J. Spottiswood, *Introduction to Mineral Processing*, Wiley-Interscience, New York, 1982.)

pH at which the zeta potential is zero is termed the isoelectric point IEP. For charging of a hydrated surface due to the reaction with OH^- or H_3O^+, raising or lowering the pH from the isoelectric point will initially increase the mobility (Fig. 10.8) and zeta potential (Fig. 10.9). However, a further addition of OH^- or H_3O^+ will cause a reduction in the zeta potential because of the compression of the double layer. When a polyelectrolyte is adsorbed on the surface, the zeta potential increases rapidly, initially, and then more slowly, as indicated in Fig. 10.10. The distance from the surface to the slippage plane may be expected to vary somewhat with the molecular size and the conformity of the adsorbed molecules. Molecules are expected to protrude farther from the surface when the relative solvent power of the liquid medium is higher.

The electrophoretic mobility of particles in a dilute suspension may be determined by measuring the migration velocity of individual particles in a known potential gradient. This technique, called microelectrophoresis, is fatiguing unless automated and is not suitable when settling occurs during the test. An average mobility of the particles in a suspension is conveniently determined using a bulk technique, in which the electrophoretic migration causes an increase in solids content in a simple cell that is large relative to the particle size. The average zeta potential is calculated from the average mobility using Eq. 10.14.

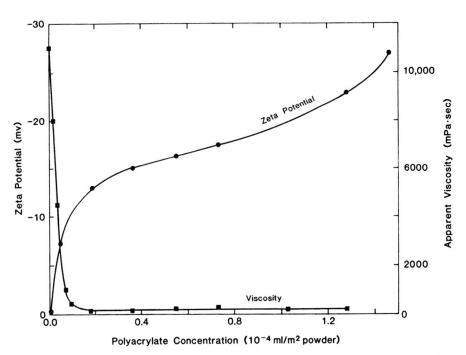

Fig. 10.10 Deflocculation of an aqueous slurry containing 50 vol % calcined alumina using an ammonium polyacrylate deflocculant.

10.4 DEFLOCCULATION AND THE STABILITY OF SUSPENSIONS

Powders suspended in a liquid spontaneously agglomerate unless they are suitably deflocculated by creating mutually repelling charged double layers or by physically preventing the close approach of particles due to the steric hindrance of adsorbed molecules (Fig. 10.6). Electrical charging may stabilize slurries in polar liquids.

The interaction of two particles with identical charged double layers was examined by Derjaguin and Landau* and Verwey and Overbeek,† and their combined theories are now referred to as the DLVO theory. The motivating force for coagulation is the ever-present Van der Waals attractive force, which is a function of the dielectric constant of the medium and the mass and separation of the particles (see Eq. 2.17). Repulsion is provided by the interaction of two electrical double layers. The form of the repulsion depends on the size and shape of the particles, the distance h between their surfaces, the double-layer thickness κ^{-1} and ϵ_r of the liquid medium. For two spherical particles of diameter a the potential energy of repulsion is‡

$$U_R = \frac{\epsilon_r a^2 \psi_0^2}{4(h + a)} \exp - \left(\frac{h}{\kappa^{-1}} \right) \tag{10.15}$$

when $a/\kappa^{-1} \ll 1$, that is, small particles with a relatively large double layer. A greater surface potential and double-layer thickness increases the repulsive force. For the situation where $a/\kappa^{-1} \gg 1$,‡

$$U_R = \frac{\epsilon_r a \psi_0^2}{4} \ln \left[1 + \exp - \left(\frac{h}{\kappa^{-1}} \right) \right] \tag{10.16}$$

The total potential energy U_T is the algebraic sum of the attractive Van der Waals potential energy U_A (Eq. 2.17) and the repulsive potential energy U_R

$$U_T = U_A + U_R \tag{10.17}$$

The dependence of U_A, U_R, and U_T on the separation of spherical particles is illustrated in Fig. 10.11. For each system there is some critical zeta potential and range of double-layer thickness for which the repulsive potential energy exceeds the attractive potential energy, producing an energy barrier to coagulation. A secondary minimum in the potential energy U_T occurs for large flat particles when the separation is of the order of the particle size, and the primary minimum occurs when the separation is approaching molecular dimensions.

*B. Derjaguin and L. Landau, *Acta. Physicochim.*, **14**, 633 (1941).
†E. Verwey and J. Th. G. Overbeek, *Theory of the Stability of Lyophobic Colloids*, Elsevier, Amsterdam, 1948.
‡M. J. Rosen, *Surfactants and Interfacial Phenomena*, Wiley-Interscience, New York, 1978.

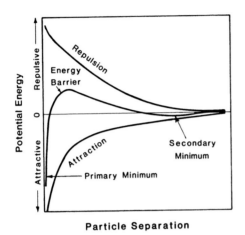

Fig. 10.11 Potential energy of interaction between two particles with electrical double layers. Steric hindrance may add to the repulsion energy and increase the energy barrier to coagulation. (The secondary minimum is not expected when $\kappa^{-1}/a < 1$.)

The DLVO interaction diagram is very useful for understanding much of the behavior of suspensions.

The kinetic energy of colloidal particles due to Brownian motion is of the order of $10k_B T$, and at $20°C$ a repulsion barrier corresponding to a zeta potential of about 25 mV is requisite to minimize coagulation by means of electrical charging. A larger repulsion barrier or other mechanism for stability is needed to retard agglomeration during pouring and mixing processes which produce a greater kinetic energy.

An adsorbed nonionic surfactant with its associated water molecules or an adsorbed polyelectrolyte may stabilize suspensions at an apparent zeta potential less than 25 mV because the adsorbed layer can provide a steric hindrance to the close approach of the particles. If the repulsive potential energy due to electrical charging is U_{Re} and steric stabilization is U_{Rs}, then

$$U_T = U_A + (U_{Re} + U_{Rs}) \tag{10.18}$$

For colloidal particles, U_A is relatively small, and with an adsorbed polyelectrolyte, the distance from the hydrated surface to the slippage plane may be greater than for deflocculation using a simple electrolyte. Deflocculants such as sodium polyacrylate can produce a very stable suspension at an apparent zeta potential of 15 mV (Fig. 10.10); a contribution of U_{Rs} to the stabilization is expected. The dissociation of the polyelectrolyte and its adsorption on positive particles increases with pH.

The Gouy–Chapman theory indicates that the thickness of the double layer is proportional to ϵ_r of the liquid and a polar liquid with a higher dielectric constant should promote dispersion. This effect of ϵ_r has been observed in preparing suspensions of MgO and CaO in alcohols where water must be avoided.

In nonpolar liquids of low dielectric constant, the specific adsorption of oriented polar molecules may provide deflocculation and stability. Tartaric,

Particle   Oleic Acid

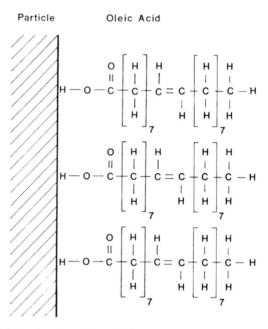

Fig. 10.12 Model for the adsorption of oleic acid deflocculant on an oxide particle. (Note that kink at C=C bond is not shown.)

benzoic, oleic, stearic, and trichloroacetic acids, glyceryl trioleate, and some natural oils produce deflocculation of oxide powders in several nonpolar liquids. Oleic acid ($C_{18}H_{34}O_2$) is a liquid at 25°C and has a dielectric constant of only about 2.5, but the molecular groups at the two ends of the molecule are very different, as is shown in Fig. 10.12. The −COOH group is strongly adsorbed on the surface of an oxide. Steric hindrance provided by the kinked molecule may produce a U_{Rs} sufficient for the deflocculation of fine particles. In general, the steric repulsion is proportional to the thickness of the adsorbed layer and the chemical nature and concentration of the adsorbed molecules.

10.5 COAGULATION AND FLOCCULATION

A suspension may coagulate because of one or more of at least three mechanisms, as shown in Fig. 10.13. Slow coagulation occurs when $\zeta < 25$ mV and steric hindrance is negligible. Particles of quartz in an aqueous suspension have a negative potential when the pH is higher than the IEP. Slow coagulation occurs for a $2 < pH < 3$ because U_{Re} is low. In the range $3 < pH < 12$ the zeta potential and U_{Re} are high and a relatively stable suspension is formed.* However, at an extreme pH, the concentration of ions is high, and the system may coagulate because the double layer is compressed.

*B. Yarar and J. A. Kitchener, *Trans. Inst. Min. Metall.*, **79**(3), 23–33 (1970).

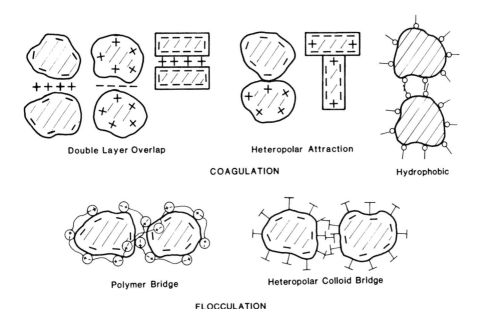

Double Layer Overlap   Heteropolar Attraction

COAGULATION

Hydrophobic

Polymer Bridge   Heteropolar Colloid Bridge

FLOCCULATION

Fig. 10.13 Two particle models showing (above) coagulation produced by adsorbed ions and small molecules and (below) flocculation produced by a binder.

Electrolytic coagulation occurs when the concentration of counterions is sufficient to reduce the thickness of the diffuse layer and reduce the zeta potential and U_{Re} (Fig. 10.14). In the Gouy–Chapman theory, the concentration N of counterions of valence Z in the double layer is in the proportion of about N^Z. Counterions of higher valence have a strong effect in causing coagulation; this observation is known as the Schulze–Hardy rule. The order of coagulation power for monovalent cations is $Li^+ > Na^+ > K^+ > Rb^+ > NH_4^+$, and for divalent cations $Mg^{2+} > Ca^{2+} > Sr^{2+} > Ba^{2+}$. This sequence is often referred to as the Hofmeister series and can be rationalized in terms of the charge/radius of the counterion, as indicated in Table 10.3. For negative counterions, the observed coagulation value is $SO_4^{2-} > Cl^- > NO_3^-$.

Negative oxide particles in an aqueous system are commonly coagulated using an additive such as $CaCl_2$, $CaCO_3$, $MgCl_2$, or $MgSO_4$. The adsorption of the more highly charged cation from the dissolved salt reduces the zeta potential. Some coagulants may also change the pH of the system in the direction of the IEP. $AlCl_3$ is a powerful coagulant and acidifier that can be used to reduce the pH to less than 4. When coagulating a clay slurry using a salt, an agglomerate with a denser particle packing may occur when the salt concentration exceeds some particular concentration. Moderately powerful coagulants are more commonly used, and the pH must be controlled, because the solubility of the fine particles is dependent on pH.

Basic organic compounds such as an amine that ionizes in solution may also be used as a coagulant in an aqueous system.

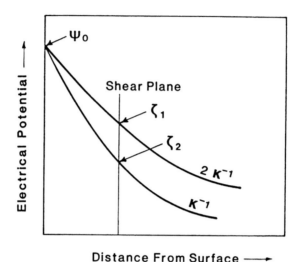

Fig. 10.14   A reduction of the thickness of the double layer reduces the zeta potential and may produce coagulation.

Adsorption of ethylamine $C_2H_5-NH_3^+$ may cause coagulation by causing the surface to become hydrophobic and less hydrated. Coagulation is motivated by contact between hydrophobic surfaces, which reduces the area of the hydrophobic surface–water interface.

Heteropolar coagulation occurs when the particles in suspension have surfaces of opposite charge. Suspensions of clay particles with negative faces and positive edges are coagulated as shown in Fig. 10.13. Admixing a deflocculated suspension of negative particles in a deflocculated suspension of positive particles may produce a coagulated system.

In general, particles in a dilute coagulated suspension settle rapidly in bulk,

TABLE 10.3   Hydrated Ionic Radii

Ion	Radius (Å)	Hydration (mol H_2O)	Hydrated Radius (Å)
Li^+	0.78	14	7.3
Na^+	0.98	10	5.6
K^+	1.33	6	3.8
Rb^+	1.49	0.5	3.6
NH_4^+	1.43	3	—
Mg^{2+}	0.78	22	10.8
Ca^{2+}	1.06	20	9.6
Ba^{2+}	1.43	19	8.8
Al^{3+}	0.57	57	—

Source: W. E. Worrall, *Clays and Ceramic Raw Materials*, Halsted Press, New York, 1975.

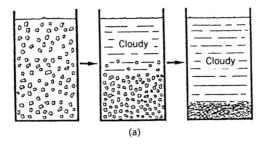

(a)

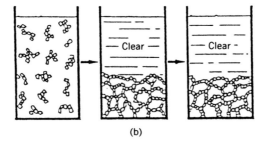

(b)

Fig. 10.15 Sedimentation behavior for (a) deflocculated and (b) coagulated suspensions.

and the supernatant solution above is clear; the sediment is of relatively low packing density and is easily redispersed. Particles in a well-deflocculated suspension settle relatively slowly, and the sediment is relatively dense and less easily dispersed. Sediment height is used as a qualitative index of coagulation (Fig. 10.15).

SUMMARY

Deflocculants and coagulants are essential additives that modify the interparticle forces, agglomerate structure, and consistency of the processing system. Agglomerating forces of the Van der Waals type may be offset by repulsive forces produced by electrical charging and steric hindrance. Particle charging and steric hindrance may produce the deflocculation of particles in polar liquids, whereas steric hindrance produces deflocculation in nonpolar liquids. Deflocculants producing particle charging may be either simple or polymer electrolytes; the type and concentration must be controlled to control the interparticle repulsion indicated by the zeta potential. The pH relative to the IEP may alter the adsorption of a polyelectrolyte. Coagulation is produced by crowding counterions in the double layer, sometimes called "overdeflocculation," or by adding counterions of a higher valence; both compress the double layer and reduce the repulsion between particles. Surfactants that produce a lyophobic surface on the particles may also produce coagulation. The bonding of oppositely charged surfaces that may occur in mixture of minerals or clay systems is called heteropolar coagulation. The role of deflocculants and coagulants in real systems is sometimes difficult to analyze because of aging caused by the effects of soluble impurities and time-dependent dissolving and adsorption processes.

SUGGESTED READING

1. Roger G. Horn, "Surface Forces and Their Action in Ceramic Materials," *J. Am. Ceram. Soc.*, **73**(5), 1117–1135 (1990).

2. Paul D. Calvert, Ellen S. Tormey, and Richard L. Pober, "Fish Oil and Triglycerides as Dispersants for Alumina," *Am. Ceram. Soc. Bull.*, **65**(4), 669–672 (1986).

3. William H. Morrison Jr., "Stabilization of Aqueous Oxide Pigment Dispersions," *J. Coatings Technol.*, **57**(721), 55–65 (1985).

4. Michael D. Sacks and Chandrashekhar S. Khadilkar, "Milling and Suspension Behavior of Al_2O_3 in Methanol and Methyl Isobutyl Ketone," *J. Am. Ceram. Soc.*, **66**(7), 488–494 (1983).

5. Robert D. Vold and Marjorie J. Vold, *Colloid and Interface Chemistry*, Addison-Wesley, Reading, MA, 1983.

6. G. D. Parfitt, Ed., *Dispersion of Powders in Liquids*, 3rd ed., Wiley-Interscience, New York, 1981.

7. R. H. Yooa, T. Salmon, and G. Donnay, "Predicting Points of Zero Charge of Oxides and Hydroxides," *J. Colloid Interface Sci.*, **70**(3), 483 (1979).

8. J. M. Horn Jr. and G. Y. Onoda Jr., "Surface Charge of Vitreous Silica and Silicate Glasses in Aqueous Electrolyte Solutions," *J. Am. Ceram. Soc.*, **61**(11), 523–527 (1978).

9. Milton J. Rosen, *Surfactants and Interfacial Phenomena*, Wiley-Interscience, New York, 1978.

10. Arthur W. Adamson, *Physical Chemistry of Surfaces*, Wiley-Interscience, New York, 1976.

11. R. F. Long, "Electrophoresis as Method of Investigating Electrical Double Layer," *Ind. Engr. Chem.*, **57**(8), 58–71 (1969).

12. Paul Sennet and J. P. Olivier, "Colloidal Dispersions, Electrokinetic Effects, and the Concept of Zeta Potential," in *Chemistry and Physics of Interfaces*, American Chemistry Society, Washington, DC, 1965.

13. H. van Olphen, *An Introduction to Clay Colloid Chemistry*, Wiley-Interscience, New York, 1963.

PROBLEMS

10.1 Where is the deflocculant phase in Fig. 4.3?

10.2 How does the surface of an hydrated oxide change when the pH is increased or decreased from the PZC?

10.3 Sketch the variation of electrical potential with pH expected for silica, alumina, and mullite.

10.4 Charged polymer ions are relative lineated. Sketch a polyacrylate molecule adsorbed on a neutral oxide particle with a hydrated surface.

10.5 Calculate the degree of polymerization and nominal size of Na-polymethracrylate deflocculant having a mean molecular weight of 13,000 g/mol.

10.6 Calculate κ^{-1} for an aqueous solution of a $1:1$ electrolyte of 0.1 M concentration at 20°C.

10.7 Show that $\psi = \psi_0/e$ when $x = \kappa^{-1}$.

10.8 What is the relative change in κ^{-1} when ethyl alcohol is substituted for water at 20°C?

10.9 For $\kappa^{-1} = 3$ nm, what is the range of size a for which $a/\kappa^{-1} > 100$?

10.10 Calculate the zeta potential for particles in an aqueous suspension at 25°C when $v_e/E = 0.23 \times 10^{-3}$ cm^2/V · s. Assume $f_H = 1$ in Eq. 10.14.

10.11 Compare the Hofmeister series with the ratio of valence/radius of the ions.

10.12 Explain why ammonium polyacrylate and sodium silicate deflocculants are especially effective when highly charged cations are present. (Clue: the addition of highly charged cations to a transparent solution of the deflocculant caused the solution to become cloudy.)

10.13 Sketch the concentration of plus and minus counterions in the diffuse layer.

10.14 Explain why an addition of AlCl$_3$ coagulant reduces the pH much more than an equal amount of CaCO$_3$.

10.15 The deflocculation of aqueous slurries of Si$_3$N$_4$ depends on the prior exposure of the powder to air. Explain.

10.16 Deflocculated slurries of silica and alumina are prepared at a pH of 7. When the two slurries are mixed, a highly coagulated slurry forms. Explain.

EXAMPLES

Example 10.1 Ammonium polyacrylate deflocculant has the molecular structure $[-CH_2-CHCOO^-NH_4^+-]_n$ and a mean molecular weight of 6,000 g/mol. Calculate the degree of polymerization n and estimate the size of the molecule. The bond length of $C-C = 1.54$ Å, $C-H = 1.09$ Å, and $C-O = 1.43$ Å. Refer to Fig. 10.5.

Solution. A mer of the polyacrylate has 3 carbon, 7 hydrogen, 2 oxygen, and 1 nitrogen atoms. The atomic weights of carbon, hydrogen, oxygen, and nitrogen are 12.0, 1.0, 16.0, and 14.0 g/mol, respectively. Summation gives 89 g/mol mers. The degree of polymerization n is

$$n = 6000 \text{ (g/mol molecules)}/89 \text{ (g/mol mers)} = 67$$

For a linear molecule, the chain length L is calculated as

$$L = 67 \, [2(C-C)/mer][1.54 \text{ Å}/C-C] = 210 \text{ Å} = 21 \text{ nm}$$

The width W of the molecule is calculated to be

$$W = C-H + C-C + C-O + \text{radius } O^- + \text{diameter } NH_4^+$$
$$W = 1.09 \text{ Å} + 1.54 \text{ Å} + 1.43 \text{ Å} + 0.66 \text{ Å} + 2.86 \text{ Å}$$
$$= 7.58 \text{ Å} = 0.76 \text{ nm}$$

The mean thickness of the layer of adsorbed molecules will be several nanometers. The actual thickness will be lower when the adsorption affinity and degree of ionization of the molecule are higher and the degree of hydration (affinity for solvent) of the molecule is lower.

Example 10.2 Estimate the electrical double layer thickness κ^{-1} for an aqueous slurry deflocculated with 0.01 M HNO_3 at 20°C. Compare your answer to the size of a water molecule which is approximately 0.2 nm.

Solution. The double layer thickness may be estimated from Eq. 10.13:

$$\kappa^{-1} = [(\epsilon_r\epsilon_0 N_A k_B T)/(F^2 \Sigma N_i Z_i^2)]^{0.5}$$
$$\kappa^{-1} = [80(8.85 \times 10^{-12} \text{ coul/J m})(8.31 \text{ J/K mol})(298 \text{ K})/$$
$$(9.65 \times 10^4 \text{ coul/mol})^2 [0.01 \, (1)^2$$
$$+ 0.01 \, (1)^2] \text{ mol/L } (10^3 \text{ L/m}^3)]^{0.5}$$
$$\kappa^{-1} = 3 \text{ nm}.$$

This thickness is equivalent to many layers of adsorbed water molecules.

Example 10.3 Estimate the length of an oleic acid molecule $CH_3(CH_2)_7C_2H_2(CH_2)_7COOH$ adsorbed on a particle, as is illustrated in Figure 10.12. Assume the angle at the $C=C$ double bond is 120°. Compare this value to the double-layer thickness and adsorbed polymer electrolyte thickness calculated in the preceding examples.

Solution. The length of two segments joined by the $C=C$ bond is

Bond Type	Number	Length (Å)	Sum (Å)
C—H	1	1.09	1.09
C—C	8	1.54	12.32
C=C	1	1.34	1.34
			$L_1 = 14.75$ Å
C—C	8	1.54	12.32
C—O	1	1.43	1.43
H—O	1	0.96	0.96
			$L_2 = 14.71$ Å

$$L = L_1 \cos 60° + L_2 \qquad L = 14.75 \text{ Å } (0.5) + 14.71 \text{ Å} \qquad L = 2.2 \text{ nm}$$

The calculated length is of the same order but slightly smaller than the thicknesses calculated for the adsorbed polymer electrolyte and the electrical double layer in the preceding examples.

Example 10.4   If the relative bonding strength of an ion is proportional to its valence/ionic radius, calculate and compare the bonding strengths of NH_4^+ and Na^+ used in deflocculation and Mg^{2+} and Ca^{2+} used for coagulation. Consult Table 10.3.

Solution. The valence/radius is calculated as follows:

Ion Type	Valence	Radius (Å)	Valence/Radius
NH_4^+	1	1.43	0.7
Na^+	1	0.98	1.0
Ca^{2+}	2	1.06	1.9
Mg^{2+}	2	0.78	2.6

The order of the values clearly indicates the use of Ca^{2+} and Mg^{2+} for coagulation and NH_4^+ and Na^+ for deflocculation of negative particles.

CHAPTER 11

FLOCCULANTS, BINDERS, AND BONDS

Polymer molecules and coagulated colloidal particles that are adsorbed and bridge between ceramic particles provide interparticle flocculation and a binding action. However, these additives may provide several additional practical functions in ceramic processing. Often a workable system is produced on using only one particular flocculating additive, but in some cases two or more different types or molecular weights are used. In ceramic processing these additives are referred to as binders more commonly than as flocculants, and the term binder is used here as a general term for these additives.

Clay binders which may contain organic matter and natural gums and waxes have been used as binders for centuries. Synthetic high-polymer molecular binders in the vinyl, cellulose, and polyethylene glycol families are also widely used. Other types of binders are resins (commonly called plastics), gels, low-temperature reaction bonds, and hydraulic cements. Films of wax or resin or an additive that polymerizes into a gel may bind the particles together. Additives that react with the ceramic particles and polymerize or crystallize are called reactive bonds. Hydraulic cements are inorganic bonds that react with water and form very fine needlelike, hydrated calcium silicate or calcium aluminate minerals which coat the particles and form a bonded network.

In this chapter we examine the compositions and structure of both organic and inorganic binder materials. Parameters for characterizing binders are considered. The dissolving of binders and their influence on viscosity are described.

11.1 USES OF BINDERS

Binders may provide many functions in ceramic processing and special names are sometimes used when a particular function is emphasized:

1. *Wetting Agent.* An adsorbed binder may improve the wetting of the particles.
2. *Thickener.* The binder commonly increases the apparent viscosity of the processing system.
3. *Suspension Aid.* The settling of particles in a suspension or slurry may be reduced on adding a binder.
4. *Rheological Aid.* The binder type and concentration can be selected to control the flow properties of a paste or slurry.
5. *Body Plasticizer.* A matrix of binder in a pressing powder, extrudate, or coating may provide plastic deformation behavior to the system containing brittle particles.
6. *Liquid Retention Agent.* Commonly the matrix of binder reduces the liquid migration rate in the system.
7. *Consistency Aid.* The binder may alter the amount of liquid required to produce a particular type of flow.
8. *Binder.* The most important function of the binder is to improve the strength of the as-formed product to provide strength for handling (green strength) before the product is densified by firing.

11.2 TYPES OF BINDERS

Colloidal clay minerals and sometimes animal dung were the binders used in processing ancient ceramics. Clay binders are used today in processing a wide variety of traditional ceramics and in more advanced ceramic systems when alumina and silica are acceptable in the composition. Examples of clay binders are fine kaolin, ball clay, and bentonite (Table 11.1). Clay binders and bentonite are colloidal particle-type binders which are now refined and characterized to produce consistent behavior. Microcrystalline cellulose is an organic colloidal particle binder. It is manufactured from high-purity cellulose pulp and may be used when submicron pores are desired or are not a problem.

Molecular binders range widely in composition and may be natural or synthetic substances (Table 11.1). The refined natural materials are more expensive than clay binders, but less expensive than the refined and synthetic organic polymers. The more popular organic molecular binders may be purchased in grades, which differ in the range of molecular sizes offered, and molecular binders provide a wide flexibility in modifying rheological behavior. When decomposed under oxidizing conditions, synthetic organic binders introduce relatively little inorganic impurity. Inorganic molecular binders are used when the inorganic component is compatible with the particle composition, because the residue on firing is large.

Molecular organic binders are often used in concert with refined clay binders when alumina and silica are present in the ceramic composition. Ball clay contains both clay minerals and natural organic matter of both a colloidal and molecular nature which influences its properties.

TABLE 11.1 Binder Materials

Colloidal Particle Type	
Organic	Inorganic
Microcrystalline cellulose	Kaolin
	Ball clay
	Bentonite

Molecular Type			
Organic	Examples	Inorganic	Example
Natural gums	Xanthan gum, gum arabic	Soluble silicates	Sodium silicate
Polysaccharides	Refined starch, dextrine	Organic silicates	Ethyl silicate
Lignin extracts	Paper waste liquor	Soluble phosphates	Alkali phosphates
Refined alginate	Na, NH_4 alginate	Soluble aluminates	Sodium aluminate
Cellulose ethers	Methyl cellulose, hydroxyethyl cellulose, sodium carboxymethyl cellulose		
Polymerized alcohols	Polyvinyl alcohol		
Polymerized butyral	Polyvinyl butyral		
Acrylic resins	Polymethyl methacrylate		
Glycols	Polyethylene glycol		
Waxes	Paraffin, wax emulsions, microcrystalline wax		

11.3 CLAY BINDERS

Coagulated clay colloids may be adsorbed and bridge the larger ceramic particles, effectively flocculating them (Fig. 10.13). The edge-face agglomeration of colloidal clay particles with positive edges and negative faces is a special case of heterocoagulation that occurs at a pH less than the PZC of the edges of the particles. Kaolin is not deflocculated at a low pH using simple acids because the faces remain negative. The coagulated clay agglomerates in the suspension are usually called clay flocs.

If a sufficient concentration of highly charged cations such as Ca^{2+} is present in the suspension prior to the coagulation, a measurable degree of parallelism in the packing of small groups of clay particles is observed, termed salt-type coagulation (see Fig. 11.1). Agglomerates formed in deflocculated suspensions due to double-layer overlap have a high degree of particle orientation.

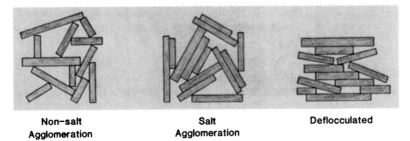

Non—salt	**Salt**	**Deflocculated**
Agglomeration	**Agglomeration**	

Fig. 11.1 Particle structure of clay agglomerates formed by edge-face heteropolar attraction, salt-assisted coagulation, and double layer overlap on sedimentation from a deflocculated slurry.

Finer clays containing more anisometric particles and a low alkali content and clays containing more of the mineral montmorillonite are more powerful flocculants and binders, as indicated by their higher MBI index and plastic strength (Table 5.5). Montmorillonite is a colloidal mineral of very high specific surface area and is a scavenger for cations. Clays containing montmorillonite tend to coagulate naturally, and their behavior is less sensitive to fluctuations in concentrations of electrolytes. Bentonite is a clay-like material containing a relatively high percentage of the mineral montmorillonite and is a relatively powerful flocculant and binder.

11.4 MOLECULAR BINDERS

Molecular binders are low- to high-molecular-weight polymer molecules which may adsorb on the surfaces of particles and bridge them together, as shown in Fig. 10.13, or form a polymer–polymer bonded network (film) among the particles. The functionality of the polymer molecule may be nonionic, anionic, or cationic (Fig. 11.2). Most of the polymer binders used in ceramic processing

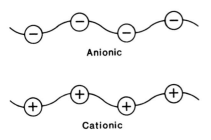

Anionic

Cationic

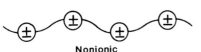

Nonionic

Fig. 11.2 Nonionic binder with polar side groups and ionic binders with anionic and cationic functionality after dissociation of charge compensating ion from molecule.

are nonionic or mildly anionic. We first consider adhesion binders, which are adsorbed on particles. Film binders, such as the waxes, are considered later.

Vinyl Binders

Consider the molecular structure of a very common binder, polyvinyl alcohol PVA, which is illustrated in Fig. 11.3. The basic repeating structural unit within the brackets is called the mer. The number of mers in the molecule, indicated by the subscript n, is the degree of polymerization. The molecular weight of a polymer increases as the degree of polymerization increases.

In the PVA structure, the $-C-C-$ linkage is referred to as the vinyl backbone, and the $-H$ and $-OH$ are referred to as the side groups. Vinyl binders with the $-C-C-$ backbone are very flexible binders. The polar $-OH$ side group is hydrophilic, which promotes initial wetting and dissolving of the PVA in a polar liquid such as water. Hydrogen bonding of the $-OH$ side group to the surface of a particle provides adhesion, and the dipolar attraction of $-OH$ side groups produces intermolecular bonding. Polyvinyl alcohol is a binder which has a strong affinity for adsorption on oxide particles dispersed in water, as is indicated in Fig. 11.4.

Polyvinyl alcohol is manufactured by the hydrolysis of polyvinyl acetate in the presence of a catalyst. Partially hydrolyzed PVA contains less than 20% acetate groups. It is commonly purchased as a powder and may be dissolved

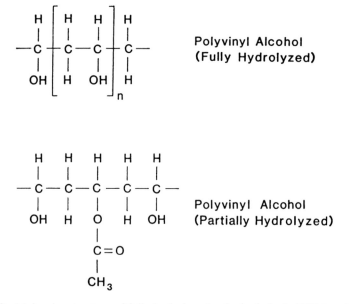

Fig. 11.3 Molecular structure of fully hydrolyzed polyvinyl alcohol PVA and partially hydrolyzed polyvinyl alcohol which retains acetate groups.

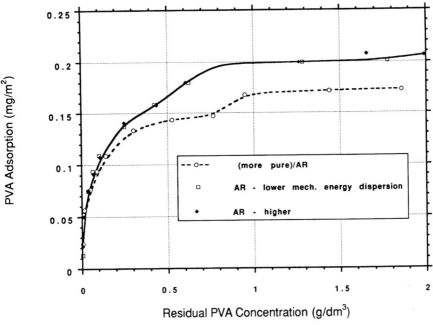

Fig. 11.4 Adsorption isotherm for fully hydrolyzed polyvinyl alcohol onto two alumina powders. (Note that the adsorption is somewhat higher for the less pure powder.)

in cold water. Fully hydrolyzed PVA contains less than 2% residual acetate and is dissolved in hot water. Other vinyl-type binders containing different side groups are indicated in Table 11.2. Polyvinyl butyral and polymethacrylate binders dissolve in nonaqueous solvents. Polyvinyl butyral is derived from PVA and some $-OH$ groups remain. Polymethacrylate is a very pure binder.

TABLE 11.2   Side Groups ($-R$) in Vinyl Binders

Type	Side Group	Functionality
Soluble in polar liquids		
Polyvinyl alcohol	$-OH$	Nonionic
Polyacrylamide	$-CONH_2$	Nonionic
Polyvinyl pyrrolidone	$-NC_2H_4C_2OH_2$	Nonionic
Carboxylic polymer	$-COOH$	Anionic
Soluble in nonpolar liquids		
Polymethyl methacrylate	$-CH_3$	Nonionic
	$-COOCH_3$	
Polyvinyl butyral	$\begin{matrix} H \\ \vert \quad \vert \quad \vert \\ O-C-O \\ \vert \\ C_3H_7 \end{matrix}$	Nonionic

Cellulose Binders

Cellulose is a natural carbohydrate that contains the basic repeating structure of anhydroglucose units (Fig. 11.5 (top)). The cellulose ethers are a second important family of binders that have the polymeric backbone of cellulose, but modified side groups indicated by $-R$ in Fig. 11.5 (middle). The cellulose backbone is much less flexible than the vinyl backbone. Methyl cellulose MC is manufactured by treating cellulose with caustic solution and then methyl chloride to produce the methyl ether of cellulose. This treatment replaces some of the $-H$ in the $-OH$ of the side groups with a $-CH_3$, as is indicated in Fig. 11.5 (bottom). Hydroxyethyl cellulose HEC is produced by reacting cellulose with ethylene oxide which replaces $-H$ in some of the side groups with the larger $-CH_2-CH_2OH$ (see Fig. 11.6). Other forms of cellulose ethers are produced by substituting other molecular groups. Table 11.3 indicates other substituent groups and the name of the cellulose material.

The manner in which the substitution occurs is described by two parameters: degree of substitution DS and molar substitution MS. The parameter DS equals the average number of OH positions on the anhydroglucose unit that have been reacted. The range of DS is 0–3. For methyl cellulose, a DS of about 1.6–2.0 provides water solubility and is typical of commercial refined cellulose binders used in ceramic processing. In hydroxyethyl cellulose, optimum water solubility is obtained at a DS of about 1.0.

In HEC binders the MS indicates the average number of ethylene oxide molecules $-CH_2CH_2O-$ that have reacted with each anhydroglucose unit. When two different types of groups are substituted, the MS is the average molar fraction of the secondary group. For example, in hydroxypropyl methyl cellulose the MS is the molar fraction of substituted hydroxypropyl. The type and DS and MS, of the side groups produce different behavior when dissolved in solution or adsorbed on ceramic particles.

Some binders may contain ionizable side groups, such as sodium alginate and sodium carboxymethyl cellulose (see Table 11.3). The ionized group can change the adsorption behavior and flocculation. A charged molecule is less twisted and effectively longer. Sodium carboxymethyl cellulose is an anionic binder used in slurries for slip casting and for glazes and sometimes to increase viscosity and control filtration properties of more traditional ceramics. In general, the behavior of ionic binders is more sensitive to electrolytes.

Starch is a natural binder that is cold-water-insoluble; it is relatively active biologically and subject to breakdown when heated and agitated. Refined starch, generically called polysaccharide, is more easily dissolved and more controlled and is used as a binder in spray-dried clay bodies. Natural gums are relatively unrefined substances derived from plants. Many types are available and many are used as thickeners in food products. They are very complex branched polymers. Many natural gums are ionic in nature. Lignosulfonates are complex polymeric by-products from paper processing using the sulfite process. They are prepared as sodium and calcium salts with a molecular weight range of

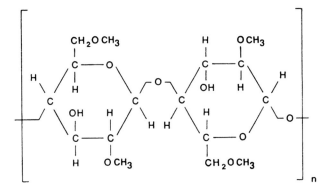

Fig. 11.5 Molecular structure of (top) a mer of cellulose, (middle) substituent positions R_1, R_2, R_3 on each anhydroglucose unit, and (bottom) methyl cellulose with DS = 2.

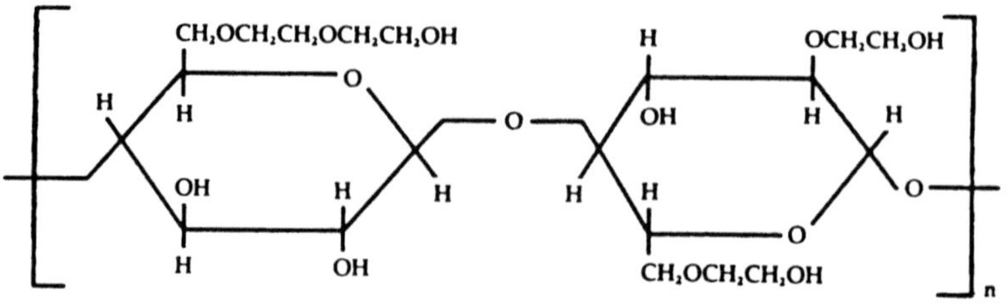

Fig. 11.6 Molecular structure of a mer of hydroxyethyl cellulose with DS = 1.5 and MS = 2. (Note that carbon atoms are not shown at 5 corners of ring.)

about 1,000–20,000 g/mol, are used as inexpensive dispersants, and can serve as binders on drying.

Polyethylene Glycol Binders

Polyethylene glycol PEG is polymerized ethylene oxide,

$$HO-[CH_2-CH_2-O]_n-H$$

The $-H$ in the OH may be substituted with another group and a general formula is $RO-[CH_2-CH_2-O]_n-H$, where R may be a group such as $-CH_3$. It is commercially available in molecular weights ranging from 200 to about 8000 g/mol.

Low molecular weight grades are relatively heat-stable liquids. As the degree of polymerization n increases, the viscosity increases and at a molecular weight of about 1000 g/mol the binder becomes a soft solid with a wax-like consistency. Higher molecular weight grades are waxy solids with a high degree of crystallinity. A PEG of much higher molecular weight in the range 15–20,000

TABLE 11.3 Substituent Groups $(-R)$ in Cellulose Binders[a]

Type	Group	Functionality
Soluble in water		
Hydroxyethyl	$-CH_2CH_2OH$	Nonionic
Methyl	$-CH_3$	Nonionic
Sodium carboxymethyl	$-CH_2OCH_2COONa$	Anionic
Hydroxypropyl methyl	$-CH_2CHOHCH_3$	Nonionic
	$-CH_3$	
Sodium alginate	$-COONa$	Anionic
Starch, dextrine	$-CH_2OH$	Nonionic
Soluble in nonpolar liquids		
Ethyl cellulose	$-CH_2OCH_2CH_3$	Nonionic

[a]Some unsubstituted OH remains.

$$\begin{matrix} & \text{H} & \text{H} \\ \text{HO[-C-C-O]}_{182}\text{-H} \\ & \text{H} & \text{H} \end{matrix}$$

$$\begin{matrix} & \text{H} & \text{H} & & \text{H} & \text{H} \\ \text{HO[-C-C-O]}_{182}\text{-L-[C-C-O]}_{182}\text{-H} \\ & \text{H} & \text{H} & & \text{H} & \text{H} \end{matrix}$$

Fig. 11.7 Molecular structure of PEG of molecular weight of 8000 and 20,000 g/mol.

g/mol is made* by compounding PEG of MW = 8000 with a special epoxide linking molecule L (see Fig. 11.7). This grade is a more amorphous material and is more suitable as a binder in pressing and extrusion. The PEG binders are very pure, are soluble in water, and have limited solubility in a wide range of solvents.

11.5 FILM-FORMING BINDERS (WAXES)

Common waxes used as film-type binders are paraffin derived from petroleum, candelilla and carnauba waxes derived from plants, and beeswax of insect origin. Paraffins are mixtures of straight-chain saturated hydrocarbons

$$\left[\begin{matrix} \text{H} \\ -\text{C}- \\ \text{H} \end{matrix} \right]_n$$

which tend to crystallize as plates or needles. Microcrystalline waxes, also derived from petroleum, are branched saturated hydrocarbons that are less crystalline, stronger, and somewhat tougher than paraffins. Nominal properties of paraffin wax are summarized in Table 11.4.

The plant waxes are more complex mixtures of straight-chain hydrocarbons, esters, acids, and alcohols that are relatively hard and have a relatively high melting point of 85–90°C. Beeswax is a complex mixture of esters and saturated and unsaturated hydrocarbons and has a much lower melting point of 60–80°C. The mechanical properties of waxes are directly related to secondary bonding between the molecules. Intermolecular bonding is weaker between straight and branched hydrocarbons, and they are softer and more plastic. Well below the hardening point, waxes are brittle. Heating disrupts the bonding, and at a higher temperature plastic flow occurs under a relatively small stress. Internal stress is frozen into the wax during cooling; it is relieved by molecular

*PEG Compound 20M, Union Carbide Chemicals and Plastics Co., Danbury, CT.

TABLE 11.4 Nominál Properties of Polymer Resins and Paraffin Wax at 25°C

	Density (Mg/m^3)	Strength[a] (MPa)	Elongation[a] (%)	CTE (ppm/K)	HDT[b] (°C)
LDPE[c]	0.91	6	100	100	40
HDPE[c]	0.95	28	30	120	85
PP	0.90	35	100	90	80
PS	1.04	41	2	75	90
Paraffin wax	0.91	4	—	400	—

Source: R. B. Seymour, *Polymers for Engineering Applications*, ASM International, Cleveland, OH, 1987.
[a]Tensile strength and elongation at failure.
[b]Temperature at which deflection occurs under stress of 1.8 MPa.
[c]LD—low density. HD—high density.

rearrangement on heating which causes distortion. This distortion is minimized by using a microcrystalline wax.

11.6 DISSOLVING AND ADMIXING BINDERS

As mentioned above, binders with polar side groups tend to be soluble in polar liquids such as water and binders with largely nonpolar side groups are soluble in organic solvents ("like likes like"). Common water-soluble binders such as polyvinyl alcohol and the cellulose types are purchased as powders and must be dissolved at some point during processing. Methyl cellulose is atypical in that it is more soluble in cold than in hot water.

In slurry processing the binder may be predissolved and added as a solution, or with the raw materials and dissolved during wet mixing. The dissolving of the particles of binder depends on overcoming bonding between molecules. Binders with polar side groups attract water molecules, and water molecules penetrate into the binder particles and hydrate the molecules. The dissolving process is faster in the less densely packed, partially hydrolyzed PVA. It is common procedure to disperse and wet the binder particles in the liquid at the temperature of lowest solubility and then alter the temperature to speed dissolving. Some binders are surface treated to improve dissolving. Inorganic molecular binders are commonly introduced into a processing system as an aqueous solution or colloidal sol.

Molecular binders with nonpolar side groups are dissolved in nonpolar solvents. Solvent blends may dissolve binders that have only weakly polar side groups. Binders of a nonpolar nature used in ceramic processing include polymethyl methacrylate, polyvinyl butyral, and ethyl cellulose. Their structural groups are indicated in Tables 11.2 and 11.3.

Polyethylene glycols containing both vinyl $-C-C-$ and ether

$-C-O-C-$ groups in the structure are soluble in water and a variety of organic solvents including alcohols and trichloroethylene.

Waxes may first be dissolved into a nonpolar solvent before adding into the processing batch, or melted and admixed into hot powder. Aqueous, stabilized wax emulsions with a wax content of about 40% are available and may be added directly into an aqueous processing system.

11.7 BINDER MOLECULAR WEIGHT AND VISCOSITY GRADE

Binders are commonly marketed by viscosity grade or a specific average molecular weight. Each grade contains molecules with a particular range and distribution of molecular weight. The number average molecular weight M_N of a polymer binder is determined experimentally from the osmotic pressure, melting point depression, or boiling point elevation of extremely dilute solutions. The mass average molecular weight M_W is determined from the slope of viscosity versus concentration of dilute solutions of less than 1% concentration, using a very sensitive viscometer. The molecular weight distribution in a binder may also be determined by gel permeation chromatography.

For the more popular binders, a series of grades are commonly available.

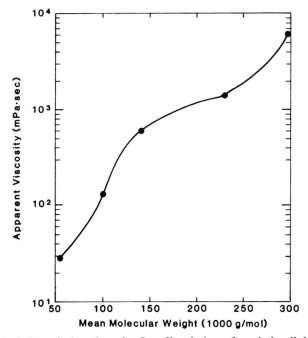

Fig. 11.8 Variation of viscosity of a 2 wt% solution of methyl cellulose increases with the molecular weight ($20°C$, $\dot{\gamma} = 10 \text{ s}^{-1}$). (Data courtesy of Dow Chemical Co., Midland, MI.)

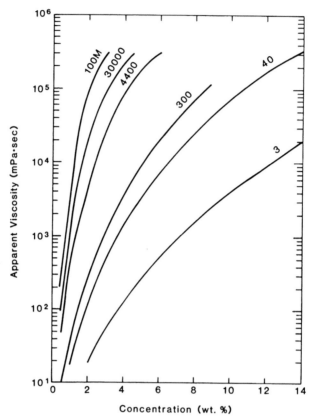

Fig. 11.9 The viscosity of aqueous solutions of hydroxyethyl cellulose increases with concentration and viscosity grade of the binder (20°C, $\dot{\gamma} = 14\ s^{-1}$). (Cellosize, Union Carbide Corp., New York, NY.)

However, molecular-weight information is seldom presented in specifications of commercial binder except for the polyethylene glycols. The apparent viscosity of a binder solution is easily measured and increases with the mean molecular weight of the dissolved binder (Fig. 11.8). The viscosity grade of a binder is its apparent viscosity at some standard concentration in solution. It is commonly used as an index of the mean molecular weight. As indicated in Fig. 11.9, the apparent viscosity of a 2 wt% solution at 20°C at a shear rate representative of pouring conditions is commonly used as the basis for designating the viscosity grade of cellulose binders.

11.8   GELATION

For some types of binders, a thermal or chemical change or a loss of solvent, increasing the concentration of the binder in solution, may cause the polymer

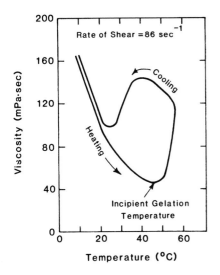

Fig. 11.10 Change in viscosity and gelation of an aqueous solution of 2 wt% methylcellulose on heating at 0.25°C/min. (Courtesy of Dow Chemical Co., Midland, MI.)

molecules to crosslink into a three-dimensional network with interspersed liquid, which is called a gel:

$$x \text{ polymer}_{(solution)} \rightarrow [\text{polymer}]_{n(gel)} + (x - n) \text{ polymer}_{(solution)} \quad (11.1)$$

where x is the concentration of dispersed polymer molecules before gelation.

A thermally reversible gel may be formed in an aqueous solution of methyl cellulose or hydroxypropyl methyl cellulose (Fig. 11.10).* Initially the viscosity of the binder solution decreases on heating. But above some particular temperature T_{gel}, additional heating causes the molecules to dehydrate and become hydrophobic and a gel is formed. The three-dimensional bonding causes the viscosity to increase rapidly.

Shearing of the gel can occur only if the shear stress exceeds the yield strength of the gel. The gel strength and elasticity increase with the concentration and molecular weight of the binder and attain a maximum value at a particular temperature. Electrolytes and glycerol, which have an affinity for water, reduce the gelation temperature; ethanol and propylene glycol raise the gelation temperature. This thermal gelation is reversible and disappears on cooling.

Gelation may occur with no loss of liquid when the liquid is occluded in capillary pores or in micelles. Syneresis is the spontaneous shrinking of a gel to an equilibrium volume with the expulsion of liquid. Imbibition of liquid or delayed condensation that produces liquid in the gel may cause the gel to expand.

Gelation may also be produced by chemical means. Using a simple acid or base, ionic binders may be coagulated at their PZC. Highly charged cations

*N. Sarkar and G. K. Greminger, Jr., *Am. Ceram. Soc. Bull.*, **62**(11), 1280–1283 (1983).

may coagulate (cross link) anionic polymers into a gel. Partially hydrolyzed polyvinyl alcohol is gelled by soluble carbonates, sulfates, and borates. Sodium alginate is used as a binder and dental impression material. The aqueous alginate sol may be caused to gel in the presence of alkaline earth ions such as Ca^{2+}:

$$Na_n \text{ alginate}_{(sol)} + \frac{n}{2} Ca^{2+}_{(solution)} \rightarrow n \text{ Na}^{+}_{(solution)} + Ca_{n/2} \text{ alginate}_{(gel)} \quad (11.2)$$

A controlled slow set is achieved by introducing the calcium as a low-solubility salt such as calcium sulfate and a retarder such as PO_4^{3-} that is more mobile than the bulky alginate and forms an insoluble tricalcium phosphate. Chemical gelation is temperature-sensitive, and the temperature of the liquid should be controlled to control the setting time. The reversal of chemical gelation in an undisturbed system occurs very slowly, and chemical gelation is often practically irreversible.

The dehydration and polymerization of esters of silicon, titanium, and zirconium, as described in Chapter 4, may produce a binding phase. Soluble and colloidal silicates are popular gel-type binders used in construction and refractory products. Aqueous solutions of sodium silicate with a molecular formula $Na_2O \cdot 2.5SiO_2$ to $Na_2O \cdot 4SiO_2$ are basic (pH 11–13) and contain polymerized silicate anions. They are important bonds because a gel structure forms on loss of a small amount of water and because of their high wetting power. The proportion and size of the larger ring structures increase as the ratio of SiO_2/Na_2O increases. A gel structure is also produced in a concentrated silicate solution on adding polyvalent cations and by the addition of an acid or the hydrolysis of an organic ester such as ethyl acetate, alcohol, or glycerol diacetate, which produces an acid. Antimigration agents such as alkali carbonates, phosphates, perchlorates, or fluorides, which cause local polymerization, are added when alkali silicate binders are to be air dried. Dry alkali silicate binder films must be gelled to prevent rehydration and dissolving in the presence of moisture.

The chemically assisted coagulation of colloidal clay minerals and colloidal cellulose particles is also a form of chemical gelation. The gelling time is inversely proportional to the concentration of colloidal particles and coagulant and is dependent on the particle size and shape and the type of coagulant.

11.9 GENERAL EFFECTS OF BINDERS

Flocculation

On adding a nonionic polymer to a suspension of particles, it is observed that the mode of adsorption changes with concentration. At first the polymer is adsorbed on individual particles and tends to stabilize the suspension. But on increasing the concentration of polymer, its adsorption between particles increases and bridging flocculation occurs. On still further increasing the concentration with continued mixing, bridging flocculation is reduced and the additional polymer again stabilizes the dispersion.

Ionic binders may flocculate particles in suspension under particular conditions. When the pH is greater than 4, negative quartz particles in an aqueous suspension are flocculated by a variety of cationic binders of high molecular weight.* With an increasing concentration of adsorbed binder, the zeta potential of the agglomerates becomes positive, and the binder may then function as a protective colloid. Suspensions of negative quartz particles are observed to be flocculated by anionic acrylamide polymers if assisted by activating cations such as Ca^{2+} that reduce the zeta potential. Flocculation occurs at a lower pH when the concentration of Ca^{2+} ions is higher in the suspension. For a constant concentration of Ca^{2+} ions, the pH for the onset of flocculation is lower when the anionic character of the polymer is higher. Highly charged cations can coagulate many anionic polyelectrolytes and apparently reduce the repulsion between negatively charged particles and negative polymer chains. In suspensions of mixtures of minerals, specific adsorption and flocculation of only one mineral type may sometimes occur.

Most oxide particles are flocculated by a variety of nonionic polymers owing to hydrogen bonding or an attraction between the hydrated surface and polar groups on the polymer chain. However, when the binder has a much higher affinity for the solvent than for the surface, adsorption of the binder may not occur. Polyvinyl alcohol is observed to adsorb on most oxides in an aqueous suspension. Methyl cellulose and polyethylene glycol are observed not to adsorb on alumina particles in water but do adsorb on kaolin and talc particles. Heteropolar coagulation of colloidal particles on coarser particles will cause the system to imitate the adsorption behavior of the colloidal particles. Flocs formed by bridging adsorption of polymer or colloid particles may be considerably stronger than when coagulated by simple ions.

Thickening, Suspending, and Rheological Control

The substitution of a viscous binder solution for a simple liquid increases the mean viscosity and effectively thickens the system. In a more concentrated suspension, flocculation may cause the suspension to have the consistency of a paste. The settling rate is reduced in the higher-viscosity liquid and may become zero when the viscosity or yield point is sufficiently high. The presence of the binder solution or gel between the particles can greatly alter the flow properties of the system. The important topic of rheology control is discussed in several sections in subsequent chapters.

*B. Yarar and J. A. Kitchener, *Trans. Inst. Min. Metall.*, **79**(3), 23–33 (1970).

Liquid Requirement, Elasticity, and Liquid Retention

The steric hindrance provided by colloidal or molecular bridges between particles decreases the particle packing density and increases the liquid required to saturate the body. The steric hindrance produced by adsorbed polymer molecules generally increases with the molecular weight. Nonionic binders adsorb on particles as trains, loops, and tails (Fig. 11.11). Charged polymers tend to lie more flat on the surface when adsorbed. An adsorbed binder having a strong affinity for the solvent will tend to have longer loops and tails.

Elastic behavior at low pressures may be produced by adsorbed flocculated colloids or polymer molecules between particles. Cellulose binders are less flexible than the vinyls, glycols, and waxes.

A matrix of polymer molecules or colloidal particles between particles greatly reduces liquid migration rates. Binder migration due to capillary effects or applied pressure is lower for binders of higher molecular weight, as indicated in Fig. 11.12, and is extremely low when the binder is gelled.

Lubricity and Film Properties

When adsorbed, a binder may reduce the apparent surface tension. Binder adsorbed on a surface may reduce the apparent surface roughness and the coefficient of friction. In water the lubricity of the surface is dependent on the hydrophilicity of the film, adsorbed layer displacement, surface tension, and molecular orientation.* The mechanical properties of binder films are very dependent on adsorbed plasticizing molecules, and this topic is addressed in Chapter 12.

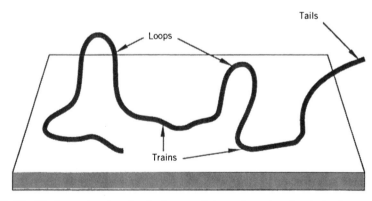

Fig. 11.11   Binder molecule adsorbed on particle surface showing train, loop, and tail segments. (Courtesy of Cynthia Incorvati, Alfred University.)

*N. Sarkar and G. K. Greminger, *Am. Ceram. Soc. Bull.*, **62**(11), 1280–1283, 1288 (1983).

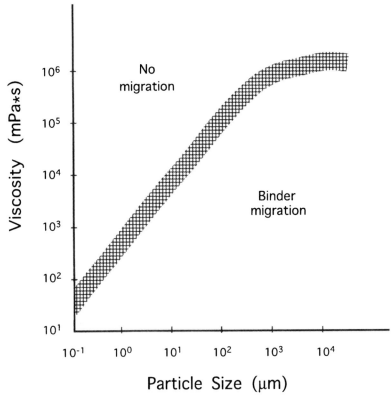

Fig. 11.12 Migration of nonadsorbed binder solution through pores in packed particles is higher when the solution is of lower viscosity and the intersticial size is larger. (Based on pore size equal to 0.5 × particle size, wetting and capillary suction; migration = migration >0.1 mm in 1 min; see Example 2.5.)

11.10 POLYMER RESINS

Polyethylene PE, polypropylene PP, and polystyrene PS, used in injection molding systems, are polymer resins. Their chemical structure is shown in Fig. 11.13. These materials are familiar to everyone and are commonly called plastics. In ceramic bodies they are usually added to form a bonding matrix, which requires that they must be moldable at an early stage and later capable of being hardened to form a structural bond.

Resins such as polyethylene and polystyrene are formed by addition polymerization, which occurs by the addition of mers to an infant polymer molecule, increasing its size. Reaction of the monomer with the chain involves the breaking of an unsaturated $C=C$ bond and requires free radicals. Free radicals are intermediate molecules that have a highly reactive, unpaired electron and are usually formed by the fracture of an initiator molecule produced by heating, light, or a chemical reaction. The polymerization of ethylene occurs by reaction

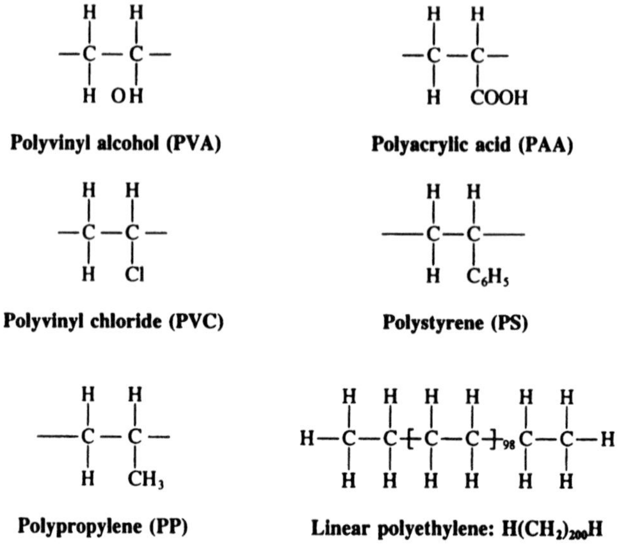

$$
\begin{array}{cc}
\text{H} & \text{H} \\
| & | \\
-\text{C}-\text{C}- \\
| & | \\
\text{H} & \text{OH}
\end{array}
\qquad
\begin{array}{cc}
\text{H} & \text{H} \\
| & | \\
-\text{C}-\text{C}- \\
| & | \\
\text{H} & \text{COOH}
\end{array}
$$

Polyvinyl alcohol (PVA)   **Polyacrylic acid (PAA)**

$$
\begin{array}{cc}
\text{H} & \text{H} \\
| & | \\
-\text{C}-\text{C}- \\
| & | \\
\text{H} & \text{Cl}
\end{array}
\qquad
\begin{array}{cc}
\text{H} & \text{H} \\
| & | \\
-\text{C}-\text{C}- \\
| & | \\
\text{H} & \text{C}_6\text{H}_5
\end{array}
$$

Polyvinyl chloride (PVC)   **Polystyrene (PS)**

$$
\begin{array}{cc}
\text{H} & \text{H} \\
| & | \\
-\text{C}-\text{C}- \\
| & | \\
\text{H} & \text{CH}_3
\end{array}
\qquad
\text{H}-\overset{\text{H}}{\underset{\text{H}}{\text{C}}}-\overset{\text{H}}{\underset{\text{H}}{\text{C}}}\text{-}\!\!\left[\overset{\text{H}}{\underset{\text{H}}{\text{C}}}-\overset{\text{H}}{\underset{\text{H}}{\text{C}}}\right]_{98}\overset{\text{H}}{\underset{\text{H}}{\text{C}}}-\overset{\text{H}}{\underset{\text{H}}{\text{C}}}-\text{H}
$$

Polypropylene (PP)   **Linear polyethylene: $H(CH_2)_{200}H$**

Fig. 11.13   Molecular structure of several polymer materials.

of the monomer CH_2CH_2 with the free radical $A^{\cdot}$:

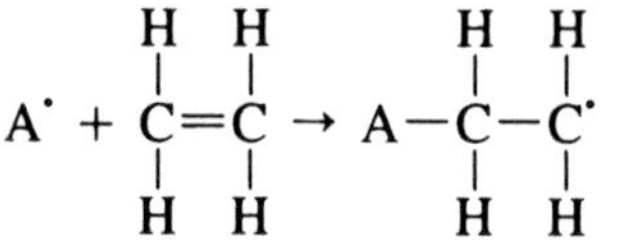

$$
A^{\cdot} + \begin{array}{cc}\text{H} & \text{H}\\ | & |\\ \text{C}=\text{C}\\ | & |\\ \text{H} & \text{H}\end{array} \rightarrow A-\begin{array}{cc}\text{H} & \text{H}\\ | & |\\ \text{C}-\text{C}^{\cdot}\\ | & |\\ \text{H} & \text{H}\end{array} \tag{11.3}
$$

followed by the activated monomer reacting with additional monomer:

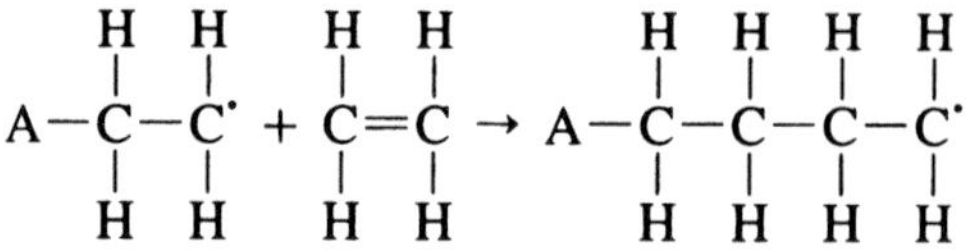

$$
A-\begin{array}{cc}\text{H} & \text{H}\\ | & |\\ \text{C}-\text{C}^{\cdot}\\ | & |\\ \text{H} & \text{H}\end{array} + \begin{array}{cc}\text{H} & \text{H}\\ | & |\\ \text{C}=\text{C}\\ | & |\\ \text{H} & \text{H}\end{array} \rightarrow A-\begin{array}{cccc}\text{H} & \text{H} & \text{H} & \text{H}\\ | & | & | & |\\ \text{C}-\text{C}-\text{C}-\text{C}^{\cdot}\\ | & | & | & |\\ \text{H} & \text{H} & \text{H} & \text{H}\end{array} \tag{11.4}
$$

until the chain propagation is terminated. Termination can occur by the reaction of two growing chains, which destroys free radicals. A high concentration of free radicals will tend to limit chain propagation. Chain transfer is a termination process that occurs when an activated polymer transfers the free radical to another specie for growth of a new polymer molecule. A retarder is a substance that can react with a free radical to form a product incapable of polymerizing.

Vinyl polymers are formed from vinyl monomers having the structure CH_2CH-R, where the $-R$ group determines the specific compound (Table 11.2). Polystyrene is polyvinyl benzene and the $-R$ group is a benzene ring. A copolymer is produced when two different functional R groups are substituted along the chain.

Polymethyl methacrylate may be formed by reacting the monomer methyl methacrylate $CH_2=C(CH_3)-COOCH_3$ in the presence of an initiator such as benzoyl peroxide $C_6H_5COO-OOC_6H_5$, which forms free radicals when heated or exposed to ultraviolet light. Heating may be produced by the addition of a tertiary amine such as dimethyl-*p*-toluidine. Spontaneous polymerization is inhibited by adding a small amount of an antioxidant such as hydroquinone. The methacrylate resin softens on heating and can be molded at 125°C. It is soluble in a variety of organic solvents.

In stepwise polymerization, both the monomer and growing chains are active. Stepwise polymerization is referred to as condensation polymerization when a nonpolymerizable molecule such as water is a by-product of the reaction. Silicone polymers have the siloxane backbone $-Si-O-Si-$ and are formed by stepwise polymerization.

Thermoplastic materials soften on heating and stiffen on cooling. Linear polymers are thermoplastic. The strength and softening temperature increase with the average molecular weight. A larger $-R$ group increases the stiffness. Material containing shorter chains can crystallize, but material with longer chains is more amorphous. Generally the more amorphous material is less strong but is more ductile. Above the glass transition temperature the viscosity of the material varies directly with the molecular weight. For materials with a high degree of polymerization, orientation of the molecules produces anisotropy of properties. Ordering of the $-R$ groups along the chain generally increases the crystallinity and brittleness. Nominal properties of several polymer resins are summarized in Table 11.4.

Chain branching (see Example 11.4) connecting linear chains and cross-linking between chains having unsaturated carbon bonds strengthen and stiffen the polymer structure. Epoxy resins have an epoxide

$$\overset{\displaystyle O}{\underset{\displaystyle -C-C-}{\diagup\diagdown}}$$

linkage and polymerize by both addition and condensation polymerization. These materials must be molded to shape before or during polymerization and are referred to as thermosetting resins. The behavior is not reversible.

Elastomers such as rubbers and silicones are coherent elastic solids composed of intertwined chains with limited cross linking. Under a load, the chains straighten, lengthen, and then slip; they return to the unloaded state when the load is removed.

11.11 REACTION BONDS

An important family of bonds used in the refractories and construction industries are formed by the reaction of alumina with orthophosphoric acid H_3PO_4

or an acidic aqueous solution of monoaluminum phosphate $Al(H_2PO_4)_3$ or monomagnesium phosphate $Mg(H_2PO_4)_2$. The reaction of alumina or hydrated alumina with orthophosphoric acid is exothermic and produces monoaluminum phosphate, $Al(H_2PO_4)_3$, or an amorphous bond:

$$Al_2O_3 + 6H_3PO_4 \rightarrow 2Al(H_2PO_4)_3 + 3H_2O \tag{11.5}$$

$$Al(OH)_3 + H_3PO_4 \rightarrow \text{amorphous bond} \tag{11.6}$$

Heating of the bond causes dehydration and the formation of the metaphosphate $Al(PO_3)_3$, but the availability and further reaction with alumina produce a phosphate with a higher Al_2O_3/P_2O_5 ratio:

$$Al(H_2PO_4)_3 + Al_2O_3 \rightarrow 3AlPO_4 + 3H_2O \tag{11.7}$$

The monoaluminum phosphate bond is an acid solution in equilibrium with $AlPO_4$ and $AlPO_4 \cdot H_3PO_4$:

$$Al(H_2PO_4)_3 \rightleftarrows AlPO_4 \cdot H_3PO_4 + H_3PO_4 \tag{11.8}$$

$$Al(H_2PO_4)_3 \rightleftarrows AlPO_4 + 2H_3PO_4 \tag{11.9}$$

Reaction with a base that consumes free H_3PO_4 or heating drives the reaction to the right, and aluminum phosphate phases precipitate in a cross-linked polymeric structure. For monomagnesium phosphate $Mg(H_2PO_4)_2$ bonds, the reactions are

$$Mg(H_2PO_4)_2 \rightleftarrows MgHPO_4 + H_3PO_4 \tag{11.10}$$

$$3MgHPO_4 \rightleftarrows Mg_3(PO_4)_2 + H_3PO_4 \tag{11.11}$$

Reaction with a base also produces precipitates having a polymeric structure.

Phosphate bonds exhibit aspects of both gel-type and reaction bonds. The viscosity of the phosphoric acid or aluminate solution decreases on heating, but very little migration of the bond on particle surfaces occurs on drying, because of gelling. The reaction bond may be formed at a temperature as low as 250–300°C, but heating above 500°C is required for complete dehydration and the formation of a stable aluminum phosphate phase.

The reaction of phosphorous acid and zinc oxide has been used to form a dental cement at room temperature. Phosphoric bonds have been produced with a wide variety of metal oxides. The reaction of a phosphoric acid solution with silicate compounds also produces a gel structure which forms silicyl phosphates such as the metaphosphate $SiO(PO_3)_2$ on heating above 250°C. Patented cements for dental ceramics have been prepared using zinc oxide and an acid such as polyacrylic acid and polycarboxylic acid. The polyacrylic acid $CH_2CHCOOH$ is soluble in water but reacts with zinc oxide to form a nearly insoluble, hard bond. Setting occurs by the cross-linking of the $-$COOH side groups by zinc ions:*

$$
\begin{array}{cccc}
\text{H} & \text{H} \quad \text{H} & \text{H} \\
| & | \quad | & | \\
-\text{C} & -\text{C}-\text{C} & -\text{C}- \\
| & | \quad | & | \\
\text{COOH} & \text{H} \quad \text{COOH} & \text{H} \\
& \diagdown \qquad \diagup & \\
& \diagdown \quad \diagup & \\
& \text{Zn} & \\
& \diagup \quad \diagdown & \\
\text{COOH} & \text{H} \quad \text{COOH} & \text{H} \\
| & | \quad | & | \\
-\text{C} & -\text{C}-\text{C} & -\text{C}- \\
| & | \quad | & | \\
\text{H} & \text{H} \quad \text{H} & \text{H}
\end{array}
\qquad (11.12)
$$

Un-cross-linked $-$COOH groups may bond to an oxide surface such as the surface of a tooth by chemisorption or chelation. The properties of the bond depend on the degree of polymerization of the acid, the particle size of the zinc oxide, and the presence of other ions such as magnesium or calcium in the zinc oxide or added as a second solid phase.

11.12 HYDRAULIC CEMENTS

Hydraulic bonds commonly called hydraulic cements are prereacted compounds of calcium oxide and silica, alumina, and iron oxide that have been ground to a fine powder. When mixed with water, the compounds harden into a strong material in air or under water, with relatively little change in volume. Cements containing natural hydrating materials such as volcanic ash or shale are called pozzolana cements.

The principal inorganic phases in Portland cement are listed in Table 11.5. Water is mixed with the cement powder to produce a paste or slurry, and the cement sets when dissolving and reaction with water occur at the surfaces, and a gel-like hydrated precipitate phase is formed. The hydration reaction is exothermic, and the water bound chemically as a hydroxide produces a hard product.

Portland cements are multiphase systems, and the reactions leading to their setting and hardening are very complex. Silicate and aluminate phases higher in calcium tend to dissolve and react more rapidly. Tricalcium aluminate $3\text{CaO} \cdot \text{Al}_2\text{O}_3$ reacts rapidly in water with or without the presence of calcium hydroxide, causing a fast set:

$$3\text{CaO} \cdot \text{Al}_2\text{O}_3 + 6\text{H}_2\text{O} \rightarrow 3\text{CaO} \cdot \text{Al}_2\text{O}_3 \cdot 6\text{H}_2\text{O} \qquad (11.13)$$

*E. H. Greener et al., *Materials Science in Dentistry*, Williams and Wilkins, Baltimore, 1972.

TABLE 11.5 Phases in Type II Portland Cement and Calcium Aluminate Cement

Phase	Oxide Formula	Amount (wt %)
Type II Portland cement		
Tricalcium silicate	$3CaO \cdot SiO_2$	46
Dicalcium silicate	$2CaO \cdot SiO_2$	28
Tricalcium aluminate	$3CaO \cdot Al_2O_3$	11
Tetracalcium		
aluminoferrite	$4CaO \cdot Al_2O_3 \cdot Fe_2O_3$	8
Gypsum	$CaSO_4 \cdot 2H_2O$	3
Magnesia	MgO	3
Calcium oxide	CaO	0.5
Sodium, potassium oxides	Na_2O, K_2O	0.5
Calcium aluminate cement		
Tricalcium aluminate	$3CaO \cdot Al_2O_3$	0-minor
Monocalcium aluminate	$CaO \cdot Al_2O_3$	major
	$12CaO \cdot 7Al_2O_3$	minor
Calcium dialuminate	$CaO \cdot 2Al_2O_3$	major

Source: H. W. Hayden et al., *The Structure and Properties of Materials*, Wiley-Interscience, New York, 1965.

and

$$3CaO \cdot Al_2O_3 + Ca(OH)_2 + 12H_2O \rightarrow 4CaO \cdot Al_2O_3 \cdot 13H_2O \quad (11.14)$$

Gypsum anhydrite is added to reduce the rate of these reactions; a fine needlelike crystalline film of the complex mineral ettringite is deposited on the aluminate particles (Fig. 11.14) which retards further hydration and causes the cement to be more plastic:*

$$3CaO \cdot Al_2O_3 + 3(CaSO_4 \cdot 2H_2O) + 26H_2O$$
$$\rightarrow 3CaO \cdot Al_2O_3 \cdot 3CaSO_4 \cdot 32H_2O \quad (11.15)$$

Hydration of $3CaO \cdot SiO_2$ and $2CaO \cdot SiO_3$ by several of many possible reactions causes hardening, for example,

$$3CaO \cdot SiO_2 + xH_2O \rightarrow 2CaO \cdot SiO_2 \cdot xH_2O + Ca(OH)_2 \quad (11.16)$$

$$2CaO \cdot SiO_2 + xH_2O \rightarrow 2CaO \cdot SiO_2 \cdot xH_2O \quad (11.17)$$

*W. Baumgart, A. C. Dunham, and G. C. Amstutz, Eds., *Process Mineralogy of Ceramic Materials*, Elsevier, New York, 1984.

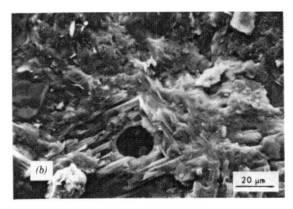

Fig. 11.14 (*a*) Portland cement paste 1 day after mixing with water; fine amorphous calcium silicate hydrate covers cement particles, and long thin needles of ettringite have formed in water-filled pores. (*b*) After 64 days, pores from air bubbles remain, and the matrix is a mixture of calcium silicate hydrates of indistinct morphology, large striated calcium hydroxide crystals, and hydrated cement particles (far left and top center). (Photos courtesy of J. F. Young, University of Illinois, Urbana-Champaign.)

Studies of the hardening of cement indicate that the hydration of $3CaO \cdot SiO_2$ continues beyond 30 days and that the hydration time for $2CaO \cdot SiO_2$ exceeds 1 year. The setting time can be reduced by grinding the cement finer or increasing the temperature of the water. Free calcium and magnesium oxides in the cement powder must be minimal because of the large volume expansion that accompanies their hydration.*

Calcium aluminate cements are used in castable refractory products. These cements set more rapidly than Portland cement and may be ready for service

*W. Baumgart, A. C. Dunham, and G. C. Amstutz, Eds., *Process Mineralogy of Ceramic Materials*, Elsevier, New York, 1984.

in 24 h. The hydration process is exothermic and sensitive to the temperature of hydration.*

$$CaO \cdot Al_2O_3 + 10H_2O \xrightarrow{T < 22°C} CaO \cdot Al_2O_3 \cdot 10H_2O \quad (11.18)$$

$$2(CaO \cdot Al_2O_3) + 11H_2O \xrightarrow{22-35°C} 2CaO \cdot Al_2O_3 \cdot 8H_2O$$
$$+ Al_2O_3 \cdot 3H_2O \quad (11.19)$$

$$3(CaO \cdot Al_2O_3) + 12H_2O \xrightarrow{T > 35°C} 3CaO \cdot Al_2O_3 \cdot 6H_2O$$
$$+ 2(Al_2O_3 \cdot 3H_2O) \quad (11.20)$$

Cement mixed and cured between 22 and 35°C is stronger and more explosion-resistant (more permeable) apparently because of the particular hydrated phase formed and the more crystallized alumina gel. Calcium aluminate cements dehydrate on heating and lose most of their strength by about 800°C. They are used in many refractory applications, because a dimensionally stable refractory material of low permeability can be formed in place and be ready for use in about 24 h.

Other hydraulic cements include gypsum anhydrite $CaSO_4$ and plaster of Paris $CaSO_4 \cdot 0.5H_2O$ which hydrate rapidly and are used in building materials and for gypsum molds. Fine CaO mortars harden by the reaction of the hydroxide $Ca(OH)_2$ with carbon dioxide in air or dissolved in water, forming calcium carbonate. Magnesium oxychloride cement $3MgO \cdot MgCl_2 \cdot 11H_2O$, known as sorrel cement, formed in a solution of magnesium chloride $MgCl_2$ containing magnesia MgO, is used as a temporary bond in some refractories.

SUMMARY

Colloidal clay minerals and organic or inorganic polymers are commonly used as binders in ceramic processing. Bond clays contain a relatively high concentration of coagulated colloidal particles which coat and bridge between larger ceramic particles. A large selection of organic polymer binders is available and these binders are widely used in ceramic processing. Families of the vinyl, cellulosic, and polyethylene glycol type have a chain-like molecular structure with a particular backbone, but different side groups. The length of the molecules is indicated by the degree of polymerization and the molecular weight. The solubility of vinyl and cellulosic binders is controlled by the type and concentration of side groups indicated by the DS and MS parameters. Properties of solutions of these polymers may be varied by changing the binder type, the substituent side groups, and the molecular weight. The viscosity grade is an

*W. Baumgart, A. C. Dunham, and G. C. Amstutz, Eds., *Process Mineralogy of Ceramic Materials*, Elsevier, New York, 1984.

index indicating the relative molecular weight of a binder in a particular family. Gelation in a binder solution increases the cohesive strength and viscosity of the system. Waxes and polymer resins are also used as binders in ceramic processing. Reaction bonds and hydraulic cements are widely used as binders in construction materials and for refractories.

SUGGESTED READING

1. R. M. German, *Powder Injection Molding*, Metal Powder Industries Federation, Princeton, NJ, 1990.

2. Wolfgang Baumgart, A. C. Dunham, and G. C. Amstutz, Eds., *Processing Mineralogy of Ceramic Materials*, Elsevier, New York, 1984.

3. Nitis Sarkar and G. K. Greminger Jr., "Methylcellulose Polymers as Multifunctional Processing Aids in Ceramics," *Am. Ceram. Soc. Bull.*, **62**(11), 1280–1283, 1288 (1983).

4. R. J. Fessenden and J. S. Fessenden, *Organic Chemistry*, Willard Grant Press, Boston, 1982.

5. F. J. Gonzalez and J. W. Hallorn, "Reaction of Orthophosphoric Acid with Several Forms of Aluminum Oxide," *Am. Ceram. Soc. Bull.*, **59**(7), 727–731 (1980).

6. G. Y. Onoda Jr., "The Rheology of Organic Binder Solutions," in *Ceramic Processing Before Firing*, G. Y. Onoda Jr. and Larry Hench, Eds., Wiley-Interscience, New York, 1978, pp. 235–251.

7. W. E. Worral, *Clays and Ceramic Raw Materials*, Halsted Press Div., Wiley-Interscience, New York, 1975.

8. E. H. Greener, J. K. Harcourt, and E. P. Lautenschlager, *Materials Science in Dentistry*, Williams and Wilkins, Baltimore, 1972.

9. P. Hopfner, "Water Soluble Cellulose Ethers as Binding and Plasticizing Agents," *Interceram*, (2), 149–151 (1970).

10. K. R. Lang, "Properties of Soluble Silicates," *Ind. Eng. Chem.*, **61**(4), 29–44 (1969).

11. J. F. Wygant, "Cementitious Bonding in Ceramic Fabrication," in *Ceramic Fabrication Processes*, W. D. Kingery, Ed., MIT Press, Cambridge, MA, 1963.

PROBLEMS

11.1 What is the MS in partially hydrolyzed PVA when the residual acetate is 12%?

11.2 Draw the structure of a mer of methyl cellulose with DS = 1.0.

11.3 Draw the structure of a mer of sodium carboxymethyl cellulose with DS = 0.75 and an ionized side group. (Note: Substitution occurs for R in CH_2OR group.)

11.4 Draw the structure of polyvinyl butyral.

11.5 Draw the structure of hydroxypropyl methylcellulose with DS = 1.0 and a molar substitution of hydroxypropyl MS = 0.25.

11.6 Compare the average degree of polymerization and average chain length for the five viscosity grades of methylcellulose in Fig. 11.8. Assume that DS = 1.0 and that the length of one mer is 1.16 nm. (Note: Actual chains in solution are not linear.)

11.7 Would you expect an aqueous solution of partially hydrolyzed PVA to have a lower surface tension than pure water? Explain.

11.8 Calculate the nominal length of a PVA molecule having MW = 53,000 g/mol.

11.9 From the molecular structure, explain why ethyl cellulose is soluble in a nonpolar liquid but hydroxyethyl cellulose is soluble in water?

11.10 Explain why the aqueous solubility of methyl cellulose varies with the DS.

11.11 A ceramic body is prepared using a solution containing 5 wt% organic binder in amounts of 4 and 8 wt% of the solution in the mix. What is the concentration of organics in each mix after drying?

11.12 Compare the plastic strength and MBI of the clays in Table 5.5 and explain your results.

11.13 Without lifting your pencil, draw the structure of a polymer resin and show regions of amorphous and crystalline structure.

11.14 Compare the coefficient of thermal expansion of polyethylene and paraffin wax to that of alumina.

11.15 Write the structural reaction between methyl methacrylate monomer and benzoyl peroxide to form polymethyl methacrylate.

11.16 Monoaluminum phosphate is used to form a protective coating on steel used in aqueous environments. Write the reaction assuming that the oxide Fe_2O_3 on the steel reacts more rapidly than the metal phase.

11.17 Explain how the particle size, alumina content, and uniformity of mixedness can influence the amount of $Al(PO_3)_3$ and $AlPO_4$ formed on reacting alumina with H_3PO_4.

11.18 Calculate the weight percent water needed to hydrate calcium aluminate cement using Eq. 11.18.

EXAMPLES

Example 11.1 The popular binder PVA, when partially hydrolyzed, contains about 10% residual acetate groups. Illustrate its structure. Explain why this binder is soluble in water.

Solution. The average structure corresponds to MS $= 0.1$ for the acetate.

$$-(CH_2-CHOH)_9-CH_2-CHOCOCH_3-(CH_2-CHOH)_9-CH_2-CHOCOCH_3-$$

The binder is soluble in water because the polar $-OH$ side groups of the molecule have a strong affinity for polar water molecules.

Example 11.2   A methylcellulose binder used for extrusion has mean MW $= 300,000$ g/mol and DS $= 1.5$. What is the structure of a mer of this binder and what is its nominal length. (The length of a mer is 1.16 nm.)

Solution. The DS $= 1.5$ indicates that on average 1.5 H in OH groups of anhydroglucose ring have been replaced by 1.5 CH_3 groups; this is equal to 1 substitution in one ring and 2 substitutions in the second ring in the mer.
  The molecular weight of a mole of mers is calculated as follows:

Element	Atomic Weight	Number/(mer)	Weight (g/mol)
C	12.0	15	180
H	1.0	26	26
O	16.0	10	160
			366 g/mol

The degree of polymerization $n = \dfrac{300,000 \text{ g/mol molecules}}{366 \text{ g/mol mers}} = 820$

The nominal length $L = (1.16$ nm/mer$) \, 820$ mers $= 950$ nm.

Note: The nominal length of the binder is 430 times the length of the surfactant molecule calculated in Example 10.3. But note that binder molecules are not linear in solution or when adsorbed.

Example 11.3   What is the best procedure for dispersing a binder received in particulate form when mixing it into the ceramic processing system.

Solution. The objective is to first disperse the particles of binder and then dissolve these to produce polymer molecules in the liquid. Several procedures are possible. A binder in powder form may first be dispersed and wetted in the solvent (water for a binder with polar groups) *at the temperature where the binder is least soluble.* This will retard formation of a viscous gelatinous mass which is difficult to disperse in a low-viscosity liquid. After dispersing, the temperature of the suspension is altered to promote dissolving of the particles of binder. Heating generally promotes dissolving, but for methylcellulose binders cooling promotes dissolving. Sometimes the binder is dispersed in a portion of the liquid and then the remaining liquid is added to suddenly change the temperature for dissolving. The molecular solution may be characterized and its viscosity checked before adding it to the processing batch.

In the dry-blending technique, the dry binder powder is commonly mixed with dry ceramic granular materials or powders in a ratio of more than four parts of inorganic particles to one part of binder. Attrition during dry mixing reduces the size of the binder particles and may coat ceramic particles. After dispersing the dry binder in the system, the ceramic + binder powder mixture is added to the solvent with continuous agitation to dissolve the binder.

For a batch with a plastic consistency, the solvent may be sprayed as uniformly as possible into the continuously agitated powder mixture. Dissolving rate depends on the relative particle sizes and the agitation energy before and after adding the solvent. For a relatively dry batch containing both granular and fine ceramic powder, the binder may be dispersed with agitation on damp coarse particles, coating them, before adding the fine ceramic powder which will adhere to the binder-coated granular material.

Example 11.4   What are the differences in structure of linear, branched, and cross linked polymer resins?

Solution. Polymer chains are able to twist and coil because of the free rotation about the $C-C$ bond. In addition to the linear configuration, the chains may have a branched structure as shown. Polyethylene $-[CH_2-CH_2-]_n$ material having the linear structure is called high-density polyethylene (HDPE) and material having the branched structure is low-density polyethylene (LDPE). Some polymers may cross link to form a continuous network having a yield strength, which is also shown.

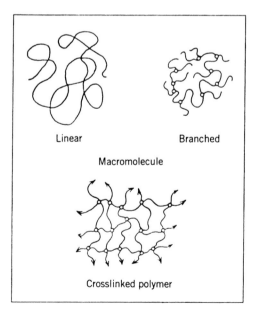

Linear   Branched

Macromolecule

Crosslinked polymer

Example 11.4   Linear, branched, and cross-linked polymer structures.

CHAPTER 12

PLASTICIZERS, FOAMING AND ANTIFOAMING AGENTS, LUBRICANTS, AND PRESERVATIVES

The processing engineer often uses small amounts of other types of additives, in addition to deflocculants, coagulants, and binders, to develop a satisfactory processing system. A plasticizer is added to modify the viscoelastic properties of a condensed binder-phase film on the particles. The tendency for bubbles to persist in a slurry may be reduced by adding a small amount of an antifoaming agent. A foaming agent may increase the stability of gas bubbles. A lubricant is a surfactant that is strongly adsorbed and especially effective in reducing the coefficient of friction of the surfaces of ceramic particles and metal dies. A preservative is used when enzymatic degradation of a binder must be controlled.

12.1 PLASTICIZERS

The moldability of a binder is very dependent on temperature. At 20°C polyvinyl alcohol is elastic and brittle, and two separate films bond very poorly when pressed together; movement of the molecules is very limited, and the brittle binder is said to be in a glassy state. At about 90°C, the thermal energy is sufficient to enable segments of the molecules to flow and realign when compressed, and bonding between films occurs. This is called the rubbery state. The temperature between these two states at which the deformation changes from elastic behavior (fracture at a strain greater than about 5%) to time-dependent viscoelastic deformation (no fracture at a strain exceeding 100%) is called the glass transition temperature (T_g) and is shown in Fig. 12.1. Polymer films exhibit a change in resistance to mechanical deformation, thermal expansion, and specific heat at the glass transition temperature, and mechanical deformation, dilatometry, and calorimetry can be used to determine the T_g of

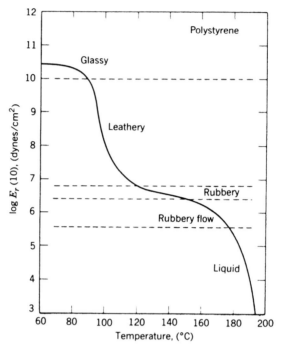

Fig. 12.1. Variation of relaxed elastic modulus of polystyrene with increasing temperature and regions of different viscoelastic behavior. (From H. W. Hayden et al., *The Structure and Properties of Materials*, Vol. III, Wiley-Interscience, New York, 1965.)

a binder material.* On heating to a temperature above the rubbery state, molding produces viscous flow behavior.

Ceramic systems containing a binder are commonly molded above the glass transition temperature of the binder. Small molecules distributed among the larger polymer molecules cause the polymers to pack less densely and reduce the Van der Waals forces binding the polymer molecules together. The presence of the small "plasticizing" molecules softens and increases the flexibility of the binder but also reduces its strength. The plasticizer effectively reduces the T_g of the polymer.

Common water-soluble binders are hygroscopic, as is indicated for polyvinyl alcohol in Fig. 12.2. Adsorbed water acts as a plasticizer, and polyvinyl alcohol plasticized by the adsorbed water has a lower Young's modulus of elasticity and strength and a greater elongation at rupture, as is indicated in Table 12.1. When stored at 50% relative humidity, the partially hydrolyzed polyvinyl alcohol, which contains about 12% acetate groups and is less densely packed, has a lower strength and elastic modulus, as indicated in Table 12.2. At each

*J. P. Polinger and G. L. Messing, in *Advances in Materials Characterization II*, R. L. Snyder et al., Eds., Plenum, New York, 1985.

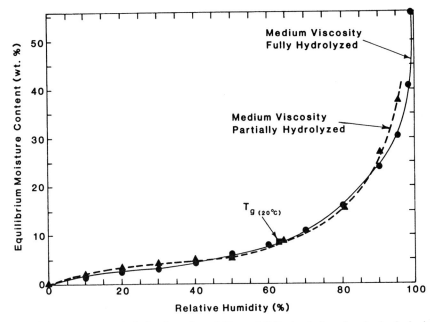

Fig. 12.2. Hygroscopic behavior of partially and fully hydrolyzed polyvinyl alcohol and the relative humidity producing a $T_g = 20°C$ in fully hydrolyzed material. (Courtesy of Du Pont Inc., Wilmington, DE.)

moisture content, T_g is 3–10°C lower for the partially hydrolyzed material. The glass transition temperature is higher for binders with stiff side groups that resist the movement of binder chains and for binders with more polar side groups that bond more strongly together. The glass transition temperature is also generally higher for a binder of higher molecular weight and when the amount of cross linking between molecules is greater.

A liquid with low vapor pressure at the molding temperature is usually selected as a plasticizer and may be used in combination with water or a nonaqueous liquid. Most plasticizers also increase the hygroscopicity of the binder system, causing a greater concentration of adsorbed moisture at a particular storage humidity. An organic plasticizer may decrease the sensitivity of the plasticizing action to relative humidity.

TABLE 12.1 Effect of Storage Humidity on Properties of Polyvinyl Alcohol Binder Film[a]

Property	40% RH	60% RH	80% RH
Tensile strength (MPa)	67	44	35
Elongation at rupture (%)	17	195	240
Young's modulus (MPa)	435	175	88

[a]88% hydrolyzed, medium viscosity grade. (Data courtesy of du Pont Inc., Wilmington, DE.)

TABLE 12.2   Typical Properties of Binder Films at 50% RH and 25°C

Property	Polyvinyl Alcohol[a] (Fully Hydrolyzed)	Polyvinyl Alcohol[b] (Partially Hydrolyzed)	Methyl Cellulose[c]
Density (Mg/m³)		1.2–1.3	1.39
Tensile strength (MPa)	60–85	50–60	60–80
Elongation (%)	150–190	140–190	10–15
Glass transition temperature (°C)	31	23	
Young's modulus (MPa)	430–500	270–370	
Thermal expansion coefficient (cm/cm/°C)		1×10^{-4}	

[a]Du Pont Inc., Wilmington, DE.
[b]Monsanto Inc., St. Louis, MO.
[c]Dow Chemical Co., Midland, MI.

Liquids used to plasticize common binders are listed in Table 12.3. Ethylene glycol is very effective in lowering the glass transition temperature and is also relatively inexpensive. The plasticizing action of ethylene glycol decreases as its molecular weight increases. A plasticizer with a relatively high boiling point, such as polymerized ethylene glycol or glycerol, is important when the system is exposed to elevated temperatures, such as in spray dying. The effect of glycerol and adsorbed water for plasticizing polyvinyl alcohol is shown in Fig. 12.3.

Plasticized viscoelastic binder films have a lower tensile strength but a greater impact strength, as is indicated for hydroxyethyl cellulose in Table 12.4. The stability of the properties of the binder film with time depends largely on the retention of the plasticizer. Viscoelastic binder films show elastic springback when the stress-producing deformation is removed. However, when the molding temperature is considerably above the glass transition temperature, the binder may exhibit viscous behavior with essentially no elastic springback.

TABLE 12.3   Common Plasticizers

Plasticizer	Melting Point (°C)	Boiling Point (°C)	MW (g/mol)
Water	0	100	18
Ethylene glycol	−13	197	62
Diethylene glycol	−8	245	106
Triethylene glycol	−7	288	150
Tetraethylene glycol	−5	327	194
Polyethylene glycol	−10	>330	300
Glycerol	18	290	92
Dibutyl phthalate		340	278
Dimethyl phthalate	1	284	194

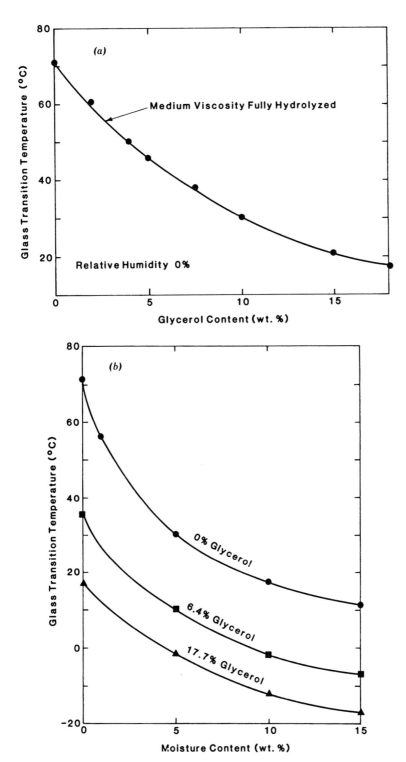

Fig. 12.3. Dependence of T_g of fully hydrolyzed polyvinyl alcohol on content of glycerol plasticizer for (a) no absorbed moisture and (b) absorbed moisture. (Courtesy of Du Pont Inc., Wilmington, DE.)

TABLE 12.4   Impact Strength of Hydroxyethyl Cellulose (HEC) Films (20°C)[a]

	10% RH	50% RH
Composition	Impact Resistance (10^3J/m^2)	
HEC (low-MW)	15	27
80% HEC, 20% diethylene glycol	46	99

[a]Courtesy of Union Carbide Corp., New York.

Plasticizers for thermoplastic polymers such as polyethylene and polystyrene include oils and waxes that are molten at the molding temperature. A solvent type of plasticizer is retained more and resists exudation during deformation. Additives such as stearic and oleic acid are plasticizers for waxes.

Many of the general effects of plasticizers apply to colloidal particle binders as well as molecular binders. Glycerine and ethylene glycol may be added to plasticize clay bodies and to extend the plastic behavior below the freezing point of water in the system.

12.2 FOAMING AND ANTIFOAMING AGENTS

A foam is air or some other gas enclosed with a thin film of liquid. Gas introduced into a pure liquid does not cause a foam, because the liquid drains from the lamellae when bubbles make contact, and the bubbles rupture. A foaming agent should reduce the surface tension of the foaming solution, increase the film elasticity, and also prevent localized thinning. An aqueous foaming additive may contain a surfactant that makes the particles hydrophobic and reduces the surface tension, and a stabilizing component. Tall oil or sodium alkyl sulfate and polypropylene glycol ether are effective aqueous foaming agents. Foaming systems are used in fabricating lightweight concrete and in the beneficiation of some minerals.

When it is desirable to eliminate bubbles from a slurry, an additive called an antifoam, or defoaming aid, is used. Commercial antifoams are used in a wide variety of industries.

An effective antifoam is a surfactant of low surface tension. It must displace any foam stabilizer and have a positive spreading coefficient S_{AL} on the liquid or existing film:

$$S_{AL} = \gamma_L - (\gamma_A + \gamma_{AL}) \tag{12.1}$$

where γ_L, γ_A, and γ_{AL} are the surface tensions of the liquid, antifoam, and antifoam–liquid interface, respectively.

Commercial aqueous defoaming surfactants include fluorocarbons, dimethylsilicones, higher-molecular-weight alcohols and glycols, and calcium and aluminum stearate. Tributyl phosphate, which reduces surface viscosity, may aid the antifoam. Many polyelectrolyte deflocculants have no significant effect on the surface energy of water, but some binders and lubricants have a moderate effect. A change in temperature, which decreases the solubility of the antifoam surfactant, may increase its effectiveness.

12.3 LUBRICANTS

An interfacial phase that reduces the resistance to sliding is an effective lubricant. Fluid lubrication is provided by a thick film of a low-viscosity liquid such as water or oil, but these may migrate rapidly from an interface under compressive stress. Boundary lubrication is provided by an adsorbed film of high lubricity which improves the surface smoothness and minimizes adhesion between surfaces.

A plasticized binder film may act as a boundary lubricant between ceramic particles, and a gelled binder solution or suspension of colloidal particles may act as a fluid lubricant. Effective boundary lubricants must have a high adhesion strength but a low shear strength. Common boundary lubricants are listed in Table 12.5. Surfactants such as stearic acid $(CH_3)(CH_2)_{16}COOH$ or its salts are particularly effective lubricants, because the carboxyl end of the molecule may be strongly bonded to an oxide surface, and the shear resistance between the first oriented adsorbed layer and successive layers is low. Molecular boundary lubricants are of a higher molecular weight than plasticizers, and they are more effective below their melting point.

Solid lubricants are fine particles with a laminar structure and smooth surfaces. Solid lubricants are effective on rough surfaces and are particularly effective at high pressure. Solid lubricants are sometimes mixed with molecular boundary lubricants that must be used at high temperatures.

12.4 PRESERVATIVES

Binders not derived from polysaccharides are biologically inert. Natural organic binders and some cellulose derivatives are subject to biological degradation because of the enzymes produced by bacteria and fungi present in an industrial environment. This degradation usually causes a decrease in the viscosity of the binder solution.

TABLE 12.5 Common Lubricants

Paraffin wax	Stearic acid
Aluminum stearate	Zinc stearate
Butyl stearate	Oleic acid
Lithium stearate	Polyglycols
Magnesium stearate	Talc (platey form)
Sodium stearate	Graphite, boron nitride

Biological activity is minimized by using sterile materials and a closed environment or by adding a chemical preservative. Many preservatives are toxic, and their use and disposal must be carefully controlled.

SUMMARY

Binders used in ceramic processing must be plasticized to produce a moldable composition. A water-soluble binder is plasticized by adsorbed moisture, and the relative humidity must be controlled to control the plasticizing effect. An organic liquid with a lower vapor pressure than water is commonly used as the primary plasticizer. The plasticized binder is of lower strength but is more deformable and resistant to failure on impact. The strength of the binder may increase when the plasticizer is lost during drying. Special surfactants may be added to promote foaming or to reduce the stability of bubbles and eliminate a surface foam. A lubricant is a special surfactant added to reduce the friction between ceramic particles and the surface of a die used in forming. A molecular boundary lubricant must be capable of being strongly adsorbed but of such a molecular size and structure that a smoother, low-friction surface is produced. A chemical preservative is used to reduce enzymatic activity in binder systems.

SUGGESTED READING

1. Errol G. Kelly and David J. Spottiswood, *Introduction to Mineral Processing*, Wiley-Interscience, New York, 1982.
2. Milton J. Rosen, *Surfactant and Interfacial Phenomena*, John Wiley, New York, 1978.
3. L. E. Nielsen, *Mechanical Properties of Polymers and Composites*, Marcel Dekker, New York, 1974.
4. P. Meares, *Polymers: Structure and Bulk Properties*, Van Nostrand, New York, 1965.
5. Ted Morse, *Handbook of Organic Additives for Use in Ceramic Body Formulation*, Montana Energy and MHD Research and Development Institute, Butte, MT, 1979.

PROBLEMS

12.1 The T_g of polystyrene is 100–105°C. What is the correlation of this value to Fig. 12.1?

12.2 Construct a graph approximating the stress–strain behavior of the polyvinyl alcohol binder in Table 12.1. Comment on the difference produced by a change in relative humidity.

12.3 Describe the change in strength and elasticity when deformable polyvinyl alcohol films with and without polyethylene glycol plasticizer are dried, eliminating adsorbed water.

12.4 Comment on the reduction of T_g when the glycerol content is 0 and 6.4% for a loss of moisture from 2.5 to 0% in Fig. 12.3.

12.5 Estimate the nominal length of a stearic acid molecule and compare this to the length of the tetraethylene glycol plasticizer in Example 12.1.

12.6 What stearate should you select as a lubricant for a ZnMn ferrite pressing powder and for a whiteware pressing composition.

12.7 Can a lubricant act as a plasticizer for a binder?

12.8 Can the order of addition of processing additives influence their adsorption and effectiveness? Consider the following sequences of addition: (1) deflocculant, binder and (2) binder/lubricant, deflocculant.

12.9 Sketch a particle with a film containing deflocculant, binder, and plasticizer molecules.

EXAMPLES

Example 12.1 Contrast the size range of low molecular weight tetraethylene glycol molecules $HO-[CH_2-CH_2-O]_n-H$ where $n = 4$ with the molecular size of the binder in Example 11.2.

Solution. The molecule contains 8 $O-C$ bonds, 4 $C-C$ bonds, and 2 $O-H$ bonds.

Bond Type	Number	Bond Length (Å)	$n \times L$ (Å)
C−O	8	1.43	11.44
C−C	4	1.54	6.16
O−H	2	0.96	1.92
			Total = 19.52

The nominal length $L = 2$ nm for the polyethylene glycol plasticizer of low molecular weight is much smaller than the nominal length of binder molecules.

Example 12.2 What is the general effect of the concentration of plasticizer on the glass transition temperature T_g and the deformation behavior at 25°C?

Solution. The temperature dependence of the shear modulus of a binder for four different plasticizer contents is shown below. At 25°C, binder compositions 1 and 2 containing less plasticizer are in the glassy state and are brittle because T_g is $\gg$ 25°C. For composition 3 containing more plasticizer, $T_g =$

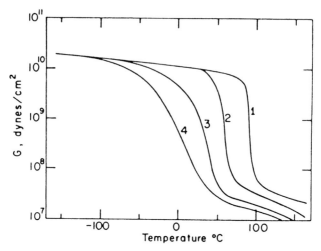

Example 12.2. A higher plasticizer content reduces the shear modulus G and T_g of the binder.

25°C and the binder is in the leathery state. Composition 4 is in the rubbery state because $T_g < 25$°C.

Example 12.3 What causes a persistent foam in a slurry and how is the foam eliminated?

Solution. Foam persistence is caused by resistance to the loss of liquid and gas in the foam even under conditions of stirring or mechanical shock. The foam is supported by the liquid lamellae of the foam. Elimination of the foam

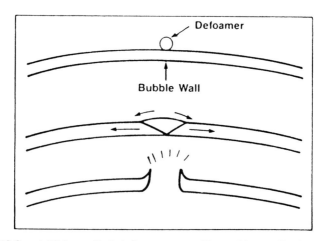

Example 12.3. Additive called defoamer or antifoam thins wall of lamellae where adsorbed and causes local thinning and bursting.

depends on thinning of the liquid lamellae and then their spontaneous rupture when a critical thickness of approximately 5–10 nm is reached. An additive which decreases foam stability when adsorbed on the walls of the lamellae by promoting drainage and gas diffusion through the lamellae is called a defoamer or antifoam (see Example 12.3).

Drainage producing thinning is caused by both gravity and the differential pressure caused by surface tension. Thinning by gravity drainage initially depends on the viscosity of the bulk liquid, but becomes limited by oriented monolayers of surfactant molecules on the surface of the lamellae, which may effectively increase the viscosity of the remaining liquid. The lower pressure at the triple point with a small negative curvature relative to that in the wall of more gradual curvature will cause migration of liquid. This effect, driven by surface tension, is larger when the bubble size is larger. At some critical thickness, the differential pressure between adjacent bubbles varying in size will cause diffusion of gas between the bubbles. As is indicated by the equation, $P = (2\gamma/r_1) - (2\gamma/r_2)$, gas will flow from the smaller bubble into the larger bubble. This can further increase a surface-tension-induced drainage effect nearby. Growth and bursting of the large bubbles commonly occurs rapidly and eliminates the foam.

PART V

PARTICLE PACKING, CONSISTENCY, AND BATCH CALCULATIONS

Several different types of particle size distributions are used when forming ceramics. Chapter 13 describes these distributions and presents principles of particle packing and parameters which influence particle packing behavior. The amount of liquid and additives introduced in processing and particle packing factors strongly influence the consistency of the system for processing. Proportions of constituents in the system for processing are commonly presented in a table called the batch composition. Chapter 14 introduces general concepts used to describe the consistency of the process material and the calculation of the proportions of ingredients commonly called batch calculations.

CHAPTER 13

PARTICLE PACKING
CHARACTERISTICS

Processing systems with particular distributions of particle sizes and shapes are produced by selecting and blending raw materials with different initial characteristics and by subsequent crushing, grinding, dispersion, classification, and granulation operations. The geometrical characteristics of the particle system have a significant impact on the particle arrangements and the packing density, size and shape of pore interstices, resistance to permeation by a fluid, bulk flow and deformation behavior, drying behavior, and microstructure development during firing.

Powders for high-density, fine-grained electronic and technical ceramics are usually submicron or finer than a few microns and approximately log-normal in size distribution to avoid exaggerated grain growth in sintering (Fig. 13.1). Systems for whitewares and some technical products sintered in the presence of a liquid phase contain particles ranging from a maximum size of 10 to 100 μm down to a submicron size and have a distribution that packs relatively densely. Conventional dense refractory shapes and concrete contain particles ranging from about 1 cm down to a submicron size and are formulated using discrete sizes to achieve a high packing density. In refractory insulation where a high volume of fine pores is required, flocculated powders mixed with refractory fibers are commonly used.

Principles of particle characterization were discussed in Section III. In this chapter we examine models for the packing of particles and experimental results for both coarse and fine systems. The packing of particles varying in size and shape, as is common in most ceramic systems, can be very complex. Many different arrangements are possible, and several parameters must be defined to consider the packing and porosity. We will see that common working models are based on spherical particles and serve best to approximate the average

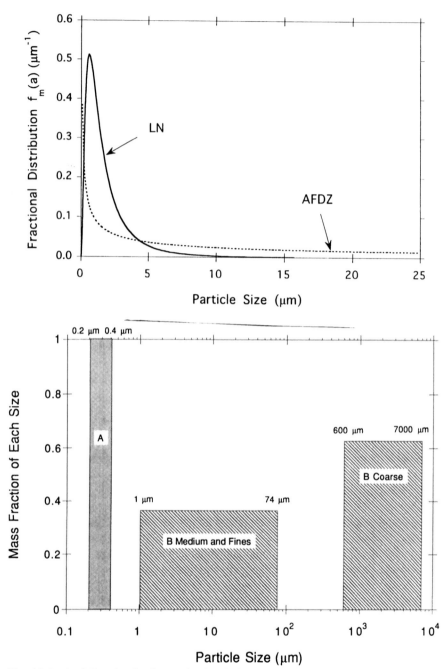

Fig. 13.1 (top) Powder for fine-grained technical ceramics such as alumina substrates, barium titanate, and ferrites have a log-normal particle size distribution LN. The more densely packing distribution for a whiteware body AFDZ is of a different form and contains relatively more colloidal size particles. (bottom) Nearly monosize colloidal particles (A) are used for specialized coatings and advanced ceramics having a sub-micron sintered grain size. System (B) with coarse, medium, and fine sizes and a gap between the coarse and medium sizes is commonly used for heavy refractories and concrete.

packing density and pore size and to suggest changes in these when the distribution of sizes is changed. Particle and binder parameters hindering particle motion and packing are also briefly discussed.

13.1 CHARACTERISTICS OF PACKINGS OF UNIFORM SPHERES

The packing of nonporous spheres of uniform size is studied in beginning courses on the structure of solids. Those results provide an approximation of the geometrical characteristics of packed, monosize ceramic particles and provide reference values and a basis for comparing experimental results for real-particle systems.

Uniform spheres may pack in five different ordered arrangements, as shown in Fig. 13.2. The fraction and percent of the bulk volume occupied by the spheres are referred to as the packing fraction PF and the packing density PF (%), respectively. As seen in Table 13.1, the packing density ranges from 52% for simple cubic packing to 74% for tetrahedral and pyramidal close packing and is independent of the size of the spheres. The packing density of nonordered arrangements of uniform spheres larger than 1 mm has been determined experimentally to be about 60% without vibrations and 64% with vibrations.*

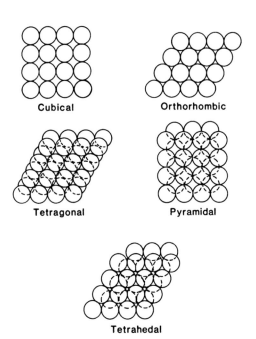

Cubical   Orthorhombic

Tetragonal   Pyramidal

Tetrahedal

Fig. 13.2 Ordered packing arrangements for spheres of uniform size.

*G. D. Scott and D. M. Kilgour, *Br. J. Appl. Phys. Ser. 2*, **2**, 863 (1969).

TABLE 13.1   Packing Density of Regular Configurations of Uniform Spheres

Configuration	Coordination Number	Packing Density (%)
Cubic	6	52.4
Orthorhombic	8	60.5
Tetragonal	10	69.8
Tetrahedral	12	74.0
Pyramidal	12	74.0

The interstitial pore fraction ϕ, or interstitial porosity $\phi(\%)$, indicates the portion of the bulk volume present as interstices among the nonporous spheres. The pore size is a function of both the size of the spheres and the packing arrangement. In a tetrahedral packing, the cross-sectional area A of the interstices ranges from $0.04a^2$ to $0.21a^2$, and for simple cubic packing $A = 0.21a^2$, as indicated in Table 13.2. For dense packings, the cross-sectional area of the interstices is a fraction of the cross-sectional area of a sphere, and the area decreases as the diameter of the spheres decreases.

The center line of a continuous pore through the packing may be tortuous, and the length of the center line relative to the edge length of the unit cell is the tortuosity T_0. For a cubic packing, $T_0 = 1$, and in a tetrahedral packing, $T_0 = 1.3$; displacement of the particles into a less dense packing would reduce the mean tortuosity.

The number of particles per unit bulk volume N_p in a system of packed spheres of uniform size is

$$N_p = \frac{6(PF)}{\pi a^3} \tag{13.1}$$

The number of particle contacts per unit bulk volume N_c is the product of N_p and half the coordination number CN:

$$N_c = \frac{3(PF)(CN)}{\pi a^3} \tag{13.2}$$

TABLE 13.2   Pore Size in Cubic and Tetrahedral Packings of Uniform: Spheres[a]

Parameter	Cubic	Tetrahedral
Entry pore area	$0.21a^2$	$0.04a^2$
Entry pore area/$\pi a^2/4$	0.26	0.05
Entry pore diameter/a	0.51	0.22
Entry sphere diameter/a	0.41	0.15
Void fraction	0.48	0.26
Vol. void/vol spheres	0.92	0.34
Radius ratio primary sphere/interstitial sphere	1.37	4.44

[a]Sphere diameter $= a$.

For regular packing arrangements, N_c is very dependent on the packing geometry.

For nonregular packing arrangements, the number of contacts is a function of the arrangement of local groups of spheres and the porosity. The *CN* is observed to range between 6 and 10; an empirical approximation of the average coordination number is $\overline{CN} = \pi/\phi$.* For random packing, the average number of contacts $\overline{N}_c$ is approximated by the equation

$$\overline{N}_c = \frac{3(1 - \phi)}{\phi a^3} \tag{13.3}$$

McGeary studied the packing of coarse (>37 μm), monosize spherical particles experimentally using axial vibrations, and observed packing mostly in an orthorhombic arrangement and a packing density of 62.5%. This arrangement appeared to be stabilized by slight lateral spreading and nesting of spheres in the vibration direction. Hindrance of the container wall was insignificant when the container diameter/particle diameter exceeded 10.

The packing density of nearly monosize spherical particles of silica and alumina of colloidal size packed by filter pressing a deflocculated slurry is about 65–70%. More densely packed domains in which the particles are ordered are separated by interdomain boundary regions containing a less ordered arrangement and larger pores.

13.2 PACKING IN INTERSTICES AMONG COARSER PARTICLES

Smaller particles introduced and distributed in the interstices of packed larger particles will reduce the porosity and pore size. Large particles added to finer particles displace fines and pores and reduce the porosity. This model of packing is called the Furnas model† and is shown in Fig. 13.3.

When the large particles in a packing are in contact, the theoretical maximum packing fraction PF_{max} for a mixture of coarse, medium, and fine particles is

$$PF_{max} = PF_c + (1 - PF_c)PF_m + (1 - PF_c)(1 - PF_m)PF_f \tag{13.4}$$

where PF_c, PF_m, and PF_f are the packing factors of the coarse, medium, and fine particles, respectively. The weight fraction f_i^w of each size at PF_{max} is

$$f_i^w = \frac{W_i}{W_{total}} \tag{13.5}$$

---

*W. O. Smith et al., *Phys. Rev.*, **34**, 1272 (1929).
†C. C. Furnas, *U.S. Bur. Mines Rep. Invest.*, **2894** (1928).

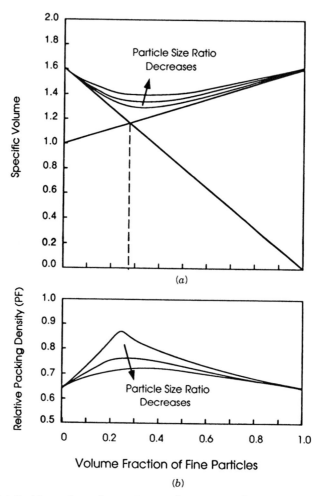

Fig. 13.3 (*a*) Packing volume for a mixture of coarse and fine particles. The straight lines correspond to ideal packing when a fraction of one size is replaced with the other size and the size ratio is infinite. Curved lines indicate packing behavior for real systems having a limited size ratio. (*b*) Corresponding relative packing fraction PF is indicated on diagram below. The maximum packing fraction PF_{max} occurs when fines just fill the interstices between touching coarse particles. (Courtesy of Jingmin Zheng, Alfred University.)

where W_i is the weight of each size and $W_{total} = \Sigma W_i$. For a three-component system the weight of coarse (W_c), medium (W_m), and fine (W_f) particles is

$$W_c = PF_c D_c \tag{13.6}$$

$$W_m = (1 - PF_c) PF_m D_m \tag{13.7}$$

$$W_f = (1 - PF_c)(1 - PF_m) PF_f D_f \tag{13.8}$$

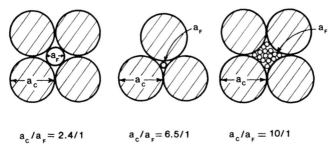

$$a_c/a_F = 2.4/1 \qquad a_c/a_F = 6.5/1 \qquad a_c/a_F = 10/1$$

Fig. 13.4 Packing of fine spheres in a planar interstice among coarse particles.

For this model, the volume fraction of particles required decreases with particle size. In practice, the maximum packing fraction is achieved when the ratio between nearest sizes is greater than about 7 and the finer particles are dispersed uniformly. Finer particles must be small enough to enter into all regions of the interstices, as indicated in Fig. 13.4. Following this packing scheme, McGeary experimentally achieved a packing density of 95% for a quaternary system of vibrated steel spheres with the sizes indicated in Table 13.3.

The blending of discrete sizes to achieve a high packing density is common practice in formulating coarse-grained refractories and concrete that have very little dimensional change on firing. Refractory batches for "common heavy refractories" composed of particles of equal density contain approximately 60–65 wt% coarse particles about equally divided between 0.7–0.17 cm and 0.17–0.06 cm, and 35–40 wt% fines <74 μm of which about one-half are finer than 44 μm; the packing density achieved on vibrating dry particles is about 65–70% (Fig. 13.5) and on pressing this commonly increases 5–10%. The packing density increases as the size ratio increases. A smaller maximum particle size is used for some refractory products such as kiln furniture, where the flexural strength must be higher. When using particles of different density, the proportions by weight for maximum packing are different from the proportions stated above. The particle size distribution of castable refractories and concrete is similar to that for heavy refractories. Hydraulic cement composes

TABLE 13.3 Packing Density of Mixed Spheres of Different Size

Diameter (cm) (Weight Fraction of Spheres)				Packing Density (%)	
1.28	0.155	0.028	0.004	Calculated	Experimental
1.000	—	—	—	60.5	58.0
0.726	0.274	—	—	84.8	80.0
0.647	0.244	0.109	—	95.2	89.8
0.607	0.230	0.102	0.061	97.5	95.1

Source: R. K. McGeary, *J. Am. Ceram. Soc.*, **44**(10), 513–522 (1961). Size ratio 320/39/7/1.

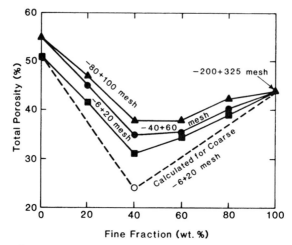

Fig. 13.5 Experimental and calculated total porosity for packings of different size coarse tabular alumina particles mixed with a common fine tabular alumina. [Note: tabular alumina particles have internal porosity (Fig. 8.3) included in the porosity indicated.]

a portion of the fines. The liquid requirement for working and the cured strength depend directly on the particle packing density.

In the packing model described above, coarse particles are assumed to be in contact; this type of microstructure is important in refractories where dimensional stability under a compressive load is requisite. Other compositions of coarse, medium, and fine sizes with the coarse particles dispersed can produce a relative maximum in packing density; these systems have received relatively little attention.

13.3 PACKING OF CONTINUOUS SIZE DISTRIBUTIONS

Most ceramics are produced from a material having a continuous distribution of particle sizes between some maximum and a finite minimum size. Calcined and milled calcined powders for fine-grained technical ceramics are often log-normal to a good approximation (Fig. 13.6). Typical values of the geometric standard deviation σ_g are in the range 1.4–2.4. The calculated minimum pore fraction for the random packing of a log-normal distribution of spheres increases as the geometric standard deviation of sizes increases, but for an increase in σ_g from 1.4 to 2.4, the maximum packing density for spherical particles is calculated to increase from 65% to only 69%. The maximum packing fraction of three milled aluminas with a log-normal distribution of sizes, determined by mechanically pressing (deflocculated) filter cake, are listed in Table 13.4. The experimental maximum packing fraction is slightly higher for the powders of larger geometric standard deviation, as predicted. The packing density values are lower than is calculated for an LN distribution of spheres, because the

TABLE 13.4   Experimental PF of Milled Calcined Aluminas Having a Log-Normal Size Distribution

Characteristic	Alumina Type		
$\bar{a}_{gM}$ (μm)	8.0	0.8	1.3
σ_g	1.8	2.2	2.5
PF (%)	62	64	66

larger particles in the coarse calcined aluminas are porous aggregates with a nonspherical shape.

One calcined alumina shown in Fig. 13.7 that is not log-normal in size distribution has a higher maximum packing fraction. The packed density depends significantly on the form of the size distribution as well as the range of particle size. In ceramic whitewares, the clay, quartz, and feldspar minerals, each having a continuous size distribution, are blended in proportions that produce a continuous distribution that packs more densely. The size distribution of the blend is approximated by the Andreasen* equation for dense packing

$$F_M(a) = \left[\frac{a}{a_{max}} \right]^n \tag{13.9}$$

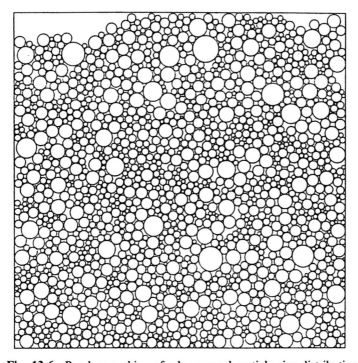

Fig. 13.6   Random packing of a log-normal particle size distribution.

*A. H. M. Andreasen and J. Anderson, Z. *Kolloid*, **50**, 217–228 (1930).

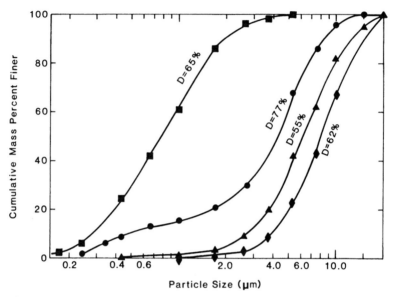

Fig. 13.7 Cumulative particle size distributions and maximum experimental packing density for several calcined Bayer process aluminas.

and

$$f_M(a) = \frac{n}{a}\left[\frac{a}{a_{max}}\right]^n \tag{13.10}$$

where a_{max} is the maximum particle size and $1/n$ is the distribution modulus.

In this model, the amount of a particular size is a constant fraction of the total amount finer than that size; that is, the proportion of fines increases as the particle size decreases, as indicated in Fig. 13.1. Andreasen recognized that for a particular a_{max}, the porosity of the packed particles would decrease as n decreased; his experiments indicated that a practical range of n was 0.33–0.50. Equation 13.9, with n equal to 0.33 and 0.50 and size data for an electrical porcelain body that has a maximum packing density of 75%, is shown in Fig. 13.8. The experimental data for sizes larger than 0.3 μm fall within the Andreasen bounds, but at smaller sizes the distribution becomes curved.

Andreasen assumed that the smallest particles would be infinitesimally small. Dinger and Funk* recognized that the finest particles in real materials are finite in size and modified the Andreasen equation using a minimum size a_{min}. Zheng† derived the equation from a fundamental model. The AFDZ equation for dense

*D. R. Dinger and J. E. Funk, *Particle Packing: Review of Packing Theories*, Fine Particle Society, 13th Annual Meeting, Sp. 1982, Chicago.
†See Reference 1.

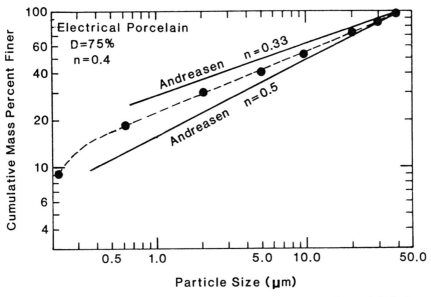

Fig. 13.8 Log–log plot of particle size distribution of an electrical porcelain body and Andreasen distributions having the same maximum size of 44 μm.

packing is

$$F_M(a) = \frac{a^n - a_{min}^n}{a_{max}^n - a_{min}^n} \tag{13.11}$$

In Fig. 13.8, the curvature of the size data is of the form predicted by Eq. 13.11. In general, for size distributions approximated by Eqs. 13.10 and 13.11, the packing density may increase as $1/n$ and the range of sizes increase, and the packing density may exceed 80%. The calcined alumina in Fig. 13.7 with a packing density of 77% is roughly approximated by Eq. 13.11, but with $n > 0.5$.

13.4 HINDERED PACKING

In real-particle systems, the movement of particles into the minimum porosity configuration during processing is hindered by both external and internal factors. Bridging of particles and agglomerates with rough surfaces between the walls of tooling is quite significant when the spacing between external surfaces is less than about 10 particle diameters. Bridging is especially a problem when the surfaces of particles and aggregates are rough (Fig. 3.2) and when friction is high at the interface between the particles and the tooling. Mechanical vibration may momentarily reduce bridging forces and facilitate the rolling and

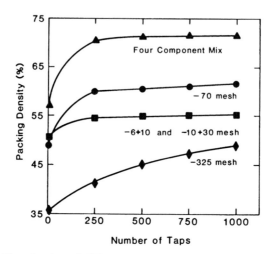

Fig. 13.9 Packing density of different size fractions of alumina and dependence on number of low-frequency vibration cycles.

sliding of dry particles into a denser configuration (Fig. 13.9). Lubrication of all surfaces may aid in achieving a higher packing density. However, mechanical forces causing fracture of some particles may be required to approach the minimal porosity configuration. The vibrated packing density of coarse angular particles of uniform size is 50–60%, in contrast to the 65% density for smooth spheres.

Coagulation, flocculation, and adhesion forces, which retard particle motion, may also hinder packing. Random arrangements of anisometric particles typically have a higher porosity and a wider range of pore sizes than for an ordered arrangement, as is shown in Fig. 13.10. A higher porosity is generally observed when the aspect ratio of the particles is higher and when a higher content of fibers is added into a powder containing isometric particles (Fig. 1.2).

Adsorbed binder molecules and colloidal particles may hinder particle move-

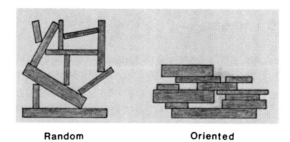

Random   Oriented

Fig. 13.10 A random packing of anisometric particles produces a lower packing density.

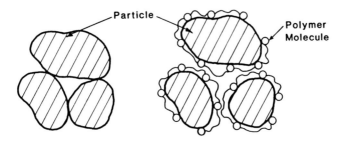

Fig. 13.11 Adsorbed binder molecules reduce the packing density in the absence of a mechanical stress.

ment and reduce the packing density a few percent when the hindrance is not overcome by applied stress (Fig. 13.11). The steric hindrance increases as the molecular weight of the binder increases and when the molecules bridge particles. Fines mixed with a sticky binder may coat larger particles rather than entering interstices. The packing density in an undispersed agglomerate may be higher than in an agglomerate of dispersed particles with a film of binder and liquid. These packing inhomogeneities tend to be more pronounced when the particles are of colloidal size and when binder concentrations are higher.

SUMMARY

The distribution of particle sizes in a processing system has a significant effect on the particle packing and pore structure and the behavior during forming, drying, and firing. Models for the packing of monosize particles indicate that the packing density increases as the coordination number increases and that the mean pore size decreases as the particle size and porosity decrease. The packing density of about 62% for a monosize system can be increased above 75% by adding a specific proportion of a finer size that packs efficiently in the interstices among the coarse; this type of sizing is used to fabricate relatively dense, coarse-grained systems. Continuous particle size distributions of the log-normal type used for ceramics have a maximum packing density of about 65%. A continuous distribution of the Andreasen type containing a greater amount of finer particles may pack to a density exceeding 80%. The packing density of a continuous distribution of sizes varies with the form and breadth of the distribution.

Rough surfaces and adsorbed molecules which prevent particle motion reduce the packing density. In packings of platelike and fibrous particles, the packing density and pore structure are very dependent on the particle orientation and arrangements. Agglomeration and the addition of fibrous particles in a system of isometric particles tend to produce inhomogeneous porosity and a higher porosity.

SUGGESTED READING

1. Jingmin Zheng, Paul F. Johnson, and James S. Reed, "Improved Equation of the Continuous Particle Size Distribution for Dense Packing," *J. Am. Ceram. Soc.*, **73**(5), 1392–1398 (1990).
2. N. G. Stanley-Wood, *Enlargement and Compaction of Particulate Solids*, Butterworths, London, 1983.
3. P. J. Sherrington and R. Olivier, *Granulation*, Heyden, Philadelphia, 1981.
4. A. N. Patankar and G. Mandel, "The Packing of Solid Particles," *Trans. Indian Ceram. Soc.*, **39**(4), 109–117 (1980).
5. G. P. Bierwagen and T. E. Saunders, "Studies of the Effects of Particle Size Distribution on the Packing Efficiency of Particles," *Powder Tech.*, **10**, 111–119 (1974).
6. H. Y. Sohn and C. Moreland, "The Effect of Particle Size Distribution on Packing Density," *Can. J. Chem. Eng.*, **46**, 162–167 (1968).
7. R. K. McGeary, "Mechanical Packing of Spherical Particles," *J. Am. Ceram. Soc.*, **44**(10), 513–522 (1961).
8. A. E. R. Westman and H. R. Hugill, "The Packing of Particles," *J. Am. Ceram. Soc.*, **13**(10), 767–779 (1930).

PROBLEMS

13.1   Estimate the change in average coordination number of randomly packed particles when the packing fraction increases from 0.55 to 0.65.

13.2   Contrast the number of particles/volume for a material whose size is 10 μm, a material whose size is 0.1 μm, and a mixture of 80% coarse and 20% fines. Assume orthorhombic packing.

13.3   For Fig. 13.3, explain why one line extrapolates to zero volume but the other does not.

13.4   The model for Eq. 13.4 assumes that fines enter interstices between coarse and medium particles in the mixture. Can some of the medium particles be replaced with fines without reducing the packing density? Explain.

13.5   Calculate the theoretical PF_{max} for a two-component system containing coarse SiC particles and alpha alumina fines. Assume a $PF = 0.50$ for each component.

13.6   Calculate the composition of SiC and alumina particles in wt% for the results in Problem 13.5.

13.7   Derive a relationship between ϕ_{min} and the ϕ_i of the components for the model used in deriving Eq. 13.4.

13.8 The packing density for coarse and medium particles is 0.62 and for the fines 0.50. Calculate the theoretical PF_{max} and the composition (wt%) when all particles have the same density.

13.9 Refractory tile is prepared by combining coarse, medium, and fine tabular alumina having an apparent density of 3.60, 3.68, and 3.86 Mg/m^3, respectively. Calculate the theoretical PF_{max} and the amount (wt%) of each size. What is the interstitial porosity and the total porosity of the ideal tile?

13.10 Estimate the PF_{max} for the powders in Table 7.6.

13.11 What are the weight proportions of gravel, sand, and cement powder used in concrete for construction purposes.

13.12 How would the workability of concrete differ at PF_{max} and with a few percent of additional fines?

13.13 Contrast the matrix phase for two-component systems when the portions correspond to PF_{max}: (1) ideal Furnas model, (2) dry fines are mixed with hygroscopic binder powder and mixed with wetted coarse particles, (3) ideal Andreasen model.

13.14 Explain why the tap density is achieved more rapidly for the <70 mesh material than for the <325 mesh material in Fig. 13.9.

EXAMPLES

Example 13.1 What is the number of particles per cubic centimeter of bulk material when the particle size $a = 1$ μm? Assume orthorhombic packing of the particles.

Solution. The number of particles per unit bulk volume N_p is

$$N_p = 6(PF)/\pi a^3$$

For orthorhombic packing, $PF = 0.605$, and

$$N_p = 6(0.605)/\pi(1 \times 10^{-4}\text{ cm})^3$$

$$N_p = 1.2 \text{ trillion/cm}^3$$

This very large number indicates that dispersing and coating all particles is a powder system is a difficult task requiring careful controls.

Example 13.2 For orthorhombic packing of spheres, calculate the diameter of the equivalent circular area of the entry pore, the size of a sphere which can enter through the interstice, and the void fraction and diameter of an interstitial spherical particle.

Solution. Note Figure 13.1 showing orthorhombic packing of spheres of diameter a in a plane. The area A_i of the planar interstitial pore is

$$A_i = a^2 - 4(1/4)(\pi a^2/4) = 0.21a^2$$

The diameter dia_i of an equivalent circular area is

$$\pi \, dia_i^2/4 = 0.21a^2 \quad \text{or} \quad dia_i = 0.51a$$

The void fraction ϕ_i is

$$\phi_i = a^3 - \pi a^3/6 = 0.476$$

The diameter dia_s of an interstitial sphere is the diagonal in the packing minus the diameter of one sphere, that is, $dia_s = (3^{1/2} - 1)a$ and $a/dia_s = 1.37$.

Example 13.3 Calculate the PF_{max} and the particle volume percent for a mixture of coarse and fine particles of the same particle density which pack with $PF = 0.60$. The size ratio is > 7.

Solution. $PF_{max} = PF_c + (1 - PF_c)PF_f = 0.60 + (1 - 0.60)0.60$

$$PF_{max} = 0.60 + 0.24 = 0.84 \quad \text{or} \quad 84\%$$

The volume of each type is

$$\text{Volume coarse} = 0.60/0.84 = 0.71 \quad \text{or} \quad 71\%$$
$$\text{Volume fines} = 0.24/0.84 = 0.29 \quad \text{or} \quad 29\%$$

Since all particles have the same density, the proportion of particle types by volume percent and weight percent is the same.

Example 13.4 If in Example 13.3 the coarse particles were silicon carbide with a particle density of 3.21 Mg/m^3 and the fines alumina with a particle density of 3.98 Mg/m^3, what would be the proportions of particles in weight percent?

Solution. The weight percent of each is calculated from the volume fractions.

$$\text{Weight coarse (\%)} = 0.71(3.21)\,100/[0.71(3.21) + 0.29(3.98)] = 66\%$$
$$\text{Weight fines (\%)} = 0.29(3.98)\,100/[0.71(3.21) + 0.29(3.98)] = 34\%$$

CHAPTER 14

BATCH CONSISTENCY AND FORMULATION

Each particular batch (processing system) used for forming a product will consist of particular inorganic powder, granular material, and/or colloid and some combination of the processing additives discussed in Part IV. The processing additives are selected on the basis of chemical type, physical characteristics such as molecular weight, viscosity, and thermal properties, and their general effectiveness as determined by experiment. The batch for processing is proportioned of powders and additives to produce a particular consistency for processing and forming.

In this chapter we consider different processing consistency states and the general structure of systems of different consistency. Then the calculation of a batch of ingredients for processing using a total batch or a powder basis proportioned by weight or volume, is described.

14.1 PROCESSING CONSISTENCY STATE

The consistency of a processing system is commonly described using a term such as bulky powder, agglomerated powder, granular material, rubbery consistency, cohesive plastic body, thick viscous paste, thin sticky paste, thick concrete-like slurry, smooth pourable slurry, or thin suspension. These descriptions suggest something about the nominal particle sizes in the batch, the apparent liquid content, and the apparent mechanical behavior. The consistency of a particular batch is a function of many parameters:

1. The amount, distribution, and properties of the liquid phase.
2. The amount, sizes, and packing of the particles.

3. The types, amounts, and distribution of additives adsorbed on the surface of the particles.
4. The interparticle forces between the particles which may be attractive or repulsive.

Clearly, ceramic processing systems may be quite complex. A small change in one of the parameters mentioned above, produced by a change in composition, temperature, pressure or mechanical agitation, may sometimes change the consistency and forming behavior rather significantly.

Five consistency states may be identified when mixing a wetting liquid or binder solution into a powder:

1. Bulky powder (no liquid)
2. Agglomerates (granules)
3. Plastic body
4. Paste
5. Slurry

These consistency states are indicated in Fig. 14.1. A dry bulky powder consists of discrete particles and random agglomerates produced by attractive Van der Waals, electrostatic, or magnetic interparticle forces. The bulk density and flow resistance depend on random particle attraction and agglomeration and both tend to decrease as the particle size decreases. The decrease occurs because the surface force relative to the gravitational force increases with decreasing particle size, as discussed in Chapters 2 and 5.

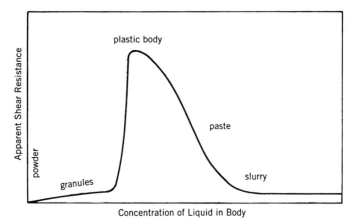

Fig. 14.1 Processing consistency states produced on mixing a powder with a wetting liquid or binder solution.

Bulky dry powders do not flow well when poured, but offer little flow resistance when stirred. When a wetting liquid of low viscosity is sprayed into a stirred dry powder, agglomerates form. Dry particles adhere to particle groups with wet surfaces and their size increases. Bonding forces in the agglomerates are capillary forces of the wetting liquid, coagulation forces between particles, and flocculation forces provided by an adsorbed binder. The degree of pore saturation DPS is defined as

$$DPS = \text{volume of liquid/pore volume} \qquad (14.1)$$

Within agglomerates, DPS < 1. Further spraying and stirring produces additional agglomerates at the expense of the dry powder. The resistance to stirring remains low. Purposefully made powder agglomerates are commonly called granules and the free-flowing material is referred to as being granular. A granular material is commonly used as a feed material when forming a ceramic product using a pressing operation.

On increasing the liquid to some particular content, nearly all of the powder will have been transformed into discrete granules with dry surfaces. A small amount of additional liquid will now cause nearly total agglomeration of the agglomerates into a cohesive mass. The cohesive mass is very resistant to shearing (Fig. 14.1), and may have plastic-like properties if a significant fraction of the powder is submicron in size and the particles are coagulated. The plasticity may be especially well developed when a clay or organic binder coats the particles. In the plastic state, DPS is < 1, as shown in Fig. 14.2. The

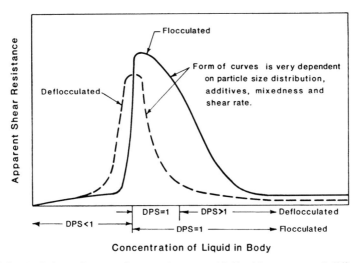

Fig. 14.2 Variation of processing consistency with liquid content and differences in DPS when system is deflocculated and when flocculated.

plastic state may extend over 10 vol% liquid content when flocculated particles rearrange into a more open packing configuration to accommodate the additional liquid. The resistance to shear is observed to decrease with increasing liquid content. Material having a plastic consistency is commonly used for the forming processes of extrusion, jiggering, and plastic transfer pressing.

Granules and the plastic materials will consolidate when a moderate compressive stress causes the agglomerates to deform and rearrange into a more dense packing configuration. The consolidation of the powder granules may be quite significant, but the frictional resistance of the closely packed particles limits the extent of densification, as indicated in Fig. 14.3. For the plastic material, the consolidation at a moderate stress is halted when DPS = 1 because the relatively incompressible liquid becomes pressurized. The mechanical compressibility X of the system is

$$X = (\Delta V/V_0)/\Delta P \qquad (14.2)$$

where ΔV is the volume change for the increment of pressure ΔP and V_0 is the initial volume. On decompression, these materials will commonly exhibit a measurable volume expansion accompanying the slight rearrangement of the particles, which is called elastic springback SB. The elastic springback may be measured in a linear or volumetric manner, and is defined as

$$SB_{linear} = \Delta L/L_0 \qquad (14.3)$$

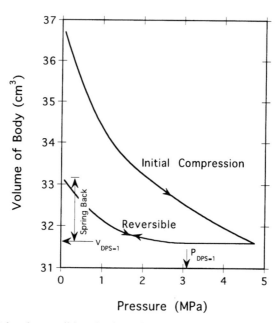

Fig. 14.3 Initial and reversible (elastic) volume change with pressure for a flocculated porcelain body. (Note $P_{DPS=1}$ and elastic spring back on decompression.)

where L_0 is the original gage length, or

$$SB_{volume} = \Delta V / V_0 \qquad (14.4)$$

where V_0 is the volume before springback.

On adding still more liquid into the system with coagulated particles, the fully saturated material (DPS = 1) with a rather open particle packing and moderate to low cohesive strength becomes more like a paste in consistency (Fig. 14.4). The flow behavior of the paste depends on the content of colloidal particle sizes and coagulation forces. The presence of a flocculating binder can significantly increase its cohesion and resistance to shear (see Fig. 14.2). The apparent consistency is influenced by the powder dispersion produced during mixing and may change with time when the interparticle forces change. The compressibility of a paste is very low at moderate pressures because DPS = 1 and the particles are not able to arrange into a denser packing. Paste materials are commonly used for printing thick films onto substrates for electronic ceramics, ceramic decorations, and as mortars and neat cements.

Adding additional liquid or a deflocculant to a paste can cause the shear resistance to decrease and a change of consistency to that of a slurry. In the

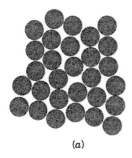

(a)

(b)

Fig. 14.4 A more open particle packing is produced on transition from plastic body (A) to paste (B), but DPS remains 1.

 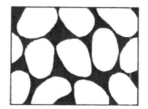 

Fig. 14.5 Distribution of liquid when DPS < 1 (granule and uncompressed plastic states), DPS = 1 (compressed plastic state and paste), and DPS > 1 (slurry state).

slurry state, coagulated/flocculated structure becomes discontinuous, particles are dispersed in the liquid, and DPS > 1. The material may flow under its own weight. A concentrated suspension that is relatively stable over time is called a slurry. Dilute systems are commonly called suspensions. The term slip is traditionally used in the whiteware industries for a slurry that contains clay. Material having a slurry consistency that is stable over time is used for casting processes and is often a temporary consistency state when the material is processed, such as in mixing.

The nominal distribution of liquid among particles in the different consistency states is shown in Fig. 14.5. Differences in the microstructural character for a uniform material are summarized in Table 14.1. In the agglomerate, plastic and paste consistency states, shear stress may be transmitted by particles in contact. Unlike the granules and plastic bodies, the paste and slurry are nominally incompressible. In our description, particles with a binder coating, which transmit force directly to each other, are considered to be in contact. In granules, the gas phase is continuous and bulk liquid if present is discontinuous. Bulk liquid is continuous in the plastic, paste, and slurry states. Because of the different liquid contents and structure, the consistency states differ significantly in particle packing factor PF, DPS, and compressibility (X).

The consistency also depends on the presence of binder between the particles (Fig. 14.6). When a system contains a significant content of a liquid or flocculant and the viscosity is strongly dependent on temperature or time, the system may appear to change from one consistency to another with a change

TABLE 14.1 Nominal Characteristics of Different Consistency States[a]

Consistency	Particle Structure	Gas	Liquid	PF	DPS	X
Powder	C	C	D	L	<1	H
Granules	C	C	D	L-M	<1	M
Plastic	C	D	C	M-H	<1	L
Paste	C	D	C	M-L	= 1	0
Slurry	D	D	C	M-L	>1	0
Suspension	D	D	C	L	>1	0

[a]C = continuous; D = discontinuous; H = high; M = moderate; L = low.

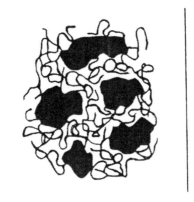

Fig. 14.6 (top) An added binder in liquid between particles influences consistency and flow of a slurry or paste and (bottom) compliant binder films on particles provides deformability needed for compaction and plastic extrusion.

in temperature or with time. Time-dependent coagulation or chemical gelation of the binder in a slurry may change the consistency to that of a paste. An internal chemical reaction in a plastic body which causes a change of inter-particle forces from attractive to repulsive, may cause the transition to a paste consistency. A material containing considerable thermoplastic binder of very high viscosity may appear to be nearly rigid at 20°C even though the particles are not touching each other. However, on heating to $> 150°C$, its flow behavior will change and become similar to that of a material with a plastic consistency, even though the microstructure of the material as processed is that of a slurry.

The changes caused by substitution of additives, internal changes in the adsorption, association, or distribution of additives due to changes in chemical behavior or temperature-dependent properties can cause many curious behaviors and complicate the understanding of the system. But these changes also present opportunities for designing processing systems which have novel forming possibilities.

14.2 BATCH CALCULATIONS

The types and proportions of the primary material components and additives for a processing system are commonly presented in a table as the "batch composition." Different bases and units may be used when listing the batch. Each of us must be familiar with different batch formulation schemes used for industrial and laboratory purposes. In stating the batch composition, it is common to first list the primary inorganic materials, followed by the minor inorganic materials, the liquid, and then the processing additives in the order of deflocculant, coagulant, polymer binder, and other additives such as a surfactant or lubricant.

The batch composition for a slurry for spray drying is listed in Table 14.2. The proportions of all ingredients are commonly listed in both weight percent and the actual weights used in the batch for processing. Here a total batch basis is used and the weight percents in Table 14.2 add up to 100%. The batch for a processing system containing a significant quantity of liquid is commonly calculated and expressed in this way.

Sometimes, especially when the amount of liquid added is small, the liquid and other additives are calculated as a percent of the dry inorganic materials. Table 14.3 presents a typical composition for granules for pressing using a total batch basis and an inorganic solids basis for comparison.

The proportions of particular additives are often based on certain ratios, such as the amount of liquid/amount of inorganic material, amount of binder/amount of inorganic material, amount of deflocculant/amount of surface area of powders, or amount of plasticizer/amount of binder, which have been determined

TABLE 14.2 Batch Composition of a Slurry for Spray Drying

Component	Weight (%)	Weight (g)
Alumina powder	73.22	2994
Water	23.23	950
Polyacrylate deflocculant[a]	2.45	100
Polyvinyl alcohol binder	0.98	40
Ethylene glycol plasticizer	0.12	5
	100	4089

[a]40% solids in water.

TABLE 14.3   Composition of Spray-Dried Granules for Pressing

Component	Total Batch Basis Weight (%)	Inorganic Basis Weight (%)
Alumina powder	96.49	100
Polyacrylate deflocculant	1.29	1.34
Polyvinyl alcohol binder	1.29	1.34
Ethylene glycol plasticizer	0.15	0.16
Moisture (plasticizer)	0.78	0.81
	100	

to be required for a particular forming process and product. These aspects are described in more detail in Chapter 21 and subsequent chapters.

For research and development purposes and when the batch constituents are measured directly by volume, the batch composition is often presented in proportions by volume rather than by weight. The volume fraction f_i^v of a component is its volume/volume of all components and may be calculated from the weight composition using the equation

$$f_i^v = (M_i/D_i)/(\Sigma M_i/D_i) \qquad (14.5)$$

where M_i and D_i are the mass and the apparent density of a component, respectively. When substituting an equal volume of one material for another, the portions by weight will be different unless the densities of the two materials are the same. Batch proportions and substitutions are commonly compared more easily using a volumetric basis. Table 14.4 presents a batch composition calculated on both a mass basis and a volume basis for a slurry for spray drying into a pressing powder. The weight proportions of the components may be calculated from the volume proportions using the equation

$$W_i(\text{wt}\%) = (f_i^v \times D_i)\,100/\Sigma(f_i^v \times D_i) \qquad (14.6)$$

Note that the weight and volume batch proportions differ considerably when the densities of the components are significantly different.

TABLE 14.4   Casting Slurry Batch in Both Weight and Volume Percent

Component	Volume (%)	Weight (%)
Zirconia powder	41.52	80.58
Water	52.33	16.92
Polyacrylate deflocculant	4.42	1.79
Polyvinyl alcohol binder	1.73	0.71
	100	100

SUMMARY

Ceramic processing systems commonly contain powders and/or granular inorganic materials mixed with colloids and various processing additives. The proportions and distribution of the particles, liquid, and additives and the interparticle forces determined the consistency state of the material. The processing and forming of the particulate systems depends importantly on the consistency state. Common consistency states are bulky powder, powder granules, plastic body, paste, or slurry. The processing batch composition may be expressed in weight or volume proportions using either a total batch or a dry powder basis.

SUGGESTED READING

1. Soil Mechanics, in *Encyclopedia of Science and Technology*, Vol. 12, 5th ed., McGraw Hill, New York, 1982.

PROBLEMS

14.1 A body is to be prepared with PF = 0.56 and DPS = 1. What is the required volume fraction of water?

14.2 The particles in Problem 14.1 are alumina with D_a = 3.98 Mg/m^3. What is the composition in percent by weight?

14.3 Explain why the PF for the granules in Table 14.1 may be only moderate and not always high.

14.4 A compact is pressed in a thick-walled die of diameter 2.00 cm. Its diameter is 2.05 cm when ejected. Calculate the percent diametral springback.

14.5 The bulk solids fraction of feed for an injection molding operation is 0.44. The pelletized feed granules are completely dense. What is the apparent compressibility when the warm feed is compressed to 40 MPa and the pore fraction is 0 for the bulk material?

14.6 Show by drawing several particles how the addition of a coagulant or flocculant may cause a relatively concentrated slurry to become a paste.

14.7 Show granules with a high and with a moderate PF by drawing several particles in contact.

14.8 Would you expect the compressibility and springback for a system of platelike of fiber particles to be different than that for a system of isometric particles? Explain.

14.9 Calculate the proportions by volume for the batch in Table 14.3.

14.10 A batch for extrusion contains 73.5 wt% inorganic particles (D_a = 2.85 Mg/m^3), 22.0 wt% water, and 4.5 wt% organic additives (D_a = 1.36

Mg/m^3). Calculate the volume proportions of each component. Is the PF of the particles in this batch high or moderate?

14.11 Calculate a batch by weight when an equal volume of silicon nitride powder is substituted for zirconia powder, for the casting slurry composition in Table 14.4.

14.12 A batch for a ceramic composite contains 85 vol% alumina and 15 vol% SiC fibers. Calculate the proportion by weight of each component. How many grams of SiC fibers are required for a 2000-g batch?

14.13 An injection molding material contains 60 vol% AlN powder and 40.0 vol% organic plasticizing material. The organic system consists of 35 wt% polypropylene binder (D_a = 0.90 Mg/m^3), 55 wt% paraffin wax (D_a = 0.91 Mg/m^3), and 10 wt% stearic acid (D_a = 0.85 Mg/m^3). Calculate the total composition in weight percent.

EXAMPLES

Example 14.1 An extrusion body with PF = 0.60 has DPS = 1 when compressed at 10 MPa. On decompression, an isotropic linear springback of 1.0% is measured. What is the DPS with no external load imposed (unconfined)?

Solution. The isotropic volume expansion is

$$\Delta V/V_0 = (1 + \Delta L/L_0)^3 - 1 = (1 + 0.01)^3 - 1 = 0.03$$

$$PF_{loaded} = V_{particles}/V_{bulk} = 0.60 \text{ cm}^3/1 \text{ cm}^3$$

$$PF_{unloaded} = 0.60 \text{ cm}^3/1.03 \text{ cm}^3 = 0.58$$

$$\phi_{loaded} = 1 - PF = 1 - 0.60 = 0.40 \qquad \phi_{unloaded} = 1 - 0.58 = 0.42$$

$$DPS_{unloaded} = 0.40/0.42 = 0.95$$

Example 14.2 Contrast the apparent reversible compressibility of the preconsolidated material in Example 14.1 with that of the same material in granular form with D_{bulk} = 1.12 g/cm^3. The granule density is 2.79 g/cm^3.

Solution

$$X = (\Delta V/V_0)/\Delta P = (1 - V/V_0)/\Delta P = (1 - D_{bulk}/D_{compressed})\Delta P$$

$$D_{compressed} = \text{mass/volume} = 2.79/(1 - 0.03) \text{ g/cm}^3 = 2.88 \text{ g/cm}^3$$

$$X_{granulated} = (1 - 1.12/2.88)/10 \text{ MPa} = 0.061 \text{ (MPa)}^{-1} = 6.1\%/\text{MPa}$$

$$X_{preconsolidated} = 0.03/10 \text{ MPa} = 0.003 \text{ (MPa)}^{-1} = 0.3\%/\text{MPa}$$

Note that the reversible compressibility of the preconsolidated material is only 5/100 of the compressibility of the granulated feed.

Example 14.3 The following batch for a slurry is to be spray dried into a granulated pressing powder. What is the composition proportioned by volume?

Material	weight (%)	D_a (Mg/m^3)
Alumina	73.22	3.98
Water	23.23	1.00
Polyacrylate deflocculant	2.45	1.25
Polyvinyl alcohol binder	0.98	1.27
Ethylene glycol plasticizer	0.12	1.13
	100%	

Solution. The volume proportion of each component is calculated from mass/density of each, as indicated by Eq. 14.5. For alumina,

$$f^v = \frac{(73.32/3.98)}{(73.32/3.98 + 23.23/1.00 + 2.45/1.25 + 0.98/1.27 + 0.12/1.13)}$$

$$f^v_{alumina} = 18.40 \ m^3/44.47 \ m^3 = 0.4138 \quad 41.38 \ vol\%$$

The volume fractions of the other additives are calculated in a similar manner:

$$f^v_{water} = 23.23 \ m^3/44.47 \ m^3 = 0.5224 \qquad 52.24 \ vol\%$$

$$f^v_{defloc} = (2.45/1.25) \ m^3/44.47 \ m^3 = 0.0441 \qquad 4.41 \ vol\%$$

$$f^v_{binder} = (0.98/1.27) \ m^3/44.47 \ m^3 = 0.0173 \qquad 1.73 \ vol\%$$

$$f^v_{plasticizer} = (0.12/1.13) \ m^3/44.47 \ m^3 = 0.0024 \qquad 0.24 \ vol\%$$

$$\overline{100 \ vol\%}$$

Example 14.4 What is the composition by weight for a slurry of B_4C powder which has the same proportions by volume as the slurry in Table 14.4?

Solution. The weight proportion of each component is obtained by multiplying each volume by its apparent density. The mass of B_4C is (41.52 m^3)(2.50 Mg/m^3) = 103.80 Mg. The (component weight/total batch weight) 100 = component (wt%). For B_4C (103.80 Mg/163.86 Mg) 100 = 63.35 wt%. The proportions of other components may be calculated in a similar manner:

Component	vol%	D_a (Mg/m^3)	Weight (Mg)	wt%
B_4C powder	41.52	2.50	103.80	63.35
Water	52.33	1.00	52.33	31.94
Polyacrylate	4.42	1.25	5.53	3.37
Polyvinyl alcohol	1.73	1.27	2.20	1.34
			163.86	100

The proportion of B_4C powder by weight is significantly lower than for alumina in a comparable slurry because the density of the B_4C particles is much lower than the density of alumina.

Example 14.5   How are the weight proportions on a total batch basis converted to an inorganic basis, as is indicated in Table 14.3?

Solution. In the total batch, the weight portion of alumina powder is 96.49 wt%. Using an inorganic basis, the proportion of alumina powder is 100 wt%.

The proportion of each component on an inorganic basis is calculated by multiplying the total batch proportion by the ratio 100/96.49. For the binder, the batch proportion is 1.29 wt%. But relative to the alumina powder, the proportion is $1.29(100/96.49) = 1.34$ wt%.

PART VI

PARTICLE MECHANICS AND RHEOLOGY

The mechanical behavior of a processing system strongly influences the manner in which the system can be processed and the selection of the equipment used for processing. Rheology is the science of deformation and flow and for particulate systems is a very complex topic. Chapter 15 presents concepts used to describe particle interactions in systems containing a minor amount of liquid and in cohesive materials nearly saturated with liquid. Valuable insights from the field of soil mechanics are introduced. Experimental techniques for determining mechanical behavior and properties are described. Chapter 16 presents continuum models which describe the flow of slurries, pastes, and saturated, compressed plastic bodies and serve to define rheological properties. Experimental techniques for determining rheological properties of different systems and the dependence of these properties on characteristics of the particles and additives are described. More particular descriptions of the mechanical behavior of systems for forming are presented in Part VIII.

CHAPTER 15

MECHANICS OF UNSATURATED BODIES

The capillary effect of a wetting liquid in the region of particle contacts and the bridging action of adsorbed binder produce cohesive strength in a consolidated powder mass. Ceramic bodies containing a pore phase incompletely saturated with a liquid may compress when deformed. For these materials the resistance to deformation is dependent on the cohesive strength and the frictional resistance between particles.

In this chapter we consider the general principles of friction and lubrication and the mechanical properties of consolidated powders and plastic bodies. The very complex topic of the deformation and flow behavior of unsaturated particulate systems is then examined using principles from the field of soil mechanics to provide theoretical insights.

15.1 CONCEPT OF EFFECTIVE STRESS

When an unsaturated particle system is compressed, the imposed pressure is supported by the mechanical stress at particle contacts and by the pore fluid pressure. The difference between the applied pressure P_a and the fluid pressure u is called the effective stress $\bar{\sigma}$:

$$\bar{\sigma} = P_a - u \tag{15.1}$$

The effective stress is sometimes called the interparticle stress, but should not be confused with the local stress σ_c at contacts between particles. In general, the contact stress is a function of the effective stress, the pore pressure, and the net pressure due to repulsive and attractive electrical interactions $(R - A)$,

$$\sigma_c = [\bar{\sigma} - (R - A)]A_T/A_c + u \qquad (15.2)$$

where A_c is the average area of interparticle contacts in the cross section of area A_T. When the electrical interactions between particles and the fluid pressure are insignificant, the contact stress is expressed by the simplified equation

$$\sigma_c = P_a A_T/A_c \qquad (15.3)$$

In an unsaturated system, the elastic strain, permanent strain, and yield strength of the body are dependent on the effective stress. The effective stress may vary with time or temperature when the pore pressure varies with time or temperature.

15.2 PARTICLE INTERACTIONS

Consider the uniaxial compression of a system of particles confined in a rigid die, as shown in Fig. 15.1. The force at contacts between particles may be resolved into normal N and tangential T components. Particles under load are nonuniformly compressed, and particles that are fibers or platelets may bend. The translation and rotation of particles into a different configuration may decrease or sometimes increase the bulk volume and distort the shape. A very intense stress at a particle contact may cause local plastic deformation and fracture.

Adsorbed surfactants and pore fluids alter the compaction and flow of particle systems. Particle sliding is enhanced by an interparticle phase which reduces the coefficient of friction. The effective stress is lower when the pore phase is

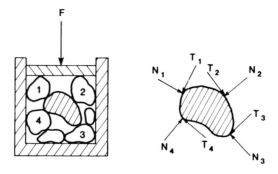

Fig. 15.1 Loading produces normal and shear forces at contacts between particles.

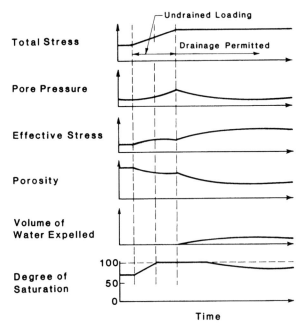

Fig. 15.2 Variation of stresses and porosity on loading a compressible, partially saturated particle system. (After T. W. Lambe and R. V. Whitman, *Soil Mechanics*, Wiley-Interscience, New York, 1969.)

saturated with a relatively incompressible fluid that becomes pressurized, as is indicated in Fig. 15.2. The flow of viscous surfactants and pore fluids is time-dependent. Resistance to flow may depend on the loading rate when the effective stress $\bar{\sigma}$ varies with time during the period of deformation.

The mean interparticle contact stress σ_c varies with the shape and orientation of the particles. For oriented platelets, the contact area ratio A_c/A_T is about 1, and σ_c is approximately equal to $\bar{\sigma}$. But as shown in Fig. 15.3, A_c/A_T may be much less than 1 in a packing of spherical or angular particles, and σ_c may be much greater than $\bar{\sigma}$.

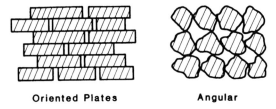

Oriented Plates   **Angular**

Fig. 15.3 The contact area between oriented platey particles is greater than between angular particles.

15.3 INTERPARTICLE FRICTION

Interparticle sliding is resisted by chemical adhesion and physical resistance produced by microscopic steps and asperites on particle surfaces (see Fig. 3.2). The classical law of Amonton states that the resistive force T_R is proportional to the normal force N and independent of the contact area:

$$T_R = fN \tag{15.4}$$

where f is the coefficient of friction. Alternatively, the concept of an angle of friction ϕ_f is used, where

$$\tan \phi_f = T_R/N = f \tag{15.5}$$

Sliding may occur when $T > T_R$ (see Fig. 15.4). For a particle in contact with several particles, sliding may not occur until $T > T_R$ at each contact or until translation or rolling reduces the shear force at a critical contact. Bulk powder flow involves the collective motion of several particles.

The surface roughness of particles can be estimated from scanning electron micrographs of particles or for coarse particles by using a profilometer. In the profilometer technique, a stylus connected to a electromechanical transducer is dragged along the surface, and the center-line average CLA roughness is calculated from the mean amplitude of the fluctuations of the amplified electrical signal. The surfaces of most particles are rough on a microscopic scale, and contact occurs at high points called asperites (Fig. 15.5). The contact stress on these asperites is much higher than the effective stress, even for small loads. Asperites of some rather brittle materials and ductile materials such as metals, alkali halides, and polymers may deform under load. The apparent contact area A_c is

$$A_c = N/\sigma_Y \tag{15.6}$$

where σ_Y is the yield strength of the deformed material for the imposed state of stress. The high contact stress may cause chemical bonding between two very clean surfaces, but less adhesion between unclean surfaces. If τ_c is the

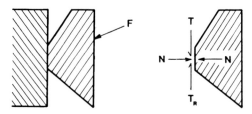

Fig. 15.4 Shear force T and friction force T_R at an interface.

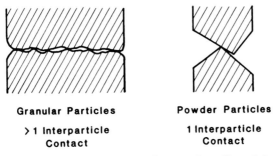

Fig. 15.5   Contacts between large and small particles.

shear strength of the contact, then

$$T_R = \tau_c A_c = \tau_c N / \sigma_Y \tag{15.7}$$

and

$$f = \tau_c / \sigma_Y \tag{15.8}$$

Generally friction is greater between clean surfaces of the same material. For a contact between a hard and soft material, the shear strength and friction are dominated by the mechanical properties of the softer material.

The static shear force at the onset of sliding is somewhat greater than the shear force required to maintain sliding. This difference may be due to a time dependence of the bonding in contact regions and the viscous flow of surface contaminants from the contact region. Rolling friction is much lower than kinetic sliding friction. During rolling, the bonding at contacts is broken in tension rather than shear. Rolling friction is generally quite independent of surface contamination. Sliding friction may sometimes be reduced by intense mechanical vibration.

Layer silicate minerals and pristine amorphous solids have relatively smooth surfaces with surface irregularities finer than about 100 nm. Crushed minerals have surfaces with steplike features coarser than a few microns, and aggregates have very rough surfaces with irregularities approaching the crystallite size. Equation 15.8 predicts that friction varies directly with the shear strength of the contact. At a contact between brittle materials, the apparent friction angle is dependent on the surface roughness and the interspersion of asperites.

An adsorbed material can reduce the coefficient of friction quite significantly if it is tenaciously bonded and resists extrusion under the high contact stress. The friction coefficient for highly polished surfaces of quartz (CLA = 70 nm) was about 0.9 (friction angle = 40°) for chemically cleaned dry or wet surfaces, about 0.5 for nominally cleaned surfaces, and only about 0.2 (11°) for dry surfaces having a residual film of oil. But for surfaces with a CLA roughness of about 1 μm, the friction coefficient was about 0.5 (27°) for both cleaned

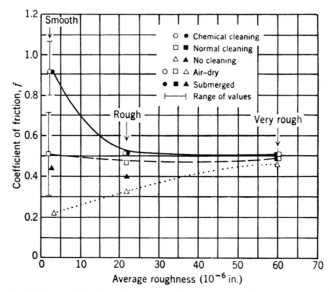

Fig. 15.6 Coefficient of friction of quartz surface varies with roughness and lubrication produced by light oil (no cleaning). (After T. W. Lambe and R. V. Whitman, *Soil Mechanics*, Wiley-Interscience, New York, 1969.)

and oil-contaminated surfaces (see Fig. 15.6). For rough surfaces, the effect of the oil film was nullified by the high contact stress.

The contact stress between oriented smooth sheet minerals is much lower than between angular particles. Surface-adsorbed contaminants and water are not completely extruded from the interface at an applied pressure below 500 MPa. The friction coefficient is observed to be about 0.3–0.4 for dry mica, but only 0.15–0.2 for wet mica, because the water acts as a lubricant.

Results for clay systems suggest that the frictional behavior of oriented, smooth kaolinite platelets and clays containing a significant content of montmorillonite and illite is similar to that for oriented mica. The behavior of clays with rough surfaces (cleavage steps) or edge-face contact is probably more similar to that for angular particles.

15.4 LUBRICATION

Materials having the same chemical composition tend to bond more strongly when in contact. A lubricant reduces the bonding at the interface. Lubrication is produced by an adsorbed boundary layer of low friction or an interfacial fluid (Fig. 15.7). In fluid lubrication (sometimes called hydrodynamic lubrication), the interfacial lubricating film is thicker than irregularities on the surface of the particle and the friction is dependent on the viscosity of the fluid medium. Fluid lubrication is produced by liquids such as water, oil, aqueous

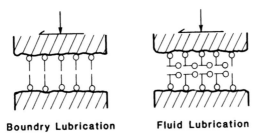

Boundry Lubrication   **Fluid Lubrication**

Fig. 15.7 Molecular configuration for a boundary lubricant and for fluid lubrication.

emulsions, a colloidal suspension, or a binder solution. Parameters affecting fluid lubrication are indicated in Fig. 15.8. When displacement is slow and the load is high, fluid escapes, and a transition to boundary lubrication occurs. The coefficient of friction rises, because a low-viscosity fluid is not an effective boundary lubricant.

Oriented adsorbed surfactant compounds of the type described in Chapter 12 are efficient boundary lubricants. An adsorbed lubricant reduces the contact between hard surfaces, which reduces friction and wear. Boundary lubricants may be tenaciously adsorbed, but of low shear resistance in the temperature range of interest. Evidence indicates that near the melting point of the lubricant, the long molecules may become disoriented, causing the friction to increase.

Finely divided solids with a laminar molecular structure can also act as lubricants between abrasive particles. Their platelike structure provides oriented smooth surfaces and a low friction coefficient. They can be admixed dry or in suspension form. Examples are platey talc, graphite, and graphitic boron nitride. Adsorbed water vapor or other gases may alter their lubricity. Solid

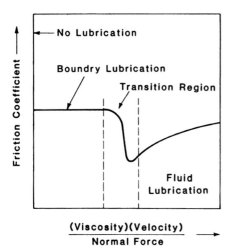

Fig. 15.8 Variation of apparent friction coefficient during kinetic loading.

lubricants resist breakdown under a high compressive stress and are good lubricants at high pressures and temperatures.

15.5 AVERAGE STRESSES IN A POWDER MASS

Ceramic systems are usually considered to be macroscopically homogeneous and isotropic for purposes of describing their deformation and flow. If the stress varies little over the dimension of the largest particle in the system, the average stress at some position is visualized as the sum of the forces between particles per unit area of the system at that position.

The analysis of plane stress and strain is discussed in undergraduate tests on the mechanics of materials. When the magnitudes and directions of normal stresses σ_x and σ_x on an element are known, as indicated in Fig. 15.9, the normal σ_θ and shear τ_θ stresses on a plane whose normal is at an angle θ are calculated from the equations

$$\sigma_\theta = \frac{\sigma_x + \sigma_y}{2} + \frac{\sigma_x - \sigma_y}{2} \cos 2\theta + \tau_{xy} \sin 2\theta \tag{15.9}$$

and

$$\tau_\theta = \frac{-(\sigma_x - \sigma_y)}{2} \sin 2\theta + \tau_{xy} \cos 2\theta \tag{15.10}$$

The average stress at a point is a function of the orientation of the plane used to define the stress. At any point, there are three orthogonal planes called the principal planes of stress on which the maximum and minimum normal stresses occur and for which the shearing stress is zero, as shown in Fig. 15.9. The largest normal stress is termed the major principal stress σ_1 and the smallest the minor principal stress σ_3. The quantity $(\sigma_1 - \sigma_3)$ is called the deviator stress.

Squaring and combining Eqs. 15.9 and 15.10 yields

$$(\sigma_\theta - a)^2 + \tau_\theta^2 = b^2 \tag{15.11}$$

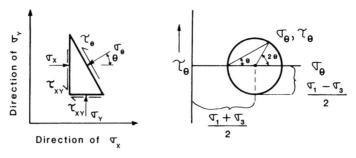

Fig. 15.9 Stress element and a general state of stress used to derive equations for the Mohr circle of stress.

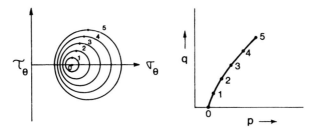

Fig. 15.10 Diagrams showing stress path produced during loading.

where

$$a = \frac{(\sigma_1 + \sigma_3)}{2} \quad \text{and} \quad b = \frac{\sigma_1 - \sigma_3}{2}$$

Equation 15.11 is the familiar equation of a circle of radius b centered at $(+a, 0)$ as shown in Fig. 15.9, and is known as the Mohr circle of stress. A plane with a normal oriented at an angle θ is indicated by a point on the Mohr circle.

The maximum shear stress τ_{max} is equal to $(\sigma_1 - \sigma_3)/2$ and occurs on planes with $\theta = \pm 45°$ from the direction of the planes of principal normal stress. A normal stress equal to $(\sigma_1 + \sigma_3)/2$ acts on each plane of maximum shear stress.

When a material is loaded, the successive states of stress can be represented by a series of Mohr circles, as indicated in Fig. 15.10. A simpler alternative is to plot the maximum shear stress versus mean principal stress for each loading at the stress points $q = (\sigma_1 - \sigma_3)/2$ and $p = (\sigma_1 + \sigma_3)/2$, respectively. The curve connecting the points on the $q - p$ diagram indicates the stress path produced during loading.

15.6 SHEAR RESISTANCE OF GRANULAR MATERIALS AND POWDERS

The shear strength of unsaturated particulate systems may be measured using the triaxial compaction and direct shear tests shown in Fig. 15.11. In the triaxial compaction test, an elongated specimen enclosed in a flexible, impermeable membrane is first subjected to a confining fluid pressure; axial pressure is applied mechanically at constant fluid pressure to produce shear stresses and deformation. A split mold box is used in the direct shear test. Vertical pressure is applied using a piston. Mechanical displacement of the mobile portion of the shear box produces a plane of shear in the compressed material.

The shear resistance of granular materials and nearly dry powders is very dependent on the packing density and shape and surface characteristics of the

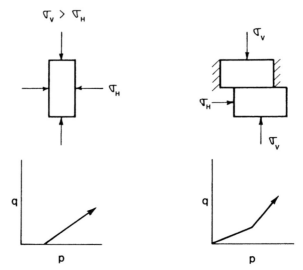

Fig. 15.11 Loading and stress path diagrams for triaxial and direct shear loading tests.

particles. Pore pressures are generally insignificant except when the material is very compressible and the compaction rate is extremely fast. Differential shear between planes of particles requires either volume dilation or fracture of particles when the particles are close-packed, because of the effective inter-locking of nested particles. In loosely packed arrays, rolling and sliding may enable shear flow without an increase in volume. Studies of the flow of sand indicate that sands that pack more densely exhibit a greater volume dilation and a greater shear resistance during flow (Fig. 15.12). Superimposed vibra-tions may momentarily reduce the contact stress between some particles and aid in particle sliding.

The strength of a granular material for a particular situation of loading may be defined at small strains by the maximum shear stress, as indicated by point P in Fig. 15.12, and after considerable straining by the ultimate shear stress

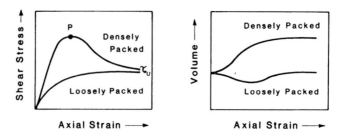

Fig. 15.12 General dependence of shear stress and specimen volume on axial strain during triaxial compaction.

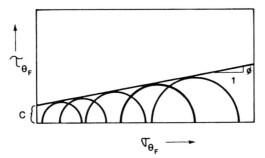

Fig. 15.13 Sequence of Mohr circles for the triaxial compaction of a cohesive body and the Mohr–Coulomb failure line.

τ_u. The principal stresses at failure σ_{1f} and σ_{3f} may be used to construct the Mohr circle. The principal stresses are similarly determined for other states of confining stress, and a series of Mohr circles are plotted on a single graph as shown in Fig. 15.13. The line tangent to the Mohr circles is called the Mohr failure envelope. The significance of the failure envelop is that only states of stress lying outside the envelope will cause flow. The Mohr envelope is often linear to a good approximation and is commonly expressed as

$$\tau_f = c + \sigma_f \tan \phi_f \tag{15.12}$$

where c is the cohesion intercept, ϕ_f is the angle of internal friction which indicates the shear resistance of the material, and the subscript f indicates flow conditions. Equation 15.12 is commonly called the Mohr–Coulomb failure law.

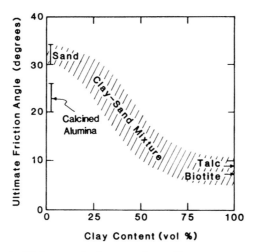

Fig. 15.14 Ultimate friction angle for wet clay $<2\ \mu m$, wet sand, wet sand–clay mixtures, alumina powder, and friction angle of talc and mica surfaces. (After T. W. Lambe and R. V. Whitman, *Soil Mechanics*, Wiley-Interscience, New York, 1969.)

Note that for materials described by Eq. 15.12, the shear strength increases as the confining stress increases when $\tan \phi_f > 0$.

The parameters $\tan \phi_f$ and c may be considered to be properties of a particular material in a particular state. The friction angle for rounded, loosely packed sand is about 30° (Fig. 15.13). When the interlocking of particles increases owing to an increase in particle size or angularity, at constant porosity, or for a decrease in porosity for the same particle system, the friction angle increases to 40–45°. Interparticle lubricants can lower the friction angle. The shear resistance of fine powders depends on the compressibility of the powder and the drainage of pore fluid as well as particle and packing parameters.

15.7 STRESS–STRAIN BEHAVIOR DURING COMPRESSION

The compression of an unsaturated plastic ceramic body or powder may cause permanent consolidation from the sliding and rearrangement of particles; the elastic compression of particles, binders, lubricants, and liquid; the compression or dissolving of gas in the liquid; particle orientation; and the migration of liquid and gas from interstices. The compressibility may be determined under isostatic or uniaxial confined compression conditions, as we see in Fig. 15.15. In the isostatic compression test, the specimen is enclosed in a thin, impermeable membrane and compressed by a fluid. True isostatic compression causes a decrease in volume but, as indicated on the $q - p$ diagram, no shear ($\sigma_1 - \sigma_2 = 0$) and no distortion. In uniaxial compression, the specimen is compressed while confined in a smooth, rigid die; compression is the dominant source of strain, but die wall friction causes shear stresses to occur as indicated on the $q - p$ diagram. The ratio of horizontal stress σ_h to vertical stress σ_v, $K_{h/v}$,

$$K_{h/v} = \sigma_h/\sigma_v \tag{15.13}$$

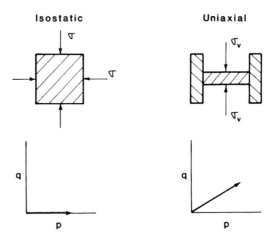

Fig. 15.15 Loading and stress path diagrams for isostatic and confined compression.

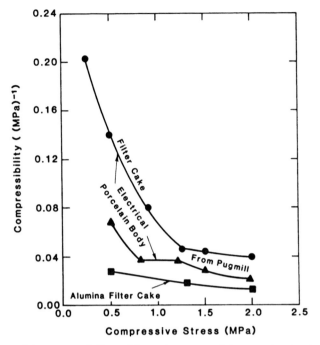

Fig. 15.16 Axial compressibility of flocculated porcelain and alumina bodies during confined compression.

may be determined as a function of pressure by using a pressure transducer in the die wall. Permeable pistons may be substituted to determine the effect of drainage of the pore fluid on the mechanical behavior.

The rolling, sliding, and rearrangement of particles during isostatic or uniaxial compression of loosely packed powders may cause large volumetric compaction. Although the average shear stress is zero during isostatic compression, shear forces at individual particle contacts may be quite large.

Dense packings of sand exhibit less than 1% volumetric strain on uniaxial compression to 3 MPa.* Initial strain occurs owing to the collapse of more loosely packed arrays and the elastic compression of particles. Above about 15 MPa, particle fracturing at contacts may enable sliding into a denser configuration. The ratio $K_{h/v}$ is about 0.4–0.5. Elastic springback on decompressing from about 50 MPa is less than 2%. Similar results are observed for packings of other hard materials. The stress causing fracture will tend to increase as the particle strength increases.

The volumetric strain on uniaxially compressing filter cake of a flocculated electrical porcelain body (50% clay, 50% quartz and feldspar) between permeable pistons is shown in Fig. 15.16. Both permanent compaction and the re-

*T. William Lambe and Robert V. Whitman, *Soil Mechanics*, Wiley-Interscience, New York, 1969.

versible axial strain (springback) up to about 10% are apparent. The primary compression of filter cake up to about 1 MPa is typical of the collapse of the coagulated clay coating hard particles, which reduces the spaces between particles. Above 1 MPa, the behavior is typical of clay with a more oriented structure. Elastic springback is very significant below 1 MPa, but is only a fraction of the total volume change due to consolidation. Shear flow of the body in the vacuum pug mill mixer increases the orientation and packing density of clay in microscopic regions, which reduces the average compressibility and springback. The reversible compressibility of filter cake from a deflocculated suspension is lower and similar to the behavior of the pugged material. The clay matrix is also partially oriented in material cast from a deflocculated suspension. When liquid is unable to drain from the pores during compression, the volumetric strain will be similar up to the pressure causing pore saturation. Bodies close to saturation will show little total compression.

On initial compression, behavior similar to that exhibited in Fig. 15.16 may be expected for highly porous material containing a significant amount of a molecular binder. Green ceramic tape containing 30 vol% binder may show an initial compressibility of 20%. However, the elastic springback is relatively low and less than about 2% for ordinary cast tape and organically plasticized extrusion bodies. A dense cast of micron-size alumina pressure cast from a deflocculated slurry exhibits relatively little compressibility (see Fig. 15.16). In general, the stress ratio $K_{h/v}$ is higher for a saturated but highly porous body, and $K_{h/v}$ approaches 1 when the friction angle ϕ_f approaches 0°. For confined compression and elastic behavior,

$$K_{h/v} = \nu/(1 - \nu) \qquad (15.14)$$

where ν is the effective Poisson's modulus of the porous compact.

15.8 STRENGTH OF AGGLOMERATES AND COMPACTS

Agglomerates may form spontaneously in a dry, nonmagnetic powder owing to either Van der Waals forces or electrostatic attraction. Electrostatic forces occur because of surface-adsorbed ions or in transfer of electrons between particles in regions of contact. For particles that are nonconductors, charge mobility is limited to the surface, and charges persist until contact with an electrically conducting material occurs. Electrostatic charges may also be diminished by an adsorbed film which enhances that charge mobility. Electrostatic and Van der Waals forces produce relatively fragile agglomerates. However, these fragile agglomerates can impede the flow and packing of dry powder and colloids.

Agglomeration is also caused by the capillary force of a wetting liquid at contacts between particles. We can estimate the effect of capillary forces on the agglomerate strength using a model of packed spheres of equal size containing a liquid distributed as shown in Fig. 15.17. The mean number of

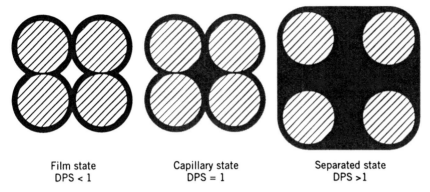

Film state   Capillary state   Separated state
DPS < 1   DPS = 1   DPS > 1

Fig. 15.17 Liquid distribution in film, capillary, and separated particle states.

contacts per unit area N_A is $(aN_c)/3$ where N_c is defined by Eq. 13.2. For a random packing, $CN = \pi/(1 - PF)$, and

$$N_A = PF/[(1 - PF)a^2] \qquad (15.15)$$

The tensile strength (S_t) is the product of the number of contacts (bonds) per unit area and the capillary force per contact b:

$$S_t = [PF/(1 - PF)a^2]b \qquad (15.16)$$

For randomly packed uniform spheres separated by an adsorbed liquid, Rumpf* has derived the nearly identical equation by other means ($\phi = 1 - PF$):

$$S_t = 9/8[(1 - \phi)/\phi a^2]b \qquad (15.17)$$

The capillary force b is equal to the force produced by surface tension and the capillary pressure (Eq. 2.6) multiplied by the area of the liquid in the neck. The force b increases as the liquid content is reduced. When the neck radius is very small compared to the radius of the circular liquid in the plane of contact, and the wetting angle is $0°$, the force $b = \pi\gamma_{LV}a$. Equation 15.17 indicates that the tensile strength from capillary liquid bonding varies directly with the surface tension and packing density and inversely with the particle size, as is observed experimentally.

The formation of a hard, brittle mass from a soft mass on drying from the capillary state to the film state is also commonly observed. Capillary liquid bonding can produce a tensile strength ranging up to about 350 kPa for water in an oxide powder of about 1 μm particle size. When pores are saturated, internal capillary liquid bonding disappears. Agglomerates in the capillary state are usually of lower strength and more plastic.

*H. Rumpf, in *Agglomeration*, W. A. Knepper, Ed., Wiley-Interscience, New York, 1962.

Stronger agglomerates and cohesive bodies are produced by flocculation bonding, as was discussed in Chapter 11. Onoda* calculated the bond strength b between two uniform spheres assuming that the binder was either uniformly concentrated at particle contacts (pendular model) or uniformly coating the particles (film model). Combining the expression for b with the Rumpf equation, the Onoda equation for the pendular model is

$$S_t = \frac{3}{8} \left(\frac{3\pi}{\phi} \right)^{1/2} (1 - \phi) S_0 \left(\frac{V_b}{V_p} \right)^{1/2} \tag{15.18}$$

and for uniformly coated particles,

$$S_t = \frac{3\pi}{16} \left(\frac{1 - \phi}{\phi} \right) S_0 \frac{V_b}{V_p} \tag{15.19}$$

The volumes of binder and particles are V_b and V_p, respectively, and S_0 is the strength of the binder in the plane of fracture. The ratio V_b/V_p is an important parameter when considering the strength of ceramic systems and will be referred to many times when discussing forming techniques and defects in green ceramic parts. Note that in Eqs. 15.18 and 15.19 the particle size does not appear as a parameter. The pendular distribution of binder concentrates the binder in the contacts and S_t is potentially higher for the same V_b/V_p.

The tensile strength of dried powder compacts and green bodies exhibiting elastic behavior may be determined using standard procedures such as the simple tension, flexural, and diametral compression tests. The simple tension test may be used when the specimen exhibits some yield which permits alignment of the specimen into a true tensile test. This test may be used for ceramic tape and extruded and injection molded materials. The engineering tensile strength is calculated from the load at fracture P_f and the cross-sectional area A,

$$S_t = P_f/A \tag{15.20}$$

The compression of a cylinder on its diameter produces a tensile failure which begins near the center and propagates toward the loading points. Soft pads are used to reduce stress concentrations between the loading platen and the specimen. For simple and triple-cleft fracture modes (see Fig. 15.18) the tensile strength is

$$S_t = 2P_f/\pi \text{ Dia } H \tag{15.21}$$

where Dia and H are the diameter and thickness of the specimen respectively. The tensile strength of bar or rod specimens determined in flexure (modulus

*G. Y. Onoda Jr., *Am. Ceram. Soc. Bull.* **58**(5–6), 236–239 (1978).

Fig. 15.18 Diametral compression test showing loading and fracture mode. (Photo courtesy of William Walker, Alfred University.)

of rupture test) as shown in Fig. 15.19 may be calculated from the appropriate formula in Table 15.1. In flexure, the tensile stress created is a maximum on some portion of the surface of the specimen between the bottom supports. Friction at the supports and loading contacts should be minimal (no indentation) and the specimen should be of a regular geometry so that loading does not produce twisting. Flexural strength results depend on the span between supports, the specimen geometry, and the condition of the surface and surface defects. A specimen with a polished tensile surface produced during forming will have a higher flexural strength than when the surface is roughened. The pattern of off-center failures in three-point tests provides information about the location of critical surface defects. Strengths determined in four-point tests are more representative of the average surface material. Comparison of strength results from the different tests for unfired ceramics indicate good agreement when the strength is controlled by internal stress-concentrating defects. Flexural strengths may be 2–4 times higher when the surface is of higher quality.

The simple tensile and diametral compression tensile strengths are more

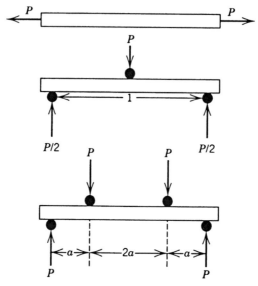

Fig. 15.19 Loading in simple tension (top) and modulus of rupture tests (middle and bottom).

sensitive to internal heterogeneity which limits strength. The flexural strength is much more sensitive to heterogeneity near or at the surface.

Lukasiewicz* observed that a low-viscosity binder solution which wetted the particles initially formed a thin coating on the particles and then concentrated in the contact regions. The dependence of the diametral tensile strength on V_b/V_p was closely approximated by Eq. 15.18 (coated particles) when $V_b/V_p < 0.02$ and by Eq. 15.19 (pendular model) when $V_b/V_p > 0.05$. The strength is influenced by the intrinsic strength S_0 of the binder and additives which decrease S_0, such as a plasticizer (Fig. 15.20). An inhomogeneous distribution of the binder may reduce the apparent strength and alter the dependence on V_b/V_p. Achieving homogeneity is more difficult when the viscosity

*S. Lukasiewicz, PhD thesis, Alfred University, Alfred, NY, 1983.

TABLE 15.1 Modulus of Rupture Equations*

Specimen Geometry	Three-Point Loading	Four-Point Loading
Rectangular bar	$3P_f L/2bh^2$	$6P_f a/bd^2$
Breadth (b), depth (h)		
Maximum stress on	Line	Surface
Circular rod	$8P_f L/\pi d^3$	$32P_f a/\pi d^3$
Diameter (d)		
Maximum stress on	Point	Line

*See Fig. 15.19

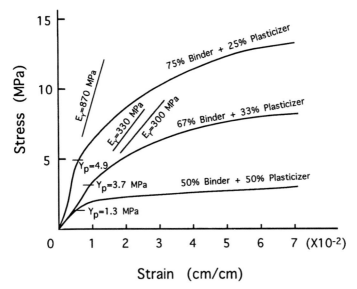

Fig. 15.20 Stress–strain behavior in simple tension for dried ceramic tape for electronic substrates showing effect of increasing plasticizer proportion on yield point Y_p, elastic modulus E, and ultimate strength [14 wt% (binder + plasticizer)/alumina]. Source: R. A. Gardner and R. W. Nufer, *Solid State Technol.*, **5** 38–43 (1974).

of the binder solution is high and when the particle size is smaller. A nonlinear increase of tensile strength with V_b/V_p is probably due to an inhomogeneous distribution of the binder solution of high viscosity. Compaction at a higher pressure may not reduce the porosity, but the higher contact stress may cause an increase in the adhesion strength S_0.

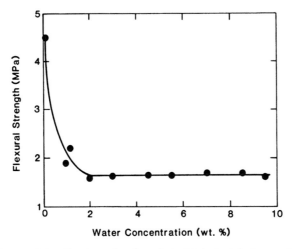

Fig. 15.21 Apparent tensile strength of an industrial electrical porcelain body as a function of liquid content.

The tensile strength of clay-bonded materials is very dependent on the liquid content (Fig. 15.21) because the strength is dependent on the coagulating forces between particles and the capillary effect. Green ceramics with a tensile strength > 10 MPa may be produced by introducing common binders into the material.

The apparent strength of individual agglomerates in shear or compression may be higher for smaller agglomerates which are often of lower porosity. The strength of a powder agglomerate is described by Eq. 15.17. However, the force to produce rupture of an agglomerate varies directly with the product of its strength and the square of its diameter. The apparent yield strength of agglomerates containing a binder varies directly with V_b/V_p and S_0 of the binder phase.

15.9 SHEAR RESISTANCE OF NEARLY SATURATED COHESIVE BODIES

On shearing a nearly saturated plastic clay body, the shear stress causing flow has the general form shown in Fig. 15.12. If the clay body has been consolidated at greater than 10 MPa, the maximum shear resistance is usually greater than the ultimate shear resistance. Highly consolidated bodies exhibit a greater initial resistance to the formation of shear zones containing an oriented clay structure, facilitating shear, although the ultimate shear stresses for sustained deformation may be comparable. Orientation of platey clay particles occurs at an angle of about $45° + \phi_f/2$ relative to the plane of the major principal stress. For nearly saturated sand–clay mixtures, the internal friction angle calculated from the ultimate stress conditions decreases as the content of submicron clay increases, as shown in Fig. 15.14, and a friction angle below 15° is achieved when the clay content exceeds about 40 vol%. The friction angle for clay may vary from 10° to as low as 4–5° for a clay with a relatively high content of montmorillonite. For an alumina body plasticized with a methylcellulose binder solution, a low friction angle of 5° is observed when the binder concentration is only 2 vol%. This indicates that much of the clay acts as a filler and only a portion is a body plasticizer (see Fig. 15.14). For these plastic bodies, the ratio $K_{h/v}$ is approximately $0.95 - \sin \phi_f$. The shear resistance and deformation are very dependent on the liquid pressure u in pores, which reduces the effective stress $\bar{\sigma}$ carried by the particle skeleton. Pore pressure effects are especially important in systems containing submicron particles and pores because the liquid permeability of these systems is very low and liquid migration occurs very slowly.

When molding ceramic materials, the term plasticity is commonly used. The term plasticity refers to a particular mode of mechanical behavior. A plastic ceramic material will exhibit permanent deformation without rupture when an applied compressive load produces a shear stress that exceeds the apparent shear strength, commonly called the yield strength, of the material. For a plastic ceramic body, measurable elastic behavior is observed before yield occurs, and when the applied load is removed. The mechanical behavior of a real plastic ceramic body is elastic-plastic in nature as illustrated in Fig. 15.22. The area under the stress–strain curve is the toughness of the material and a plastic body is a relatively tough material.

The compressibility of unconsolidated fine powders and plastic bodies containing a significant fraction of colloidal particles is significantly higher than the compressibility of liquid water. For a nearly saturated body, an increase in the applied stress first increases the effective stress and the packing density of the particles. As the degree of pore saturation increases, liquid may become pressurized in microscopic regions. Eventually, the pore structure becomes saturated if the system is sufficiently compressible (Table 15.2). Air in the pores can dissolve in the pore liquid at a liquid pressure of about 500 kPa. Once the DPS $= 1$, an increment in the applied stress will increase the pore pressure but not the effective stress. The ultimate shear strength, as is shown in Fig. 15.23, varies directly with the effective stress. When the pore pressure is not zero, the Mohr–Coulomb equation is

$$\tau_f = \bar{c} + \bar{\sigma}_f \tan \bar{\phi}_f \tag{15.22}$$

The material parameters $\bar{c}$ and $\bar{\phi}_f$ are effective parameters calculated using the effective stress $\bar{\sigma}_f$. If the pore pressure is not known, c and ϕ_f determined using the total stress are total stress parameters.*

General compositions of plastic bodies are presented in Table 15.3. In plastic clay bodies, the submicron coagulated colloidal clay coats the larger particles and forms a matrix, as shown in Fig. 15.24. Under load the body densifies and DPS $= 1$. During deformation, flow occurs between larger particles in shear zones containing oriented clay and finer nonplastic particles. Pressurized water separates the particles and acts as an internal lubricant. Liquid migration from a shear zone will cause the effective stress and the shear resistance to increase there. A concentrated stress may cause the nucleation of a crack. Plastic deformation requires a high concentration of shear zones which are of uniform shear resistance.

In bodies plasticized with a molecular binder, the macromolecules play a role similar to that of the clay. Binders of very high molecular weight contain molecules which range up to 0.5 μm in length. Hydrophilic groups attract water molecules. Particles have a film of adsorbed, hydrated binder molecules which are surrounded by a binder solution. Stiffer and higher-molecular-weight molecules may provide relatively more initial compressibility needed to reduce the effective stress. Binder molecules of high molecular weight resist migration and are effective in reducing the liquid migration. A critical volume of fine particles and plasticized molecular binder is required for plastic flow.

Properties determined for plastic extrusion bodies are presented in Table 15.4. For the clay body, increasing the water content reduced the cohesive strength, the internal friction, and the ultimate shear resistance. For the alumina

*T. W. Lambe and R. V. Whitman, *Soil Mechanics*, Wiley-Interscience, New York, 1969.

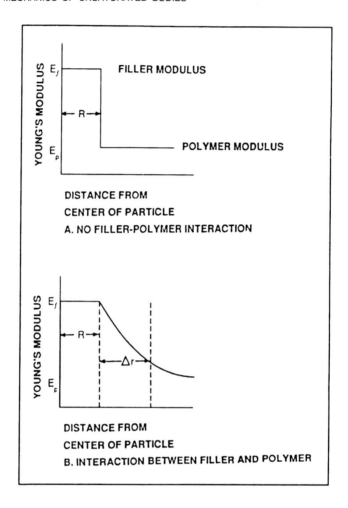

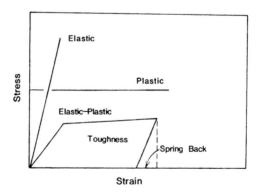

Fig. 15.22 (Top) Adsorbed polymer of lower elastic modulus on particle modifies near-surface mechanical properties. (Bottom) With deformable polymer coating, system containing hard particles becomes elastic-plastic and exhibits toughness.

TABLE 15.2   Parameter Values for a Plastic Body

Pressure Range	DPS	X	$\bar{\sigma}$	tan ϕ_f
Shear resistance climbs	<1	>0	Increasing	Moderate
Shear resistance constant	$=1$	$=0$	Constant (τ_u)	Very low

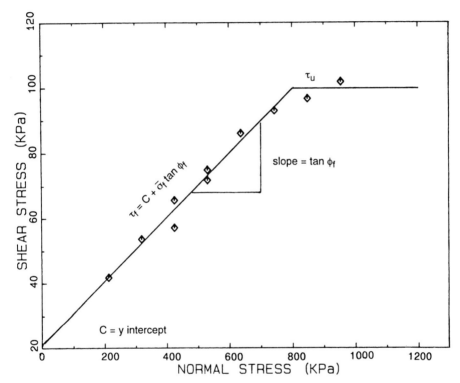

Fig. 15.23   Shear stress causing flow in alumina body plasticized with methylcellulose solution first increases with imposed pressure (Mohr–Coulomb behavior), but shear resistance remains constant when body becomes saturated (direct shear test results).

TABLE 15.3   General Composition of Plastic Bodies

Clay Plasticized	Component	Organic Plasticized
25–38 (vol%)	Filler	20–30 (vol%)
10–15[a]	Body plasticizer	1–12
3–8	Pores	3–8
Balance	Liquid	Balance

[a]Coarser clay is considered to be a filler.

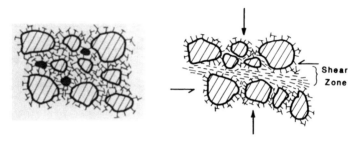

| Unstressed | Shear with Consolidation |

Fig. 15.24 Model of distribution of filler, finer clay particles of high aspect ratio/binder molecules, pores (dark regions) in plastic body, and formation of shear zone during plastic deformation.

body, increasing the content of methylcellulose binder above 2 vol% increased the ultimate shear resistance but did not further reduce the angle of internal friction.

The ultimate shear resistance τ_u of the plastic material may be considered the shear strength of the saturated plastic body. An equation proposed for the dependence of the shear strength on the volume fraction of liquid f_L^v for a clay plasticized body is*

$$\tau_u = A \exp - [\alpha(f_L^v - f_{cr}^v)] \qquad (15.23)$$

where f_{cr}^v is the critical concentration of liquid at which particle flow becomes blocked and A and α are constants that depend on the particle packing factor PF and the content and aspect ratio of colloidal particles. As is seen in Fig. 15.25, the dependence of the shear strength of industrial clay bodies on the

*R. H. Weltmann and H. Green, *J. Appl. Phys.*, **14**(11), 569–576 (1943).

TABLE 15.4 Coulomb Parameters for Industrial Extrusion Bodies

Binder (vol %)	Water (vol %)	c (kPa)	ϕ_f (deg)	τ_u (kPa)
Porcelain[a]				
31 (clay)	39	4	4	39
33 (clay)	34	39	12	122
Alumina[b]				
6.0	40		5	129
4.0	40		5	92
2.0	40		5	76

[a]Basis includes liquid; total clay content is indicated.
[b]Methy cellulose binder (R. Locker PhD thesis, Alfred University, Alfred NY, 1985).

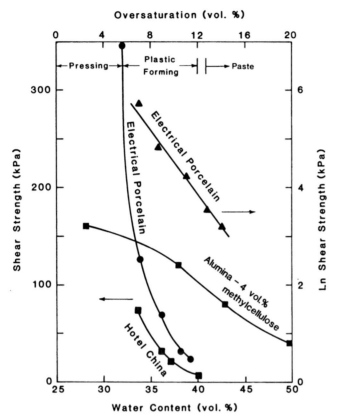

Fig. 15.25 Variation of ultimate shear strength (direct shear test) on liquid content for several plastic bodies. The range of plastic forming is indicated for the electrical porcelain body.

parameter $(f_L^v - f_{cr}^v)$ is approximated well by Eq. 15.23. However, the dependence of the shear strength of the alumina body plasticized with methylcellulose binder on the liquid content is not described well by this equation.

A variety of empirical tests for determining indices of plasticity have been devised and are discussed in the ceramics literature. Most involve imposing some simple or complex strain and measuring the resulting stress. These can be categorized as torque of mixing, indentation, tension, torsion, unconfined compression, and simple shear tests. The torque used to rotate mixing blades through a mixture is a quick, convenient way to note qualitative trends on varying the liquid content, chemical additives, content of colloidal particles, and so on. Penetrometers indicate the resistance to deformation of a relatively small volume of material. Simple torsion tests do not produce a uniform strain, and as normally practiced the specimen is not compressed. Tests in which the body is compressed during deformation, simulating industrial processing, should provide results that correlate with industrial workability. Studies of well-characterized systems using tests that provide a well-defined state of stress are needed to develop a science of plasticity.

SUMMARY

When a ceramic material is compressed, the pressure of fluid in the interstices between particles reduces the effective stress carried by the particles. The contact stress between particles is high relative to the effective stress when the contact region between particles is small relative to the particle size and the pore pressure is small. Interparticle sliding is restricted by friction at particle contacts and is very dependent on surface roughness and lubrication.

Compression of a particle system produces both compressive stress and shear stress. Compression may produce both elastic and permanent volume strain. Volume expansion called springback occurs on decompression. The compressive strength of an agglomerate and the tensile strength of a compact depend on the PF of the particles, the volume and strength of the binder, and the uniformity of the binder distributed among the particles. Cohesive particulate bodies are elastic-plastic in behavior. Their shear behavior is described by the Mohr–Coulomb equation. Mechanical parameters of the ceramic body are the cohesion, internal friction, and ultimate shear resistance. The shear resistance of a particle system is very dependent on the imposed compressive load when the effective stress varies with the load. Mechanical properties of the particle system depend on particle parameters, packing, interparticle forces, adsorbed additives, and DPS.

SUGGESTED READING

1. M. Takahashi and S. Suzuki, "Deformability of Spherical Granules under Uniaxial Loading," *Am. Ceram. Soc. Bull.*, **64**(9), 1257–1261 (1985).

2. Soil Mechanics, in *Encyclopedia of Science and Technology*, Vol. 12, 5th ed., McGraw Hill, New York, 1982.

3. G. Y. Onoda and M. A. Janney, "Application of Soil Mechanics Concepts to Ceramic Processing," in *Advances in Powder Technology*, G. Y. Chin, Ed., American Society for Metals, Metals Park, OH, 1982.

4. H. Rumpf and H. Schubert, "Adhesion Forces in Agglomeration Processes," in *Ceramic Processing Before Firing*, G. Y. Onoda and L. L. Hench, Eds., Wiley-Interscience, New York, 1978.

5. G. Y. Onoda, "Theoretical Strength of Dried Green Bodies With Organic Binders," *J. Am. Ceram. Soc.*, **58**(5–6), 236–239 (1978).

6. W. M. Price, "Green Strength of Ceramics: Techniques for Measuring the Tensile Strength of Unfired Clays and Ceramic Bodies," *Interceram.*, **3**, 197–200 (1974).

7. W. O. Williamson, "Strength of Dried Clay—A Review," *Am. Ceram. Soc. Bull.*, **50**(7), 620–625 (1971).

8. N. Claussen and J. Jahn, "Green Strength of Metal and Ceramic Compacts as Determined by the Indirect Tensile Test," *Powder Metal. Int.*, **2**(3), 87–90 (1970).

9. T. W. Lambe and R. V. Whitman, *Soil Mechanics*, Wiley-Interscience, New York, 1969.

10. A. H. Cottrell, *The Mechanical Properties of Matter*, Wiley-Interscience, New York, 1964.

PROBLEMS

15.1 A tank truck carrying a dry cement powder is loaded and unloaded pneumatically by pumping air into the lightly packed powder. Explain the ease of flow using Eq. 15.1.

15.2 A pressure of 70 MPa is applied to concrete and the pore pressure is measured to be 3 MPa. What is the effective stress carried by the particles.

15.3 Calculate the contact stress in Problem 15.2 for $A_c/A_T = 0.01$. What is the contact stress if the liquid drains from the system?

15.4 For an applied stress of 10 MPa, calculate the contact stress for the two cases in Fig. 15.3. Assume A_c/A_T is 0.9 for the oriented plates and 0.05 for the angular particles; $(R - A)$ and u may be neglected.

15.5 The CLA roughness of industrial minerals is of the order of 0.5 μm. Explain why a low molecular weight surfactant is an unsatisfactory lubricant.

15.6 Derive Eqs. 15.9, 15.10, and 15.11.

15.7 Illustrate an element of material with $\sigma_y = 300$ kPa, $\sigma_x = 100$ kPa, $\tau_{xy} = 0$. Calculate the principal normal stresses, the maximum shearing stress, and the angle of the plane of τ_{max}.

15.8 Construct a Mohr circle diagram for the results of Problem 15.7.

15.9 In confined compression, $\sigma_y = 10$ MPa and σ_x is determined to be 6 MPa. What is $K_{h/v}$?

15.10 Construct a stress path diagram for isostatic compression and for confined compression. The range of applied pressure for each is 125 MPa. For confined compression, $K_{h/v} = 0.4$ and $\tau_{xy} = 0.21$ σ_x for all pressures.

15.11 Estimate $K_{h/v}$ for an alumina compact with $\nu = 0.23$.

15.12 The strength of ceramic tape is determined in simple tension. The load at yield is 4 N and the specimen is 50 $\mu m \times 40$ mm in cross section. Calculate the yield strength.

15.13 A bar specimen is broken in three-point loading at a load of 11.7 N. The span $L = 5$ cm, $b = 1$ cm, and $h = 0.5$ cm. Calculate the tensile strength S_t.

15.14 A cylindrical specimen is fractured in diametral compression at a load of 6.6 N; Dia = 2 cm and thickness = 1 cm. Calculate the tensile strength.

15.15 Estimate the tensile strength of a compact when PF = 0.6, V_b/V_p = 0.08, and S_0 = 70 MPa.

15.16 The diameter d of the contact area between two elastic spheres is $d = [(12(1 - \nu^2)N/E)(R_1 R_2)/(R_1 + R_2)]^{1/3}$. N is the normal load, E is Young's modulus, and R_1 and R_2 are the radii of the two spheres. Derive an equation expressing the dependence of the ratio of contact stress/applied stress as a function of ν, E, and applied stress. Assume $R_1 = R_2$ and orthorhombic packing.

15.17 From your result in Problem 15.16, derive an equation for the elastic strain as a function of the applied stress. Then calculate the elastic springback for $R_1 = R_2 = 1$ μm, $\nu = 0.25$, $E = 350$ GPa, and an applied stress of 70 MPa.

15.18 The shear stress at failure τ_f is equal to one-half the normal stress σ_f at failure for a noncohesive powder. What is the angle of internal friction?

15.19 For a plastic ceramic body, the friction angle $\phi_f = 5°$. Estimate the stress ratio $K_{h/v}$.

15.20 Estimate the constants A, α, and f_{cr}^{ν} for the data in Fig. 15.25.

EXAMPLES

Example 15.1 For the case of $(R - A) = 0$, how will the effective stress $\bar{\sigma}$ vary with time t when the pore pressure $u = 0$ and when $u > 0$?

Solution. For those systems where $u = 0$ during fabrication, $\bar{\sigma} = P_a$ and $d\bar{\sigma}/dt = dP_a/dt$, that is, the change of effective stress varies directly with the rate at which the applied pressure is increased. For the case where $u > 0$, $d\bar{\sigma}/dt = d(P_a - u)/dt$ and the effective stress depends directly on the magnitude of u and its dissipation, du/dt. For saturated, compressible systems the effective stress is very dependent on u and gradients in the pore pressure du/dx produce gradients in the effective stress and the shear resistance which depends directly on $\bar{\sigma}$.

Example 15.2 What is the mean contact stress between particles when the contact area/cross-sectional area $A_C/A_T = 0.01$, $P_a = 10$ MPa, pore pressure $u = 3$ MPa, and $(R - A) = 0$.

Solution. The effective stress is calculated using Eq. 15.1,

$$\bar{\sigma} = P_a - u = 10 \text{ MPa} - 3 \text{ MPa} = 7 \text{ MPa}$$

The contact stress may then be calculated from Eq. 15.2,

$$\bar{\sigma}_C = [\bar{\sigma} - (R - A)]A_T/A_C + u = 7 \text{ MPa } (100) + 3 \text{ MPa } = 703 \text{ MPa}$$

The contact stress is considerably higher than the applied pressure. If the pore fluid were to drain and $u = 0$, then $\bar{\sigma} = 10$ MPa and $\sigma_C = 1000$ MPa.

Example 15.3   The addition of a boundary lubricant onto the surface of a die reduces the shear resistance τ_R for sliding from 4.2 MPa to 2.1 MPa. What is the change in the coefficient of friction produced by the boundary lubricant?

Solution. The percent change in the shear resistance is $(\Delta f/f_0)\% = ((\tau_R/N)_0 - (\tau_R/N)_{\text{final}}) 100/(\tau_R/N)_0$. When N is the same,

$$(\Delta f/f_0)\% = (4.2 \text{ MPa} - 2.1 \text{ MPa}) 100/4.2 \text{ MPa} \qquad \Delta f/f_0 = 50\%$$

Example 15.4   For the volume–pressure results shown in Fig. 15.16, calculate the relative percent change in volume for both initial compression and subsequent compression to 2.8 MPa, the reversible engineering compressibility X, and the elastic springback on decompression. The volume of the material before initial axial compression is 25.0 cm^3 and at a pressure of 2.8 MPa is 20.5 cm^3. On decompressing, the material expands to a volume of 22.5 cm^3 and the subsequent compression and decompression behavior is reversible.

Solution. On compressing to 2.8 MPa, the relative change in volume is

$$\Delta V/V_0 \times 100 = (20.5 \text{ cm}^3 - 25.0 \text{ cm}^3) 100/25.0 \text{ cm}^3 = -18\%$$

The reversible expansion on decompression (elastic springback) is

$$\Delta V/V_0 \times 100 = (22.5 \text{ cm}^3 - 20.5 \text{ cm}^3) 100/20.5 \text{ cm}^3 = 10\%$$

and the reversible compression is

$$\Delta V/V_0 \times 100 = (20.5 \text{ cm}^3 - 22.5 \text{ cm}^3) 100/22.5 \text{ cm}^3 = -9\%$$

The compressibility is defined for a particular increment of pressure. The compressibility on compression from 0 to 1 MPa is much higher than from 1 to 2 MPa. For compression from 0 to 1 MPa,

$$X = (\Delta V/V_0)/\Delta P$$

$$X = ((21.0 \text{ cm}^3 - 22.5 \text{ cm}^3)/21.0 \text{ cm}^3)/(2 \text{ MPa} - 0 \text{ MPa})$$

$$X = -0.04 \text{ (MPa)}^{-1}$$

Example 15.5. Estimate the maximum green strength produced by a small amount of water—wetting particles of 1 μm diameter packed randomly in a compact with a pore fraction of 0.6. Compare this to a tensile strength of 1 MPa needed for handling strength. Refer to Example 2.6 for a model.

Solution. The tensile strength S_t is given by the equation

$$S_t = PF\ CN\ b/(\pi a^2) = PF\ (\pi/(1 - PF))b/(\pi a^2)$$

The maximum strength is obtained for a small amount of wetting liquid in the neck region of the particle contacts (pendular case) as explained in Example 2.7. The bonding force per contact b is

$$b = -(\Delta P_{neck})(A_{neck}) = -(-\gamma_{LV}/r_n)(\pi x^2) = (\gamma_{LV}/r_n)(\pi 2 r_p r_n)$$
$$b = 2\pi r_p \gamma_{LV} = \pi a \gamma_{LV} = (10^{-6}\ m)(73 \times 10^{-3}\ N/m) = 229 \times 10^{-9}\ N$$

The tensile strength S_t is

$$S_t = PF(\pi/(1 - PF)(b)/\pi a^2)$$
$$= (1 - 0.4)(\pi/0.4)(229 \times 10^{-9}\ N)/\pi(10^{-6}\ m)^2$$
$$S_t = 344\ kPa$$

The tensile strength is considerably less than the strength required for industrial handling. A binder is commonly required to develop adequate green strength.

Example 15.6. The results shown in Figure 15.23 are for a deaired plastic extrusion body. What is the significance of the information shown?

Solution. The cohesion $c = 20$ kPa and indicates the resistance to shear deformation of unsaturated free-standing material. The angle of internal friction is

$$\tan \phi_f = (100\ kPa - 20\ kPa)/(800\ kPa - 0\ kPa) = 0.1 \qquad \phi_f = 5.7°$$

The positive slope indicates that DPS < 1, the body is compressible and the effective stress increases with applied pressure. The low friction angle indicates that the body is very well plasticized. Above about 800 kPa, the shear resistance (ultimate shear strength) of 100 kPa is independent of the normal stress. This constant shear resistance indicates the effective stress $\bar{\sigma}$ does not increase with an increasing normal stress which is consistent with DPS = 1.

CHAPTER 16

RHEOLOGY OF SATURATED SYSTEMS (SLURRIES AND PASTES)

Rheology is the science of deformation and flow. Knowledge of rheological behavior is essential when designing or selecting equipment for storing, pumping, transporting, milling, mixing, atomizing, and forming a ceramic system. Rheological measurements are an integral part of the research and development of slurry systems, and rheological tests are used in programs for monitoring and controlling the consistency and behavior of slurries for casting, spray-drying, or glazing, and pastes for surface printing and decorating.

Ceramic slurries and pastes are commonly multicomponent systems that are relatively complex in structure and often poorly characterized. Particles may range from granular sizes to colloids. Added electrolytes and polymers may significantly change interparticle forces and the state of dispersion. The interparticle spacing depends directly on the concentration of the particles (solids loading), the state of dispersion, and the particle packing. Particle coagulation and polymer flocculation may produce a structural linkage, a gel structure, that varies with the mixing procedure and the shear history prior to measuring the rheological properties. Changes in the microstructure of the system produced during flow may be indicated by the rheological behavior.

In this chapter we discuss fundamentals and techniques used to determine the rheological properties of these important systems. Rheological behavior observed is then described in terms of models for the effects of individual components and their mixtures.

16.1 EFFECTIVE STRESS AND SHEAR IN A SATURATED SYSTEM

Suspensions, slurries, and pastes are systems containing particles not in close packing and DPS ≥ 1. When these systems are compressed, the imposed pressure is supported by the liquid medium, and the system is relatively incompressible at common processing pressures, if free from gas bubbles. The difference between the applied pressure and the liquid pressure, called the effective stress $\bar{\sigma}$, was defined by Eq. 15.1. In a suspension of completely dispersed particles (Fig. 16.1), the effective stress is

$$\bar{\sigma} = R - A \tag{16.1}$$

where $(R - A)$ is the effective pressure due to repulsive R and attractive A electrical interactions between particles. The coagulation and flocculation of particles can alter $(R - A)$ and the resistance to shear. However, the effective stress is independent of pressure imposed in processing, which simplifies the

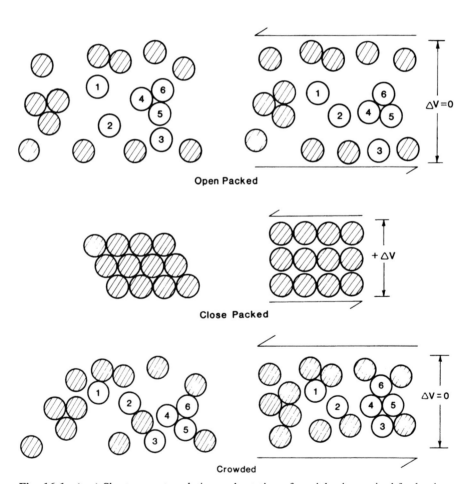

Fig. 16.1 (top) Short-range translation and rotation of particles is required for laminar flow in a slurry; (bottom) crowding restricts laminar flow; and (middle) volume dilation may be required for shear.

interpretation of the flow behavior. The shear resistance of these systems depends primarily on the viscosity of the liquid and the interparticle forces.

In a more crowded slurry, shear may be momentarily blocked by neighboring particles. The resistance to shear flow is dependent on particle translation away from the plane of shear, which is time dependent. The shear resistance will be very shear rate dependent. It also depends on mechanical interactions between particles. In a nearly close-packed system, initial flow produces significant contact stress between particles. Volume dilation of the system must occur to accommodate shear flow. For these systems the effective stress is described by Eq. 15.1 and the shear resistance is not independent of the confining pressure.

16.2 RHEOLOGICAL MODELS AND PROPERTIES

A shear stress is required to initiate and maintain laminar flow in a simple liquid. When a shear stress τ is linearly dependent on the velocity gradient $-dv/dy$ (Fig. 16.2), the liquid is said to be Newtonian, and

$$\tau = \eta_L \, (-dv/dy) \tag{16.2}$$

The constant of proportionality is the coefficient of viscosity η. The coefficient of viscosity indicates the resistance to flow due to internal friction between the

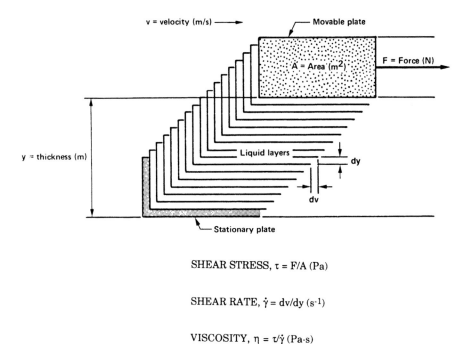

SHEAR STRESS, τ = F/A (Pa)

SHEAR RATE, $\dot{\gamma}$ = dv/dy (s^{-1})

VISCOSITY, η = $\tau/\dot{\gamma}$ (Pa·s)

Fig. 16.2 Model indicating viscous flow and definitions of shear stress, shear rate, and coefficient of viscosity.

molecules of the liquid. The velocity gradient $-dv/dy$ is the shear rate $\dot{\gamma}$; for simplicity, $\dot{\gamma}$ is used in equations that follow.

In liquids and solutions containing large molecules and suspensions containing nonattracting anisometric particles, laminar flow may orient the molecules or particles. When orientation reduces the resistance to shear, the stress required to increase the shear rate by an increment diminishes with increasing shear rate. This behavior is often described by an empirical power law equation,

$$\tau = K\dot{\gamma}^n \tag{16.3}$$

where K is the consistency index and $n < 1$ is the shear thinning constant which indicates the departure from Newtonian behavior. The apparent viscosity decreases with increasing shear rate, and the behavior is said to be shear thinning or pseudoplastic (see Fig. 16.3). The apparent viscosity of a power law material is

$$\eta_a = K\dot{\gamma}^{n-1} \tag{16.4}$$

The power law with $n > 1$ may also approximate the flow behavior of moderately concentrated suspensions containing large agglomerates and concentrated, deflocculated slurries. With an increase in the shear rate, particle interference and the apparent viscosity increases. This dependence on shear rate is called shear thickening behavior. Power law materials have no yield point.

Slurries containing a linkage of bonded molecules and particles require a finite stress called the yield stress τ_Y to initiate flow. A material is called a Bingham plastic when the flow behavior is described by the equation

$$\tau = \tau_Y + \eta_p\dot{\gamma} \tag{16.5}$$

The apparent viscosity η_a of a Bingham material is higher when the yield stress is higher and decreases with increasing shear rate:

$$\eta_a = (\tau_Y/\dot{\gamma}) + \eta_p \tag{16.6}$$

Equation 16.6 indicates that the Bingham material is shear thinning and η_p is the viscosity limit at a high shear rate. Note that for both shear thinning systems, the apparent viscosity decreases with the shear rate and two parameters are required to characterize the viscous behavior.

A more general model is the Herschel–Bulkley model:

$$\tau = \tau_Y + K\dot{\gamma}^n \tag{16.7}$$

This equation provides shear thinning behavior after the stress exceeds the yield stress, like the Bingham equation, but provides for a nonlinear dependence of shear stress on shear rate as described by the power law equation. Another

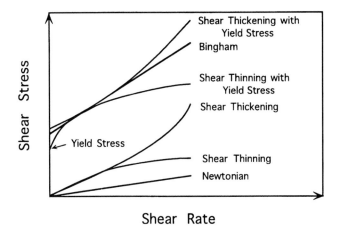

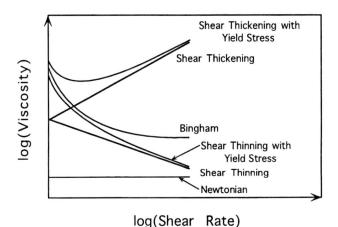

Fig. 16.3 (top) Variation of shear stress with shear rate and (bottom) variation of log (viscosity) with log (shear rate) for the different models of flow behavior.

equation often used successfully to describe the stress–shear rate behavior of a system containing weakly bonded particles is the Casson equation.*

$$\tau^{0.5} = k_1 + k_2 \dot{\gamma}^{0.5} \tag{16.8}$$

where k_1 and k_2 are structure-dependent constants for the system. In deriving this equation, Casson assumed that the suspension contained chain-like units which controlled the viscosity. The group size is dependent on the disruptive

*N. Casson, in *Rheology of Dispersed Systems*, C. C. Mill, Ed., Pergamon Press, London, 1959, pp. 84–104.

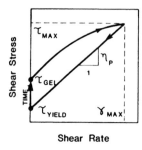 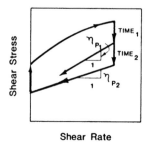

Fig. 16.4 Shear stress decreases with shear flow at constant shear rate, which indicates thixotropic behavior.

stresses produced during flow, and the viscosity varies with the shear rate. A generalized form of the Casson equation is

$$\eta_a^m = \eta_\infty^m + (\tau_Y/\dot\gamma)^m \tag{16.9}$$

where η_∞ is the viscosity limit when structure is broken down at a very high shear rate, and m is a constant indicating the deviation from linearity. Note in Fig. 16.3 that for materials with a yield point, the variation of log η_a with log (shear rate) is nonlinear.

The rheological behavior discussed above was assumed to be independent of the shear history and shearing time. For some materials the apparent shear resistance and viscosity at a particular shear rate may decrease with shearing time (Fig. 16.4). This behavior, called thixotropy, is commonly observed for shear-thinning materials when the orientation and bonding (coagulation) of molecules or particles change with the time during shear flow ($\dot\gamma \cdot t$). Thixotropic behavior in ceramic slurries is often reversible. For a thixotropic material with a yield stress, the apparent yield strength is higher after the suspension has been at rest and a particle structure has reformed. This higher apparent yield stress after a period of rest is often called the gel strength. A few materials exhibit an increase in shear resistance with time when sheared at a constant rate. This behavior is called rheopectic and may be expected when agitation enhances bonding between structural units in the suspension.

16.3   DETERMINATION OF VISCOSITY

Ranges of shear rate encountered in ceramic processing operations are indicated in Fig. 16.5. Viscosity data are more relevant when the viscosity is determined in the range of shear rate of the process.

The most widely used viscometer for ceramic suspensions and slurries is the variable-speed rotating cylinder viscometer. Depending on the design, either the inner or outer cylinder of length L rotates at an angular frequency ω, and the torque T produced by the viscous medium is recorded (Fig. 16.6). At a

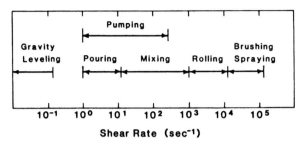

Fig. 16.5 Different processing operations occur in different ranges of shear rate.

position r in the viscous material in the annulus, $\dot{\gamma} = -r(d\omega/dr)$, and

$$\tau = \frac{T/r}{2\pi rL} \tag{16.10}$$

The apparent viscosity is obtained using Eq. 16.2 and integrating between the radii of the two cylinders, a and b,

$$\eta_a = \frac{T}{\omega_a} \frac{b^2 - a^2}{a^2 b^2 4\pi L} \tag{16.11}$$

The length L in Eq. 16.11 may be increased to correct for the end effect. For suspensions, it is assumed in deriving Eq. 16.11 that the spacing $(b - a)$ of the annulus is much greater than the maximum particle size, the viscous material makes continuous contact with the surfaces of the cylinders, no slippage occurs, and laminar flow is fully developed. The shear rate varies with position

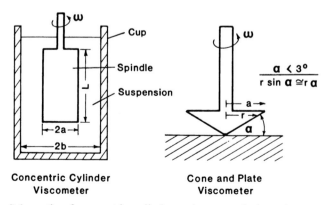

Fig. 16.6 Schematic of concentric cylinder and cone and plate viscometers showing parameters used in equations.

in the annulus, and

$$\dot{\gamma} = \frac{2\omega_a}{r^2} \frac{a^2b^2}{b^2 - a^2}$$

(16.12)

The range of shear rate $(\dot{\gamma}_a - \dot{\gamma}_b)$ in the annulus is reduced by using a viscometer with a narrow annulus (<1 mm).

A concentric cylinder viscometer with a rotating cup is conventionally used for measuring the viscosity of a Bingham plastic material. Slip at the wall of the inner cylinder may occur before the onset of flow. Laminar flow begins at the inner cylinder when $T = T_a = \tau_Y 2\pi a^2 L$ and progresses outward to the point where $T = T_b = \tau_Y 2\pi b^2 L$, as is shown in Fig. 16.7. Continued rotation at increasing speed causes the torque to increase to a value that is dependent on the rheological properties of the material. The plastic viscosity and yield strength are calculated using the equations

$$\eta_p = \frac{(T - T_o)}{\omega_b} \frac{(b^2 - a^2)}{b^2 a^2 4\pi L}$$

(16.13)

and

$$\tau_Y = T_o \frac{b^2 - a^2}{b^2 a^2 4\pi L \, \ln(b/a)}$$

(16.14)

where T_o is the torque when ω is reduced to zero. The effective shear rate may be calculated from η_p and τ_Y,

$$\dot{\gamma} = (1/\eta_p)[(T/2\pi r^2 L) - \tau_Y]$$

(16.15)

The apparent viscosity η_a is calculated using Eq. 16.6.

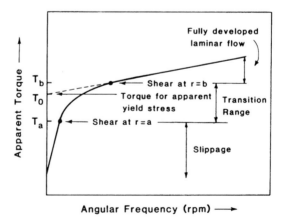

Fig. 16.7 Flow development in the annulus of a concentric cylinder viscometer for a Bingham material.

Rotating cylinder viscometers are widely used for determining the rheological properties of slips and slurries. An apparent viscosity determined at one shear rate is not sufficient for comparing the rheological behavior. The properly designed concentric cylinder viscometer provides a means for determining the rheological properties for a controlled range of shear rate. Rotational viscometers are commonly used at shear rates up to 1200 s^{-1}; at higher rotational speeds, centrifugal effects cannot be neglected.

The cone and plate viscometer shown in Fig. 16.6 is used for determining the viscosity of molecular liquids and colloidal suspensions. Interchangeable cones provide a selection of radius and cone angle α (from 0.3 to 3 degrees). For a small α, the shear rate is independent of the radial position r, and

$$\dot{\gamma} = -dv/dy = -d(\omega s)/d(r\alpha) \cong \omega/\alpha \tag{16.16}$$

The viscosity (no yield point) is calculated from the torque T using the equation

$$\eta = \frac{3\alpha T}{2\pi a^3 \omega} \tag{16.17}$$

Slightly truncated cones are used for suspensions containing colloidal-size particles.

The viscosity of a Newtonian liquid may be determined from the pressure drop ΔP during laminar flow at a volumetric rate Q in a capillary tube of radius R and length L. It is assumed that no slippage at the wall occurs. The viscosity is calculated from the Poiseuille equation

$$\eta = \frac{\pi \Delta P R^4}{8QL} \tag{16.18}$$

and

$$\dot{\gamma} = \frac{\Delta P r}{2L\eta} \tag{16.19}$$

For a shear-thinning liquid, the velocity profile in a capillary viscometer is somewhat flattened toward the center (Fig. 16.8), and the deviation increases

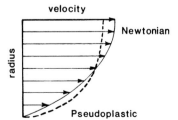

Fig. 16.8 The velocity profile of a shear thinning liquid is relatively flattened.

when the flow constant n is lower. The laminar flow of a power law shear thinning (pseudoplastic) liquid in a capillary tube (no slip at wall) is described by the equations

$$K = \frac{1}{2}\left[\frac{n}{3n+1}\right]^n \frac{R^{n+1}\Delta P}{\bar{v}^n L} \tag{16.20}$$

$$n = d[\ln(R\Delta P/2L)]/d[\ln(4\bar{v}/R] \tag{16.21}$$

where $\bar{v}$ is the mean flow velocity. The constant n is determined from a graph of log K versus log $\bar{v}$. A capillary viscometer is commonly used for determining the viscosity of a Newtonian or pseudoplastic liquid at a shear rate beyond the limit for a rotational viscometer.

The description of the flow behavior of a Bingham plastic fluid or a yield-pseudoplastic fluid in a capillary tube is much more complex. The analysis of the flow behavior of these systems is presented in References 3 and 10 and in Chapter 23.

When some critical shear rate is exceeded, turbulent flow begins and is characterized by local fluctuations in the velocity and direction of flow. For flow in a pipe, the conditions causing a transition to turbulent flow can be predicted using the dimensionless Reynolds number Re:

$$Re = 2\bar{v}D_L R/\eta_L \tag{16.22}$$

where $\bar{v}$ is the average flow velocity of a liquid of density D_L and viscosity η_L in a tube of radius R. The transition to turbulent flow begins when Re is approximately 2100 for a Newtonian liquid; flow is fully turbulent when Re > 3000. The critical Re for the onset of turbulent flow is higher for pseudoplastic liquids and generally increases as the shear thinning constant n decreases. Turbulent flow is of importance in some mixing operations and atomization such as spray drying.

The rate at which a heavy sphere falls through a suspension may also be used to determine the viscosity of a Newtonian liquid. Stokes' law (Eq. 7.2) is used for calculating the coefficient of viscosity from the time required to fall a fixed distance. The viscosity of very viscous liquids and pastes can be determined by measuring the stress on a thin sheet or foil pulled through a slit filled with material. The viscosity may be calculated using Eq. 16.2. A variation of this is the axial displacement of a cylinder in a tube filled with material. If carefully controlled, the coefficient of viscosity of the material can be determined for a controlled displacement (shear) rate.

A relative index of viscosity may be determined using experimental techniques which include the viscous damping of a sonic probe, the torsional resistance to stirring, and mixing under conditions simulating industrial conditions. The time for flow through an orifice in a cup is also used as a quality control index of viscosity. As always, the shear rate of the test should be considered when relating results to industrial behavior.

For a partially coagulated slurry, the yield strength of undisturbed material, often called the gel strength τ_{gel}, is dependent on the time at rest following agitation. When using a concentric cylinder viscometer to determine the gel strength, the material is initially sheared in the viscometer at a high shear rate and then permitted to rest for a specified period of time (quiescent time). The gel strength is the maximum shear stress before the gel breaks, when the viscometer rotates at a specific slow rate (≈ 3 rpm). The relative gel strength may also be determined experimentally using a special torsion viscometer. An inner cylinder or disk of high mass supported by a wire is initially twisted through some specific angle and immersed in the thixotropic material. After a particular rest period, the cylinder is released, and the angular rotation is measured as the torque in the spindle decreases. Less angular rotation occurs in a material with a higher gel strength.

Indices of gelation and thixotropic behavior may be conveniently determined using a concentric cylinder or cone and plate viscometer.

An index of structural buildup B_{gel} is the difference of the yield stress obtained after two quiescent times (such as 6 and 30 min):

$$B_{gel} = (\tau_{Y2} - \tau_{Y1})/\ln(t_2/t_1) \tag{16.23}$$

When the plastic viscosity is determined after shearing for two different times at a constant shear rate or two shear rates and a constant time, indices of structural breakdown can be calculated from the difference in plastic viscosity or extrapolated yield strength relative to the differential time or rate of shear flow (Fig. 16.4):

$$B_{thix} = (\tau_{Y2} - \tau_{Y1})/\ln(t_2/t_1) \tag{16.24}$$

or

$$B_{thix} = (\eta_{p1} - \eta_{p2})/\ln (t_1/t_2) \tag{16.25}$$

For viscoelastic materials, two or more properties are required to characterize the flow of the material. Elasticity implies memory. A viscoelastic material poured onto a flat surface will not smoothly cover the surface and will retain part of its shape before pouring. This behavior is not commonly desired in a processing system. Viscoelasticity of ceramic processing systems has not been widely studied. It may be determined using a dynamic mechanical test or a stress relaxation test. In the dynamic test, the phase angle is $0°$ for an elastic material, $90°$ for a viscous material, and $0-90°$ for a viscoelastic material. The closer the angle is to $90°$, the closer the material is to a viscous material. A relaxation test subjects the test sample to a constant rate of shear for a specified time before instantly stopping the shear. The decay of the stress built up during shear deformation is monitored. The stress after shearing will decay instantly for a strictly viscous material, remain constant for an elastic material, and decay with time for a viscoelastic material (Fig. 16.9). Alternatively, a constant stress

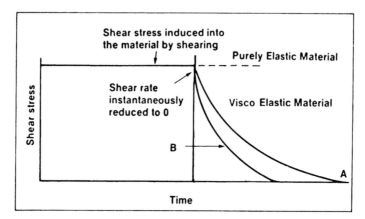

Fig. 16.9 Shear stress relaxes when shear is stopped for a viscoelastic material, but not for an elastic material (material A has a higher elastic component than material B).

may be imposed and held for a period of time before it is removed. The deformation response provides an insight into the viscoelastic deformation behavior.

16.4 VISCOSITY OF LIQUIDS

The viscosity of simple liquids used in ceramic processing is Newtonian in behavior. For a series of liquids of similar type, the viscosity increases with molecular size, as indicated in Table 9.2. With only a few exceptions, heating increases the specific volume and reduces the viscosity of simple liquids. The variation of viscosity over a limited temperature range is approximated by the equation

$$\eta_L = A \exp(Q/RT) \tag{16.26}$$

where A and Q are constants for a particular liquid. For water $A = 1.25 \times 10^{-3}$ mPa·s and $Q/R = 1970$ K. The viscosities of several liquids at 0–70°C are listed in Table 9.3.

In liquids containing large molecules, the molecules tend to be randomized by thermal vibrations and oriented in laminar flow. Thermal energy increases the free volume which facilitates orientation during flow, and the viscosity decreases with temperature and shear rate.

Most linear polymers such as polystyrene and polyethylene are non-Newtonian shear thinning materials. The viscosity increases with the mean molecular weight. A branched polymer that is more symmetrical in structure and orients less when sheared exhibits less shear thinning. Polymers with a wider molecular-weight distribution generally exhibit a lower melt viscosity. Degradation of polymers during high shear mixing may reduce their apparent viscosity.

16.5 VISCOSITY OF BINDER SOLUTIONS

Binder molecules have a large sphere of influence relative to their size, and dissolved binder molecules may significantly increase the viscosity of a simple liquid. As shown in Fig. 16.10, binders with a higher average molecular weight are especially effective in increasing the viscosity. At low concentrations the dependence of viscosity of the solution η_s at low shear rates on the concentration C is often approximated by the equation

$$\eta_s/\eta_L = \exp(SC) \qquad (16.27)$$

where S is an effective sphere of influence for molecules (hydrated) of a particular molecular weight and is determined from the slope of $[\ln(\eta_s/\eta_L)$ vs. $C]$ as $C \rightarrow 0$. Interaction between molecules causes a departure from Eq. 16.27 at concentrations greater than about 2–10%. Solvation of the molecules may reduce their interaction and extend the range described by Eq. 16.27.

The random orientation of dispersed binder molecules due to Brownian motion is offset by shear stresses in laminar flow, and binder solutions exhibit shear thinning behavior, as shown in Fig. 16.11. The reduction in viscosity due to shear alignment is greater for solutions of binders of higher molecular

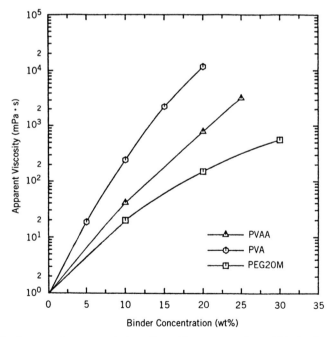

Fig. 16.10 Increase in apparent viscosity with concentration of binder in solution for PVA, PVAA, and PEG binders (MW of PVA and PVAA = 50,000; MW of PEG = 20,000 g/mol). (Courtesy of W. J. Walker, Jr., Alfred University.)

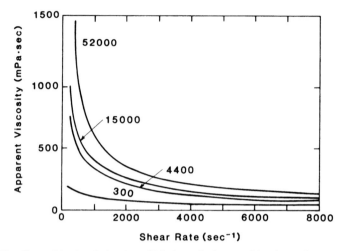

Fig. 16.11 Shear thinning behavior of binder solutions and its dependence on viscosity grade of the hydroxyethyl cellulose binder. (Cellosize, Union Carbide Corp., New York.)

weight. Shear may also reduce the average solvation of the molecules. When a binder is available in several viscosity grades, different grades can be blended to modify the viscosity and shear thinning behavior of the solution.

The viscosity of solutions of dispersed binder molecules decreases with increasing temperature, as shown in Fig. 16.12. Temperature control is essential to obtain a reproducible viscosity. For binders that resist gelation, the

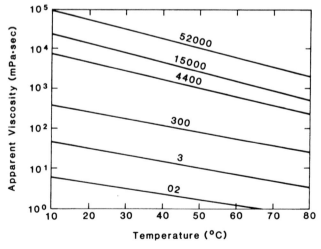

Fig. 16.12 Viscosity of hydroxyethyl cellulose binder solutions depends on viscosity grade and temperature. (Cellosize, Union Carbide Corp., New York.)

variation of viscosity with temperature is often reversible between the freezing and boiling points of the solvent.

A plasticizer reduces the flow viscosity of a polymer. Plasticizers used may be Newtonian or non-Newtonian. An example of a Newtonian plasticizer is mineral oil for polystyrene. A low molecular weight resin may be a non-Newtonian plasticizer for the higher molecular weight form. Plasticizers may also be divided into compatible types, which do not exude during flow, and noncompatible, which do exude. The shear rate behavior of the polymer is modified to a degree by the shear rate behavior of the plasticizer and its concentration.

16.6 VISCOSITY OF SUSPENSIONS AND SLURRIES OF DISPERSED POWDERS AND COLLOIDS

The viscosity of a suspension η_S is greater than the viscosity of the liquid medium η_L in the suspension, and the ratio is referred to as the relative viscosity η_r. For a very dilute suspension of noninteracting spheres in a Newtonian liquid, the viscosity for laminar flow is described by the Einstein equation:

$$\eta_r = \eta_S/\eta_L = 1 + 2.5\,f_p^v \qquad (16.28)$$

where f_p^v is the volume fraction of dispersed spheres. The higher viscosity is caused by the dissipation of energy as liquid flows around the particles. A more general equation for the viscosity of a suspension of nonspherical particles is

$$\eta_r = 1 + K_H f_p^v \qquad (16.29)$$

where K_H is the apparent hydrodynamic shape factor of the particles. In suspensions of particles with an aspect ratio greater than 1, particle rotation in the velocity gradient during flow produces a larger effective hydrodynamic volume (Fig. 16.13) and $K_H > 2.5$.

As the volume fraction of particles increases above about 5–10 vol%, the interaction between particles during flow causes the viscosity to increase at the

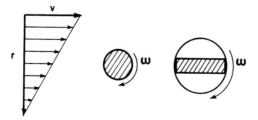

Fig. 16.13 Anisometric particle has the same cross-sectional area of the sphere, but a larger time-average hydrodynamic shape factor K_H.

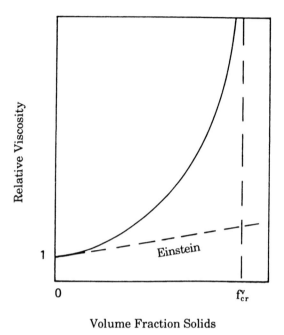

Fig. 16.14 Typical form of increasing relative viscosity when increasing the powder concentration dispersed in the slurry.

rate shown in Fig. 16.14. Rheological data for suspensions of uniform spherical colloidal particles are often approximated by the Dougherty–Krieger equation*

$$\eta_r = \eta_S/\eta_L = [1 - f_p^v/f_{cr}^v]^{-K_H f_{cr}^v} \qquad (16.30)$$

where f_{cr}^v is the packing factor at which flow is blocked. The variation of relative viscosity with volume fraction of dispersed particles for different values of K_H and f_{cr}^v is shown in Fig. 16.15. The graph of relative viscosity shows the effects of the particles on the viscosity. Effects of both the liquid viscosity and the particle loading are shown in the figure of apparent viscosity η_S with powder loading (Fig. 16.16). The graph of apparent slurry viscosity will also reveal the effect of changing the temperature which changes the liquid viscosity. In suspensions of particles of anisometric shape both K_H and f_{cr}^v depend on particle orientation produced during laminar flow. Brownian motion somewhat randomizes colloidal particles and increases the effective hydrodynamic shape factor. Woods and Krieger† obtained values of K_H and f_{cr}^v indicated in Table 16.1 for aqueous suspensions of dispersed monosize latex sphere finer than 0.7

*I. M. Krieger and T. J. Dougherty, *Trans. Soc. Rheol.* 3, **137** (1959).
†M. E. Woods and I. M. Krieger, *J. Colloid Interfac. Sci.*, **34**(1), 91–99 (1970).

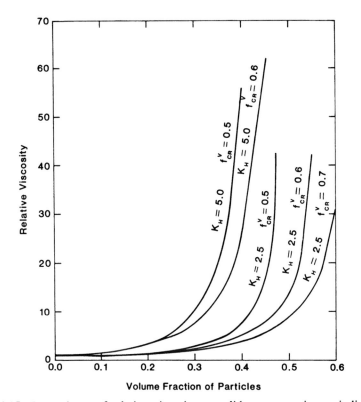

Fig. 16.15 Dependence of relative viscosity on solids concentration as indicated by the Dougherty–Krieger equation.

μm. The volume fraction of dispersed particles f_p^v was calculated from the volume fraction of spheres f_s^v, their mean size $\bar{a}_{V/A}$, and the effective thickness of the adsorbed surfactant Δ (4.5 nm in Table 16.1). The effective volume fraction of the particles with adsorbed additive is calculated from the equation

$$f_p^v = f_s^v [1 + 2\Delta/\bar{a}_{V/A}]^3 \qquad (16.31)$$

This correction becomes increasingly significant when the particle size decreases below 1 μm, as is seen in Fig. 16.17. Agglomeration reduces the effective f_{cr}^v (Table 16.1). Suspensions were observed to be shear thinning for $f_p^v < 0.50$, apparently because of the formation and persistence of agglomerates at low shear rates.

Several other empirical equations with two adjustable parameters have been proposed to explain the dependence of slurry viscosity on the particle content when the slurry becomes crowded. Particle interactions at some shear rate and the relative viscosity diminish as the volume fraction of liquid increases above that at which flow is blocked, indicated by the difference $f_{cr}^v - f_p^v$. The Mooney

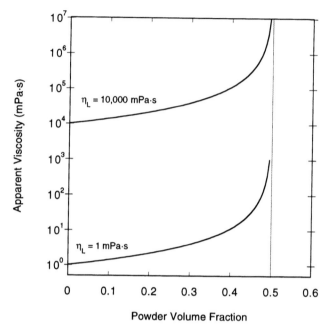

Fig. 16.16 The apparent viscosity depends on viscosity of the liquid medium and particle parameters described by the Dougherty–Krieger equation.

equation*

$$\eta_r = \exp[K_H f_p^v/(1 - k_2 f_p^v)] \tag{16.32}$$

may sometimes approximate the effect of solids loading on the relative viscosity of deflocculated casting slurries and slips of fine powders (Fig. 16.18).† The equation of Chong et al.‡ in general form is

$$\eta_r = [(f_{cr}^v - k f_p^v)/(f_{cr}^v - f_p^v)]^2 \tag{16.33}$$

*M. Mooney, *J. Colloid. Sci.*, **6**, 162–170 (1951).
†M. Rajala, *Am. Ceram. Soc. Bull.*, **71**(12), 1817–1819 (1992).
‡J. S. Chong, E. B. Christiansen, and A. D. Baer, *J. Appl. Polym. Sci.*, **15**, 2007–2202 (1971).

TABLE 16.1 Dougherty–Krieger Constants for Aqueous Suspensions of Latex Spheres

Shear Rate	K_H	f_{cr}^v
Low	2.7	0.57
High	2.7	0.68

Source: Reference 9.

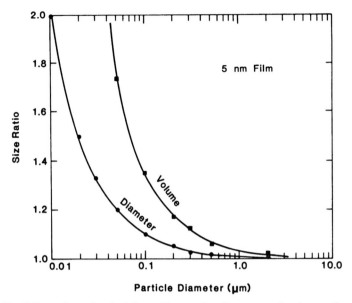

Fig. 16.17   Effect of an adsorbed 5-nm film on the diameter and volume of a spherical particle.

and has been shown to approximate the viscous behavior of injection molding materials containing a highly viscous polymer liquid.* These equations predict a viscosity increase similar to that of Eq. 16.30. A viscosity sensitivity index S calculated as $d \ln \eta_S / df_p^v$ will indicate the viscosity effect at some constant powder loading f_p^v on substituting an alternative material. For the Mooney description,

$$S_{f_p^v} = K_H / (1 - k_2 f_p^v)^2 \tag{16.34}$$

More efficient particle packing reduces the maximum size of the interstices among particles. For a size distribution which packs more efficiently, less of the liquid is segregated in large interstices and more of the liquid is effectively mobilized in flow, as shown in Fig. 16.19. For a particular solids loading, a powder with a higher PF will have a higher f_{cr}^v and the relative viscosity will be lower. For mixtures of two particle sizes, the slurry viscosity is a function of the proportions of coarse and fines, the size ratio, and the volume fraction of total solids in the slurry. Slurries with relatively more mobilized liquid and a matrix of colloids with a compliant, adsorbed additive may be relatively fluid and pseudoplastic over a wide range of shear rate.

In crowded slurries, the movement of particles around one another to accommodate shear flow requires time. Above a particular shear rate, the hindered

*T. Zhang and J. R. G. Evans, *J. Euro. Ceram. Soc.*, **5**, 165–172 (1989).

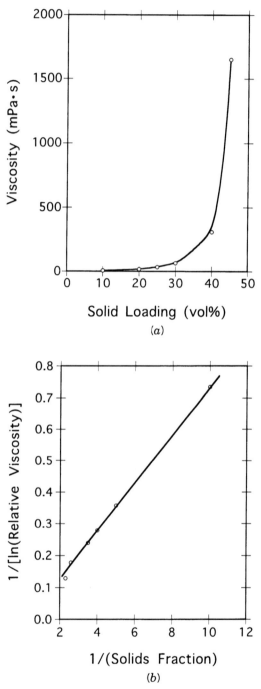

Fig. 16.18 (a) Variation of apparent viscosity with concentration of dispersed alumina of 0.7 μm mean particle size (shear rate = 12 s^{-1}) and (b) approximation using Mooney equation.

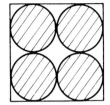 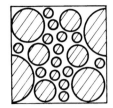

Fig. 16.19 Relatively more liquid is mobilized on flow of a slurry containing a powder with a higher packing factor (PF) and the apparent viscosity is lower. (Equal volume fraction of particles in the two illustrations.)

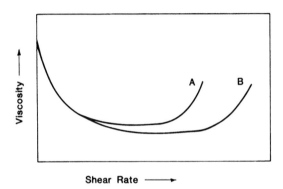

Shear Rate ⟶

Fig. 16.20 General variation of viscosity with shear rate for two slurries indicating shear thickening behavior at a high shear rate. The less well-dispersed system A exhibits shear thickening behavior at a lower shear rate.

rotation and particle interference may cause the appearance of shear-thickening behavior (Fig. 16.20). Increasing the solids loading in the slurry generally decreases the shear rate at which dilatant behavior begins.* Dispersion of agglomerates or modification of the particle size distribution to produce additional fines and a higher f_{cr}^{v} may extend the range of shear rate before shear thickening behavior is observed in the slurry. The absence of shear thickening behavior is very important in the pumping and spraying of slurries.

16.7 RHEOLOGY OF COAGULATED SYSTEMS

As seen in Fig. 10.10, the viscosity of the slurry varies inversely with the zeta potential on adding a polyelectrolyte deflocculant. In suspensions of oxides

*R. L. Hoffman, *J. Colloid. Interfac. Sci.*, **46**(3), 491–506 (1974).

deflocculated using simple electrolytes, the viscosity also varies indirectly with the zeta potential. When the pH corresponds to the IEP of the particles, co-agulation forces produce large, relatively porous agglomerates which may form a continuous gel structure when the solids content is sufficiently high. Smaller, less porous agglomerates form at an intermediate zeta potential corresponding to a pH higher or lower than the IEP. The viscosity of a slurry is very sensitive to the coagulation forces and the coagulated structure, as shown in Fig. 16.21. The viscosity of the coagulated structure is relatively high and commonly shear thinning and thixotropic in behavior. [A binder may change pH range of de-flocculation (Fig. 16.22).] Slurry that is only partially coagulated may be of lower viscosity, but exhibit shear-thickening behavior owing to the interaction of agglomerates when sufficiently concentrated. The minimum viscosity occurs when the zeta potential is high and the agglomerates are dispersed, and the f_{cr}^{v} is a maximum. Over-deflocculation which actually reduces the zeta potential may cause the viscosity to increase because agglomerates are formed.

The maximum viscosity of a clay slip occurs when edge-face heterocoagu-lation produces large porous agglomerates, called flocs. A continuous flocced

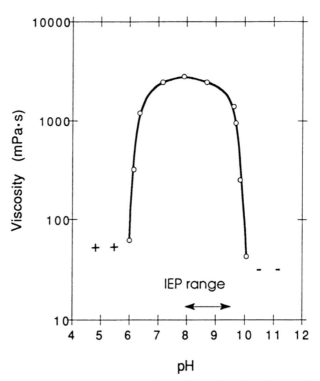

Fig. 16.21 Dependence of viscosity of slurry of reactive alumina powder on solution pH. Coagulation of weakly charged particles occurs near IEP of alumina.

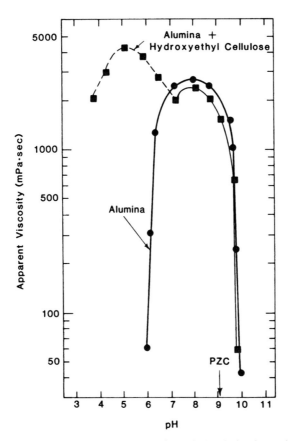

Fig. 16.22   Adsorbed binder alters deflocculation behavior at low pH.

structure is formed when the clay is sufficiently concentrated. Increasing the pH reduces the floc size and eventually produces a dispersion of clay platelets with both negative edges and negative faces, and the apparent viscosity is relatively low, as is shown in Fig. 16.23. At low pH, the edges of the platelets become positive, but the faces remain negative, the coagulated slip is not dispersed, and the viscosity remains relatively high. The relative maximum in the apparent viscosity and the yield stress are higher in a coagulated clay slip of finer particle size (Fig. 16.24). The dependence of the apparent yield stress on particle size depends on the number of particle contacts, as is approximated by Eq. 15.16. Substitution of clays containing more of the mineral montmorillonite and a higher clay loading can also produce a higher maximum apparent viscosity.

The yield stress and viscosity of a coagulated clay slurry may decrease or increase on heating. Heating reduces the viscosity of an aqueous medium, but increases both the double layer thickness and the electrochemical activity. The

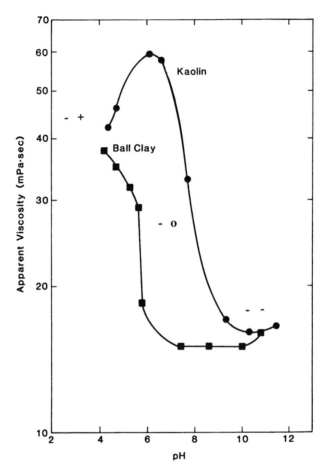

Fig. 16.23 Viscosity of clay slips decreases with increasing pH above IEP of edges of clay particles, but viscosity remains high at low pH because faces remain negative.

apparent yield strength may increase at temperatures approaching 100°C when lime is present because Ca^{2+} may precipitate in the contact region.

The flow properties of coagulated slurries and pastes are very dependent on the shear history. In the absence of an energy barrier, the coagulation rate is a function of the frequency of particle collisions. The number fraction of dispersed particles (N/N_0) remaining in a suspension of attracting particles after time t is given by the von Smoluchowski equation*

$$N/N_0 = [1 + t/t_c]^{-1} \qquad (16.35)$$

*Paul Sennett and J. P. Oliver, Chapter 7 in *Chemistry and Physics of Interfaces*, D. E. Gushee, Ed., American Ceramic Society, Washington, DC, 1965.

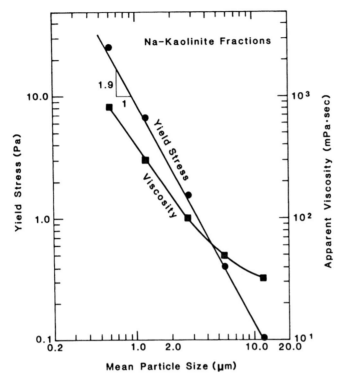

Fig. 16.24 Yield stress and viscosity increase with decreasing particle size in a co-agulated suspension of kaolinite. (Data courtesy of W. G. Lawrence, Alfred University.)

where t_c is the period for $N/N_0 = 0.5$. When t/t_c can be expressed as $f/\dot{\gamma}$, where f is the coagulation frequency, a shear rate dependent coagulation index λ is

$$\lambda = c/(1 + \dot{\gamma}/f) \tag{16.36}$$

where c is a constant. Pseudoplastic and thixotropic behavior is expected when $\dot{\gamma}/f > 0$. The constant f is dependent on the interparticle separation and at-tractive forces and is proportional to N^2T/η. An equation used by Worrall* to describe the viscosity of partially coagulated clay slurries is

$$\eta_a = \tau_Y/\dot{\gamma} + (\lambda + \eta_p) \tag{16.37}$$

The change in the structure and the rheological properties of a slurry with time is called aging. Aging is caused by time-dependent dissolving, chemical

*W. E. Worral, *Clays and Ceramic Raw Materials*, Halsted Press, New York, 1975.

reactions, and mechanical agitation which change the coagulated structure. Aging is usually accelerated by increasing the temperature and the intensity of agitation. In clay slurries, aging is expected and must be controlled to reproduce the flow behavior and thixotropy.

16.8 RHEOLOGY OF SUSPENSIONS AND SLURRIES CONTAINING BINDERS

Systems containing dispersed particles, electrolytes, and an adsorbed binder are relatively complex and wide variations in rheological behavior can be produced by changing the proportions and types of components. When the binder flocculates particles in the suspension, the viscosity and yield strength may be increased in a range of pH where particle charging and dispersion would otherwise occur (Fig. 16.22). The effects of the binder may be profound at low shear rates. But with an increasing rate of shear, the viscosity behavior becomes less dependent on flocculation effects relative to effects of the particle size distribution, particle shape, and particle loading.

An ionic binder may act as either a flocculant or a deflocculant. An adsorbed low-molecular-weight ionic binder that does not flocculate the particles may increase the viscosity of the solution but decrease the viscosity of the slurry when particle dispersion is stabilized. Solvated nonionic molecular binders of

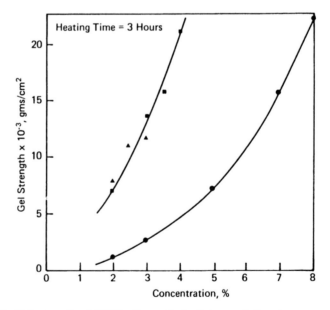

Fig. 16.25 Gel strength at 65°C of a methylcellulose solution increases with binder concentration (lower curve is for low MW; upper curve is for medium MW). (Dow Chemical Co., Midland, MI.)

low molecular weight may act as a flocculant at very low concentrations when they bond and bridge between particles. On increasing the concentration, they may adsorb predominantly on individual particles and act as a deflocculant. When the concentration exceeds that for a monolayer coverage, flocculation is produced by bonding and bridging between particles. Nonionic binders of high molecular weight which are adsorbed are usually flocculants. Flocculation forces and the presence of the solvated, pseudoplastic binder between the particles significantly alter the rheological properties of the particle system. The viscosity of nonionic binder solutions is usually stable over a wider range of pH or concentration of highly charged ions than solutions of ionic binders. On gelation of a binder, the gel strength increases with concentration and is higher for a binder of higher molecular weight (Fig. 16.25).

Figure 16.26 shows the shear thinning viscous behavior preferred for slurries that must be mixed, stored, pumped, sprayed, and so on. Specific examples are shown in chapters that follow. Steric hindrance of the binder resists the formation of hard agglomerates, and the yield stress provides a mechanism to prevent product separation during storage. However, shear thinning (pseudoplasticity) at a high shear rate provides the necessary flow during mixing, pumping, and spraying. Similarly, pseudoplastic coatings with a yield stress resist running (creep flow) due to gravitational forces or mild mechanical vibrations. Gelation of the binder solution may cause a significant increase in the viscosity of a slurry (Fig. 11.10). This characteristic can be valuable to obtain a strong coating without runs when a thermogelling slurry is applied on a hot surface. Gelation may also be produced by chemical means.

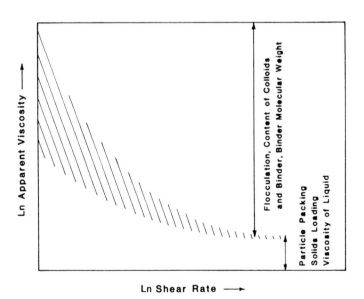

Fig. 16.26 Nominal rheological behavior of many industrial slurries at low to moderate shear rate and important variables.

A slurry of powder having a viscosity at a low shear rate less than about 1000 mPa·s is commonly called a paste when the yield strength is in the range of about 0.1–10 kPa. It is described as being buttery, creamy, and watery as the yield strength decreases below 0.1 kPa. Intuitive descriptions of the general rheological behavior of a complex slurry may sometimes be insufficient or imprecise because of shear thinning or thixotropic behavior during flow and because of variation in the shear rate when the material is used.

One of the problems in interpreting the behavior of suspensions and slurries is determining the character of the system under flow conditions. Minimal characterization must include determining the characteristics of the ingredients and the state of dispersion. The agglomerate structure can be analyzed indirectly in a carefully diluted suspension using standard wet size analysis techniques. Agglomerates may also be separated by filtration or freeze drying and examined microscopically. Coagulants and flocculants may be doped with a salt easily identified by microscale chemical analysis, prior to admixing into the system, to aid in analysis. The concentration of dissolved or added electrolytes and surfactants in the liquid surrounding particles may change when the proportion of powder/liquid is increased, and simple extrapolation of behavior from that of dilute systems may be inaccurate.

SUMMARY

Rheological control depends on determining flow properties under conditions providing a controlled shear rate, as a function of shear rate, for characterized systems. The shear rate dependence of the shear stress and viscosity of a slurry and paste is commonly determined using a rotating viscometer or by measuring the flow velocity through a tube when the boundary conditions are appropriate. The structure of the system may change with time, temperature, and shear flow. Physical properties and interactions between the liquid, colloid, and particle components may vary widely. Colloids may be dispersed or agglomerated macromolecules and/or isometric and anisometric particles. In dispersed systems, the viscosity depends on the viscosity of the liquid medium and the hydrodynamic shape factor and volume fraction of the dispersed particles with adsorbed surfactant. The proportion of liquid required for flow varies directly with the packing efficiency of the particles. Slurries for processing may contain particles of variable character dispersed in a liquid–colloid matrix and linked by coagulation forces and flocculating additives. Partial coagulation and flocculation produce slurries having a yield strength and shear thinning behavior which is commonly preferred in industrial processing.

SUGGESTED READING

1. J. K. Wright, M. J. Edirisinghe, J. G. Zhang, and J. R. G. Evans, "Particle Packing in Ceramic Injection Molding." *J. Am. Ceram. Soc.*, **73**(9), 2653–2658 (1990).

2. C. S. Khadilkar and M. D. Sacks, "Effect of Polyvinyl Alcohol on the Properties of Model Silica Suspensions," in *Ceramic Transactions*, Vol. 1, G. L. Messing, E. R. Fuller, and H. Hausner, Eds., American Ceramic Society, Westerville, OH, 1988, pp. 397–409.

3. G. R. Gray and H. C. H. Darley, *Composition and Properties of Oil Well Drilling Fluids*, Gulf Publishing, Houston, 1980.

4. Temple C. Patton, *Paint Flow and Pigment Dispersion*, Wiley-Interscience, New York, 1979.

5. W. E. Worrall, *Clays and Ceramic Raw Materials*, Halsted Press, New York, 1975.

6. R. L. Hoffman, "Discontinuous and Dilatent Viscosity Behavior in Concentrated Suspensions. II. Theory and Experimental Tests," *J. Colloid. Interfac. Sci.*, **46**(3), 491–506 (1974).

7. Rex W. Grimshaw, *The Chemistry and Physics of Clays*, Wiley-Interscience, New York, 1971.

8. J. S. Chong, E. B. Christiansen, and A. D. Baer, "Rheology of Concentrated Suspensions," *J. Appl. Polym. Sci.*, **15**, 2007–2202 (1971).

9. M. E. Woods and I. M. Krieger, "Rheological Studies on Dispersions of Uniform Colloidal Spheres," *J. Colloid. Interfac. Sci.*, **34**(1), 91–99 (1970).

10. A. H. P. Skelland, *Non-Newtonian Flow and Heat Transfer*, Wiley, New York, 1967.

11. M. Mooney, "The Viscosity of a Concentrated Suspension of Spherical Particles," *J. Colloid. Sci.*, **6**, 162–170 (1957).

12. H. Eilers, "The Viscosity of Emulsions Made of Highly Viscous Materials as a Function of the Concentration," *Kolloid-Z.*, **97**, 313–321 (1941).

PROBLEMS

16.1   Identify the planes of shear for the open packed, close packed, and crowded systems in Fig. 16.1.

16.2   Calculate the viscosity of water at 10 and at 40°C.

16.3   Graph the variation of apparent viscosity/K with shear rate for a power law pseudoplastic material for $n = 0.2$, 0.5, and 0.9.

16.4   Graph the viscosity/K as a function of shear rate for power-law materials with $n = 0.25$ and 4.0.

16.5   For a coagulated clay suspension exhibiting Bingham behavior, the yield stress = 2.0 Pa and the plastic viscosity = 10 mPa·s. Calculate and graph the apparent viscosity with shear rate. Comment on the behavior.

16.6   A concentric cylinder viscometer has a rotating inner cylinder of length 4.0 cm and radius 1.5 cm. The annulus $(b - a) = 0.1$ cm. Calculate the range of shear rates during a test when the spindle speed is 60 rpm. Compare your answer to the range when $(b - a) = 1$ cm.

16.7 Calculate the viscosity and shear rate using the geometrical parameters in Problem 16.6 when the speed is 100 rpm, $T = 30$ Nm, $b - a = 0.1$ cm, and $L = 4.0$ cm.

16.8 A concentric cylinder viscometer is used to determine the properties of a Bingham plastic material ($L = 4.0$ cm, $a = 1.8$ cm, $b - a = 0.1$ cm). Calculate the shear rate and the flow properties. $T_o = 4.07 \times 10^{-4}$ N·m and $T = 7.26 \times 10^{-4}$ N·m at 300 rpm.

16.9 Show graphically why knowledge of the shear stress at one shear rate is insufficient when comparing the behavior of two different non-Newtonian materials.

16.10 Show that Eq. 16.20 becomes the Poiseuille equation when $n = 1$.

16.11 The relative velocity v/v_{max} of a power-law shear-thinning material as a function of radial position r in a tube of radius R is $v/v_{max} = [(3n + 1)/(n + 1)] [1 - (r/R)^{(n + 1)/n}]$. Graph v/v_{max} as a function of r/R for $n = 0.2$ and compare to a material with $n = 1$.

16.12 The gel strengths for a casting slip were 3.5 Pa after 10 min and 8.0 Pa after 60 min. Calculate the gelation index B_{gel}.

16.13 Estimate the sphere of influence S for the PVA and PEG in Fig. 16.10.

16.14 Calculate the viscosity of a 50–50 blend of the two 10 wt% binder solutions in Fig. 16.10 (note: $\log \eta_{blend} = A \log \eta_a + B \log \eta_b$).

16.15 Calculate and compare the hydrodynamic volumes of rectangular particles $1 \times 1 \times 6$ μm and $1 \times 1 \times 2$ μm. Assume that the particles may rotate in three dimensions. Compare these values to the actual volume.

16.16 Estimate the shear thinning index for the binder solutions of grade 300 and 15,000 in Fig. 16.11.

16.17 Calculate and graph the variation of relative viscosity with solids loading when $f_{cr}^v = 0.55$ and $K_H = 3.5$ using the Dougherty–Krieger equation.

16.18 Agglomeration in a slurry causes a change in f_{cr}^v from 0.6 to 0.45 and a change in K_H from 3.5 to 2.5. Calculate and compare the viscosity for each slurry when $f_p^v = 0.40$.

16.19 Calculate the effective increase in volume fraction when particles of 0.05 μm adsorb an organic polyelectrolyte film of 5 nm thickness. The volume fraction of inorganic particles is 0.40.

16.20 For the suspension of alumina in Fig. 16.21, sketch the shear rate dependence of viscosity when pH = 8.5 and 9.8.

16.21 A ceramic slurry exhibiting rheological behavior similar to that of house paint is described by the Casson equation with $\eta_\infty = 200$ mPa·s, yield

yield stress = 0.7 Pa, and m = 0.5. Graph the variation of apparent viscosity with shear rate.

EXAMPLES

Example 16.1   What is the influence of the value of the consistency index K and the exponent n on the viscosity of a material described by the power law model?

Solution. The behavior is best examined by considering separately the shear thinning (pseudoplastic case) for which $n < 1$ and the shear thickening case for which $n > 1$. For $n < 1$, the slope of viscosity with increasing shear rate is negative and the slope diminishes as $n \to 1$. Increasing K has the effect of displacing the curve upwards. For $n > 1$, the slope of viscosity with shear rate is positive and increases as n increases. A greater K increases the viscosity.

Example 16.2   What is the general effect of molecular size on the viscosity of liquids?

Solution. The increase in viscosity with molecular size for alcohols is shown in the figure below. In general, the smaller molecules exhibit less resistance to flow. On heating, the increase in specific volume has a greater effect in reducing the viscosity of the higher molecular weight alcohols, and the difference in viscosity diminishes.

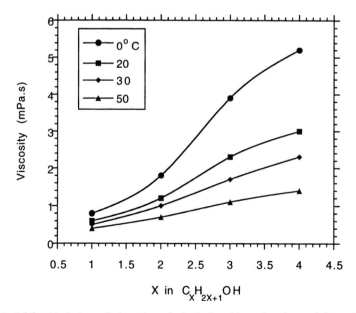

Example 16.2   Variation of viscosity of alcohols with molecular weight and temperature.

Example 16.3 A concentric cylinder viscometer with a spindle diameter of 26.0 mm and an outer diameter of 28.0 mm is used to determine the viscosity of a well-deflocculated ceramic slurry. The effective length is 50.0 mm. When the spindle is rotating at 60 rpm the torque is 2.4×10^{-3} Nm. Calculate the apparent viscosity of the slurry and the shear rate at the surface of the spindle and the cup.

Solution. The viscosity is calculated using Eq. 16.11 recognizing that the angular velocity in radians/s is $60(2\pi/60) = 2\pi$ rad/s.

$$\text{Viscosity} = [(2.4 \times 10^{-3}\ \text{Nm})/(4\pi(50.0\ \text{mm})2\pi\ \text{s}^{-1}]$$
$$\cdot\ (14.0^2 - 13.0^2)\text{m}^2/(14.0^2)\ (13.0^2)\text{m}^4$$

$$\text{Viscosity} = 497\ \text{mPa·s}$$

The shear rate is calculated using Eq. 16.12, at spindle

$$\dot{\gamma} = 2(2\pi\ \text{s}^{-1})(14.0^2/14.0^2 - 13.0^2) = 91\ \text{s}^{-1}$$

at cup

$$\dot{\gamma} = 2(2\pi\ \text{s}^{-1})(13.0^2/14.0^2 - 13.0^2) = 79\ \text{s}^{-1}$$

Note that the range of shear rate is quite small, that is, the shear rate is well defined for this viscometer with an annulus of 1 mm.

Example 16.4 A flocculant (binder) of low molecular weight may be preferred for a slurry of very high powder concentration and fluidity. But a higher molecular weight flocculant is more effective for increasing the viscosity and shear thinning behavior (pseudoplasticity) of a slurry, paste, mortar, or grout when using just a minimum amount of the flocculant. Explain this difference by calculating the sphere of influence for two aqueous binder solutions of viscosity grades 3 and 4400 (specified at a concentration $C = 2$ wt%).

Solution. The sphere of influence S is calculated using Eq. 16.27 and assuming linear dependence up to a concentration of 2 wt%,

$$S = [\ln(\eta_s/\eta_L)]/C \qquad S_{4400} = \ln(4400/1)/0.02 \qquad S_{4400} = 419$$

$$S_3 = \ln(3/1)/0.02 \qquad S_3 = 55$$

Each molecule of the 4400 grade is much longer and more hydrated with water molecules. These larger molecules provide nearly an $8\times$ greater sphere of influence than an equal weight concentration of smaller molecules.

Example 16.5 How is the critical solids volume loading in a slurry dependent on the ultimate packing factor PF_{max} of the particles and the volume of adsorbed processing additives?

Solution. The PF_{max} of the particles will have a direct influence on the amount of powder causing flow to be blocked. When the PF_{max} is higher, interstice spaces are smaller and a greater fraction of the liquid is mobilized in flow (less immobilized liquid). At the PF_{max}, the particles are touching and dilation would be required for flow. The critical fraction f_{cr}^v blocking flow will be somewhat smaller than the PF_{max}; the value for uniform spheres in Table 16.1 is 0.68, which is smaller than the $PF_{max} = 0.74$ for the tetrahedral close packing. For particles finer than about 1 μm, the volume of adsorbed additives becomes a significant portion of the volume of a particle. An increased thickness of the adsorbed film and a less compliant film may cause the viscosity to increase.

PART VII

BENEFICIATION

Processing includes operations designed to improve both the microscopic characteristics and the macroscale uniformity of the system, and also to modify the liquid content and rheology for forming. These processes, called beneficiation processes, are essential both for the control of the process rheology and for control of the microstructure. Chapter 17 presents the fundamentals of particle size reduction which is called comminution. In Chapter 18 we discuss dispersion and mixing processes for both dry and wet systems. Chapter 19 describes processes for the removal of particles of a particular size, the removal of liquid, and the removal of soluble impurities. Chapter 20 discusses the preparation of powder agglomerates of controlled character which is called granulation.

CHAPTER 17

COMMINUTION

The comminution processes of crushing and milling are widely used in ceramic processing to reduce the average particle size of a material, to liberate impurities and reduce the porosity of particles, to modify the particle size distribution, to disperse agglomerates and aggregates, to reduce the maximum particle size, to increase the content of colloids, and to modify the shape of particles. Some milling processes also provide effective dispersion and mixing, which is discussed in Chapter 18, and are used to provide communication and mixing simultaneously. The ceramic processing engineer must be familiar with these very important processes, which can have such a large impact on the rheology, fabrication behavior, sintering behavior, and ultimate microstructure of the product.

17.1 COMMINUTION EQUIPMENT

Primary crushers such as jaw crushers and cone crushers (Fig. 17.1), which produce compression and shear stresses (nipping), are commonly used individually or in series to reduce the size of coarse feed to an average size ranging down to about 5 mm or larger in size. Crushing rolls may be used to reduce less coarse feed to below 1 mm. In a hammer mill, rotating hammers pulverize particles of a brittle but relatively soft material and force the fines through the openings in a circular, wear-resistant screen (Fig. 17.2). A hammer mill is capable of producing a large reduction in size, down to about 0.1 mm. One or more of a variety of mills may then be used to further reduce the average particle size, as indicated in Fig. 17.3. Common mills used for grinding ceramic

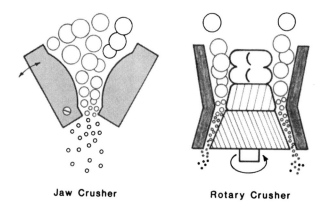

Fig. 17.1   Jaw and rotary crushers. (Courtesy of Sturtevant Inc., Boston.)

materials are ball mills, vibratory mills, attrition mills, fluid energy mills, and roller mills.

A ball mill is a hollow rotating cylinder or conical cylinder partially filled with hard, wear-resistant media having the shape of rods, short cylinders, balls, or pebbles, and granular feed material (Fig. 17.4). The mill may be as simple as a plastic bottle or porcelain jar, or a steel cylindrical shell with a hard, wear-resistant ceramic or hardened steel lining. Industrial mills range in diameter from several centimeters with a capacity of several grams to several meters in diameter with a capacity of several tons. The tumbling media in a rotating mill produce a grinding action by impacting and shearing the particles on their surfaces. Operating variables include the size and angular velocity of the mill, the size of the media relative to the size of the feed material, the loading of the mill, the relative volumes of media and feed material, the physical characteristics of the media, agglomeration of feed or product, and, in wet milling,

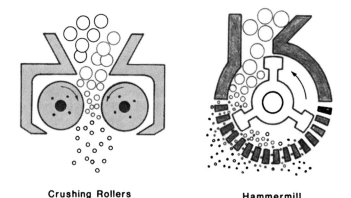

Fig. 17.2   Roll crusher and hammer mill. (Courtesy of Sturtevant Inc., Boston.)

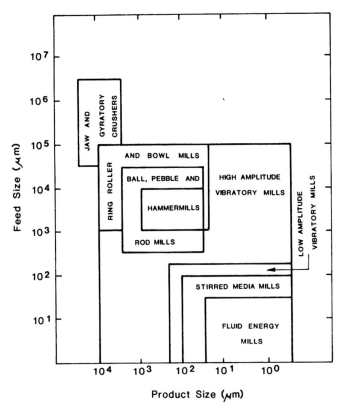

Fig. 17.3 Nominal feed and product mean size capabilities of industrial equipment. [From W. Summers, *Am. Ceram. Soc. Bull.*, **62**(2), 213 (1983).]

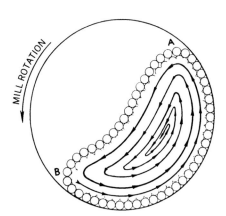

Fig. 17.4 Schematic diagram of ball mill showing media cascading from A to B and subsequent media movement.

the viscosity of the slurry during milling. In continuous ball milling, dry feed material is continually added, and the forced convection of air through the mill entrains and removes fine particles. Ball mills are typically used to produce <200 and <325 mesh materials with a wide size distribution, more narrow distributions with a mean size of several microns, and to deagglomerate and mix slurries and powders.

Industrial vibratory mills are either of the horizontal tube type or the vertical torus type (Fig. 17.5). Low-amplitude mills are used for wet grinding submillimeter feed material to a submicron size with a minimum of wear and contamination. High-amplitude mills are used for the wet or dry grinding of coarse feed and to obtain a wider size distribution. The high-amplitude mill is also a satisfactory mixer. The vertical-type vibratory mill is nearly completely filled with cylindrical media and the feed slurry. Weights mounted eccentrically control the vibration pattern and reduce the power to maintain the vibration. Vibratory mills used in wet grinding usually have a rubber or ceramic lining and are supported on rubber or metal springs. Media acceleration produces an impact energy that is significantly greater than the energy in ball milling.

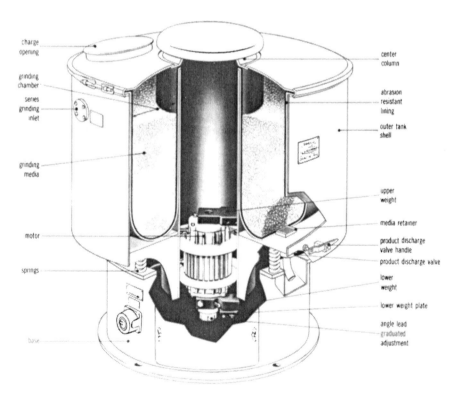

Fig. 17.5 Schematic of vibratory wet grinding mill. (Courtesy of SWECO Inc., Florence, KY.)

Milling may be continuous in some models. With vibrations, discharge is quick even when the product is a pseudoplastic suspension. Vibratory mills are used industrially for milling a wide variety of ceramic materials to a particle size that is a few microns or smaller. The grinding capacity of a large vibratory mill is limited to about 2 tons per hour, which is much smaller than can be achieved using a large ball mill.

A planetary attrition mill is a stirred media mill. A central shaft with arms rotating at 1–6 Hz continually stirs the slurry of particles and spherical media 0.3–1.0 cm and provides a means to vary the grinding energy (Fig. 17.6). Intense rolling and in-line impacts are produced by the differential velocity of media moving around the agitator arm into the cavity behind the trailing edge. The stirred media mill is used industrially for wet grinding to below 1 μm and for the dispersion of agglomerates of submicron particles. A pumping system in production models maintains circulation and uniformity of the slurry and is used for discharge. Oxidizable particles such as carbides and nitrides can be milled under an inert gas atmosphere, and a thermal jacket can be used to control the slurry temperature. Slurry capacities are quite large and continuous models are available. In a high-speed attrition mill, small grinding media and granular material are intensely agitated at 20–30 Hz to produce turbulence between a cylindrical rotor and a stator. The intense milling action produces considerable heat, and water cooling is required. Contamination due to wear of the mill lining and media is considerable and is commonly eliminated by chemical leaching, sedimentation, or magnetic separation. Attrition grinding is used for producing submicron powders of hard refractory oxides, carbides, nitrides, titania pigments, and paper-grade kaolin.

Fluid energy mills grind and classify in a single chamber. Feed material finer than about 5 mm may be either hard or soft in nature. Impacts and attrition between high-speed particles entrained in a fluid and moving in a circular orbit provide the grinding action. Centrifugal force causes oversize particles to remain in the peripheral grinding zone, but fines are drawn off in a central

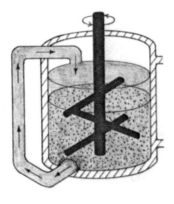

Attrition Mill

Fig. 17.6 Recirculating stirred media attrition mill. (Courtesy of Union Process Inc., Akron, OH.)

collector. The particle size and output of the product are controlled by the propellant pressure and the material feed rate. Mill capacities range from a few grams to several tons per hour. Mill linings may be wear-resistant hard materials or expendable plastic liners. Product heating in a fluid energy mill is lower than in other dry grinding operations, and an inert or nonoxidizing atmosphere such as N_2 may be used.

A roller mill in which feed material passes between a rotating table and arm supported rollers, and the disk mill in which material passes between a stationary and a rotating grinding disk are used for the comminution of moderately hard clays and porous calcined aggregates which break down under moderate stress.

Precious-metal inks of a relatively high viscosity used for electrodes and conduction paths in electronic ceramics and for decorating china and glass are commonly milled in a small, three-roller mill. The viscous paste adheres to the rolls. Differential roller speeds produce high shear stresses for dispersion and a precisely controlled particle size.

17.2 LOADING AND FRACTURE OF PARTICLES

Crushing and milling operations produce both compressive and shear loads on particles. Falling or vibrating media may produce a compressive in-line impact (Fig. 17.7). Shear is produced when a particle is seized between two surfaces moving with different velocities. Attrition is produced by frictional stresses. In milling situations using large media or rollers a combination of in-line and rolling impacts can occur. The role of attrition increases as the media size and impact force decrease and as the frequency of rubbing contacts increases.

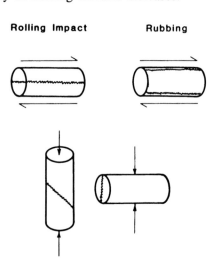

Rolling Impact   **Rubbing**

In-Line Impact

Fig. 17.7 Shear stresses in particles produced by rolling and rubbing actions and shear and tensile stresses produced by compressive loading.

The grinding energy produced during milling is proportional to the mass (m) and the change in velocity (v) of the media on impact:

$$\text{Energy} = \Delta(\tfrac{1}{2}mv^2) \tag{17.1}$$

The mass is increased by using media of a larger size or density (Table 17.1). Media with a high elastic modulus can produce a high dv/dt; media must also be hard, fine-grained, and nonporous to resist abrasive wear. Using the tetragonal to monoclinic phase transition in partially stabilized zirconia particles as an index of the grinding stress on milling, we see in Fig. 17.8 that vibratory milling with dense media produces a much greater stress than is produced in ball milling. Viscous flow and anelastic deformation reduce the grinding stress. When grinding an anelastic material the impact load on the particle is lower, because the anelastic deformation reduces dv/dt. Deformable dense materials and porous agglomerates and aggregates are somewhat anelastic.

During milling, shear and tensile stresses are produced by in-line compressive loads. Shear is also produced by rolling loads, and attrition is produced by the sliding and rubbing of particles between hard surfaces. The initial tensile and shear fracture of large particles is expected in microscopic regions of intensified stress produced by point loads and preexisting defects, as indicated in Fig. 17.9. The reduction of the fracture strength S_f of a material due to the presence of stress-intensifying flaws of depth c is described by the equation

$$S_f \sim \frac{K_{Ic}}{y\sqrt{c}} \tag{17.2}$$

where y is a constant that depends on a flaw geometry and K_{Ic} is the fracture toughness parameter that includes the work required to extend the crack. Microfissures at the edges and surfaces of particles, surface pores, and internal microcracks and pores which increase c reduce the fracture strength of brittle particles. Fracture fragments containing defects that produce a smaller stress intensification are more resistant to grinding. Dense particles with a finer grain

TABLE 17.1 Density of Industrial Grinding Media

Material	Density (Mg/m^3)
Flint pebbles	2.4
Porcelain	2.3
Steatite porcelain	2.7
High-density alumina	3.6
Zircon	3.7
Zirconia	5.5
Steel (hardened or carburized)	7.8
Tungsten carbide	15.6

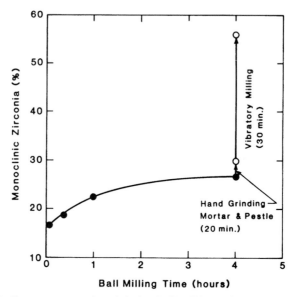

Fig. 17.8 Grinding stresses produced during ball milling a zirconia slurry with alumina media are relatively low and less than produced by hand grinding; when vibratory milling in a high-energy vibratory shaker with tungsten carbide media, the stress is much more intense. (Determined from tetragonal to monoclinic phase transformation in zirconia powder.)

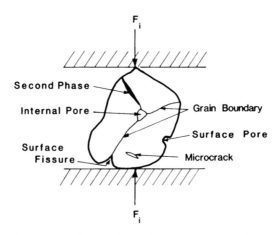

Fig. 17.9 Microstructural defects reduce the strength and grinding resistance of a particle.

size and higher fracture toughness are more resistant to attrition. Abrasion grinding may increase the concentration of edge and surface flaws, which can aid in fracture. Particles exhibiting anelastic deformation are tougher particles. Attrition is more important for the grinding of anelastic particles and particles finer than a few microns. Particles having a glassy matrix are less tough below their glass transformation temperature.

In general, the mean size of fracture fragments is smaller when the impact force is higher, but the reduction ratio is also very dependent on the microstructure of the particle. Microstructures, intergranular fracture paths, and material defects that cause crack branching may increase the apparent reduction ratio. As seen in Fig. 17.10, the milling rate of coarse-grinding tabular alumina is midway between that of single-crystal fused alumina particles and a calcined, porous, finer-grained alumina aggregate.

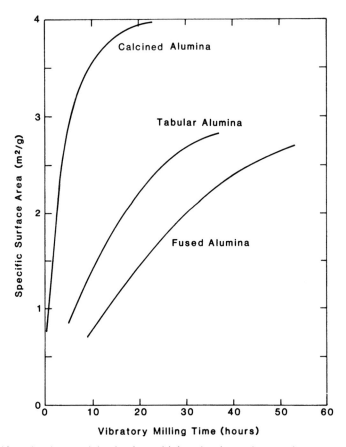

Fig. 17.10 Alumina particles having a higher density and strength are more grinding resistant. (Data courtesy of SWECO Inc., Florence, KY.)

Milling also produces atomic-scale lattice defects in particles, as revealed by an observed increase in the dislocation density, a change in the index of refraction, a phase transition, and a reduction of the coercive force of hard ferrite particles. These effects are usually less pronounced in wet ball milling, where the impact stresses are lower. When grinding calcined aggregates that contain fine crystals dispersed in a matrix phase, chemical leaching of a less resistant matrix phase by the milling liquid significantly reduce the milling time and lattice damage.

Chemical mechanisms other than leaching may improve or retard comminution in a particular mill. Water and other chemical species may be adsorbed in cracks and reduce the strength of the material, which should aid grinding. However, water vapor often causes agglomeration in dry milling. Agglomerates absorb some of the impact energy, reducing the energy for particle fracture, and a "hard pack" of agglomerates on the lining of the mill isolates particles from the milling process. Surfactants such as alcohols, oleic acid, glycols, and silicones added at a concentration less than 1% can minimize agglomeration and powder packing. Control of agglomeration is required for efficient and

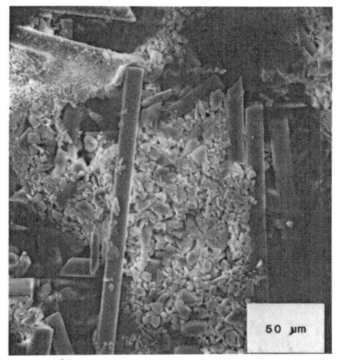

Fig. 17.11 Ball milling for 0.5 h using 1 cm media produces relatively isometric fracture fragments from silica fibers of high-aspect ratio.

Fig. 17.12 Large angular particles of refractory-grade zirconia are rounded by attrition during ball milling for 3 h.

reproducible grinding. In wet milling, a deflocculant is added to disperse agglomerates and to increase the solids loading in the slurry. The deflocculant used should not degrade during the milling process.

It is generally observed that mineral particles from compression crushers and hammer mills are more angular and anisometric than particles produced by grinding. Particles with a low aspect ratio are produced rapidly on ball milling very uniform glass fibers, as is shown in Fig. 17.11. After 1 h no fibers are observed and the preponderance of particle surfaces are fracture surfaces. Both fresh angular fracture surfaces and rounded surfaces due to attrition are observed in ball-milled, stabilized refractory zirconia with a feed size of 200–500 μm (Fig. 17.12).

17.3 MILLING PERFORMANCE

Although the choice of a mill may depend on the ultimate capital cost, capacity, period of a milling cycle, particle size distribution, and material factors, the mill selected must be used efficiently commensurate with wear and maintenance expense. We can gain insight into factors effecting milling performance by examining Eq. 17.3.

$$\frac{\text{Particles generated}}{\text{Time}} = \frac{\text{Media collisions}}{\text{Time}} \cdot \frac{\text{Particle impacts}}{\text{Collision}} \cdot \frac{\text{Particles}}{\text{Impact}} \qquad (17.3)$$

Media Collision Frequency

The probability of fracture is higher statistically when the frequency of collisions is higher (Fig. 17.13). The collision frequency per unit volume of the mill increases rapidly with a decrease in the size of the media (Eq. 13.2) and an increase in the media velocity, which is higher in attrition mills and vibratory mills. In ball mills, the frequency of impacts is limited by the tumbling velocity and the one-dimensional nature of impacts, and because the mill is only partially filled. During ball milling, collisions causing grinding occur primarily in the tumbling layers near the center. In vibratory and attrition mills, collisions causing grinding occur between a relatively larger fraction of media at some instant of time.

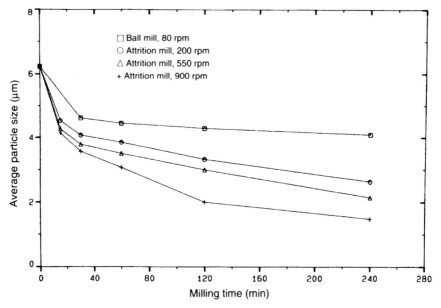

Fig. 17.13 Change in mean particle size of a deflocculated slurry containing 30 vol% of a grinding-resistant calcined alumina powder when milling in a ball mill and in a planetary attrition mill at different rotational speeds. (Courtesy of Mary Kerr, Alfred University.)

Particle Impact Frequency

The probability of hitting a particle during a collision is higher for cylindrical media, which pack more densely with finer interstices. In vibratory mills, both area and line impact zones occur, as shown in Fig. 17.14. The probability of an impact also increases as the concentration of particles on the surface of the media increases and when the particles are more uniform in size. Particles become more numerous as the particle size decreases. Agglomeration reduces the frequency of impacts. In wet milling, the slurry viscosity should be high enough to reduce the particle mobility and maintain a uniform coating on the media and retain particles in the impact zone.

Particle Fracture

The probability of fracture is high when the grinding stress is large and the strength of the particles is low. Agglomerates may disperse on impact but absorb energy needed for fracture or attrition. Collision forces are dissipated when the slurry viscosity at the shear rate of milling is too high. A combination of shear and compression stresses may enhance the fracture of tough particles of a micron size. Crack branching during fracture produces finer fragments (i.e., the reduction ratio is larger); this is enhanced in crystals of low symmetry, in flaw-loaded particles, and when the fracture velocity is higher (higher impact force, chemical interaction at the crack tip). The relative reduction ratio generally decreases as the particle size becomes smaller.

The factors in Eq. 17.3 indicate that the rate of size reduction decreases as the particles become finer and when no deflocculant is present to prevent the formation of agglomerates. For efficient milling of a slurry, the slurry should be deflocculated to disperse agglomerates, and the solids content should be increased to develop a coating of an adequate viscosity on the media to prevent the escape of particles from the grinding zone but not completely dissipate the grinding stress (Fig. 17.15); this will also minimize media wear and slurry contamination. Larger media are used for larger and stronger feed particles.

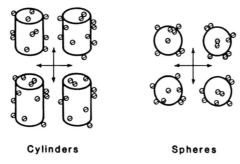

Cylinders   Spheres

Fig. 17.14 Line and area impact zones between cylindrical media produce a higher frequency of impact than for spheres.

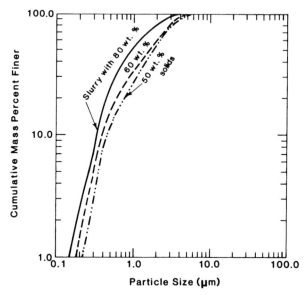

Fig. 17.15 The grinding rate of a 95% alumina body for alumina substrates is increased on increasing the solids content and viscosity of the deflocculated slurry.

Smaller media, ranging down to about 1 mm in diameter, are used for milling micron-size particles or dispersing agglomerates, because the rate of fine grinding is very dependent on the frequency of collisions. A grinding action that increases both the frequency and energy of the collisions increases the rate of milling.

17.4 MILLING PRACTICE

In ball milling, impact occurs between tumbling media; attrition also occurs between these media and other rotating media near the bottom. The height of the media before cascading is a function of the media charge, solids loading, angular speed, and the viscosity of the suspension in wet milling. The critical angular frequency ω_{cr}(Hz) causing centrifuging is

$$\omega_{cr} = 0.5R^{-1/2} \tag{17.4}$$

where R (meters) is the radius of the mill. Adhesion produced by a viscous slurry that wets the mill effectively reduces ω_{cr} and the mill is normally operated at (0.65–0.85) ω_{cr} to produce a maximum media height of 50–60° measured from the horizontal through the center of rotation. A lining with baffle bars may be used to minimize media slippage. Typically, the media charge is about 50% of the volume of the mill, and powder or slurry is added to slightly exceed the void spaces in the media; this combination provides a good compromise between grinding efficiency and rate of wear of the media and lining. Wear increases rapidly when the material does not fill the interstices between the media. The wear resistance of the media and lining should be matched. Smaller media charges and higher material loading are often used when dispersing agglomerates and mixing are the primary objectives. More wear-resistant media are used for tough materials and when wear particles are a contaminant in the product. The net cost of media and mill linings may favor low-cost, less wear-resistant media for easily milled material when the wear particles are acceptable in the milled product. Media exceeding 8 cm in diameter are available. Media 1.3 cm in diameter are very popular, and 0.6 media are extensively used for fine grinding. Media of mixed size may wear more rapidly unless the feed particles are soft. In general, the ratio of media size to feed size should exceed 25 to 1. Higher-density and/or larger media are used when the viscosity of the suspension is high. In wet milling the viscosity of the deflocculated slurry is in the range 500–2000 mPa·s at the milling temperature. The viscosity should be sufficient to form a film of slurry on the media, which holds particles in the impact zone and protects media from wear, and to minimize slippage between the media and the wall of the mill. Results indicate that when milling to a micron-size chemical deflocculation is critical, the viscosity should not be increased by partial flocculation. When milling a deflocculated slurry, the viscosity may decrease with milling time as more colloids are produced or as the packing efficiency increases.

Low-amplitude wet vibratory milling is normally limited to feed sizes smaller than 250 μm, but feed sizes may range up to about 1.25 cm in high-amplitude mills. Cylindrical media 1.3 cm in diameter are recommended for low-amplitude milling to a fine size, because smaller media do not produce a uniform distribution of vibration energy. Larger media are used in high-amplitude mills. Grinding rates in terms of the increase in the specific surface area of the product are reported to be 10–50% greater than for ball milling when the eccentric weights are properly adjusted. The viscosity of the feed suspension is normally the maximum that can be discharged from the mill with vibration. In low-amplitude milling, the feed suspension should be premixed and fill the void space between the media. Comparative results for the milling of silicon carbide and zircon sand to a submicron size, as shown in Fig. 17.16, indicate that power consumption and media wear are about seven times lower for vibratory milling.

Feed material in attrition mills is usually finer than about 50 μm, and the solids content of slurries can range from 30% to 70%. The size of the spherical grinding media is in the range of 0.5–5 mm. Attrition mills are operated at a velocity that produces turbulence and an increase in the apparent volume of the fluidized charge of slurry and media. Very rapid attrition is produced by the intense combined compression and shearing stresses, and the frequency of collisions is very high.

Compressed air, superheated steam, or compressed inert gases may be used as the dynamic medium in fluid energy milling. Feed material may range up to several millimeters in size.

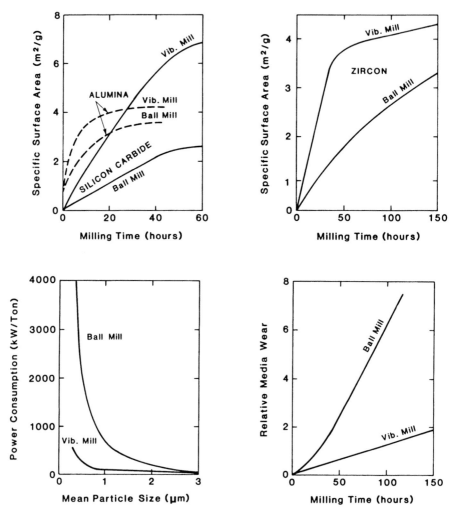

Fig. 17.16   Grinding results for calcined alumina, silicon carbide, and zircon and the power consumption and media wear for vibratory and ball milling of zircon under optimal conditions. (Courtesy of SWECO Inc., Florence, KY.)

17.5   PARTICLE SIZE DISTRIBUTIONS

The particle size distribution of the milled product will vary with the grain size and fracture behavior of particles of the feed material and the type of mill and its operating conditions. The grain size distribution in many calcined materials is approximately log-normal if regular grain growth has occurred, and the particle size distribution of the milled product is often approximately log-normal if milling disperses the grains. On grinding materials containing particles that are amorphous or single crystals, a somewhat different size distribution that is a function of their different fracture behavior may be produced.

The particle size distribution after the industrial ball milling of a < 200 mesh calcined alumina composed of particles that are aggregates of grains finer than 5 μm is shown in Fig. 17.17. The finest particle size is about 0.1 μm, and the distributions are log-normal in form. Milling preferentially reduces particles coarser than the geometric mean $\bar{a}_{gM}$ and adds to the concentration of fines smaller than $\bar{a}_{gM}$. The maximum particle size decreases slowly, and the distribution remains skewed to larger sizes. Particles as coarse as 30 μm remain after milling for 20,000 revolutions. The general effect of increasing the material loading of the mill is to decrease the rate of reduction of the maximum size, and a very broad size range is produced. Dry ball milling typically produces fewer submicron particles because of agglomeration.

The comparative log-normal distribution of sizes produced on ball milling and vibratory milling a calcined, aggregated alumina is shown in Fig. 17.18. A smaller maximum size and a narrower particle size distribution are obtained after vibratory milling, because larger particles are less likely to escape impact in three-dimensional vibratory milling using cylindrical media. Vibratory milling is typically used to grind refractory oxides such as alumina and zirconia to a size finer than 10 μm and titanates and ferrites for electroceramics and zircon opacifiers with a size range of about 0.1–1 μm.

Attrition milling produces powders having a relatively narrow size distri-

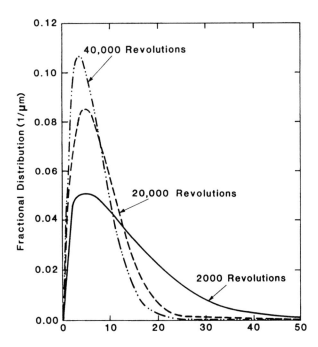

Fig. 17.17 Change in size distribution $f_M(a)$ with mill revolutions on wet ball milling a 200-mesh calcined alumina.

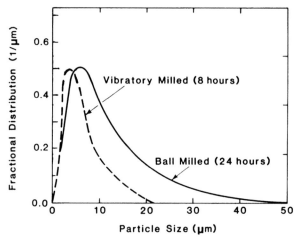

Fig. 17.18   Size distributions $f_M(\ln a)$ of 60-mesh calcined alumina after vibratory and ball milling under optimal conditions (log-normal distribution: ball mill $\bar{a}_{gM} = 5.6\ \mu m$ and $\sigma_g = 2.24$; vibratory mill $\bar{a}_{gM} = 2.4\ \mu m$ and $\sigma_g = 1.97$). (Data courtesy of SWECO Inc., Florence, KY.)

bution with a mean size ranging from a few microns to less than 0.1 μm. Fluid energy milling is used to obtain a narrow particle size distribution with an average particle size below 10 μm and relatively little submicron material.

The rate of size reduction of dense hard particles such as zircon, alumina, and silicon carbide decreases rapidly when the mean size $\bar{a}_{V/A}$ is less than about 1 μm in ball milling, less than 0.5 μm in vibratory milling, and less than about 0.1 μm in attrition milling. For each material and mill, there is a practical grinding limit that is a function of the grinding energy and the strength and toughness of the particles.

17.6   MILLING EFFICIENCY

The history of comminution is replete with equations attempting to describe energy consumption during milling. A quantitative description of milling can aid in selecting comminution equipment, in optimizing the design and operation of a particular mill, and in understanding the mechanics of material breakdown for different milling conditions. In industrial milling situations, the kinetic impact energy is only a fraction of the total energy input. Considerably more energy is consumed in accelerating the container and media, the elastic compression of media and lining materials, shearing viscous liquid, and as heat.

Charles* studied the fracture of glass cylinders of size a_o using a dropped weight crusher of constant mass and observed that the characteristic mean size

*R. J. Charles, *Trans. AIME*, **208**, 80–88 (1957).

$\bar{a}$ of the crushed product ($\bar{a} \ll a_0$) could be correlated to the kinetic energy input U_K:

$$U_K = A(1/\bar{a}^n) \tag{17.5}$$

The parameters A and n ($n > 1$) are empirical constants for a particular material and constant mode of fracture. This relation was later confirmed for the compressive fracture of glass spheres. An implication of Eq. 17.5 is that the energy for size reduction increases rapidly as the product size decreases.

In milling, the total energy input in producing a unit of product, U_T, is commonly monitored. A general empirical equation which often approximates the dependence of U_T on a characteristic mean particle size during milling is:

$$U_T = A_c[1/\bar{a}^m - 1/\bar{a}_0^m] \tag{17.6}$$

where A_c is an efficiency constant for a particular milling system and m is a fracture constant for a particular material. Equation 17.6 may be expected to describe the dependence of the energy consumption on particle size during milling when the fracture mode and general form of the size distribution remain constant, so that a single characteristic size can be used to represent the size distribution. The constant m, indicating the particle size dependence, is listed in Table 17.2 for several different materials. Higher values are observed for porous aggregates of fine grains, as would be expected from fracture theory. As is indicated in Table 17.3 for the milling of zircon to a submicron mean size, the lower value of A_c for vibratory milling indicates that the specific energy consumption is less than for ball milling. For the stirred media, mill, the exponent m is constant for both materials, but the apparent efficiency is much higher for the softer limestone.

Results such as those presented in Table 17.3 may be used to evaluate the effect of changing milling conditions or to ascertain the relative efficiency of

TABLE 17.2 Size Dependence Index m in Eq. 17.6 for Several Industrial Raw Materials[a]

Material	m
Fused alumina	1.1
Silicon carbide	1.3
Quartz	1.4
Tabular alumina	1.6
Bauxite	2.4
Calcined alumina (porous aggregate)	4.0[b]
Calcined titania (porous aggregate)	4.4[b]

[a]Ball milling and low-amplitude vibratory milling to a submicron mean size. (Data courtesy of Sweco Inc., Florence, KY.)
[b]Varies with calcination temperature.

TABLE 17.3 Values of m and A_c for Milling to a Submicron Mean Size

Mill	Material	m	$\dfrac{A_c[KW(\mu m)^m]}{(T)}$	$\dfrac{A_c[KW(\mu m)^m]}{(m^3)}$
Ball	Zircon[a]	1.8	650	3350
Vibratory	Zircon[a]	1.8	100	520
Attrition	Quartz[b]	1.8	920	2680
Attrition	Limestone[b]	1.8	500	1500

[a]Data courtesy of Sweco Inc., Florence, KY.
[b]From J. A. Herbst and J. L. Sepulveda, *Proceedings of Powder and Bulk Solids Handling Conference*, Chicago, 1978.

mills of different capacity or type. In terms of the new surface energy relative to energy input in milling processes, milling is less than 1% efficient. A small improvement in efficiency may be a very significant factor in reducing the processing cost.

SUMMARY

Crushing and grinding are important processes used to change the particle size distribution and disperse agglomerates in materials. Feed material finer than about 5 mm may be ground in a variety of mills that differ in capacity, frequency of impacts for each mill, and impact stress. The feed material must be of a specific size range and consistency. Control of the mass and size of the grinding media and the solids concentration and dispersion of the feed slurry are important to maximize the rate of grinding to a particular size with a minimum of contamination. Particle fracture is controlled by the frequency and energy of particle impacts in the mill and the fracture resistance of the particles. Relative to ball milling, vibratory milling and attrition milling produce finer particles and a narrower size distribution at a faster rate. Ball milling is used for high-capacity grinding and when a wide size distribution or mixing and dispersion of agglomerates with a minimum of damage to particle or additive phases is preferred. Control of the mill materials, mill loading, and feed consistency is required to minimize wear and chemical contamination during milling.

SUGGESTED READING

1. Sheila A. Padden and James S. Reed, "Grinding Kinetics and Media Wear During Attrition Milling," *Am. Ceram. Soc. Bull.*, **72**(3), 101–112 (1993).
2. Mary C. Kerr and James S. Reed, "Comparative Grinding Kinetics and Grinding Energy during Ball Milling and Attrition Milling," *Am. Ceram. Soc. Bull.*, **71**(12), 1809–1816 (1992).

3. Wade Summers, "Broad Scope Particle Size Reduction by Means of Vibratory Grinding," *Am. Ceram. Soc. Bull.*, **62**(2), 212–215 (1983).

4. Errol G. Kelly and David J. Spottiswood, *Introduction to Mineral Processing*, Wiley-Interscience, New York, 1982.

5. Temple C. Patton, *Paint Flow and Pigment Dispersion*, Wiley-Interscience, New York, 1979.

6. P. Somasundaran, "Theories of Grinding," in *Ceramic Processing Before Firing*, G. Y. Onoda and L. L. Hench, Eds., Wiley-Interscience, New York, 1978.

7. S. M. Wiederhorn and H. Johnson, "Effect of Zeta Potential on Crack Propagation in Glass in Aqueous Solutions," *J. Am. Ceram. Soc.*, **58**(7–8), 342 (1975).

8. D. A. Stanley et al., "Attrition Milling of Ceramic Oxides," *Am. Ceram. Soc. Bull.*, **53**(11), 813–815, 829 (1974).

9. B. Dobson and E. Rothwell, "Particle Size Reduction in a Fluid Energy Mill," *Powder Tech.*, **3**, 213–217 (1969/1970).

10. B. Steverding, "Brittleness and Impact Resistance," *J. Am. Ceram. Soc.*, **52**(3), 133–136 (1969).

11. L. D. Hart and L. K. Hudson, "Grinding Low Soda Alumina," *Am. Ceram. Soc. Bull.*, **43**, 1–17 (1964).

12. R. J. Charles, "Energy-Size Reduction Relationships in Comminution," *Trans. AIME*, **208**, 80–88 (1957).

PROBLEMS

17.1 Define grinding rate, grinding efficiency, and media wear rate.

17.2 Calculate the mass ratio for porcelain, high-density alumina, zircon, zirconia, steel, and tungsten carbide, setting alumina at a value of 1.0.

17.3 Calculate the mass ratio for spherical media of diameter 13 mm and 3 mm.

17.4 Calculate the mass ratio for spherical media of diameter a and cylindrical media of diameter a and length $L = a$.

17.5 Contrast the Young's modulus of grinding media and a lining material such as polypropylene.

17.6 Contrast the grinding surface for cylindrical media and spherical media in ball milling. Assume a 1-mm interface between rolling spherical media in contact.

17.7 State three important reasons for adding chemical additives when milling a slurry.

17.8 Beads of about 1 mm may be effective for the grinding of calcined powders containing weak aggregates, but are ineffective when grinding very dense particles. Why?

17.9   How should milling be conducted to minimize contamination of the product?

17.10   Why should the temperature be controlled when milling a slurry?

17.11   At what shear rate should the slurry viscosity be measured to correlate with ball milling, vibratory milling, and planetary ball milling?

17.12   How might the milling of slurry of talc in a ball mill differ from the milling of the talc slurry containing some 325-mesh alumina?

17.13   Consider a cylindrical particle of diameter a in contact with two crushing rolls of diameter Dia and a gap X. The angle of nip is defined by the intersection of tangents at the roll/particle contact. Draw a free body diagram indicating the compressive and shear stresses at the contact. If the coefficient of friction between the particle and the steel roll is 0.4, what is the critical roll diameter to reduce particles of $a = 1$ cm to <0.5 cm?

17.14   Mathematically the grinding rate df_N/dt can be shown to vary directly with f_N and $(a/\bar{a}_g)^n$. Does this prediction follow observations?

EXAMPLES

Example 17.1   When milling particles to reduce their size, the material to be ground is only a portion of the volume of the mill. Calculate the volume fraction of particles in a ball mill and contrast this to the volume fraction for a vibratory mill and a planetary attrition mill. Is this difference large enough to explain the faster grinding rates in vibratory and planetary attrition milling? Assume that the volume fraction of particles in the slurry is 0.40 and the PF of the media is 0.55.

Solution.   Following common practice for ball milling, the media occupy one-half of the mill volume and the slurry just exceeds the volume of interstices in the media. The volume fraction of particles in the slurry + media charge f_c^v is

$$f_c^v = (1 - PF_{media})(0.40) = (0.45)(0.40) = 0.18$$

With one-half of the mill volume filled, the effective volume fraction in the mill f^v is

$$f^v = 0.5(0.18) = 0.09 \quad \text{or 9 vol\%.}$$

Assuming that 80% of the volume is filled in the vibratory mill and planetary attrition mill, the f^v for milling this slurry would be

$$f^v = 0.8(0.18) = 0.14 \quad \text{or 14 vol\%}$$

The higher f^v indicates a more efficient use of the mill volume. As is indicated in the figure below, the grinding stress generated when vibratory milling a slurry using alumina media is very comparable to the stress produced in ball milling. The faster grinding rate of a unit volume of slurry in vibratory milling occurs because of the 3-dimensional nature of the collisions and the higher collision frequency. In planetary attrition milling, both the grinding stress and the collision frequency are commonly higher when using comparable media because of the greater media velocity.

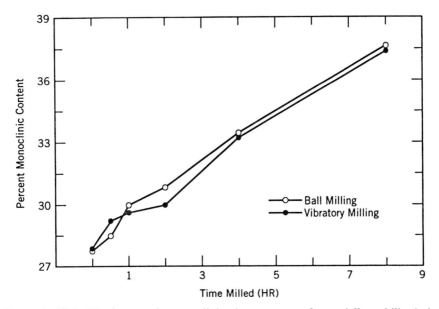

Example 17.1   The increase in monoclinic phase content of a partially stabilized zirconia powder produced by grinding stress is comparable in ball milling and vibratory milling.

Example 17.2   Smaller media are sometimes observed to more efficiently grind particle aggregates of low–moderate strength because the collision frequency per unit volume is higher when using smaller media. Calculate and compare the number of contacts per unit volume N_c for spherical media of 3, 6, 13, and 25 mm diameter. Assume a PF = 0.524 for the monosize media.

Solution.   Equation 13.2 may be used to calculate N_c,

$$N_c = 3 \text{ PF } CN/\pi a^3 = 3(0.524)(6.0)/\pi a^3 = 3.0/a^3$$

The N_c for the different media are

Media Diameter (mm)	N_c (contacts/cm^3)
3	111
6	14
13	1.4
25	0.2

When only a small grinding force is needed for comminution and the viscosity of the slurry is not high, smaller size media, which increase N_c, may improve the grinding efficiency.

Example 17.3   Calculate the operational rotation frequency for a ball mill 100 cm in diameter, operating at 75% critical speed. Compare this to the rotor speed of a planetary ball mill which may range up to 15 Hz.

Solution. The critical angular frequency is calculated using Eq. 17.4

$$\omega_{cr} \text{ (Hz)} = 0.75(0.5\ R^{-0.5}) = (0.75)[0.5(0.05\ m)^{-0.5}] = 1.7\ Hz$$

The operational rotational velocity for the ball mill is considerably less than that used for planetary ball milling.

Example 17.4   During milling it is desirable to minimize wear of the media and mill lining to minimize the contamination of the powder. What are the important factors?

Solution. Wear of the media and mill lining are lower when the hardness of the media and lining is higher than that of the particles being ground. The effective particle hardness is lower when the particles are porous aggregates. Media that toughen (K_{IC} increases) during grinding and microstructural factors such as smoothness and the absence of pores and grain boundary fracture paths in media also reduce wear rates. The volume of the slurry must exceed the PF of the media to minimize media–media abrasion. A higher particle loading in the slurry may reduce the particle mobility on the media and media–media grinding, but will produce an unacceptably slow grinding rate if the slurry viscosity is too high. The effect of slurry solids content on the wear of a standard zirconia media of 4.8 mm diameter when milling a grinding-resistant alumina at 900 rpm in a planetary attrition mill is shown in the diagram below. The higher solids loading reduced the grinding rate only slightly.

An abrasion-resistant polymer lining is sometimes used when the wear product can be tolerated or eventually eliminated from the powder. Reduction of the milling time through good general milling practice may also reduce the contamination.

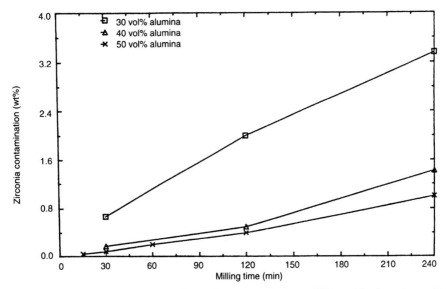

Example 17.4 Contamination of alumina slurry during milling with zirconia media depends on the alumina content in the slurry and milling time. (Courtesy of Sheila Padden, Johnson Matthey Inc.)

Example 17.5 Why are the grinding rates different for fused, tabular, and calcined aluminas. Why these differences?

Solution. Calcined alumina is in the form of porous aggregates. Tabular alumina is calcined at a much higher temperature and the aggregates are much more strongly bonded but still slightly porous (see Chapter 8). Fused alumina particles are large crystals and are nearly completely dense. The microstructural differences influence the fracture strength and mode and the apparent abrasion resistance of the particles. The porous calcined alumina aggregates of lowest strength have the least resistance to grinding, as indicated by the higher fracture constant m in Table 17.2. The resistance is much higher for the more dense, second-generation particles produced on grinding. The fused alumina particles are most dense and have the highest strength and abrasion resistance.

CHAPTER 18

BATCHING AND MIXING

Batch additives may be in the form of a granular material, a powder, a liquid, a chemical solution, an emulsion, or a slurry. Stored raw materials are transported for feeding, proportional gravimetrically or volumetrically, and then charged into a mixer to form a batch. Mixing combines, distributes, disperses, and intermingles the batch materials differing in chemical and/or physical form.

Mixing must be well controlled to reproducibly wet the particles, disperse additives, and produce a homogeneous batch of the proper consistency. Improper or nonreproducible batching and mixing is often the root cause of microscopic defects in sintered ceramics. In this chapter we consider batching and mixing equipment and processes, and the analysis of the mixedness of a material system.

18.1 BULK SOLIDS TRANSPORT AND BATCHING

Ceramic processing operations involve the transport of materials that may range widely in consistency. Bins or silos are commonly used for storing incoming dry materials and to provide "surge capacity" when there is a change in the input and output rates. The conical or wedge-shaped portion at the bottom of the bin is called a hopper. Dry solids are conveyed using a continuous belt, bucket, chain, screw, vibratory slide, or pneumatic slide type of device (Fig. 18.1). Slurries are commonly stored in agitated tanks and pumped through pipes using a positive displacement or centrifugal pump to motivate flow. A very viscous slurry or paste may be pumped using a reciprocating, progressing cavity pump.

Bin and hopper designs that produce first-in, first-out flow called mass flow produce less material segregation and supply material that is more uniform in

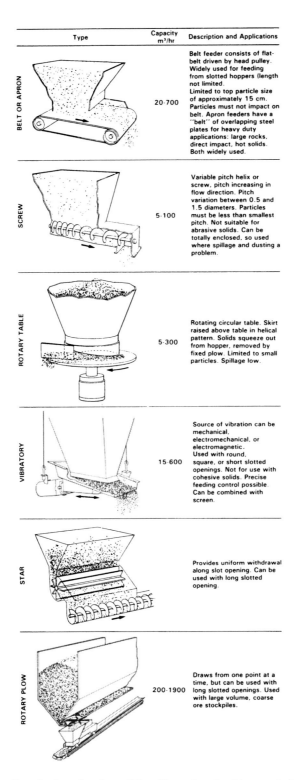

Type	Capacity m³/hr	Description and Applications
BELT OR APRON	20·700	Belt feeder consists of flat-belt driven by head pulley. Widely used for feeding from slotted hoppers (length not limited. Limited to top particle size of approximately 15 cm. Particles must not impact on belt. Apron feeders have a "belt" of overlapping steel plates for heavy duty applications: large rocks, direct impact, hot solids. Both widely used.
SCREW	5·100	Variable pitch helix or screw, pitch increasing in flow direction. Pitch variation between 0.5 and 1.5 diameters. Particles must be less than smallest pitch. Not suitable for abrasive solids. Can be totally enclosed, so used where spillage and dusting a problem.
ROTARY TABLE	5·300	Rotating circular table. Skirt raised above table in helical pattern. Solids squeeze out from hopper, removed by fixed plow. Limited to small particles. Spillage low.
VIBRATORY	15·600	Source of vibration can be mechanical, electromechanical, or electromagnetic. Used with round, square, or short slotted openings. Not for use with cohesive solids. Precise feeding control possible. Can be combined with screen.
STAR		Provides uniform withdrawal along slot opening. Can be used with long slotted opening.
ROTARY PLOW	200·1900	Draws from one point at a time, but can be used with long slotted openings. Used with large volume, coarse ore stockpiles.

Fig. 18.1 Feeding devices for dry solids. (Reproduced with permission from *Introduction to Mineral Processing*, Wiley-Interscience, New York, 1982.)

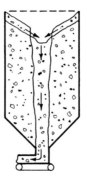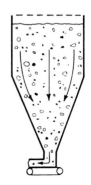

Fig. 18.2 Funnel flow and mass flow on discharging dry material from a bin.

Funnel — Flow Bin   Mass — Flow Bin

bulk density and flow rate (Fig. 18.2). A decrease in the slope of the hopper or an increase in the wall friction of the hopper or internal friction of the material increases the tendency for first-in, last-out flow called funnel flow or "rat holing" (Fig. 18.3). Funnel flow is usually acceptable only for rather coarse materials that do not segregate. An internal baffle or "insert" may decrease the central flow velocity and the tendency for funnel flow.

Powders and granular materials stored in bins and silos are subjected to the compressive load of the overlying material. The internal friction angle of the material can be determined as a function of compressive stress using the direct shear test (Eq. 15.12). A related flow property is the dynamic angle of repose, which is the slope of the cone of material formed when the material is poured and flows freely onto a horizontal surface. Flow also depends on the compressibility and cohesion of the material, as is indicated in Table 18.1. Cohesive materials gain strength when compressed. Adsorbed moisture tends to increase

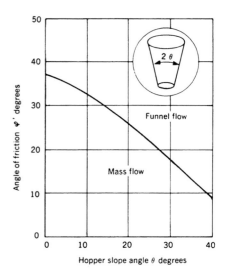

Fig. 18.3 Type of flow from a bin depends on slope of the wall and kinetic friction angle between the material and wall of the hopper. (Courtesy of Jenike and Johanson Inc., Winchester, MA.)

TABLE 18.1   Effect of Compressibility on Flow of Powders

Compressibility (%)	Unloaded	Flow	Angle of Repose
5–18	Free-flowing granules	Good to excellent	25–35
18–22	Powder and granules	Fair	35–45
22–28	Very fluid powder	Floodable[a]	33–40
>28	Cohesive powder	Very poor	>60

Source: Ralph L. Carr Jr., *Ceram. Age*, **11**, 21–28 (1970).
[a]Unstable, gushing, discontinuous flow.

the cohesion and formation of lumps in common materials, and the flowability decreases with an increase in the moisture content. Cyclic freezing and thawing may produce agglomerates or lumps that impede flow. Compaction and cohesion increase the tendency for blocked flow caused by arching and may cause the flow induced pneumatically to be nonuniform floodable flow, due to variations in the air permeability.

The flow of a powder through an opening decreases significantly when the size of the flow unit is larger than about 0.15 of the opening, because of arching. The free flow of particles and agglomerates finer than about 44 μm is retarded by particle adhesion. Ceramic particles with an adsorbed processing additive may exhibit poorer flowability when the temperature is increased and especially when the melting point is approached or the glass transition temperature of the surface film is exceeded.

TABLE 18.2   Material Storage and Transport

Consistency	Storage	Transport/Flow Assist
Granular	Bags on pallets	Belt conveyor
	Bins	Bucket conveyor
	Silos	Pneumatic conveyor
		Screw conveyor
		Rotary table
		Vibrating slide
Plastic	Plastic wrap	Auger feeder
	Polymer tubes	Piston/ram feeder
	Auger/metal tubes	
Paste	Sealed jars/bottles	Progressing cavity pump
	Covered tank	Piston pump
	Sealed hopper	
Slurry/liquid	Sealed jars/bottles	Progressing cavity pump
	Covered tank	Centrifugal pump
	Ball mill	
	Vibratory mill	
	Attrition mill	

Abrasion is a problem when transporting ceramic slurries through pipes. Valves should not restrict flow when open, and elbows in pipe systems should be of a large radius to minimize wear and contamination. Positive displacement pumps provide a volumetric flow that is less dependent on the suction or discharge pressure and are commonly used where a large pressure head is needed. Centrifugal pumps are used to produce a high flow rate.

Materials handling systems for batching ceramics are now commonly automated. A computer may be used to open and close valves, activate feeding systems, verify the gravimetric or volumetric proportioning of batch ingredients to specified tolerances, activate dust collection systems, print batching records, and maintain an inventory of raw materials. Storage and transport of materials is summarized in Table 18.2.

18.2 MIXING AND MIXEDNESS

In the initial batch, raw materials are segregated chemically or physically. Mixing is the process used to improve the chemical and physical uniformity of the mixture. The mixedness of a system refers to the state of the mixture that can be described after chemical or physical analysis.

The length, area, or volume of the largest region of each component in the mixture is referred to as the scale of segregation of that component. In a chemical solution the minimum scale of segregation is limited by the size of the largest molecules, but in a particle system it is limited by the largest particle size. A mixture is homogeneous when the composition does not vary with position. All mixtures are inhomogeneous when examined on a scale that is smaller than the size of the components. The scale of inhomogeneity is the maximum sample size which reveals a variation in composition with position. The requisite "scale of scrutiny" is the sample size requisite for determining unacceptable variations in the composition or structure of the ultimate product. During mixing the minimum scales of segregation and inhomogeneity are obtained only when agglomerates and viscous additives are well dispersed.

When two components are initially combined, an unmixed arrangement exists. Mixing caused by random flow will produce, at best, a fully randomized mixture containing incompletely dispersed components. Although randomly mixed, the degree of segregation is larger than for the fully dispersed (nonrandom) mixture, as is shown in Fig. 18.4. The intensity of segregation indicates the relative deviation of a chemical or physical characteristic when a component of one type has been intermingled with components of another type. Idealized mixtures differing in both the scale and intensity of segregation are shown in Fig. 18.5.

The degree of mixedness must be determined quantitatively to assess the homogeneity of the mix and the efficiency of a mixing device. Numerous statistical indices have been proposed to indicate the departure of a mixture from a randomly mixed state. A random mixture of particles differing only in

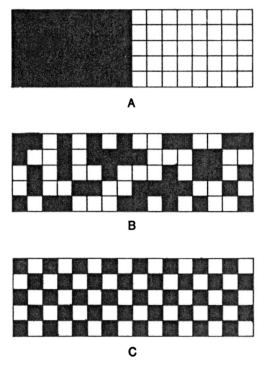

Fig. 18.4 Particle arrangements in two dimensional mixtures: (A) completely segregated, (B) completely random, and (C) completely dispersed.

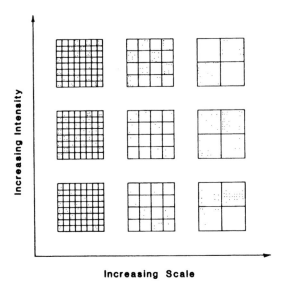

Fig. 18.5 Diagrammatic representation of mixtures varying in both intensity and scale of segregation.

composition but not physical form can be defined in statistical terms as one in which the probability that a particle drawn at random is of a particular type depends only on the proportion of that type. According to probability theory, if n particles are withdrawn from a binary random mixture one at a time and each is replaced before the next is withdrawn, the probability $p(r/n)$ that some number r would be of the first component in n trials is given by the binomial theorem

$$p\left(\frac{r}{n}\right) = \frac{n!}{(n-r)!r!} p^r q^{n-r} \tag{18.1}$$

The probability of failure $q = 1 - p$. If a sample containing n particles is withdrawn as is common in sampling, the probability given by Eq. 18.1 is correct when n is a very small fraction of the total number of particles in the mixture. If n is large and neither p nor q is close to zero, the binomial distribution is closely approximated by the Gaussian (normal) distribution.

The variance s^2 and its square root, the standard deviation s

$$s = \left[\frac{\sum\limits_{0}^{N} (C_i - \overline{C})^2}{N-1} \right]^{1/2} \tag{18.2}$$

where N is the number of samples analyzed, C_i is the fractional concentration of the component in a sample, and $\overline{C}$ is the mean fractional concentration of that component, are used as indices of mixedness but are dependent on the mean concentration of the component analyzed. Many different indices of mixedness that range from 0 to 1 have been defined in terms of the standard deviation of the original segregated mixture (σ_0), the standard deviation after mixing (s), and the ultimate standard deviation for a completely random mixture (σ_r). The index M, where

$$M = \frac{\log \sigma_0 - \log s}{\log \sigma_0 - \log \sigma_r} \tag{18.3}$$

has been recommended because it provides a clear distinction between mixtures approaching the theoretical optimum.* For a completely random two-component mixture of particles of separate identity but otherwise physically uniform, the standard deviation σ_r decreases as the number of particles in the sample n (i.e., the sample size) increases:

$$\sigma_r = \left(\frac{C_1(1 - C_1)}{n} \right)^{1/2} \tag{18.4}$$

*F. H. H. Valentin, *Chem. Eng.*, **208**(5), 99–104 (1967).

In a mixture of two different materials varying in particle size, the standard deviation of a completely random mixture may be calculated using the equation

$$\sigma_r = \left[C_1 C_2 \left(\frac{C_2 \left(\Sigma f^W \right)_1 + C_1 \left(\Sigma f^W \right)_2}{M_s} \right) \right]^{1/2} \quad (18.5)$$

where M_s is the mass of a sample taken and Σf^W is the sum of the product of the weight fraction f of particles in each size class and the mean particle weight W in the class. As defined by Eq. 18.5, the standard deviation of a statistically random mixture σ_r decreases as the proportions of the two components becomes less equal and the mean particle size of the components decreases. The standard deviation of completely segregated material is*

$$\sigma_0 = \left[\frac{C_1}{1 - C_1} \right]^{1/2} \quad (18.6)$$

A multicomponent system may be treated as a binary system by considering one component dispersed in a mixture of the other components.

Indices such as s and M are used to assess the spatial uniformity of bulk material—that is, the macroscale mixedness—as a function of parameters of a mixing operation and to compare the effects of different mixing techniques.

Analyses used to determine macroscale mixedness include chemical, mineral, and phase analysis; colorimetric analysis; and the analysis of phosphorescent or radioactive tracers. Samples should be removed from diverse regions of the mixture following a scheme that will reveal segregation and inhomogeneity. The sampling technique used should not alter the mixedness in any significant way.

Sampling and analysis errors must be kept to a minimum if the calculated variance is to provide an accurate assessment of the mixedness. The uncertainty in the standard deviation calculated for a mixture is low only when a large number of samples are taken (Table 18.3). A comparison of the mixedness of two systems may be meaningless when only a few samples have been taken.

Microscopic analyses and microscale chemical analyses may be used to determine the scale of segregation in an individual sample—that is, the microscale mixedness. Examples of microscale inhomogeneities are shown in Fig. 18.6. Microscopic examination and chemical reactivity that depends on the diffusion path between components may be used to study dispersion and microscale mixedness. Agglomerates may increase the diffusion path and the time or temperature to complete a chemical reaction. Organic inhomogeneities of low mean atomic number may be identified conveniently using the backscattered electron detector of the scanning electron microscope.

*F. H. H. Valentin, *Chem. Eng.*, **208**(5), 99–104 (1967).

TABLE 18.3 Uncertainty in the Ratio of Estimated s to True s_t Standard Deviation

	Range of s/s_t	
N	95% Confidence	99% Confidence
5	0.60–2.87	0.52–4.39
10	0.69–1.82	0.62–2.29
15	0.73–1.58	0.67–1.85
20	0.76–1.46	0.70–1.67
50	0.84–1.24	0.79–1.34
100	0.88–1.16	0.85–1.22
200	0.91–1.11	0.88–1.14
500	0.94–1.07	0.91–1.09

Source: F. H. H. Valentin, *Chem. Eng.*, **5**, 99–104 (1967).

(A)   (B)

Fig. 18.6 Segregated organic binder appears dark in (A) secondary electron image and (B) backscattered electron image. (Courtesy of Ward Votava, Alfred University.)

18.3 MIXING MECHANISMS

Mixing is produced by the mechanisms of convection, shear, and diffusion (Fig. 18.7). Convection transfers components from one region to another. Shear increases the interface between components by deforming their shape. Diffusion interchanges molecules and particles randomly between neighboring microscopic regions of the mixture.

The relative importance of each mixing mechanism depends on the design of the mixer, the consistency and rheology of the material, and the energy

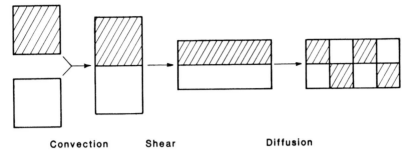

Convection   Shear   Diffusion

Fig. 18.7   Flow of two materials during mixing and mixing mechanisms.

input in mixing. Flow produced by feeding devices and stirring and baffling contribute importantly to convection. Mixer surfaces moving at different velocities will produce a shear stress gradient in intervening material if slippage does not occur. Intense shear stresses can disperse agglomerates and mix viscous materials on a microscale. Turbulent flow, cavitation produced by ultrasonication, and impact can disperse agglomerates and accelerate diffusive mixing.

The relative contribution of the mixing mechanisms is dependent on the design of the mixer and the Reynolds number during material flow. Bulky powders and liquids and suspensions of low viscosity may be mixed by convection, shear, and turbulence. In the mixing of very viscous liquids, pastes, slurries, and plastic bodies, only laminar flow commonly occurs, and the mixing mechanisms are convection and shear.

18.4 MIXING EQUIPMENT AND PRACTICE

Each material system has a particular consistency and flow resistance that may vary with the time of mixing and flow rate (see Fig. 14.2). Agglomerates in the mixture have a range of strengths, and a mixing intensity in excess of that required for bulk flow is normally required to disperse the agglomerates. Mixing at a low shear rate but a high shear flow, the product of shear rate and time is often used for the bulk mixing of low-viscosity suspensions. The shearing stress developed can be increased by initially mixing at a higher solids content and viscosity and then admixing additional liquid. Relatively high shear stresses produced at a high shear rate are usually required to completely disperse agglomerates in a low-viscosity suspension. Many commercial mixers are designed with two or more mixing elements to produce both high shear mixing in a local region of the mixture and low shear bulk mixing. When mixing a system of high viscosity, considerable torque and mixing energy are required for both the dispersion of agglomerates and bulk flow. High-torque mixers for ceramic systems must be rugged in design, abrasion-resistant, and relatively expensive. Material systems of high viscosity such as an extrusion body may

TABLE 18.4 Mixers for Different Consistencies

Consistency	Mixers
Slurry	Impeller, paddle type, planetary, flat blade turbine, anchor, ball mill, vibratory mill, planetary ball mill, microstratifier
Paste	Sigma type, three roller, muller, planetary, ribbon, microstratifier
Plastic body	Sigma type, double planetary, muller, mix mill, pugmill, twin screw auger
Granules	Planetary, pan, ribbon, mix mill granulator, twin cone blender, conical mixer

be mixed as a slurry with excess liquid which is removed later by filter pressing or spray-drying. Mixers used for mixing materials of different consistencies are listed in Table 18.4.

Slurries and Pastes

Each type of mixer is effective over a particular range of viscosity, as is shown in Fig. 18.8. Impeller mixers that produce both circulation and circumferential flow (Fig. 18.9) are used for mixing a wide variety of suspensions and slurries. In blungers and storage tanks, agitation and flow are produced by rotating paddles, a marine propeller, or a turbine impeller. The flow type in an impeller mixer can be characterized using a Reynolds number (Re),*

$$\text{Re} = \frac{ND_s(\text{Dia})^2}{\eta} \qquad (18.7)$$

where D_s and η are the density and viscosity of the slurry, respectively, N is the impeller speed (rev/time), and Dia is the diameter of the agitator. The power requirement P is approximated by the equation*

$$P = C_d N^3 D_s(\text{Dia})^5 \qquad (18.8)$$

where C_d is the drag coefficient for the impeller. An increase in those factors that increase Re also increase P, and a single impeller is not satisfactory for creating bulk volumetric flow and turbulence in a slurry of high viscosity. Small, high-speed (several 1000 rpm), serrated, or vaned impellers or an ultrasonic vibratory probe may be introduced to produce cavitation and turbulence for dispersing agglomerates and friable particles in a local region. The high-intensity mixing device may also be installed in line to continuously disperse material entering or leaving the tank. The flow of a suspension through an orifice can produce intense microscale mixing if turbulent flow occurs. Ultra-

*E. J. Kelly and D. J. Spottiswood, *Introduction to Minerals Processing*, Wiley-Interscience, New York, 1982.

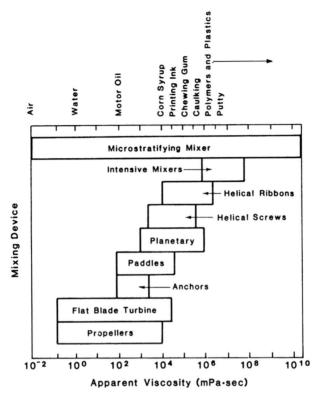

Fig. 18.8 Working ranges of mixers for mixing slurries, pastes, and plastic bodies. (Courtesy of TAH Industries Inc., Imlaystown, NJ.)

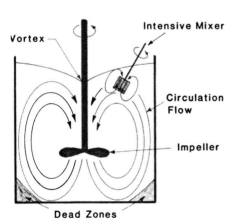

Fig. 18.9 Convection and shear flow in an impeller mixer; the intensive device may provide turbulence or cavitation.

sonic tanks, probes, and throats provide a controlled, intense dispersion of suspensions by creating cavitation. Slurries of moderate viscosity are also mixed using anchors, paddles, and planetary impellers. Mixing under a vacuum to remove air from the slurry is accomplished by using a vacuum-tight cover; the elimination of bubbles may also aid in dispersing agglomerates.

These systems are also commonly mixed in ball mills and high-amplitude vibratory mills. Media impacts disperse agglomerates, and the tumbling media provide convection and shear for bulk mixing.

More viscous slurries and pseudoplastic pastes may be mixed in a planetary mixer, helical mixer, or intensive mixers (Fig. 18.10). Mixing elements are relatively large and strong to provide the high torque needed for shear flow in bulk. These mixers are sometimes provided with a heating or cooling jacket to improve wetting, control the viscosity more precisely, or prevent overheating.

A microstratifying mixer produces a controlled stratification in the material rather than random flow. Static mixing elements fixed in a conduit divide, invert, and radially mix material received from the preceding element with a programmed precision and efficiency, as is indicated in Fig. 18.11. Two inlet streams become four alternating streams after the first element, eight after the second element, and so on. The number of strata increases exponentially with the number of mixing elements n and the striation thickness X_S is

$$X_S = \frac{\text{Dia}}{2^n} \qquad (18.9)$$

where Dia is the inside diameter of the unit. Shear force produced during circulation flow in the semicircular channel causes material to migrate from the center of the pipe to the inside wall every 5 elements. The mixer may be jacketed for control of temperature. Some axial mixing is provided by the constantly changing velocity profile. Microstratification may be used for mixing liquids, suspensions, and pastes ranging widely in viscosity (Fig. 18.8).

Plastic Bodies

High-energy mixers constructed from wear-resistant materials are required to directly mix ceramic bodies having a plastic consistency. Cylindrical pan mixers, which may vary somewhat in design, are used for mixing plastic material. Blades or plows direct the flow of material, and heavy Muller wheels rolling over material or rapidly rotating blades or vanes may be used to provide local high-shear energy to break up agglomerates (Fig. 18.12). Liquids are commonly introduced as a spray after dry solids have been mixed. Batching, mixing, and discharge can be automatically controlled, and the temperature of the pans may be controlled. Pan-mixed material is often mixed further in a pug mill. In the pug mill, angled paddle-type knives cut and shear material and then consolidate it under pressure and extrude it onto a carrier. Large deairing

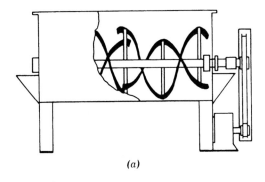

(a)

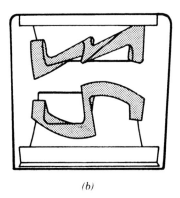

(b)

(c)

Fig. 18.10 Examples of (a) helical ribbon mixer, (b) sigma-blade intensive mixer, (c) double planetary mixer, (d) horizontal mix-mill intensive mixer, and (e) double roller mill for mixing a viscous paste at a controlled temperature. [Photo for c courtesy of Chas. Ross and Son Co., Hauppauge, NY; photo for d courtesy of Processall Inc., Cincinnati, OH; schematic e from B. C. Mutsuddy, *Am. Ceram. Soc. Bull.*, **68**(10), 1797 (1989).]

pug mills can continuously process tons of material per hour. Plastic filter cake is often preblended in a version of a pug mill called a wad mill. Intensive mixers that typically have very heavy-duty sigma blades and heavy-duty double-planetary mixers (Fig. 18.10) are also used widely for mixing plastic bodies in relatively small batches.

(*d*)

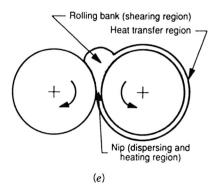

(*e*)

Fig. 18.10 (*Continued*)

Granular Materials and Powders

Granular materials and powders are typically mixed with a small amount of liquid, binder solution, and so on, using a pan mixer, conical blender, ribbon mixers, or rotating drum mixers. Muller-type wheels or rotating beaters are required to disperse agglomerates and more uniformly distribute the liquid or solution introduced using a spray nozzle. Wetted powders are not free-flowing unless they become granulated after distribution of the liquid additives. The Muller wheel or rotary disperser may be elevated to allow for granulation.

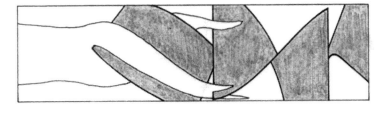

Element Number

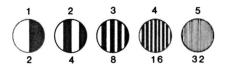

Number of Striations

Fig. 18.11 Mixing produced by flow through a microstratifying mixer. (Courtesy of TAH Industries Inc., Imlaystown, NJ.)

Conical blenders are also widely used for mixing free-flowing powders and liquid additives introduced as a spray (Fig. 18.13). Tumbling flow provides convection and shear mixing; in the double-cone blender, the intersection of the cones splits the flow once every revolution. Models with an internal intensive mixing bar provide additional energy for comminuting agglomerates in dry feed powders and those formed on admixing liquid. Tumbling without intensive mixing produces granulation. Construction may be of either plastic or stainless steel, and some models are available with a thermostatically controlled jacket.

When mixing slurries, the liquid and soluble processing additives are normally introduced into the tank prior to introducing particle materials. Automatic material handling systems convey, meter, and charge materials under computer control. Colloidal materials and recycled scrap is usually slurried separately at

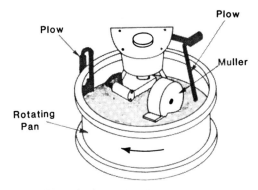

Fig. 18.12 Cylindrical pan mixer.

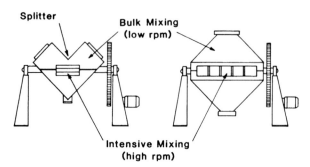

Fig. 18.13 V-type and double-cone mixers with intensifier element.

a particular specific gravity and added prior to coarser materials. When mixing more viscous slurries and pastes, solids may be blended dry before adding to the liquid or a colloidal suspension. Intense dispersers may be incorporated in the mixing tank or in the piping system following discharge.

In mixing granular materials and powders to produce pressing material or a plastic body, the extreme fines tend to adsorb the liquid and form granules. Prewetting of the coarser particles, a fine liquid spray, and the use of Muller wheels or rotary beaters to disperse agglomerates aid in producing coarse particles coated with wetted fines. Systems with porous particles or extremely fine colloids may adsorb considerable liquid. Powders are usually dry blended and deagglomerated before adding liquids. Humidified aging may improve the moisture uniformity. A viscous binder solution is not easily admixed uniformly into fine ceramic powder. Pressing powders for high-performance technical ceramics are commonly mixed as a slurry and spray dried into pressing granules to achieve a higher degree of homogeneity.

18.5 MIXING PERFORMANCE

Mixing is a complex process, and the performance of a mixing system is usually evaluated experimentally. However, a simple model can aid in visualizing the process. A general mixing law for the rate of mixing dM/dt produced in a particular mixer is

$$\frac{dM}{dt} = A(1 - M) - B\phi_u \tag{18.10}$$

where M is the mixedness parameter, A and B are mixing and unmixing constants, respectively, for a particular mixing system, and ϕ_u is the unmixing potential of the material. The constant A is larger when the mixing action is intense relative to the cohesive resistance of the material. The unmixing potential ϕ_u indicates the tendency of materials to demix owing to differences in the particle size, shape, density, and surface attraction, and B is a function of the dynamics of the mixer and the formation of dead zones.

The mixedness initially increases with the number of revolutions, but on nearing a maximum, the rate levels off (Fig. 18.14). The maximum mixedness achieved depends on the mixing and unmixing rates for the system, and the optimum mixing time depends on the unmixing potential. A change that reduces the unmixing potential such as an increase in viscosity may reduce the mixing constant A as well. For $B > 0$, an optimum mixing time exists.

The increase in mixedness with number of revolutions for the mixing of initially segregated 5-mm spherical particles in a two-dimensional double-cone blender is shown in Fig. 18.15. The rate of mixing dM/dt is dependent on both the number of rotations and the mixer loading. There is an optimum residence time for each loading condition, and the number of revolutions required to achieve a nearly random mix increases as the loading increases.

The correlation of the microstructure of a ceramic material with mixing parameters has been reported in relatively few studies. Macroscale nonuniformity arises from nonflow, biased flow, and segregation. Relatively little mixing occurs in stagnant regions, where convection is poor. The ability of a small, high-speed impeller to distribute batch components in the bulk and disperse all agglomerates decreases as the apparent viscosity increases, as is shown in Fig. 18.16. Settling of larger or denser particles is a problem when mixing a low-viscosity suspension. Chemical changes during mixing that cause agglomeration may also produce segregation. Fibrous particles or colloidal particles may adhere more strongly to mixer surfaces. When mixing dry powders ranging widely in size, vibrations may cause fine particles to percolate down through pores in the coarse particles. Corners, surfaces of mixing hardware, and wear pockets may collect poorly mixed material which may later escape into the bulk. Close tolerances between moving and static portions of the mixing hardware aid in minimizing stagnant zones.

Difficult mixing situations include the admixing of a high-viscosity solution

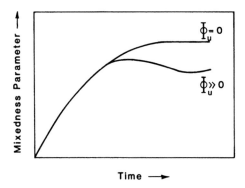

Fig. 18.14 The general dependence of mixedness on mixing time; an optimum mixing time is apparent when the unmixing parameter $\phi_u > 0$. (Refer to Eq. 18.10.)

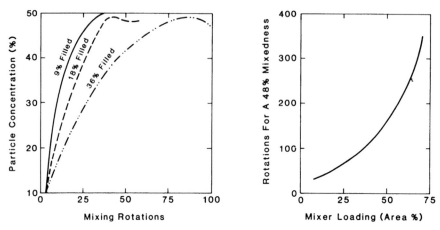

Fig. 18.15 Intermingling of initially segregated black and white spheres in a two-dimensional version of a V-blender increases with mixing revolutions and the number of rotations required increases when the volume loading of the mixer increases. (Note: the initial relative concentration of each type was 0.5.)

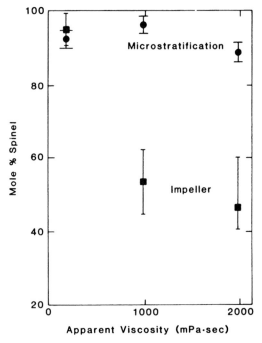

Fig. 18.16 Average amount of $ZnAl_2O_4$ spinel formed on reacting samples of mixed ZnO and Al_2O_3 powders (as slurries) depends on the scale of inhomogeneity in the mixture. For a higher viscosity, convection currents occur nearer the small impeller and convection and shear of the bulk material are incomplete.

into a fine powder or low-viscosity liquid, dispersing agglomerates in a high viscosity matrix, and dry mixing a minor additive uniformly in a powder or granular material. A sensitive additive is sometimes premixed with a component of the batch and its state checked before adding to the total mixture.

SUMMARY

Material handling and batching systems are now highly mechanized and automated, and factors that influence the flow behavior must be recognized to ensure proper flow and to minimize segregation during handling.

Mixing technology is used to reduce the scale of segregation of the components and to reduce the scale of inhomogeneity in the mixture. Mixing mechanisms are convection, shear, and dispersion. Penetration of liquid into agglomerates slows as the viscosity of the liquid increases. Steps in powder dispersion are wetting, mechanical disruption of agglomerates, and the spreading of liquid on particle surfaces. Mechanical disruption of agglomerates in slurries is produced by impact in wet milling or turbulence during high shear rate mixing, and in pastes and plastic bodies by high shear stress, which is the product of the viscosity and the shear rate.

Mixing must be controlled on both the macroscale and the microscale. The degree of mixedness may be determined from the analysis of samples. Macroscale mixedness is commonly specified using a parameter such as the standard deviation or mixedness index which indicates the variation within a sample set. Microscale mixedness is determined by microscale physical and chemical analyses. Unmixing tendencies reduce the degree of mixedness achievable. Unmixing is influenced by reducing differentials in the particle size, shape, density, and surface behavior. The rate of mixing and homogeneity achieved depend on the mechanics of the mixing device, the degree of loading of the mixer, and the chemical and physical properties of the mixture. The selection of mixing technology depends on the consistency and flow properties of the mixture, the acceptable scale of inhomogeneity, and the batch capacity required.

SUGGESTED READING

1. Errol G. Kelly and David J. Spottiswood, *Introduction to Mineral Processing*, Wiley, New York, 1982.
2. Richard Hogg, "Grinding and Mixing of Nonmetallic Powders," *Am. Ceram. Soc. Bull.*, **60**(2), 206–211, 220 (1981).
3. Milton N. Kraus, *Pneumatic Conveying of Bulk Materials*, McGraw-Hill, New York, 1980.
4. Thomas D. Croft, James S. Reed, and Robert L. Snyder, "Microstratified Mixing of Ceramic Systems: Viscosity Dependence," *Am. Ceram. Soc. Bull.*, **57**(12), 1111–1115 (1978).

5. G. Daniel Lapp, "A Numerical Index of Stirrer Performance in Continuous Flow," *Am. Ceram. Soc. Bull.*, **56**(2), 186–188, 193 (1977).

6. H. M. Sutton, "Flow Properties of Powders and the Role of Surface Character," in *Characterization of Powder Surfaces*, G. D. Parfitt and K. S. W. Sing, Eds., Academic Press, New York, 1976, pp. 107–158.

7. J. M. Pope and K. C. Radford, "Blending of UO_2–ThO_2 Powders," *Am. Ceram. Soc. Bull.*, **53**(8), 574–578 (1974).

8. M. Cable and J. Hakim, "A Quantitative Study of the Homogeneity of Laboratory Melts Using Simple Stirrers," *Glass Tech.*, **14**(4), 90–100 (1973).

9. H. Ries, "The Importance of Mixing Technology for the Preparation of Ceramic Bodies," *Interceram.*, **3**, 224–228 (1971).

10. Jerry R. Johanson, "Working Guide: Feeding and Storage Methods," *Ceram. Age*, **11**, 21–28 (1970).

11. Ralph, L. Carr Jr., "Properties of Solids: How They Affect Handling," *Ceram. Age*, **10**, 25–30, (1970).

12. R. C. Rossi, R. M. Fulrath, and D. W. Fuerstenau, "Quantitative Analysis of the Mixing of Fine Powders," *Am. Ceram. Soc. Bull.*, **49**(3), 289–292 (1970).

13. F. H. H. Valentin, "The Mixing of Powders and Pastes: Some Basic Concepts," *Chem. Eng.*, **208**(5), 99–104 (1967).

14. V. W. Uhl and J. B. Gray, *Mixing: Theory and Practice*, Vol. 1, Academic Press, New York, 1966.

15. K. Strange, "Degree of Mixing in a Random Mixture as a Basis for Evaluating Mixing Experiments," *Chem.-Eng. Tech.*, **26**(3), 331–337 (1954).

16. P. M. C. Lacey, "Developments in the Theory of Particle Mixing," *J. Appl. Chem.*, **4**(5), 257–268 (1954).

PROBLEMS

18.1   Estimate the maximum hopper angle for mass flow when the friction between the material and wall is 0.4.

18.2   Should the outlet dimension of a bin X_B depend on the moisture content of the material? Explain using the concept of $X_{arch} = 2\sigma_Y/D_b g$, where X_{arch} is the dimension for which arching is expected, σ_Y is the yield strength of consolidated material of bulk density D_b, and g is the acceleration of gravity.

18.3   Explain why vibration may cause a material stored in a bin to flow more poorly.

18.4   Explain pneumatic loading and unloading using the concept of effective stress (see Chapter 15).

18.5   Compare and contrast belt and screw feeding devices.

18.6   Illustrate mixtures containing a minor phase with both a large and a small scale of segregation dispersed with a small and a large scale of inhomogeneity.

18.7 Prepare two grids containing 90 squares numbered 10–99. On one grid shade one-half of the squares corresponding to a microstratified mixture. For the other, generate two-digit random numbers and shade these squares until one-half are shaded. Contrast the two grids in terms of scale of segregation and scale of inhomogeneity.

18.8 The weight percent concentrations of an ingredient in 10 samples removed at random from a mixture after 10 min of mixing are 20.2, 21.1, 21.0, 16.4, 20.3, 17.1, 22.0, 20.4, 23.4, and 20.1 wt%, and after 30 min 19.5, 21.0, 23.0, 18.6, 20.1, 18.2, 23.0, 19.1, 21.2, and 19.3 wt%. Calculate the standard deviation for each data set. Using a 95% confidence interval, can you conclude that the additional 20 min of mixing improve the mixedness? Explain.

18.9 If the standard deviations in Problem 18.8 had been obtained for 20 samples in each set, would your conclusion be different?

18.10 From Eq. 18.10, derive an expression for M_{max} as a function of A, B, and ϕ_u. What is the value of M_{max} when $\phi_u \rightarrow 0$?

18.11 Show that the standard deviation for a completely random mixture of black and white particles, otherwise identical, is larger when the fraction of the minor component is larger. Assume an invariant sample size.

18.12 Calculate the standard deviation for a completely random mixture of equal proportions of petalite (Table 7.2) and quartz (Table 7.7). The sample size is 5 g. Consult Eq. 18.5.

18.13 If the experimental standard deviation in problem 18.12 is 0.09 after mixing for 20 min, calculate the mixedness index M (see Eq. 18.3).

EXAMPLES

Example 18.1 When charging a material with a wide particle size distribution into the center of a bin, the coarser particles tend to segregate at a greater radius. Why does this occur and how does the segregation influence the particle size distribution of material unloaded from the bin?

Solution. The segregation is caused by the greater mass and momentum of the larger particles, which slide and roll to a greater radial position from the point of discharge. On unloading, some differential in particle size distribution for mass flow discharge can be expected for small but not large batches. On discharge by funnel flow, the mean size will increase as the bin is unloaded and a differential batch to batch size can be expected even for large batches.

Example 18.2 What are the differences in the scale of segregation and the scale of inhomogeneity for the three mixtures in Fig. 18.4.

Solution. Assessed in terms of an area parameter, the scale of segregation is 40 blocks in A, 8 blocks in B, and 1 block in C. The size of an area which can reveal inhomogeneity is 79 blocks for A, 39 blocks for B, and 3 blocks for C.

Example 18.3   In assessing the mixing time effect on the macroscale mixedness for an industrial mixing procedure, 10 samples are withdrawn at random from the mixture after mixing for 10 min and another 10 after mixing for 20 min. The samples are analyzed for component A and the standard deviation of the concentrations of the first set is 1.9 wt% and for the second, 1.7 wt%. For a 95% confidence interval, can it be concluded that the additional mixing time was beneficial?

Solution. The uncertainty in the standard deviations calculated from the limited data may be estimated using the parameters in Table 18.3. For the 10 samples and a 95% confidence interval, the factors are 0.69 and 1.82. The calculated range is

10 min	$0.69(1.9 \text{ wt}\%) = 1.3\%$ to	$0.82(1.9 \text{ wt}\%) = 3.5 \text{ wt}\%$
20 min	$0.69(1.7 \text{ wt}\%) = 1.2\%$ to	$0.82(1.7 \text{ wt}\%) = 2.1 \text{ wt}\%$

Because of the large uncertainty in the calculated experimental standard deviation, that is, the ranges for each overlap, it cannot be concluded that the additional mixing time was beneficial. Considerably more samples are needed to distinguish the time effect with certainty.

Example 18.4   When mixing a viscous slurry with a single impeller, why is it difficult to maintain a high pumping capacity for macroscale mixing and also produce turbulence for microscale mixing?

Solution. The ratio of the power requirement to the Reynolds number provides an insight into the effect of the operational variables. Combining Eqs. 18.7 and 18.8,

$$\text{Re} = ND_s \, (\text{Dia})^2/\eta \quad \text{and} \quad P/\text{Re} = C_d \eta N^2 \, (\text{Dia})^3$$

The power requirement to exceed the critical Re for turbulence increases when the viscosity increases. For mixing with a constant power, this increase may be offset by reducing the impeller diameter (note cubed dependence) and increasing the speed. But the smaller impeller may not produce flow throughout the bulk of the slurry and segregation and undispersed agglomerates may be found in poorly sheared regions of the mixture.

Example 18.5   When dispersing a powder in a wetting liquid, what is the most resistant step?

Solution. Steps in immersing and wetting a particle are shown in the figure below. Assuming that the solid surface is in equilibrium with the vapor of the liquid, the work of adhesion W_a, work of immersion W_i, and work of spreading W_s may be quickly derived by considering the change in number of wetted surfaces and the surface tension on each surface:

$$W_a = \gamma_{SL} - (\gamma_{LV} + \gamma_{SV}) = -\gamma_{LV}(\cos \theta + 1)$$

$$W_i = 4\gamma_{SL} - 4\gamma_{SV} = -4\gamma_{LV} \cos \theta$$

$$W_s = (\gamma_{SL} + \gamma_{LV}) - \gamma_{SV} = -\gamma_{LV}(\cos \theta - 1)$$

When W is negative, the step will occur spontaneously. When W is positive, work must be expended on the system for the process to occur. It may be concluded from the equations that W_a is spontaneous for $\theta < 180°$ and W_i is spontaneous for $\theta < 90°$. But W_s is positive for $\theta > 0°$ and $W_s \rightarrow 0$ as $\theta \rightarrow 0°$. Work must be done to obtain spreading for real values of the wetting angle.

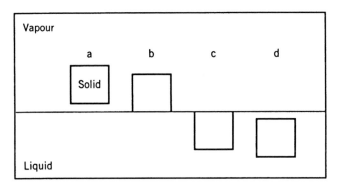

Example 18.5 Model for a–b adhesion, b–c immersion, and c–d spreading of liquid on a particle.

CHAPTER 19

PARTICLE SEPARATION, CONCENTRATION, AND WASHING PROCESSES

The microstructure of the final ceramic product is very dependent on the range of particle sizes and impurities in the processing system. The separation of particles by size is a rather common operation in ceramic processing. Impurity particles in raw materials and those introduced during processing, such as from the abrasion of materials handling and processing machinery, are sometimes removed based on a difference in size, density, surface behavior, or electromagnetic properties. Liquid separation processes are used to remove soluble impurity salts and to reduce the concentration of ions that interfere with deflocculation or coagulation processes. Separation processes are also used to remove liquid to concentrate particles and change the consistency of the system and to remove particles from recycled waste liquids.

Separation and concentration processes have long been used on an industrial scale for beneficiating industrial minerals and in the processing of traditional ceramics. These beneficiation operations should also be considered in developing processes for producing many of the newer ceramics.

19.1 PARTICLE SIZING

Continuous screening and batch-sieving operations are used to directly separate rather coarse particles. Finer particles are separated by differences in their settling rate, which is called classification. Sizing may involve the separation of particles into a relatively narrow range of size, or discrete size ranges, or it may eliminate extremely coarse or fine sizes. Cyclone and centrifugation techniques are commonly used for particle sizing in the subsieve range of size.

Screening Techniques

Screening apparatus is commonly used for the removal of extreme coarse or fine particles or for separation in size fractions. The screen size may range from about 200 mm down to about 37 μm in size. A vibratory continuous wet-screening apparatus used for multiple separations is shown in Fig. 19.1. Particle motion is assisted by mechanical vibrations which cause shear flow over the mesh and by entrainment in the liquid passing through apertures in the screen. Standard sieves ranging up to about 2 m in diameter are used, and screening rates of several thousand gallons per hour for slurries with a nominal viscosity ranging up to 5000 mPa·s are attainable.

Both dry forced-air screening and sonic sieving are used for the dry separation in the range of 850 to about 37 μm. Production models may be configured in series to obtain multiple size cuts. Disk and rotary drum sieves are used for high-capacity dry sieving.

Classification Techniques

The purpose of classification is generally separation by size, but other factors, such as particle density and shape, fluid flow, and hindered particle flow, affect the separation. Classification by differential sedimentation is widely used in minerals processing, and a wide variety of equipment is available. A detailed discussion of these operations may be found in E. G. Kelly and D. J. Spottiswood, *Introduction to Mineral Processing* (Wiley-Interscience, New York, 1982).

Cyclone separators, which are relatively simple in construction with no moving parts, are widely used for particle separation in ceramic processing. Cy-

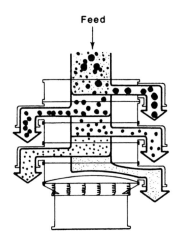

Vibratory Sieving

Fig. 19.1 Vibratory continuous wet screening separation. (Courtesy of SWECO Inc., Florence, KY.)

clones are inertial separators. The feed suspension enters the cyclone tangentially under pressure and circulates in a spiral path, as shown in Fig. 19.2. The centrifugal force on a particle is opposed by the inward-directed drag force of the fluid. Particles coarser than the cut point migrate downward in the primary spiral along the wall and are discharged. Finer particles migrate into the secondary vortex moving upward and are discharged with the majority of the fluid. The mechanics of flow is rather complex, and design is based on empirical correlations of the parameters listed above for classifiers and the feed pressure and size of the apparatus. The cut size is a function of the settling rate of particles suspended in the slurry, the gradient of the spiral, the fluid velocity, and the size of the classifying chamber. The sharpness of the cut is high when the particle density and shape and fluid currents are very uniform. Industrial models may accommodate a feed rate ranging up to about 1 ton/h and a cut size (separation) in the range of 0.5–50 μm. Laboratory models are available for classifying a few grams of material. Hydroclones are cyclones used for separating particles in liquid slurries and suspensions. Industrial models provide a cut in the range of 2–100 μm. Hydroclones are available with wear-resistant

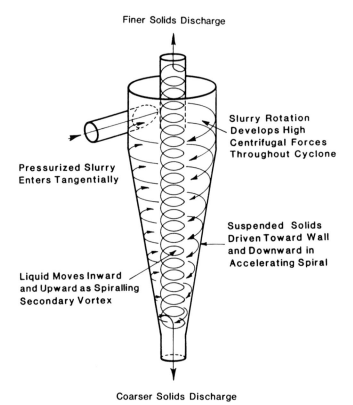

Fig. 19.2 Fluid flow in a hydroclone separates fines in an upward spiralling vortex which discharges through the top.

or expendable polymer linings. A progressing cavity pump is used to provide feed slurry under a controlled pressure. Industrial units processing up to several hundred liters per minute may be connected in parallel to a manifold to increase the capacity and efficiency. Models that process a few liters per minute are available for laboratory samples.

Two widely different size cuts may be produced by combining a screening device in series with a cyclone or a hydroclone (Fig. 19.3). Coarse particles are separated first by a screen. Finer particles entrained in the fluid are fed into a cyclone for a second size cut.

Centrifuging

Centrifuges are used for separate particles that settle under conditions described by Stokes' law (Eq. 7.1). At particle volume concentrations higher than about

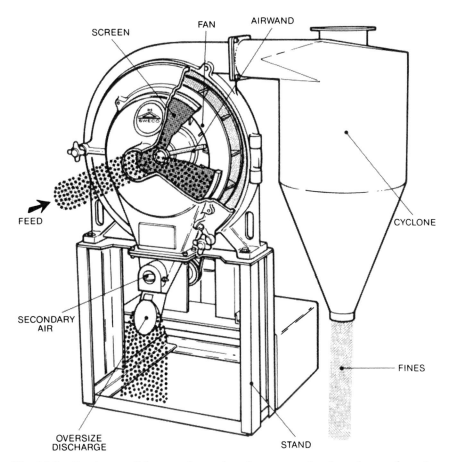

Fig. 19.3 Continuous disk screening and cyclone separation in series configuration. (Courtesy of SWECO Inc., Florence, KY.)

5%, the average density and viscosity of the medium through which the particle settles are significantly increased, and particle settling is hindered. Centrifugal classification is used for separating particles by size, as was discussed in Chapter 7.

19.2 FILTRATION AND WASHING PROCESSES

Filtration is widely used in ceramic processing to concentrate particles from a suspension or slurry. The plate and frame filter press (Fig. 19.4) is the most common filtering device used in the ceramics industry for reasons of low maintenance and flexibility. The press receives pressurized suspension and distributes it to the filter surfaces. Filtrate passing through the filter exits through an outlet duct. A solid concentrate is retained in the frames between the filter surfaces. Some models (Fig. 19.5) have separate ducts for introducing and removing wash water with dissolved impurities. Modern filter presses have alternating frames and filter plates hung on a rack which are compressed together using a hydraulic ram. The filter cloth is nylon or some other synthetic material. The frames are made of cast iron, stainless steel, high-strength polymer, or rubber. Frames are opened by hand or hydraulically, and cake drops onto a conveyor below. Equipment size is directly dependent on the surface area required to form the product at a particular rate. A common practice is filtration at a constant rate followed by constant pressure filtration to minimize the initial loss of fine particles and maintain a constant liquid uniformity.

Rotary disk, rotary drum, and horizontal vacuum filters are used for the continuous filtering of high-permeability precipitates, slurries, and so on. Vac-

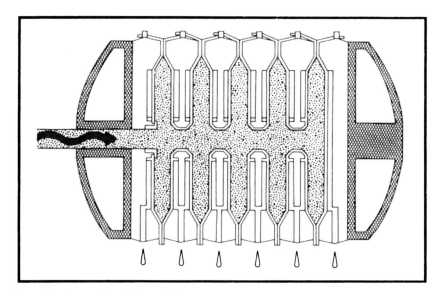

Fig. 19.4 Center feed industrial filter press. (Courtesy of Netzsch Inc., Exton, PA.)

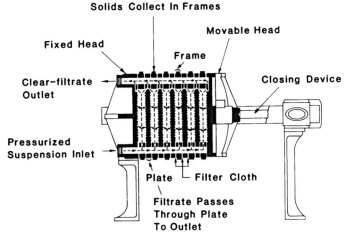

Fig. 19.5 Filter press for concentrating and washing filter cake.

uum is used to remove water, and filter cake adhering to a belt, rotating disk, or drum is removed after rising above the slurry.

Liquid Permeability

Filtration depends directly on the permeability of the filter cake, and the coefficient of permeability is an important property of any porous system. A model for the permeability of a packed bed of particles was introduced by Darcy* in 1856 to explain permeation through a sand filter. He observed that the uniaxial volume flow rate Q through the filter (Fig. 19.6) was proportional to the differential pressure ΔP across the filter of thickness L and cross-sectional area A. The Darcy equation is

$$Q = \frac{K_D A \Delta P}{L} \tag{19.1}$$

where K_D is the Darcy permeability coefficient. Normalizing Eq. 19.1 for the viscosity of the fluid η_f, the coefficient of specific permeability K_P is

$$K_P = \frac{\eta_f Q L}{A \Delta P} \tag{19.2}$$

Equation 19.2 is valid when the fluid flow through the pores is laminar, which may be estimated from the equation

$$\text{Re} = \frac{D_f \bar{v} \bar{a}_{V/A}}{\eta_f} < 1 \tag{19.3}$$

*H. Darcy, *Les Fontaines Publiques de la Ville de Dijon*, Libraire des Corps Impériaux des Ponts et Chaussées et des Mines, Paris, France, 1856.

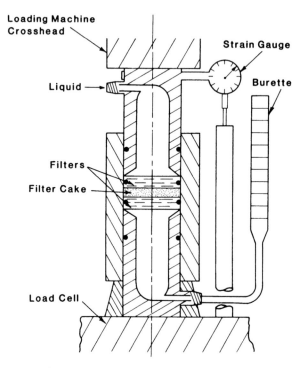

Fig. 19.6 Apparatus for determining the coefficient of permeability and compressibility of filter cake.

where $\bar{v} = Q/(A\phi)$ is the apparent velocity of the fluid of density D_f in the pore channels among particles with the mean size $\bar{a}_{V/A}$.

Capillary flow models assume that the connecting interstices among particles can be represented as a network of closed channels. Kozeny considered the porous system to be a bundle of parallel, uniform, but tortuous channels of total length (L_T) and hydraulic radius (R_H). The hydraulic radius (R_H) is defined as the cross-sectional area divided by the wetted perimeter of the channel and is calculated from the specific surface area per unit volume of the particles S_s:

$$R_H = \frac{\phi}{S_s(1 - \phi)} \tag{19.4}$$

Carmen showed that the apparent liquid velocity was also dependent on the tortuosity of the pores. The Carmen–Kozeny* equation is

$$K_P = \frac{\phi R_H^2}{K_0} \tag{19.5}$$

*P. C. Carmen, *Trans. Inst. Chem. Eng. (Lond.)*, **16**, 168–188 (1938).

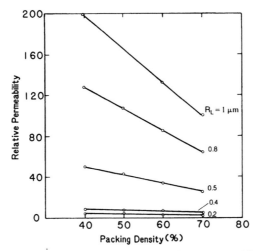

Fig. 19.7   Dependence of relative permeability coefficient on packing density and pore radius (calculated from Eq. 19.6; base condition is $R_{\mathrm{L}} = 0.1~\mu$m and PF = 0.70).

where K_0 is a constant representing the shape and tortuosity of the pores and has a value of 5 for a cylindrical capillary.

Interconnected interstices among packed particles have a varying cross-sectional area, and the permeability is very dependent on the minimum pore dimension. Lukasiewicz* has shown that the specific permeability of interstitial pores is better represented by the equation

$$K_{\mathrm{P}} = \frac{\phi \overline{R}_{\mathrm{L}}^{2}}{K_{\mathrm{L}}}$$ (19.6)

where $\overline{R}_{\mathrm{L}}$ is the volume average linear mean pore radius determined by mercury penetration porosimetry and K_{L} is a constant for a particular pore network. For a cylindrical pore, $K_{\mathrm{L}} = 8$.

Fluid migration is higher when the pressure differential and permeability are high and the viscosity of the fluid is low. The permeability is relatively more dependent on the effective pore size than the porosity (Fig. 19.7). In compacts of dry powders, agglomeration can cause a significant increase in the effective $\overline{R}_{\mathrm{L}}$ and K_{P}, although the porosity is only slightly changed.

The permeability of packings of particles with a continuous distribution of particle sizes depends significantly on the form of the size distribution. Figure 19.8 illustrates four different distributions with the same range of sizes; the permeability decreases for the sequence A through D as a result of more fines in the system packing to give smaller pore interstices. For distributions of type A and B, increasing the size range will increase K_{P}, but for those of type C

*S. Lukasiewicz, Ph.D. thesis, Alfred University, Alfred, NY. 1983.

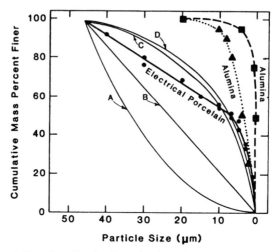

Fig. 19.8 Cumulative size distributions having the same maximum and minimum particle size but which vary greatly in packing and permeability, and curves for a reactive and a calcined alumina. [After M. K. Bo et al., *Trans. Inst. Chem. Engr.*, **43**, 228–232 (1965).]

and D, an increase in the size range will decrease the permeability. The specific permeability is sensitive to the pore structure and size influenced by the particle packing, as is indicated in Table 19.1. The presence of an organic binder in the liquid may drastically reduce K_P. The adsorbed polymer chains may cause a slight increase in ϕ, but the bridging of polymers across pores reduces the effective hydraulic radius and the permeability.

TABLE 19.1 Permeability Coefficients

System	Porosity (%)	K_p $(10^{-18}$ m$^2)$
Gypsum mold	40	28,600
Porcelain casting body (deflocculated)	35	30
Hole china body (flocculated)		170
Electrical porcelain body (50% clay)		
Filter cake (flocculated)	43	85
From pug mill	29	12
Compressed at 2.5 MPa	25	5
Alumina powder (<5 μm)		
Coagulated cast	56	1,100
Deflocculated cast	30	0.8
Deflocculated cast (0.5% binder)	36	0.2
Zirconia power (<0.05 μm)		
Deflocculated cast	59	0.6

Filtration

In uniaxial filter pressing, the increase in thickness dL of the cake with time dt can be related to the volume of filtrate dV_f using a materials balance:

$$dL = \frac{W\, dV_f}{A(D_s - \phi D_s - W\phi)} = \frac{J\, dV_f}{A} \tag{19.7}$$

where W is the mass of solids in suspension per unit volume of liquid, ϕ is the fraction of interstices among the particles in a cake of cross-sectional area A, and D_s is the density of the solids in the cake. The parameter J is the volume of cake per volume of filtrate and is a contraction for $W/(D_s - \phi D_s - W\phi)$. Substituting Eq. 19.7 into Eq. 19.2 for liquid permeability and expressing the volume flow rate in differential form,

$$\frac{1}{A}\frac{dV_f}{dt} = \frac{K_P \Delta P_c A}{\eta_L J V_f} \tag{19.8}$$

where ΔP_c is the pressure drop across the cake of thickness dL. In practice, the total pressure drop ΔP_T across the cake and filter medium is measured. Assuming the filter medium has a constant resistance R_m, and expressing the specific resistivity α_c of the cake as $\alpha_c = K_P^{-1}$,

$$\frac{1}{A}\frac{dV_f}{dt} = \frac{\Delta P_T}{\eta_L \left(\dfrac{J V_f}{A}\alpha_c + R_m\right)} \tag{19.9}$$

For an incompressible cake, $J\alpha$ is constant, and on integrating

$$L = \left[\frac{2J\Delta P_T t}{\alpha_c \eta_L} + \frac{R_m^2}{\alpha_c^2}\right]^{1/2} - \frac{R_m}{\alpha_c} \tag{19.10}$$

When $R_m/\alpha_c \ll 1$, Eq. 19.10 may often be expressed as

$$\frac{L^2}{t} = \frac{2JK_P\Delta P_T}{\eta_L} \tag{19.11}$$

The pressure drop ΔP_c across the cake produces a gradient in the effective stress. For a compressible cake, J and K_P decrease with pressure, and J and K_P in Eq. 19.11 are apparent values.

Filtration data graphed as shown in Fig. 19.9 for a well deflocculated suspension of alumina may be analyzed using the reciprocal of Eq. 19.9. The filtration behavior of a partially deflocculated clay porcelain body where

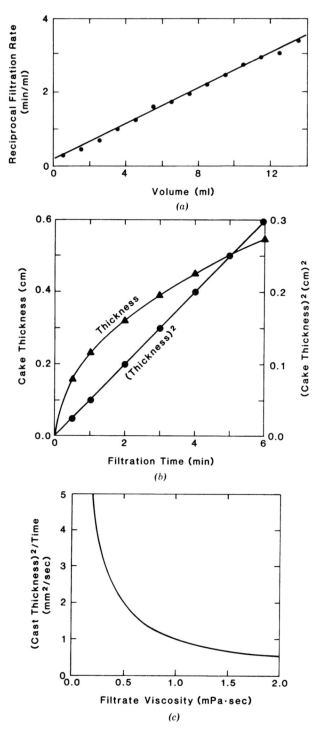

Fig. 19.9 (*a*) Filtration results for deflocculated slurry plotted in form described by Eq. 19.9. (*b*) Filtration results for a deflocculated whiteware slip. (*c*) Dependence of the filtration behavior of a deflocculated alumina slurry on filtrate viscosity calculated using Eq. 19.9.

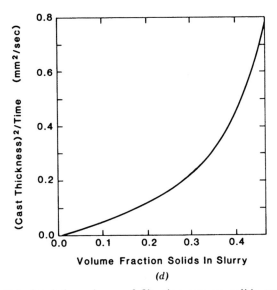

Fig. 19.9 (d) Calculated dependence of filtration rate on solids concentration in a deflocculated alumina slurry (volume fraction solids in cake = 0.60).

$R_m/\alpha_c \ll 1$ is described by Eq. 19.11, as is shown in Fig. 19.9. The parameter L^2/t is termed the filtration rate. The filtration rate is significantly dependent on the viscosity of the filtrate, and heating of the slurry or substitution of a liquid of lower viscosity may reduce the filtration time. A slurry temperature of 35–40°C is commonly used in filtering slurries in the production of electrical porcelain. Filtration also occurs more rapidly when the solids loading in the slurry and J is higher (Fig. 19.9). Filtration is very sensitive to both dispersion and agglomeration, which affect K_P. As is shown in Fig. 19.10, the filtration rate of an alumina slurry is significantly reduced by the presence of an organic binder of a higher viscosity grade (molecular weight) and by removing coagulating ions from the surfaces of the particles prior to filtration.

Washing

Washing operations may be conducted after filtering with the filter cake in place (Fig. 19.5). Washing rates are of the order of the final filtration rate dV_f/dt when wash liquid passes through the thickness of the final cake. Batch or continuous centrifugation may also be used for concentrating solids and for washing operations. Washing may significantly reduce the concentration of soluble impurities in industrial minerals, inorganic chemicals, precipitates, and so on, as indicated in Table 19.2.

Manganese zinc ferrite powders are washed during industrial processing to reduce the alkali content and thus to obtain a much more uniform grain size on sintering and improved magnetic properties. Other industrial chemicals are washed during processing. Filter pressing is widely used in preparing plastic clay bodies and removes significant quantities of soluble salts.

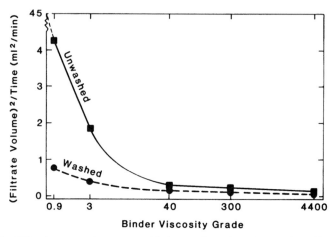

Fig. 19.10 Filtration rate decreases with increasing viscosity grade of hydroxyethyl cellulose binder in slurry and packing of particles influenced by soluble impurity in alumina (0.5 wt% binder in solution, 7 vol% solids in slurry, filtration rate = 45 ml^2/min with no binder).

TABLE 19.2   Impurity Removed by Washing

Material	Concentration (wt %)	
	As Received	Washed
Bayer process alumina powder ($<5~\mu$m)		
(Na$_2$O impurity)	0.04	0.003
Zirconia powder from chloride precursor ($<0.05~\mu$m)		
(Cl impurity)	0.79	0.15

19.3   PARTICLE CONCENTRATION AND SEPARATION PROCESSES

Particles may be concentrated using a cyclone and by gravity sedimentation, flotation, magnetic separation, and electrical precipitation. Sedimentation and flotation are widely used in processing industrial minerals. In froth flotation, surfactants are used to produce a stable froth containing suspended particles having a hydrophobic surface that has a greater affinity for air bubbles than for water; surface-specific surfactants may provide separation by mineral type. Particles differing in density may be separated by sink/float behavior in a liquid or suspension of intermediate density.

Magnetic separation is used to remove magnetic minerals and magnetic iron particles produced by the abrasive wear of processing machinery. Superconducting magnetic separators are used in the bulk processing of kaolins to remove mica and ferrous titanates of a micron size. Smaller magnets are used to remove tramp iron in wet processing. In electrical precipitation operations, fine particles charged on adsorbing gas ions are removed when they migrate to the collecting electrode.

SUMMARY

Particle separation and classification processes are used to modify the range of particle size, to remove impurities, and to concentrate particles and remove liquids. Screening and cyclone-type separators are widely used to remove extremely coarse or fine sizes and to separate particles by size in the range of 0.5 μm to several 100 mm. Hydrocloning, filtration, and centrifuging are used to concentrate particles in a slurry. The filtration rate is very dependent on the solids content of the slurry and the permeability of the filter cake. The permeability of a system of packed particles is an important property influencing the processing of ceramic systems and can range over several orders of magnitude for similar systems differing in agglomeration and pore size. Magnetic separation and other separation processes are used to remove impurity particles and to concentrate particles having a different density or surface behavior. Washing processes are used to remove soluble substances.

SUGGESTED READING

1. Errol G. Kelly and David J. Spottiswood, *Introduction to Minerals Processing*, Wiley-Interscience, New York, 1982.
2. F. A. L. Dullen, *Porous Media: Fluid Transport and Pore Structure*, Academic Press, New York, 1979.
3. J. Bear, *Dynamics of Fluids in Porous Media*, American Elsevier, New York, 1972.
4. Terrence J. Fennelly and James Reed, "Mechanics of Pressure Slip Casting," *J. Am. Ceram. Soc.*, **55**(5), 264–268 (1972).
5. D. S. Adcock and I. C. McDowall, "Mechanism of Filter Pressing and Slip Casting," *J. Am. Ceram. Soc.*, **40**(10), 355–362 (1957).
6. P. M. Heertjes, "Studies in Filtration," *Chem. Eng. Sci.*, **6**, 190–208 (1957).
7. H. P. Grace, "Resistance and Compressibility of Filter Cakes," *Chem. Eng. Prog.*, **49**(6), 303–318 (1953).
8. David Cornell and D. L. Katz, "Flow of Gases Through Consolidated Porous Media," *Ind. Eng. Chem.*, **45**(10), 2145–2152 (1953).
9. P. C. Carmen, "Some Physical Aspects of Water Flow in Porous Media," *Disc. Faraday Soc.*, **3**, 72–77 (1948).

PROBLEMS

19.1   Using Stokes' law as a qualitative description of the balance between the centrifugal force on the particles and the drag force of the liquid flowing inward in a hydroclone, predict the change in "cut size" on increasing the liquid viscosity, particle density, and the tangential velocity of flow.

19.2 Calculate the time required to centrifuge zirconia particles larger than 1 μm in diameter from a column of an aqueous suspension 6 cm high at 1000 rpm. The initial radial position is 12 cm, $T = 25°C$, and $D_P = 6.02$ Mg/m^3.

19.3 The volume flow rate of water (20°C) through a filter that is 60% porous, 0.6 cm thick, and 3.5 cm in diameter is 3.7 mL/s at a pressure head of 420 kPa. Calculate the specific permeability of the filter assuming laminar flow.

19.4 Calculate the apparent axial flow velocity in Problem 19.3. The particle size is 1 μm; calculate Re to check for laminar flow conditions.

19.5 Estimate the change in permeability coefficient of a filter cake when the cylindrical pore diameter decreases from 1.0 to 0.1 μm when the pore fraction is 0.45.

19.6 The permeability of a filter cake is $K_P = 10 \times 10^{-16}$ m^2 and the pore fraction is 0.45. Assuming cylindrical pores, estimate the pore radius R_L. Assume $K_L = 8$.

19.7 Calculate the J parameter for filter pressing a clay clip containing 300 kg of clay and 100 kg of water (25°C). The pore fraction of the cake is 0.40.

19.8 Calculate the permeability of the filter cake in Problem 19.7. The liquid flow rate is $V_f^2/t = 5$ mL^2/s and the resistance of the filter cloth can be neglected. The diameter of the filter cell is 7.8 cm and the pressure head is 170 kPa.

19.9 In Problem 19.8, what is the cake thickness after 1 hr?

19.10 Calculate and graph the cake thickness versus time from the data in Problem 19.8.

EXAMPLES

Example 19.1 How does a screening operation differ for large and small particles?

Solution. Screening efficiency depends on particle motion, the ratio of the number of apertures/number of particles, and entrainment of the particles in the fluid passing through the openings. For coarse particles which are of greater mass but are fewer in number, a greater vibration amplitude is particularly needed to maintain particle motion over the screen. For fines, the solids loading must be lower and the screening time longer because of the greater particle/aperature number ratio. For wet screening using very fine screens, it may be necessary to assist the flow of liquid through the aperatures, and particle dispersion in the slurry by chemical means is essential.

Example 19.2   A slip for filter pressing is prepared using 200 g of powder/100 mL of water. The average density of the solids (powder) D_s is 2.60 g/cm^3. The filter cake is determined to have PF = 0.60 (ϕ = 0.40). Calculate the J parameter indicating the volume of cake/volume of filtrate. Assume the cake is incompressible at this filtration pressure. What would be the value of J if only 150 g of powder had been used to prepare the slip?

Solution. The mass of solids/volume of liquid in the slurry W is

$$W = 200 \text{ g}/100 \text{ cm}^3 = 2.00 \text{ g/cm}^3$$

The parameter J indicating the cake volume/filtrate volume is

$$J = W/(D_s - \phi D_s - \phi W)$$

and

$$J = (200 \text{ g/cm}^3)/[2.60 \text{ g/cm}^3 - 0.4(2.60 \text{ g/cm}^3) - 0.4(2.00 \text{ g/cm}^3)]$$

$$J = 2.63$$

For 150 g/100 mL, W = 1.5 g/cm^3 and J = 1.56. Note that the filter cake thickness at some time varies as $J^{1/2}$ and the higher powder loading in the slip significantly reduces the filtration time.

Example 19.3   For the filtration process in Example 19.2, calculate the specific permeability constant K_P for the cake when the filtrate is collected at a (volume)^2/time rate of 0.55 mL^2/s. The filtrate temperature was 25°C and the pressure head P = 170 kPa. The area A of the laboratory filter press is 47.78 cm^2.

Solution. Equation 19.11 may be used to calculate K_P. Rearranging this equation and substituting $J(V/A)$ for L,

$$K_P = J(V^2/t)\eta_L/2A^2P$$

$$K_P = \frac{2.6(0.55 \times 10^{-12} \text{ m}^6/\text{s})(0.89 \times 10^{-3} \text{ Pa} \cdot \text{s})}{2(47.78 \times 10^{-4} \text{ m}^2)^2(170 \times 10^3 \text{ Pa})}$$

K_P = 164 × 10^{-18} m^2, which is similar to K_P for a flocculated hotel china body (see Table 19.1).

CHAPTER 20

GRANULATION

Dry bulky powders, semidry powders which spontaneously form variable agglomerates, and granular materials and powders mixed with a small amount of binder solution are not a convenient feed material. These kinds of systems do not flow well and do not fill a mold or die uniformly. For a powder system, a satisfactory semidry feed material consists of controlled agglomerates called granules which are produced by means of granulation processes.

Granulation may be achieved directly by pressing, by extruding the material through an orifice, or by spraying a wetting liquid or a binder solution into a stirred powder, which is called spray granulation. Granules are produced indirectly in the process of spray drying by atomizing a slurry of the powder into a drying chamber. Modern powder pressing operations now depend on a granule feed. Granulated material is used extensively as a feed material for calcining and melting processes and directly as a product serving as a catalyst support. Spray drying is also used as a unit operation to reduce the liquid content of a slurry to form a semidry material containing additives for use in extrusion.

We examine granulation processes and the characteristics of granules in this chapter.

20.1 DIRECT GRANULATION

Powder granules may be produced directly using pressing, extrusion, and spray granulation, which are sometimes referred to as pelletizing processes. Materials prepared in this way include technical aluminas, ferrites, clays, tile bodies, porcelain bodies, conventional refractory compositions, catalyst supports, and feed materials for glass melting and metal refining.

Fine powders premixed with only a few percent of a wetting liquid or a binder solution may be compacted in a tableting die or between briquetting rolls or belts. A mixer with a roller and a perforated bottom plate is also used to produce granulated powder. Granules produced by compaction are commonly dense, hard, and strong if compacted above about 10 MPa. An auger extruder or a kneader may be used to produce a granular product having a plastic consistency. Spaghetti-shaped material exiting from an orifice plate is cut with rotary knives. In roll extrusion, moist feed is forced through perforated rolls or through a perforated plate. Dried material is crushed to reduce the particle size.

Spray granulation is the formation of granules when a liquid or a binder solution is sprayed into a continually agitated powder. The formation of the granules may be considered to occur in two stages:

1. Nucleation of primary agglomerates at random
2. Growth by the addition of particles or small agglomerates to the central agglomerate

When stirring the powder, particles roll and slide, and fines become airborne. A minute agglomerate called a nucleus or seed is formed when a droplet of liquid or binder solution hits and is adsorbed on the surface of a group of particles. Capillary forces and binder flocculation give the nuclei strength. Nuclei are more numerous when the liquid is introduced as a fine mist and the powder is agitated vigorously. Capillary forces may cause particle sliding and rearrangement, densification, and the migration of liquid to the surface of the agglomerate nuclei.

The growth of granules by layering occurs by the contact and adhesion of particles to the nuclei. Alternatively, the agglomeration of young nuclei and fracture fragments can also produce a granule. These growth mechanisms are illustrated in Fig. 20.1. The rate of each mechanism depends on the liquid feed rate, the adsorption of liquid into the agglomerate, and the mixing action.

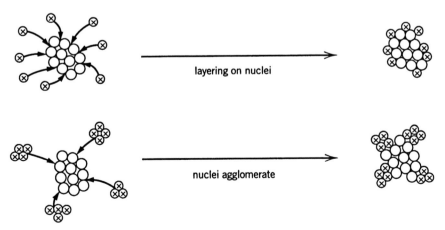

layering on nuclei

nuclei agglomerate

Fig. 20.1 Nucleation and growth of granule in spray granulation may occur by addition of primary particles or small agglomerates.

Rubbing between granules during tumbling may cause particle transfer at surfaces and surface smoothing. Fracture of granules and attrition are higher when the forces during mixing are higher and the granules lack strength. The size, shape, and surface appearance of granules formed by spray granulation are shown in Fig. 20.2.

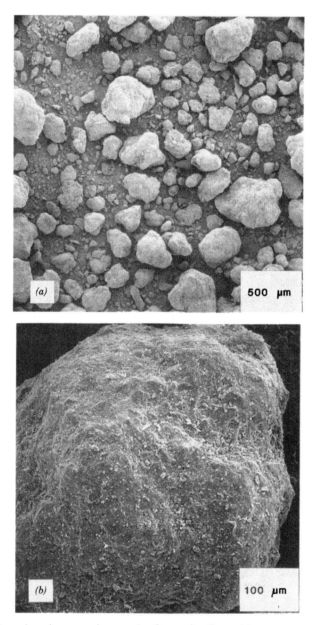

Fig. 20.2 Scanning electron micrograph of granules formed by spray granulation: (*a*) general view and (*b*) shape and surface smoothed by rubbing during tumbling.

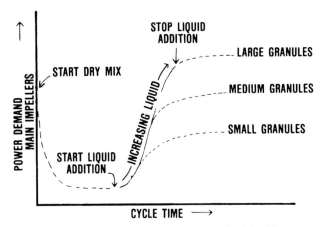

Fig. 20.3 Granule size development on adding a liquid with constant mixing.

For each system, there is a critical range of liquid content for granulation. At the low end, granules are eventually formed when the processing energy is sufficiently high. The addition of a greater amount of liquid generally increases the mean size, the size distribution, and the porosity of the granules (see Figs. 20.3 and 20.4). The liquid requirement for granulation is greater when the

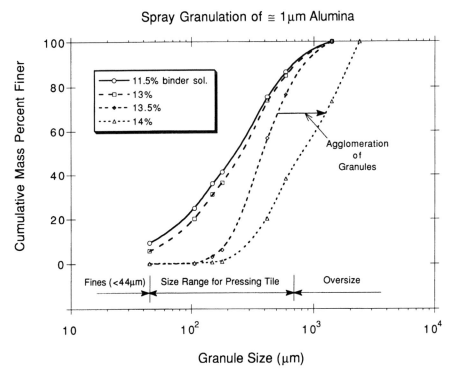

Fig. 20.4 Concentation of added binder solution must be controlled to obtain maximum yield of granules within target size range.

specific surface area of the powder is higher. Common liquid requirements in the range of 20–36 vol% indicate that during granulation the pores in granules are incompletely saturated. The spray geometry, mixing process, and liquid/ binder solution content must be well planned and controlled to produce the maximum yield of granules falling within a target size range (see Fig. 20.4).

Rotating pan-type mixers, ribbon mixers, double planetary mixers, V-blenders, and continuous rotating drum and disk pelletizers are used for directly granulating ceramic powders. Rotating drums and disks are commonly used for producing rather coarse granules 1–10 mm in size as a feed for melting operations. Mixers with a more intense mixing action produce smaller granules, but larger than the 44 μm required for flow and for pressing. A binder aids in forming granules and increases the strength of granules of coarser powders. Increasing the temperature of the powder tends to produce smaller granules.

Granules produced by spray granulation are more nearly spherical than granules produced by compaction and extrusion. Capacities of industrial granulators range up to several tons per hour. Spray-granulated material is usually partially dried to produce the powder feed for pressing operations.

20.2 SPRAY DRYING

Cocurrent and mixed-flow spray dryers are illustrated in Fig. 20.5. Spray drying is the process of spraying a slurry into a warm drying medium to produce nearly spherical powder granules that are relatively homogeneous. The dispersion of small droplets of high specific surface area in the drying air provides

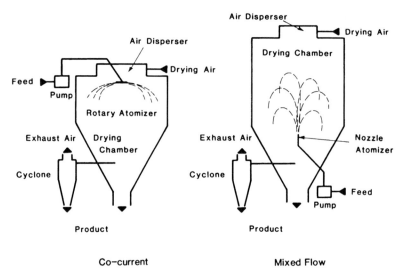

Fig. 20.5 Spray dryers: (left) cocurrent spray dryer with centrifugal atomizer, and (right) mixed-flow spray dryer with a nozzle atomizer.

a relatively high drying efficiency. Spray drying is used widely for preparing granulated pressing feed from powders of ferrites, titanates, other electrical ceramic compositions, alumina, carbides, nitrides, and porcelain bodies. A milling process is commonly used to disperse powder agglomerates and for mixing. The binder solution may be filtered to remove undissolved globules. Secondary mixing during atomization further reduces the scale of inhomogeneity in the product. When properly controlled, spray drying produces a nearly spherical, relatively dense granulated product. Granules larger than 20 μm in size flow and compact well in pressing operations. Capacities of industrial spray dryers range from less than 10 to several 100 kg/h.

Slurry Controls

The batch, which commonly includes recycled fines from the cyclone, is mixed to form a deflocculated, shear thinning slurry feed. Consideration must be given to soluble inorganic impurities retained in spray drying, and the effect of the drying on the degradation or evaporation of the organic binders and plasticizers. Aqueous systems are commonly used, but recycled organic liquids are sometimes utilized when the chemical or thermal reactivity of the powder or binder precludes using an aqueous system.

Slurry characterization includes the determination of the slurry density, the foam content, the percent solids, and the viscosity behavior (Fig. 20.6). A slurry with a high solids content is desirable for technical reasons and to in-

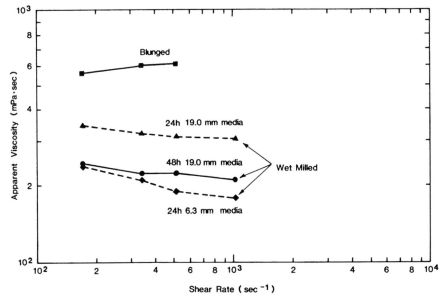

Fig. 20.6 Viscosity behavior of alumina slurries for spray-drying is shear thinning (determined using a concentric cylinder viscometer).

crease the energy efficiency in drying and the product yield (see Fig. 20.7). The slurry should be relatively free of air because air bubbles reduce the density of both the slurry and the resultant granules. The slurry density is determined by weighing a standard volume of a slurry sample. The foam index is calculated from the theoretical density of the batch D_T and the apparent experimental slurry density D_E

$$\text{Foam index } (\%) = (D_T - D_E)\,100/D_T \qquad (20.1)$$

Often an antifoam must be added to the slurry to control the foam index below a control limit.

Weighing errors in batching may cause errors in the inorganic content of the product. The percent solids is determined by heating a slurry sample to eliminate liquid and organic additives and comparing the experimental result with the value for the batch. The viscosity must be determined over a wide range of shear rate to check for reproducibility and the absence of shear thickening behavior. The shear rate during atomization may exceed 10^4 s^{-1}. Wet-milled slurries with the shear thinning viscosity behavior shown in Fig. 20.6 were satisfactory for spray drying; blunging did not produce sufficient powder dispersion.

Atomization

Slurry screened to remove coarse particles is commonly atomized using either a nozzle or a rotary atomizer. In a pressure nozzle, slurry accelerating through small channels achieves turbulence and breaks up into droplets. Pressure noz-

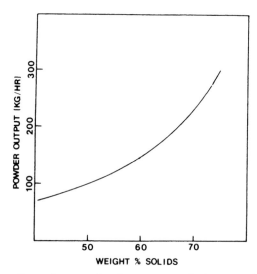

Fig. 20.7 Output of powder granules from a spray drying as a function of powder content in feed slurry. Dryer evaporative capacity of water is 100 kg/h and a yield of 100% is assumed. [From S. J. Lukasiewicz, *J. Am. Ceram. Soc.*, **72**(4) 618 (1989).]

zles are relatively inexpensive but are subject to blocked flow, and the high operational pressures, ranging up to 10 MPa, increase abrasion rates. In a two-fluid nozzle (Fig. 20.8), a high-velocity gas is mixed with the slurry to produce turbulence and droplets. Operating pressures may be reduced to less than 1 MPa. Centrifugal energy from a disk or wheel rotating at several thousand revolutions per minute produces droplets in a rotary atomizer. The larger flow channels and lower operating pressure increase the feed rate capacity and the reliability. Spray dryers using radial atomization are relatively large in diameter because of the larger radial/axial flow pattern of the spray.

Parameters that control the droplet size in the spray are listed in Table 20.1. Characteristics of spray-dried alumina granules listed in Table 20.2 are indicative of these trends. Granules ranging up to about 400 μm may be produced using pressure nozzles, and sizes up to about 150 μm using a rotary atomizer. Relative to a pressure nozzle, a two-fluid nozzle usually produces a larger fraction of sizes, < 40 μm.

Atomization–Drying Air Configuration

Drying involves simultaneous heat and mass transfer. The airflow pattern influences the droplet drying time, residence time in dryer, residual liquid content, and wall deposits. Heat is transferred by convection from the air to the droplet. Heat absorbed evaporates the liquid and the vapor is transported by convection into the drying air. A cocurrent dryer design positions the atomizer and drying-air inlet at the top of the dryer to provide cocurrent flow (see Fig. 20.5). Atomized slurry is of maximum liquid content when it encounters the laminar flow of hot incoming air. Using this design, the maximum product temperature is relatively low and the evaporation time is relatively short. However, the product exits with moist air, and the exit temperature must be relatively high to obtain a dry product. Using a countercurrent design, slurry and drying air inlets are at opposite ends of the drying chamber. Droplets encountering the incoming hot air have already been partially dried, and the product heating is greater. The mixed flow design (Fig. 20.5), which increases the residence time of the spray and provides a means for reducing the size of the chamber needed for drying, is a convenient compromise and is characteristic of many small industrial and laboratory dryers.

Two Fluid Nozzle

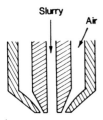

Fig. 20.8 Two-fluid nozzle with external mixing for atomizing a slurry.

TABLE 20.1 Parameters Influencing Droplet Size

Atomization	Size Proportional To
Pressure nozzle	Surface tension (viscosity)/pressure
Pneumatic (2-fluid)	Surface tension (viscosity)/mass(air/slurry)
Centrifugal	Surface tension (viscosity) (slurry feed rate)/ (speed × diameter) of wheel

Drying Droplets to Form Granules

The rate of drying varies directly with the rates of heat and mass transfer and depends on the temperature and humidity of the air, the transport properties of the air near the droplet, and transport properties within the drying droplet. Considerable evaporation must occur during the first few milliseconds of drying to prevent the formation of agglomerated granules and a coating of sticky agglomerates on the wall of the dryer. The velocity of the droplets from the atomizer is considerably higher than the velocity of the drying air. Incoming air keeps the humidity low and carries moisture away. Filtered drying air is commonly heated using natural gas or electrically to a temperature in the range of 250–600°C. The inlet air temperature depends directly on the dryer design, the product residence time, and the thermal limitations of the product. The product temperature is normally lower than the outlet air temperature and is usually maintained below 100°C when using organic binders and additives.

The evaporation rate initially increases as the droplet is heated. Volume shrinkage occurs as liquid is lost and the PF of the particles increases. Evaporation reduces the air temperature. The drying rate remains high as long as the temperature of the drying air is maintained and the surface is saturated with liquid. At some point the rate of flow of liquid to the surface becomes rate controlling. Slow capillary flow and fast evaporation (i.e., a relatively high inlet air temperature and a relatively low humidity) may form a dried surface layer almost instantaneously. Capillary flow is reduced by parameters which

TABLE 20.2 Characteristics of Spray-Dried Alumina Granules[a]

Slurry		Granule Size[c] (μm)			Granule Density (%)	Fill[d] Density (%)
Solids (vol%)	Binder (wt%)[b]	90% <	50% <	10% <		
35	1.0	172	92	40	54	30
35	0.5	144	74	27	55	32
50	0.5	202	118	54	58	34
52	0.5	216	113	41	57	34

[a]Two-fluid nozzle, utility dryer.
[b]Fully hydrolyzed polyvinyl alcohol; powder basis.
[c]Sieving data.
[d]Bulk density after pouring into graduated cylinder.

lessen the permeability, such as binder molecules, binder migration forming a surface skin, and the migration of fine particles or precipitated salts into pores near the surface. Rapid heating may cause liquid to evaporate within the droplet, forming a vapor bubble which may expand the agglomerate. The granule is nearly in equilibrium with the warm, moist air leaving the dryer, and the outlet air temperature is controlled to maintain a product with a constant moisture content.

The majority of the spray-dried product is discharged through a rotary valve into interchangeable containers attached at the base. Product fines entrained in the exhaust air are separated using a cyclone or bag filters. Product fines may be used in some pressing operations but are commonly recycled into the feed slurry. Periodic cleaning of the dryer using a liquid spray is required to remove the granulate crust. Cleaning is facilitated when the dried binder system is readily redissolved. Abnormally large granules and flakes of granulate crust from the wall of the dryer are undesirable and are commonly removed by screening.

Granule Character

Instability causes the high-velocity slurry stream to break up into droplets, as shown in Fig. 20.9. Flow required to produce a spherical shape is motivated by surface tension and resisted by the viscosity. Colliding droplets may coalesce into a larger droplet which, if viscous, will be less nearly spherical in shape or form a granule that is a cluster of smaller agglomerates that are partially merged. Air bubbles present in the feed slurry or occluded during atomization

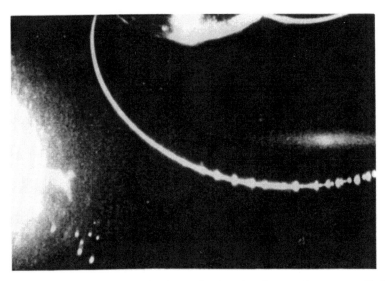

Fig. 20.9 High-speed photograph of droplet forming from ''slurry comet'' during spray drying. (Photo courtesy of Niro Atomizer Inc., Columbia, MD.)

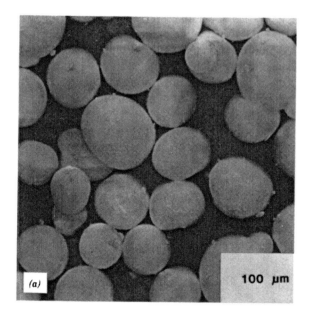

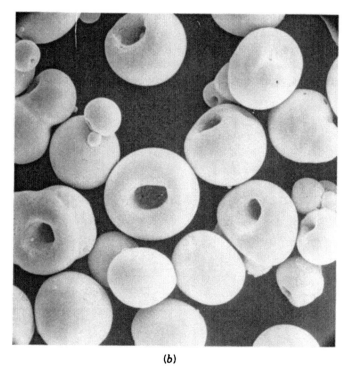

Fig. 20.10 Scanning electron micrographs of spray dried granules: (*a*) smooth, round, dense granules of alumina and (*b*) granules of zinc oxide having a "donut" shape.

using a two-fluid nozzle may persist as relatively large pores in the ultimate granule.

Granules with a large crater or a donut shape are often observed in industrial spray-dried material (Fig. 20.10), regardless of the atomization method used. The tendency for these shapes is higher when the inlet temperature is relatively high, the binder content is relatively high, and the solids loading in the slurry is relatively low. Rapid surface drying and the formation of surface material of low permeability produce a greater initial heating rate. Momentary internal evaporation may form a vapor bubble. When the bubble breaks through the last formed, binder-deficient surface regions and the agglomerate is viscous, surface tension induced flow causes a dimpled or donut shape to be formed. Granules containing an abnormally large internal pore may be produced when the migration of finer particles, salts, or binder during drying produces a surface layer of higher PF that shrinks relatively less on drying.

Characteristics of granules prepared by spray drying and spray granulation are listed in Tables 20.2 and 20.3. The PF of 0.55–0.58 for the particles in spray-dried alumina granules is a few percent lower than the PF_{max} for this powder. Relative to the spray-granulated product, the spray-dried granules are somewhat finer, of comparable bulk density, and slightly faster in flow rate owing to their more nearly spherical shape. The ratio of tap density to fill density is an index of hindered packing during gravity settling after pouring and is slightly higher for the less spherical spray-dried granules. The higher fill density for the whiteware composition reflects the higher granule density achieved for the more densely packing AFDZ particle size distribution.

TABLE 20.3   Characteristics of Granulated Whiteware Bodies[a]

Parameter	Wall Tile Spray Drying	Hotel China Spray Drying	Hotel China Spray Granulation
Moisture Adsorption (wt%)			
33% RH	2.7	3.5	2.4
92% RH	4.8	4.9	3.5
Granule size (μm)[b]			
90% <	300	446	992
50% <	151	148	403
10% <	55	99	122
Flow time (s)[c]			
33% RH		7	10
92% RH		8	12
Fill density (%)	38	40	41
Tapped density/fill density	1.15	1.13	1.20

[a]Granules larger than 1000 μm removed.
[b]Sieving data.
[c]Funnel test.

SUMMARY

Granulated powders are commonly used for pressing, calcining, and melting operations. Compaction and extrusion processes produce a relatively large granulated product. Spray granulation is used to produce granules that are satisfactory for use as a pressing powder for traditional ceramics and refractories and some fine-grained technical ceramics. Spray-drying is used to produce granules that are more homogeneous. Controls for spray-drying include slurry formulation and control, atomization, the droplet-drying air configuration, and the drying conditions. Spray-drying must be properly controlled to avoid the formation of granules with irregular shapes and large internal pores. Spray-drying is relatively efficient because the product is well dispersed in the drying medium and has a high specific surface area.

SUGGESTED READING

1. F. V. Shaw, "Spray Drying: A Traditional Process for Advanced Applications," *Am. Ceram. Soc. Bull.*, **69**(9), 1484–1489 (1990).
2. S. L. Lukasiewicz, "Spray-Drying Ceramic Powders," *J. Am. Ceram. Soc.*, **72**(4), 617–624 (1989).
3. H. B. Ries, "Methods of Preparing Bodies for the Preparation of Oxidic and Non-Oxidic Ceramics," *Keram. Z.*, **35**(2), 67–71 (1983).
4. P. J. Sherrington and R. Oliver, *Granulation*, Heyden and Sons, London, 1981.
5. K. Masters, *Spray Drying*, Wiley-Interscience, New York, 1979.
6. W. H. Engelleitner, "Pelletizing as Applied to the Ceramic Field," *Am. Ceram. Soc. Bull.*, **54**(2), 206–207 (1975).
7. D. A. Lee, "Comparison of Centrifugal and Nozzle Atomization in Spray Dryers," *Am. Ceram. Soc. Bull.*, **53**(3), 232–233 (1974).
8. R. H. Perry and C. H. Chilton, *Chemical Engineers' Handbook*, McGraw-Hill, New York, 1973.

PROBLEMS

20.1 Sketch the change in the physical character of the granules in Fig. 20.4 on increasing the content of binder solution from 11.5 to 14%.

20.2 The size data in Table 20.2 are log normal. Estimate and compare the parameters $\bar{a}_{gM}$ and σ_g for the powders.

20.3 Alumina slurries of 60, 70, and 80 wt% solids are spray dried. Calculate and contrast the relative mass of liquid removed per mass of product formed. Assume 0.4 wt% moisture in product.

20.4 Assuming that the evaporation rate (mass water/time) is a constant value during the drying of the granules in Problem 20.3, calculate and compare the relative product formation rates when the residual moisture content in the product is 0.4 wt%.

20.5 A slurry for spray drying contains 40 vol% zirconia powder ($D_p = 6.0$ Mg/m³) and 60 vol% water. The measured density of the prepared slurry is 2.60 Mg/m³. Calculate the foam index.

20.6 Assuming that in spray granulation the liquid requirement is directly proportional to the specific surface area of the powders used, calculate the ratio of liquid required for two powders having specific surface areas of 1.2 and 12 m²/g. What is the ratio for a powder containing 2/3 of the coarser powder and 1/3 of the finer powder (relative to the coarser powder)?

20.7 Alumina granules with a solids fraction of 0.48 are produced on spray drying an aqueous slurry containing 80 wt% alumina. Did the droplets shrink in size during drying?

20.8 Spray-dried granules of alumina and barium titanate powders have a granule density $D_G = 55\%$. Contrast this density to the powder bulk density for each given in Tables 5.3 and 5.4.

EXAMPLES

Example 20.1 Granules for pressing tile are prepared by spray granulation at 14 wt% water; these are dried to 10 wt% water for the pressing feed. Prepared by spray drying, the starting slurry contains 60 vol% water and the spray-dried product has 10 wt% water. Contrast the weight of water removed per weight of product for the two processes. The density of the powder is 2.60 Mg/m³.

Solution. For spray granulation, the as-formed granules contain 86 g of powder and 14 g of water. The product contains 90 g of powder and 10 g of water. Relative to 100 g of powder, the corresponding liquid contents are 16.3 g as-formed and 11.1 g after drying. The relative weight loss is

$$\Delta W/W_{product} = (16.3 \text{ g} - 11.1 \text{ g})\,100/(100 \text{ g} + 11.1 \text{ g}) = 4.7\%$$

For the spray drying, the weight percent water in the slurry is

$$\text{Water (wt\%)} = 0.60(1 \text{ g/cm}^3)\,100/[0.60(1 \text{ g/cm}^3) + 0.40(2.60 \text{ g/cm}^3)]$$

$$\text{Water}_{slurry} = 36.6 \text{ wt\%}$$

On drying, for a basis of 100 g of powder,

$$\Delta W/W_{product} = (57.7 \text{ g} - 11.1 \text{ g})\,100/(100 \text{ g} + 11.1 \text{ g})$$

$$\Delta W/W_{product} = 42\%$$

Considerably less water is removed by drying in the spray granulation process and this component of the processing cost is lower. However, granules made by spray granulation are commonly less homogeneous.

Example 20.2 When preparing a slurry for spray drying, an antifoam must often be added to reduce the foam content of the slurry. Otherwise, the granule density and the fill density may be unacceptably low. How is the foam content determined?

Solution. The foam content can be determined by comparing the measured slurry density with the calculated slurry density. For example, consider a slurry which consists of 100 g of alumina powder, 60 g of water + deflocculant, and 1 g of polyvinyl alcohol. The experimental density D_E of the slurry determined by measuring the mass and volume V_E of a sample from the bulk of the mixed slurry is 1.58 g/cm^3. The volume calculated V_T assuming no foam is

$$V_T = \Sigma W_i / D_i$$

$$V_T = 100 \text{ g}/(3.98 \text{ g/cm}^3) + 60 \text{ g}/(1.00 \text{ g/cm}^3) + 1 \text{ g}/(1.26 \text{ g/cm}^3)$$

$V_T = 85.9$ cm^3 for a 161-g batch, and the calculated density D_T is

$$D_T = 161 \text{ g}/85.9 \text{ cm}^3 = 1.87 \text{ g/cm}^3$$

The foam index FI is calculated as

$$\text{FI } (\%) = (V_E - V_T) 100/V_E = (D_T - D_E) 100/D_T$$

$$\text{FI } (\%) = (1.87 \text{ g/cm}^3 - 1.58 \text{ g/cm}^3)/1.87 \text{ g/cm}^3 = 16\%$$

100 mL of slurry will contain 16 mL of occluded air which may cause an increased porosity in the atomized droplets and the granules.

Example 20.3 Table 20.2 indicates that increasing the binder concentration and the solids content of the slurries altered the sizes and the density of the spray-dried granules. What caused these results?

Solution. Increasing the binder content will increase the viscosity of the slurry and the droplet size on spraying; increasing the powder content in the slurry will have a similar effect. The consequence is larger granules and clearly the solids content had the bigger effect. The granule density will depend on steric hindrance of particle motions during drying and the reduced K_P for fluid transport produced by the binder. Increasing the solids content of the slurry and reducing the binder content contributed a higher packing density in the droplet and granule.

Example 20.4 Compare a granule density of 58% and a fill density of 34% with the bulk density of 1.8 Mg/m³ for the capacitor grade barium titanate powder in Table 5.4. The density of barium titanate is 6.0 Mg/m³.

Solution. The bulk density D_{bulk} of the powder is

$$D_{bulk} = (1.8 \text{ Mg/m}^3)\,100/6.0 \text{ Mg/m}^3 = 30\%$$

The granule density of 58% is nearly twice the density of the dry powder. However, the interstices within the granules and larger interstices between the granules cause the fill density of the granules to be only slightly higher (34%). But the higher density and larger size of the granules endows the granules with much better flowability.

PART VIII

FORMING

Forming transforms the processed feed material into a consolidated form having a particular geometry and microstructure. Selection of a particular forming operation depends on many factors such as product size and shape, surface character, tolerance specifications, microstructural characteristics, forming productivity, business considerations such as capital investment, and safety and environmental impact. Chapter 21 presents a general description of forming processes, the merits of each process, binder requirements, feed material characteristics, pressures and shear rates in forming, and general concepts of process control. Particular compositions and forming processes are described and discussed in Chapters 22–26.

CHAPTER 21

GENERAL CERAMIC FORMING PRINCIPLES

Forming transforms the feed material into a "green product" having a controlled size, shape, and surface and a particular density and microstructure. Careful control of the density and microstructure of the green ceramic is necessary to obtain ultimate product performance, because large defects introduced in forming are commonly not eliminated when the product is fired. Surface smoothness is commonly desirable and may be essential for some products. The smoothness may be enhanced when using a particular forming technique. As-formed strength must be sufficient for handling and subsequent finishing operations. Reproducibility of the dimensions of product items is also very important in industrial production. The size and green density of the part must be controlled to maintain a constant shrinkage factor between the formed and fired product.

Productivity is greatly enhanced when the ceramic can be formed to a near-net shape so that subsequent machining and surface finishing are kept to a minimum. Ease of manufacturing, adaption to mass production, capital investment required, and production yield figure importantly into productivity. Forming directly impacts productivity and the ultimate quality and cost of the manufactured product. This chapter introduces important concepts which provide a general perspective on forming processes. Specific forming processes are discussed in detail in Chapters 22–26.

21.1 POWDER-FORMING PROCESSES

Although practiced until recently largely as an industrial art, powder forming is now much better understood as an engineering science. A feed material having a particular consistency and flow rheology is required for each forming process. This requires that the particulate materials and additives be properly selected, proportioned, and mixed together when preparing the feed material. These preforming processes have been discussed in the preceding sections.

Selection of the forming process to produce a particular product depends on several technical factors and business-related factors such as the volume of production and capital costs. Technical considerations include the shape and size of the product item, the quality of the internal microstructure and surface, tolerance specifications for size and shape, property specifications, the ease and availability of secondary forming/finishing operations, technical support of mechanization and automation, and the general sensitivity of the processing system to fluctuations in the characteristics of incoming starting materials. Financial analyses are not within the scope of this book.

Forming Techniques

Table 21.1 lists common forming processes, feed material consistency, mold/ die considerations, production features, and some examples of products made using each technique. The general schemes of forming by slip casting, injection molding, and extrusion are shown in Fig. 21.1.

Pressing begins with a prepared granulate feed (see Fig. 3.6) which is compressed at a relatively high pressure in a rigid die. Complex shapes and large items may be produced when using a rubber mold. Pressing technology is relatively mature, capital intensive, and widely used for the mass production of a wide variety of products.

Feed material having a plastic consistency may be formed by extrusion through a die when the yield strength of the material is sufficient to support the weight of the product. Extrusion is a high productivity forming process for lineated shapes. Alternatively, a plastic-like material containing either a thermoplastic or a thermosetting binder may be injection molded to shape at the temperature for which the viscosity is low and cooled or heated by the mold to increase the apparent viscosity and strength of the part before it is ejected. Injection molding is used for the high-productivity forming of small products of complex shape.

Feed having a slurry consistency may be formed into a product using a casting technique. In slip casting, the slurry is poured into a porous mold; capillary suction of the mold absorbs liquid from the slurry and the particles that consolidate on the surface of the mold form a cast. Pressure or vacuum may be applied to increase the casting rate. In gel casting, the slurry contains a binder which is polymerized after the mold has been filled to effectively "solidify" the slurry and impart strength for removal and handling. Castable concretes commonly contain a hydraulic binder which hardens in the presence of water.

Ceramic substrates may be produced by roll forming a granulate at a high pressure or a plastic material at a moderate pressure. Thin substrates with a

TABLE 21.1 Aspects of Powder-Forming Processes

Process/Feed	Mold	Features	Products
Pressing/granulate			
Metal die	Metal die, surface relief in punch	Mass production, mature technology	Small electronic, magnetic, and refractory products
Isostatic	Rubber mold, complex/simple shape	Uniform density, mechanized, large or small sizes	Chemical, structural, electronic, refractory items
Hot	Refractory ceramic/metal die	Simultaneous molding and sintering	Dense structural, electronic, material stock and shapes
Rolling	Metal roll	Continuous mass production	Thick substrates
Pressing/slurry	Porous pistons	Particle/fiber alignment, porous or high density	Oriented magnets, fiber insulation board, advanced materials
Pressing/plastic			
Porcelain	Porous molds	Low pressures, surface relief, small/large size	Table ware, electrical insulators, refractories
Thermosetting	Heated metal	Surface relief, small/large size	Refractories, insulation
Thermoplastic	Cooled metal		
Extrusion/plastic	Metal die, constant x-section	Mass production, continuous, small/large items,	Refractory/electronic, rods, tubes, cellular products
Injection molding/plastic	Metal die	Warm temperature, complex shape, dimensional accuracy	Advanced structural components, nozzles
Rolling	Metal rolls	Continuous	Bonded abrasives
Casting/slurry			
Slip	Porous gypsum/polymer	Wide variety shapes, ease of installation, mechanization available	Refractories, closed-end tubes, advanced structural items, whitewares
Gel	Metal/polymer mold	Complex shapes, good powder dispersion, uniform density	Advanced structural items, insulation, refractories
Reaction bond	Variable materials	Complex shape, large/small size	Concrete, refractories, insulation, dental fillings
Tape	Metal blade	Mass production, continuous	Thin substrates

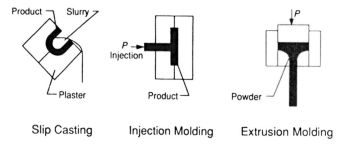

Slip Casting   Injection Molding   Extrusion Molding

Fig. 21.1 Simulation of slip casting, injection molding, and extrusion. [From L. M. Sheppard, *Am. Ceram. Soc. Bull.*, **68**(10), 1815 (1989).]

very smooth surface are produced by forming a coating of controlled thickness on a stationary or moving carrier film, and then separating it after drying to a leather-hard state. Small substrate items are produced by blanking, cutting, and punching controlled shapes.

Processing Additives

The liquid content and composition of processing additives will differ somewhat for each technique. Consistency states used in forming and the dependence on liquid content were discussed in Chapter 14. Functions of processing additives discussed in prior chapters are summarized in Table 21.2.

TABLE 21.2   Functions of Processing Additives

Additive	Functions
Deflocculant	Particle charging, aid and maintain dispersion
Coagulant	Uniform agglomeration after dispersion
Binder/flocculant	Modify rheology, retain liquid under pressure, yield strength, green strength, adhesion
Plasticizer	Change viscoelastic properties of binder at forming temperature, reduce T_g of binder
Lubricant	Reduce die friction (external), reduce internal friction, mold release
Wetting agent	Improve particle wetting by liquid, aid dispersion
Antifoam	Eliminate foam
Foam stabilizer	Stabilize foam
Antistatic agent	Charge control of dry powder
Chelating agent/sequestering agent/precipitant	Inactivate undesirable ions
Antioxidant	Retard oxidative degradation of binder

The binder molecular weight and concentration are all important to developing the necessary forming rheology and other properties. Viscosity grades and molecular weights of binders used in different forming processes are listed in Table 21.3. Granules for pressing are commonly made by spray drying a slurry. For a slurry with about 40 vol% powder and a binder concentration of 2–10 vol%, a binder of low to medium viscosity grade must be used to obtain a slurry that can be spray dried. The droplet diameter of <1 mm facilitates rapid drying. In slip casting, the content of binder must be very low when casting dense parts of fine powders because the resistance to liquid migration through the cast varies directly with the concentration of the binder. A binder of higher molecular weight will commonly migrate less with the migrating liquid, which is desirable. In gel casting, a binder of low viscosity grade (low molecular weight) is used in the initial slurry.

Tape cast films are thin and must be strong but also flexible after drying. This requires a relatively high concentration of both binder and plasticizer. Vinyl binders are relatively flexible and are popular for tape casting systems. Binders are of high molecular weight and orient during the casting process. This facilitates the flow rheology and improves the strength of the tape.

A binder of medium-high viscosity grade is required to extrude a material of high powder content which has an appropriate yield strength, apparent viscosity, and pseudoplasticity, and because the liquid permeability and binder migration in the body should be low. Extrusion bodies containing a flocculated platey clay binder are mixed in slurry form and filter pressed or spray dried to form a feed material. But bodies containing a molecular binder of medium-high molecular weight must be admixed directly into the batch having a plastic consistency. The cellulose binders, which are relatively less flexible, impart elasticity to the extrudate and are popular for extrusion.

A thermoplastic binder of medium molecular weight is commonly used for a nonaqueous injection molding material. The viscosity of a thermoplastic

TABLE 21.3 Binder Viscosity Grade and Molecular Weight Guidelines

Process	Viscosity Grade[a] (mPa · s)	MW (kg/mol)
Spray drying/pressing	Low-medium	8–50
Casting		
Slip	Medium-high	20–300
Gel	Very low-medium	3–20
Tape casting	Medium-very high	100–1000
Extrusion	Medium-high	100–500
Injection molding	Medium (average value)	20–50

[a]Based on a 2-wt % solution.

polymer binder is relatively low at the forming temperature for compounds such as polyethylene, polypropylene, and wax, but on cooling to room temperature the apparent viscosity becomes very high. These materials may also be pressure molded hot and then cooled to stiffen for handling. Because of the very high polymer concentration in the molded part, the composition of the binder system must also be designed to facilitate removal from the molded part prior to sintering.

Small parts may be produced by dry pressing, isostatic pressing, slurry pressing, injection molding, and slip/gel casting. Dry pressing would be the choice when the shape is relatively simple, a large-volume market warrants mass production, and the dimensional precision must be high. Slurry pressing and isostatic pressing might offer some improvement in microstructure, but at a loss of productivity. Injection molding and casting would be an option for a small product of complex shape. Very large shapes may be made by isostatic pressing, extrusion, and casting processes. Isostatic pressing is commonly used for forming very dense, large refractory products and extrusion for large clay-based ceramics. Slip casting is not commonly used for forming large solid products because of the long casting times. But slip casting is commonly used for forming large thin-walled shapes and large solid products when the production volume is very limited.

Table 21.4 presents insights into the microstructure of the feed and changes in the microstructure for different forming techniques. Processing the feed material as a slurry facilitates the dispersion of particles and additives, which may reduce the scale of segregation and inhomogeneity and potentially improve the properties of the product and reduce variability in a production lot. The percent volume of binder/volume of particles $100V_b/V_p$ is an important parameter of a processing system. As seen in Table 21.4, $100V_b/V_p$ is very high for an injection molding material, high for a cast tape product, moderate for extrusion and gel cast materials, and relatively low for slip cast and pressed

TABLE 21.4 System Characteristics on Forming and After Drying

Process	$100\ V_b/V_p$	Feed	Formed	Dried	ΔPF_f	$100\ V_o/V_p$
		\multicolumn{3}{c}{DPS}		
Pressing (all forms)	2–10	$\ll 1$	$\ll 1$	$\ll 1$	High	2–10
Extrusion	8–16	< 1	$= 1$	$\ll 1$	Very low	8–16
Injection molding	15–40	> 1	> 1	> 1	Very low	30–50
Casting						
Slip	0–4	> 1	$= 1$	$\ll 1$	Medium	0–6
Gel	5–25	> 1	> 1	$\ll 1$	Very low	5–25
Tape	15–25	> 1	> 1	< 1	Medium	45–70

products. The DPS of the feed is >1 for feed for casting and injection molding systems, nearly 1 for extrusion feed, and $\ll 1$ for feed for pressing. Drying significantly reduces the DPS, except in commonly processed injection molded products. The change in particle packing (ΔPF) on forming is indicative of both the densification produced on forming and the potential for density gradients due to nonuniform pressure or from the migration of processing additives. However, when ΔPF is smaller, the formed product will be of relatively higher porosity and the firing shrinkage will be higher if sintered to a high density. The volume of residual organics/volume of particles $100 V_o/V_p$ in the green product indicates the potential difficulty and time required for thermal elimination of the organics in the presintering stage of the firing cycle. Thermolysis of organics must be accomplished especially carefully for injection molded and tape cast products and for gel cast products containing a relatively high binder content.

Pressures and shear rates differ for the different processes. In general, the pressure for forming increases as the yield stress and apparent viscosity of the feed material increase. Nominal forming pressures and shear rates for the different forming processes are listed in Table 21.5. In general, pressures for isostatic and roll compaction exceed those used in metal die pressing of the granulate feed. Forming pressure is significantly lower for plastic and slurry feed and decreases in the order injection molding, extrusion, and casting. Shear rates in forming vary widely depending on the mode of delivery and batch/ continuous nature of the process.

The formed product must have a strength sufficient for removal from the die/mold and subsequent handling without distorting its shape. The strength of the product commonly increases as the product item is dried, when cracks or

TABLE 21.5 Nominal Pressures and Shear Rates in Forming

Forming Process	Pressure (MPa)	Shear rate (1/s)
Pressing		
Roll/isostatic	>150	
Metal die	<100	
Plastic forming		
Injection molding	Varies	10–10,000
Extrusion	<40	10–1000 (die)
		100–10,000 (die-land)
Casting		
Slip (mold suction)	<0.2	<10 Pouring/draining
Slip (slurry pressurized)	<10	<100 Pumping
Slip (vacuum on mold)	<0.7	<10 Filling
Gel	<0.1	<10 Filling
Tape	<0.1	10–2000

TABLE 21.6   Tensile Strength Guidelines for Forming and Machining[a]

Strength (MPa)	Surface	Edges	Threads
<0.4	Crumbles	Crumbles	—
<0.7	Pock marked	Chipped	Rough
1.0	Smooth	Sharp	Smooth
2.0	Very smooth	Sharp	Smooth
5.0	Very smooth	Sharp	Very smooth

[a]After A. R. Teter, *Ceram. Age*, 30–31 (1966).

laminations are not formed. Parts may be machined or surface finished in a subsequent operation. Strength guidelines are presented in Table 21.6.

21.2   INDUSTRIAL FORMING EQUIPMENT

Ceramic forming equipment and tooling used in production varies widely depending on the basic process, the degree of mechanization, and modernization investments. Pressing is commonly carried out using highly automated presses which press and eject small parts at a very high rate (exceeding 1/sec). Tooling is commonly hardened steel with carbide inserts in high wear sections. In isostatic pressing, molds made of a rubber material are hydraulically compressed at a rapid rate. Granulate is pressed between hard rolls into substrates in a continuous manner in roll compaction. Pressing technology is highly mechanized, capital intensive, and designed for mass production. A high degree of automation is now the standard. A constant supply of a well-controlled granulate feed is requisite. Pressed parts are commonly handled and conveyed mechanically.

Modern extrusion is also highly mechanized and automated. Extruders may be of the piston type for small-scale production or for the production of some high-performance parts. A product with a large cross section is often continuously extruded using a single auger or a twin screw feeding mechanism. Feed material may be in the form of a deaired billet for piston extrusion or a shredded or coarse granulate material for auger extrusion and some piston extrusion operations. A means for evacuation of most of the air introduced with the feed is requisite. Dies are commonly made of a hardened steel which has a well-controlled surface finish. Product from continuous extruders is commonly handled mechanically. Small product items and small-volume production items may be handled by hand.

Machines for high pressure injection molding are mechanized, highly automated, very sophisticated, and capital intensive. A well-controlled source of premixed, pelletized feed material is required. The feeding mechanism is commonly either a ram or a screw configuration. Temperature control and shear rates throughout the delivery and molding process must be carefully controlled. Tooling must be highly wear resistant and polished and is commonly made of hardened steel. Considerations in mold designs are vents, ejector pins, weld lines, and edge radii, and the molds are quite complex. Low-pressure injection molding equipment in which a low-viscosity feed slurry becomes solidified on contact with a mold of different temperature is less complex and much lower in capital cost.

The degree of mechanization, automation, and capital cost for casting processes varies widely depending on the mode of casting, product type, and volume of production. Porous gypsum molds which remove liquid from the slurry by capillary suction are widely used for slip casting large and small shapes. A large mold inventory and large mold conditioning and drying facility is required for high-volume production. Mechanized mold handling and slurry delivery is common for large shapes and high-volume production. Pressure casting machines in which a pressurized slurry is injected into a single or battery of porous molds are now available. The porous molds may be manufactured of a polymer material and do not require drying. Pressure casting conserves floor space and is less labor intensive, but the capital cost is relatively higher. Vacuum casting, in which a vacuum assist is used to form a deposit of particles on a porous mold, is used to form porous insulations of complex shapes. In gel casting, nonporous molds cast from a polymer or machined from a metal or high polymer are used, and mold life is potentially very high.

Modern machines for continuous tape casting and drying are highly mechanized and automated and of high capital cost. A very well-controlled slurry feed is required. The controlled slurry film supported on a moving belt is dried into a continuous tape on moving through a drying tunnel. Flexible tape is commonly blanked to size and shape, print coated, and laminated using very precise mechanized transport and handling systems.

21.3 PROCESS FLOW CHARTS

Selecting a forming process for a particular product requires consideration of both technical and business factors. For high-volume production, a high-productivity mass production technique may offset a high capital cost. The choice of forming method may also depend on the feed material preparation systems already in place, expertise and experience in the workforce, technical support, projected markets and competition, and environmental concerns. Here we consider alternative forming processes primarily based on technical considerations.

As seen in Table 21.1, more than one forming method may often be used to produce a product having a particular size and shape. The choice may be very limited, however, when the microstructure in the product is difficult to obtain or control, precision of dimensions must be very high, and surface smoothness is paramount. Sometimes the choice depends solely on productivity considerations.

A general process flow chart is presented in Fig. 21.2. As is indicated, forming involves operations that impart energy to a material system to alter its size, shape, and microstructural characteristics. Measurements are required both to monitor the pressures and temperature in forming and to ascertain the characteristics of the feed and the as-formed product. Flow charts for different forming processes are indicated in Figure 21.3.

Pressing and casting techniques commonly begin with a prepared slurry. Note that the feed preparation technology is significantly different for plastic

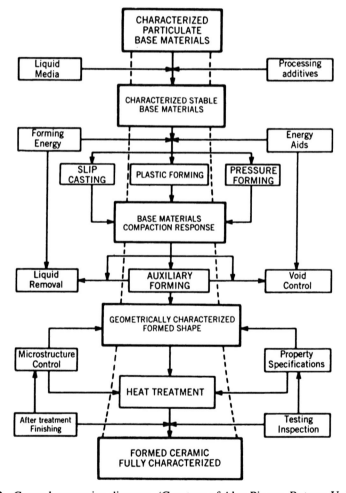

Fig. 21.2   General processing diagram. (Courtesy of Alex Pincus, Rutgers University.)

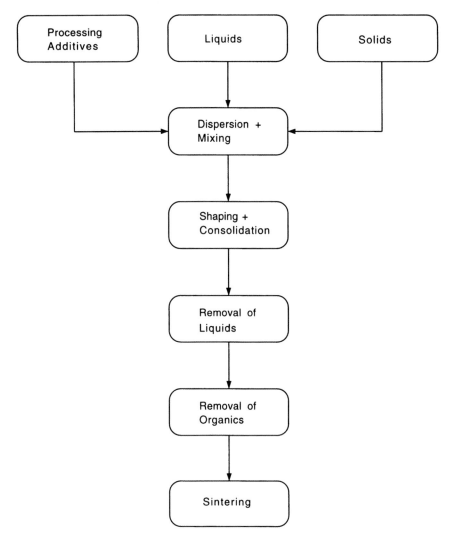

Fig. 21.3 General processing flow diagram (top) and examples of processing flow diagrams for forming by (middle) pressing and (bottom) injection molding.

forming. Control of the microstructure is more critical for advanced high-performance ceramics, and this may restrict the options for forming. The process steps for the slip casting process for forming a high-performance structural ceramic becomes very involved because of the tight controls on both dimensions and microstructure, as shown in Fig. 21.4. Clearly, ceramic processing and forming is a systems problem.

Statistical process control SPC is now widely practiced to improve product performance and to keep the product parameters within specification. In SPC,

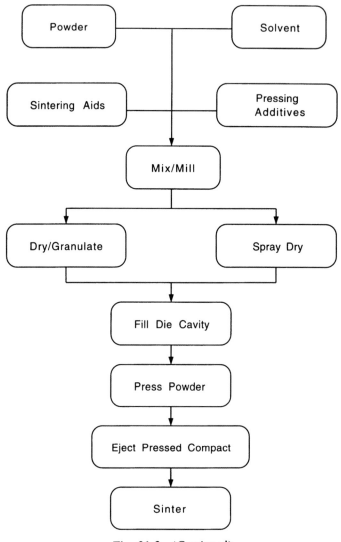

Fig. 21.3   (*Continued*)

upper and lower control limits about the mean equal to $+/-$ three times the standard deviation of data are established for key parameters specified for the process. These parameters are tracked during the processing and forming of the product, as shown in Fig. 21.5. Points falling outside of the control limits indicate lack of control. The control charts indicate natural tolerances of the process and may aid in detecting and eliminating causes of variability.

To a first approximation, the production cost using a particular process varies

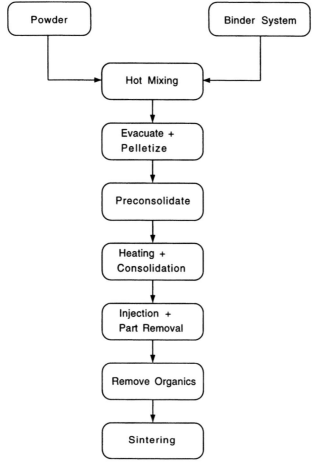

Fig. 21.3 (*Continued*)

with the production yield as

$$Cost = 1/Yield \qquad (21.1)$$

Production yields for ceramics are much lower than yields in metals and polymers processing. The lower yield may present a barrier to opening new markets. Systems exhibiting "process tolerant" behavior offer an avenue to reduce variability and increase yields, as is indicated in Fig. 21.6. Clearly, production yield must be considered when selecting a process and estimating the product cost.

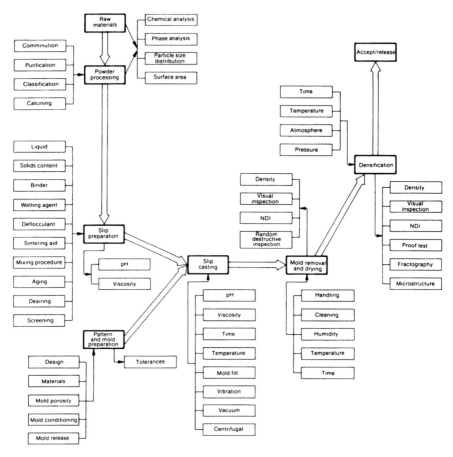

Fig. 21.4 Processing parameters, steps, and controls for the fabrication of an advanced ceramic using the slip casting process. (From D. W. Richerson, *Modern Ceramic Engineering*, Marcel Dekker, New York, 1982, p. 189.)

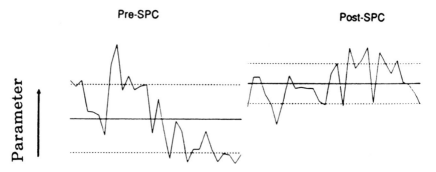

Fig. 21.5 Control chart indicating variability and parameter mean before and after statistical process control (SPC).

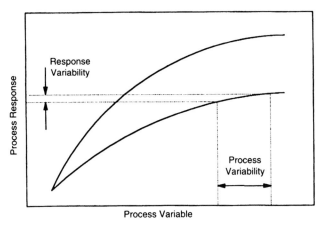

Fig. 21.6 General diagram indicating variability in a processing-tolerant process. [From M. J. Readey and C. D. Lovejoy II, *J. Am. Ceram. Soc.*, **72**(9), 77 (1993).]

21.4 DIMENSIONAL CONTROL

Control of product dimensions within specifications is essential when manufacturing a product. For products having a tight dimensional tolerance, superior process control through the sintering process may significantly improve yields within specification and reduce the product cost. Machining of the fired product to specification contributes significantly to the cost of manufacturing and is often prohibitive.

Post-forming shrinkage is commonly much higher in ceramics processing than in metals and polymer processing because of the large differential between the final density and the as-formed density. Volume shrinkage may occur during drying Sv_d, during the removal of the organic binder or reaction of bond phases producing consolidation Sv_b, and during sintering Sv_s. The total volume shrinkage Sv_{total} is

$$Sv_{total} = Sv_d + Sv_b + Sv_s \qquad (21.2)$$

where each shrinkage is specified using an as-formed volume basis. Precise control of the final product dimensions depends on control of Sv_d, Sv_b, and Sv_s. For dry pressed and injection molded parts, $Sv_d = 0$. But for extruded and slip cast parts, Sv_d is in the range 3–12%. Shrinkage on elimination of the binder (Sv_b) is relatively high for injection molded and tape cast parts. The sintering shrinkage Sv_s is commonly about 35–45% for products having a high fired density and may range from 20 to 50%. Better dimensional control can commonly be expected when the shrinkages Sv_d and Sv_b are low.

The final volume of the product depends on the as-formed volume and the total shrinkage

$$V_{final} = (1 - Sv_{total}) V_{formed} \qquad (21.3)$$

Control of the final product dimensions depends on controlling both the as-formed dimensions and the shrinkages in processing.

Figure 21.7 shows the volumes of constituent phases in an as-molded part and the relative volume shrinkage which occurs on densification. The final volume V_{final} is larger than the volume of the inorganic solids V_{solids} when the product has residual porosity; the ratio D_r is

$$D_r = V_{solids}/V_{final} \tag{21.4}$$

Accordingly, $V_{final}/V_{formed} = V_{solids}/(V_{formed} \, D_r)$, and

$$V_{final} = V_{formed} \, (PF_{formed})/D_r \tag{21.5}$$

Equation 21.5 indicates the essential controls required in processing to reproduce V_{final}. V_{formed} is controlled by the mold or die. Control of PF_{formed} requires consistent batching and dispersion of the starting materials and reproducibility of forming response. Adjustments in the proportions of processing additives or variations in the liquid content to maintain a particular flow rheology commonly alter PF_{formed} and V_{final}. Control of D_r requires control of both the microstructure of the as-formed part and the firing, as both will influence the shrinkage Sv_s. When the firing temperature–time–atmosphere conditions are well reproduced during firing, the reproducibility of D_r depends on its sensitivity to processing-related variations in the microstructure of the as-formed product.

When shrinkage is isotropic, the linear shrinkage ($\Delta L/L_0$) is calculated from the volume shrinkage as

$$\Delta L/L_0(\%) = [1 - (1 - Sv_{total})^{1/3}]\,100 \tag{21.6}$$

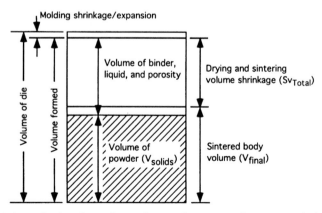

Fig. 21.7 Schematic showing volume changes in a processing system during forming, drying, and sintering.

The linear shrinkage $\Delta L/L_0$ on drying is commonly $<2\%$ and on binder elimination and sintering $<15\%$. But it should be realized that when batching and thermal processing are consistent, scatter in the dimensions of the sintered product commonly follows the scatter in the PF and dimensions of the as-formed product.

SUMMARY

Ceramic systems for forming contain the basic ceramic material and a carefully designed system of processing additives. Several forming processes may commonly be used to form a particular product. The selection of a process is based on technical forming and material factors, manufacturing factors, market factors, and production costs. Forming to a near-net shape is desirable to reduce finishing costs. Ceramic production yields are low relative to yields in metals and polymer manufacturing. Statistical process control is now widely practiced to reduce product variability and improve product yield. Designing systems that exhibit processing tolerance provides a distinct advantage for improving yields. Dimensional control depends on reproducing the density of the as-formed product and dimensional shrinkages on drying, binder elimination, and sintering. Batch adjustments to reproduce forming rheology may negatively impact dimensional control.

SUGGESTED READING

1. M. J. Readey and C. D. Lovejoy, "An Optimization Strategy for Improving Manufacturing Yields," *Am. Ceram. Soc. Bull.*, **72**(9), 75–80, 101 (1993).

2. Jingmin Zheng and James S. Reed, "The Different Roles of Forming and Sintering on the Densification of Ceramics," *Am. Ceram. Soc. Bull.*, **79**(9), 1410–1416 (1992).

3. *Engineered Materials Handbook*, Vol. 4, ASM International, Cleveland, OH, 1991.

4. Randall M. German, Karl F. Hens, and Shun-Tian P. Lin, "Key Issues in Powder Injection Molding," *Am. Ceram. Soc. Bull.*, **78**(8), 1294–1302 (1991).

5. A. C. Young, O. O. Omatete, M. A. Janney, and P. A. Menchhoofer, "Gelcasting of Alumina," *J. Am. Ceram. Soc.*, **74**(3), 612–618 (1991).

6. Andrew J. Ruys and Charles C. Sorrel, "Slip Casting of High-Purity Alumina Using Sodium Carboxymethylcellulose as a Deflocculant/Binder," *Am. Ceram. Soc. Bull.*, **69**(5), 828–832 (1990).

7. Laurel M. Sheppard, "Fabrication of Ceramics: The Challenge Continues," *Am. Ceram. Soc. Bull.*, **68**(10), 1815–1820 (1989).

8. Stanley J. Lukasiewicz, "Spray Drying Ceramic Powders," *J. Am. Ceram. Soc.*, **72**(4), 617–624 (1989).

9. Edward G. Blanchard, "Pressure Casting Improves Productivity," *Am. Ceram. Soc. Bull.*, **67**(10), 1680–1683 (1988).

10. A. Roosen, "Basic Requirements for Tape Casting of Ceramic Powders," in *Ceramic Transactions*, Vol. 1B, G. L. Messing, E. R. Fuller, Jr., and Hans Hausner, Eds., The American Ceramic Society, Westerville, OH, 1988, pp. 675-692.

11. James S. Reed, "Critical Issues and Future Directions in Powder Forming Processes," in *Ceramic Transactions*, Vol. 1B, G. L. Messing, E. R. Fuller, Jr., and Hans Hausner, Eds., The American Ceramic Society, Westerville, OH, 1988, pp. 601-610.

12. Edmond P. Hyatt, "Making Thin, Flat Ceramics—A Review," *Am. Ceram. Soc. Bull.*, **65**(4), 637-638 (1986).

11. S. J. Lukasiewicz and James S. Reed, "Character and Compaction Response of Spray-Dried Agglomerates," *Am. Ceram. Soc. Bull.*, **57**(9), 798-801, 805 (1978).

12. J. E. Fenstermacher, "Dimensional Control of Dry-Pressed Electrical Ceramics," *Am. Ceram. Soc. Bull.*, **48**(8), 775-780 (1969).

PROBLEMS

21.1 Compare and contrast dry pressing and isostatic pressing.

21.2 Compare and contrast extrusion and powder pressing.

21.3 Compare and contrast slip casting and injection molding.

21.4 Compare and contrast process material formulation, homogeneity, and microstructure when forming polymers and ceramics.

21.5 Why is the molecular weight of the binder for extrusion and tape casting relatively high.

21.6 What is the significance of $100V_b/V_p$ being higher for injection molding but $100V_o/V_p$ after drying being higher for tape casting?

21.7 What is the relatively compressibility of a substrate after drying, when made by tape casting, injected molding, and roll pressing?

21.8 Why are the binders used in injection molding not of highest molecular weight?

21.9 The wetting angle of water on a polymer mold material is $> 90°$. When can these molds be used for casting an aqueous slurry?

21.10 Why may the PF after forming and drying be as high in a slip cast part as in a dry pressed part?

21.11 Compare and contrast qualitatively Sv_d and Sv_b for dry pressed, extruded, injection molded, and tape cast products.

21.12 Explain why changes in batch $100V_b/V_p$ may cause variability in final product dimensions even when V_{formed} and D_r are reproduced.

21.13 What is the volume shrinkage when the linear shrinkage is 1.5% and isotropic?

21.14 Two products are formed with PF of 0.56 and 0.75. Both are fired to $D_r = 0.97$. Calculate the $\Delta L/L_0$ (%) for each.

21.15 An as-formed product shrinks 38 vol% on drying and firing. Calculate the linear shrinkage and final length ($L_0 = 10$ cm).

21.16 The PF of pressed products is in the range 0.57–0.60. The volume formed is 10 cm^3; $D_r = 0.97$. What is the range of Sv_{total} and the fired lengths for the dimension with $L_0 = 10$ cm?

21.17 For a production process, PF $= 0.60$ but D_r ranges from 0.95 to 0.98; $L_0 = 25$ cm. Calculate the range for Sv_{total} and L_{fired}.

EXAMPLES

Example 21.1 Contrast injection molding and casting from a strictly shaping perspective.

Solution. Each requires introducing a feed material into a mold. In injection molding, the mold is connected to the feed mechanism and repetitively closed, filled, opened to eject part, and so on. In common casting, slurry is transported to sequentially fill a set of molds or molds are conveyed to a filling station. A modern pressure casting machine imitates the shaping action of an injection molding machine except that liquid removal rather than a change of temperature is used to develop strength in the green product.

Example 21.2 Small tubes of a high-performance ceramic may be produced by isostatic pressing or by extrusion. Compare and contrast each of these forming methods for this product.

Solution. Each method is capable of mass production of small tubes of high quality. However, the processing of the powder feed is very different. A spray-dried granulate would be required for isostatic compaction and dispersion of agglomerates and processing additives in processing the precursor slurry should be possible. Isostatic pressing should produce a product with a small scale of inhomogeneity when the granules deformability is well controlled. Achieving an equivalent degree of dispersion in a high-viscosity extrusion batch will be much more difficult. Mixing will be a critical processing step in order to achieve a small scale of inhomogeneity in the green extruded tubes. Achieving a high degree of dimensional precision will be more difficult in extrusion forming because of the lower strength of the extruded tubes as formed and because of their drying shrinkage. However, if satisfactory homogeneity and dimensional control can be achieved, extrusion may offer productivity advantages.

Example 21.3 Explain why a low–medium viscosity grade binder must be used when preparing a spray-dried powder. A common batch will contain about 40 vol% powder, 55 vol% liquid, 2 vol% binder, and 3 vol% other additives. (Consult Example 14.3.)

Solution. The binder concentration $100V_b/V_p = 5$ is high enough to produce adequate green strength, and the powder content in the slurry must be as high as possible for efficient drying into satisfactory granules. The concentration of binder relative to the concentration of liquid is about 3.6 vol% or about 4.6 wt%. At this concentration, the viscosity of the solution of dissolved binder can range from 3 to 4 mPa·s (low-viscosity grade) to over 1000 mPa·s (high-viscosity grade). The viscosity with the powder added will be much higher. For atomization in spray drying, a low-viscosity slurry is required. A low-medium viscosity grade binder must be used.

Example 21.4 Compare and contrast the system characteristics during forming and drying for injection molding and slip casting.

Solution. Both processes begin with the particles dispersed in a matrix (DPS > 1), but the matrix of the injection molding feed contains mostly viscous polymers (note much high V_b/V_p). In the molded part, the DPS > 1. After casting and removal of the low viscosity liquid on drying, DPS << 1. The relatively higher increase in PF during slip casting indicates that its PF should be higher, but PF gradients may be produced when pressure gradients are significant in casting. Thermal gradients may cause small PF gradients in cooled injection molded material. The green strength of the molded parts of much higher binder content will facilitate handling. But the relatively high organic content and lack of particle contact will be unfavorable when thermally eliminating the organics prior to sintering.

Example 21.5 The same alumina ceramic product is made using two different processes. In Process A, the bulk density (inorganic basis) is 2.43 Mg/m³ green, and 3.84 Mg/m³ fired. In Process B, the bulk density is 2.74 Mg/m³ green and 3.90 Mg/m³ fired. The ultimate density for the product is $D_u = 3.96$ Mg/m³. What are the differences in volume shrinkage Sv_s, linear shrinkage, and ratio D_r for the two processes.

Solution. The relative volume change $Sv_s = (1 - D_{green}/D_{fired})$ and

$$Sv_{s(A)} = 1 - 2.43/3.84 = 0.367 \quad \text{or} \quad 36.7\%$$

$$Sv_{s(B)} = 1 - 2.74/3.90 = 0.297 \quad \text{or} \quad 29.7\%$$

The linear shrinkage $\Delta L/L_0 = [1 - (1 - Sv_s)^{1/3}]$ and

$$\Delta L/L_{0A} = 0.141 \quad \text{or } 14.1\%$$

$$\Delta L/L_{0B} = 0.111 \quad \text{or } 11.1\%$$

$$D_r = D_{\text{fired}}/D_u$$

and

$$D_{r(A)} = 3.84/3.96 = 0.97 \qquad D_{r(B)} = 3.90/3.96 = 0.98$$

CHAPTER 22

PRESSING

Pressing is the simultaneous compaction and shaping of a powder or granular material confined in a rigid die or a flexible mold. For industrial pressing operations, powder feed is in the form of granules of controlled size and deformability. The feed granules contain processing additives and are commonly prepared by spray drying or spray granulation. Feed containing coarse particles and a binder is commonly in the form of a poorly flowing, semicohesive mass. Pressing is the most widely practiced forming process for reasons of productivity and the ability to produce parts ranging widely in size and shape to close tolerances with essentially no drying shrinkage. Products produced by pressing include a wide variety of magnetic and dielectric ceramics, various fine-grained technical alumina products including chip carriers and spark plugs, engineering ceramics such as cutting tools and refractory sensors, ceramic tile and porcelain products, and coarse-grained refractories, grinding wheels, and structural clay products.

Pressing by means of punches in hardened metal dies, commonly called dry pressing, is commonly used for pressing parts thicker than 0.5 mm and parts with surface relief in the pressing direction. Isostatic pressing in flexible rubber molds, commonly called isopressing, is used for producing shapes with relief in two or three dimensions, shapes with one elongated dimension such as rods and tubes, and very massive products with a thick cross section. Large shallow products are produced using a combination of metal die and isostatic pressing and large sheets by roll pressing. Here we consider the very important topic of pressing technology.

22.1 PROCESS VARIABLES IN DRY PRESSING

Stages in dry pressing include (1) the filling of the die, (2) compaction and shaping, and (3) ejection, as shown in Fig. 22.1. Note that free-flowing granules are fed to the die by means of a sliding feed shoe and are metered volumetrically. Punch and die motions are commonly coordinated to induce vacuum assisting settling of the powder in the cavity of the die. Poorly flowing feed is usually preweighed and fed by hand or by mechanically induced flow.

Pressing modes classified in terms of the motion of the punches and die are listed in Table 22.1. Movement of the punches and die is synchronized, as shown in Fig. 22.2. Punch stops are set at a particular pressure when using hydraulic control and at a particular displacement using mechanical control. Vibrating punches are used in pressing some coarse granular products such as refractories and grinding wheels. Tooling for dies and punches is commonly constructed of hardened steel, but special steels and carbide or ceramic inserts are used in high-wear areas. The clearance between the die and punch is about 10–25 μm when pressing micron-size powders and up to 100 μm when pressing granular particles. Punch motions are commonly controlled in relation to the centroid of the pressed part. The die wall is sometimes tapered (< 10 μm/cm) to facilitate ejection. Pressed parts may be ejected with or without contact pressure from the top punch. Small ejected pieces are commonly displaced to a conveyer by the leading edge of the feed shoe. Larger pieces are lifted and moved mechanically, often using a vacuum assist or by hand.

Pressing rates range from a fraction of a second for small parts to several minutes for large parts using a single-action press. Rates exceeding 5000 ppm are achieved using a multistation rotary press. Press capacities range up to several hundred tons. The maximum pressing pressure used in dry pressing is commonly in the range of 20–100 MPa; higher pressures are used for technical ceramics than for clay-based materials. Die-set life may range up to several hundred thousand pieces for a simple die and a low pressuring pressure. Advertised tolerances for industrial pressing are now better than $\pm 1\%$ in mass and ± 0.02 mm in thickness. Dimensional tolerances and the uniformity of the microstructure are much more difficult to control when the part has several surface levels. The classification of pressing difficulty is summarized in Table 22.2.

Systems used to prepare granulated pressing powders by spray drying commonly contain a deflocculant, binder, plasticizer, and sometimes a lubricant, wetting agent, and defoamer; representative examples of the organic systems used are listed in Table 22.3. A deflocculant is used to aid powder dispersion and to reduce the liquid requirement to form granules. The binder content for pressing is low and in the range of $100 V_b / V_p = 0.02$–0.12. The plasticizer increases the deformability of the binder and reduces the moisture sensitivity of the binder. Moisture commonly acts as a secondary plasticizer and the

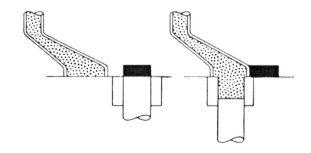

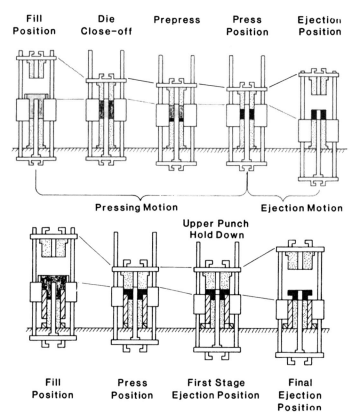

Fig. 22.1 (top) Action of feeding shoe pushes away pressed part and fills die; pressing and ejection motions using a floating die for (middle) one-level part and (bottom) two-level part. (Courtesy of Dorst Maschinen und Anlangenbau, Kochel A. See, Germany.)

TABLE 22.1 Dry-Press Modes

Type	Die	Top Punch[a]	Bottom Punch[a]
Single action	Fixed	Motion	Fixed
Double action	Fixed	Motion	Motion
Floating die	Moves	Motion	Fixed

[a]Simple or composite.

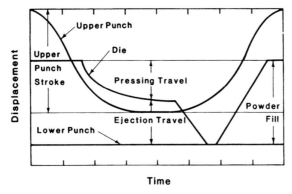

Fig. 22.2 Synchronized punch and die motions during a pressing cycle using a floating die.

amount adsorbed between granulation and pressing should be controlled. A lubricant may be introduced when forming granules or admixed later in a separate operation to coat the granules. A lubricant can reduce die wear and the ejection pressure on pressing and also improve density uniformity in the pressed part. Soft wax and polyethylene glycol binders and plasticized polyvinyl alcohol provide some lubricating action. A list of additives for a slurry for spray drying, containing a polyvinyl alcohol binder, are listed in Table 22.4. Pressing powders should be free flowing, of a relatively high bulk density, composed of deformable granules, and stable under ambient conditions. Pressed parts should not adhere to the punch and must be sufficiently strong to

TABLE 22.2 Class of Pressing Difficulty in Dry Pressing

Class	Part Geometry	Pressing
I	One level, <5 mm thick	One direction
II	One level, >5 mm thick	Two directions
III	Two level, any thickness	Two directions
IV	Multilevel, any thickness	Two directions

TABLE 22.3   Additives Used in Industrial Pressing Powders

Product	Binder	Plasticizer	Lubricant
Alumina	Polyvinyl alcohol[a]	Polyethylene glycol[b]	Mg Stearate
96% alumina substrates	Polyethylene glycol[c]	None	Talc,[d] clay[d]
Alumina spark plug insulation	Microcrystalline wax emulsion	KOH + tannic acid	Wax, talc[d] clay[d]
MnZn Ferrites	Polyvinyl alcohol[a]	Polyethylene glycol[b]	Zn Stearate
Ba Titanate	Polyvinyl alcohol[a]	Polyethylene glycol[b]	
Steatite	Microcrystalline wax, clay	Water	Wax, talc,[d] clay[d]
Ceramic tile	Clay	Water	Talc,[d] clay[d]
Hotel china	Clay, polysaccharide	Water	Clay[d]
Refractories	Ca/Na lignosulfonate	Water	Stearate

[a]Low-viscosity grade.
[b]400 molecular weight.
[c]20,000 molecular weight.
[d]Colloidal size.

TABLE 22.4   Organic Additive System for a Pressing Powder Produced by Spray Drying

Ingredient	Concentration (wt%)
Polyvinyl alcohol	83.1
Glycerol	8.3
Ethylene glycol	4.1
Deflocculant	4.1
Nonionic surfactant	0.2
Defoamer	0.2

Source: Du Pont Inc., Wilmington, DE.

survive ejection and subsequent handling. The binder content is usually as low as practicable to minimize both the higher binder cost and the amount of gas produced during binder burnout.

22.2   POWDER FLOW AND DIE FILLING

Good powder flow is essential for reproducible volumetric filling, a uniform density of the fill, and a rapid pressing rate. Flow rates are often determined by measuring the time for a specific mass or volume of powder to flow through a standard funnel or are inferred from the angle of repose of powder poured onto a flat surface. The flow rate and angle of repose for pressing granules correlates well, as shown in Fig. 22.3.

Dense, nearly spherical particles or granules with smooth, nonsticky surfaces that are coarser than about 20 μm have good flow behavior and are preferred (Fig. 22.4). The presence of more than 5% of fines <20 μm in size may sometimes stop flow altogether. Also, fines may enter the annulus between the punch and die wall, which increases friction and reduces the escape of air. Extremely large granules are commonly irregular in shape, and the bridging action between coarse granules may impede flow and the achievement of a uniform bulk density when powder flows into the die. These should be removed by sieving. When the separation between surfaces of a composite punch is small, the maximum granule size must be a fraction of the cross section available for flow. The surface of granules is smoother when composed of a finer particle size and when small granules do not adhere to larger ones (see Chapter 20). Binders become stickier when the temperature approaches the melting point or exceeds the glass transition temperature of the binder system, and control of the concentration of the binder system and plasticizer (and temperature) is important for controlled flow. The flow behavior of granules containing a water-soluble binder system may be quite sensitive to the relative humidity of the storage environment and humidity control may be required (Fig. 22.5).

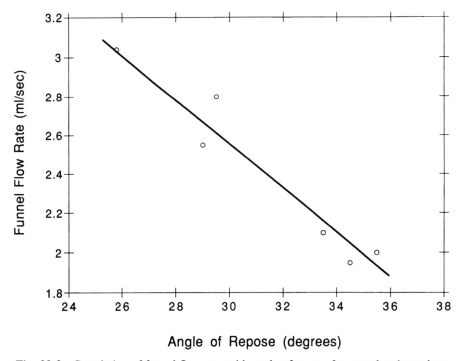

Fig. 22.3 Correlation of funnel flow rate with angle of repose for granulated powders.

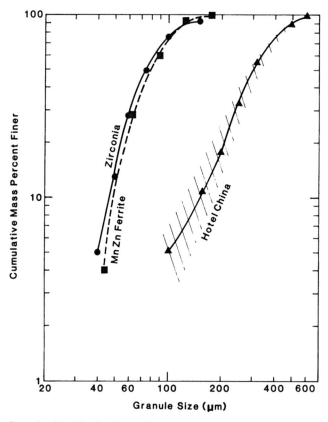

Fig. 22.4 Granule size distributions of spray-dried feed used in industrial pressing.

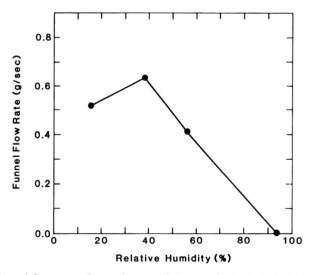

Fig. 22.5 Funnel flow rate of granules containing a polyvinyl alcohol binder depends on the moisture content controlled by ambient relative humidity (surfaces become adhesive when T_g < ambient temperature).

TABLE 22.5 Characteristics of Several Industrial Spray-Dried Pressing Powders

	Particle Size[a] (μm)	Organics (vol%)	Granule Size[b] (μm)	Granule Density (%)	Fill Density (%)
Alumina substrate	0.7	3.6	92	54	32
Spark plug alumina	2.0	13.3	186	55	34
Zirconia sensor	1.0	10.5	75	55	37
MnZn ferrite[c]	0.7	10.0	53	55[d]	32
MnZn ferrite[e]	0.2	8.5	56	31	18
Silicon carbide	0.3	20.4	174	45	29
		Water (vol%)			
Wall tile	9	13–21	134		38

[a]Volume mean.
[b]Volume geometric mean (50%).
[c]Conventional powder.
[d]Donut shapes (low value).
[e]Coprecipitated powder.

Pressing problems are reduced when the bulk density of the feed in the die, called the fill density D_{fill}, is high. A higher fill density reduces both the content of air in the powder and the punch travel. Powders composed of high-density granules that pack efficiently on flowing into the die have a higher D_{fill}. Powder characteristics that retard flow may hinder packing and produce density non-uniformities in the filled die. Because mechanical vibration may cause rearrangements of granules into a higher packing density, the ratio of the vibrated bulk density to the poured density is used as an index of hindered flow and filling (see Table 20.3).

A typical fill density for a granulated powder is in the range of 25–35%, as seen in Table 22.5. The granule size distribution for these spray-dried powders is approximately log normal, and the fill bulk density depends directly on the granule density and the packing behavior. Powders composed of granules of low density and/or granules containing large pores, donut-shaped granules, and granules with rough surfaces have a relatively lower packing density.

22.3 COMPACTION BEHAVIOR

In dry pressing, pressure produced by the moving punches compacts the granulated powder into a cohesive part having a particular shape and microstructure. The rate of densification is high initially, but then decreases rapidly for pressures above about 5–10 MPa (see Fig. 22.6). The initial pressure is transmitted by means of contacts between granules. Granule deformation occurs by the

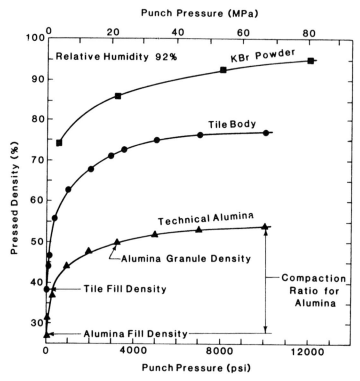

Fig. 22.6 Compaction behavior of spray-dried granules for alumina and clay tile as a function of punch pressure; compaction of KBr powder containing ductile particles at 25 °C is also shown. (Density is calculated on an inorganic basis.)

sliding and rearrangement of particles coated with the binder system. Granule deformation reduces the intergranular porosity and increases the number and area of intergranular contacts. Air compressed in pores migrates and partially exhausts between the punch and die. Relatively little densification occurs above about 50 MPa, but the die wear is higher at these pressures when pressing hard ceramic particles. Industrial pressing pressures are commonly less than 100 MPa for high-performance technical ceramics and less than 40 MPa for whiteware and tile compositions.

The final compact density is less than the PF_{max} of the particles because the frictional forces at particle contacts retard particle sliding. The compaction ratio CR in pressing the product is

$$CR = V_{fill}/V_{pressed} = D_{pressed}/D_{fill} \qquad (22.1)$$

For ceramic pressing powders, a low compaction ratio will reduce both the punch displacement and compressed air in the compact, and a CR < 2.0 is desired. Ceramic pressing powders contain deformable granules but brittle par-

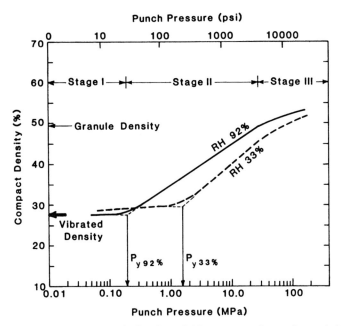

Fig. 22.7 Compaction diagram indicating yield pressure of granules and three stages of compaction for cases of good plasticizing (92% relative humidity RH) and poor plasticizing (33% RH) of granules. (Note Stage II begins when $P > P_Y$ and ends when $D_{compact} \rightarrow D_G$.)

ticles; a high fill density assures a low CR. Powders containing ductile particles such as Al, Cu, and KBr may be pressed to nearly 100% density at a pressure of about 50 MPa at room temperature; for these materials the CR $\gg$ 2.0.

A graph of compact density versus log(punch pressure) provides important information about the compaction of granules. Three compaction stages may be identified when examining the behavior, as shown in Fig. 22.7:*

Stage I—Granule flow and rearrangement

Stage II—Granule deformation predominates

Stage III—Granule densification predominates

In Stage I, a small amount of sliding and rearrangement of granules at a low pressure when the punch contacts the powder feed may produce a slight densification above the fill density.

Stage II begins with granule deformation into neighboring interstices when the pressure exceeds the apparent yield pressure P_Y of the granules. A reduction in the volume and the size of the relatively large interstices occurs with granule deformation, as shown in Figs. 22.8 and 22.9. The P_Y of the granules is less

*R. A. DiMilia and J. S. Reed, *J. Am. Ceram. Soc.*, **66**(9), 667–672 (1983).

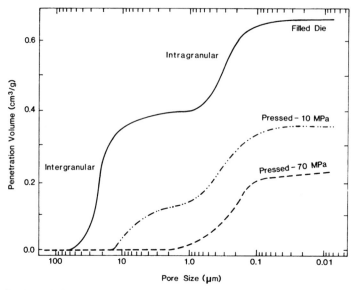

Fig. 22.8 Cumulative pore size distributions indicating intergranular and intragranular porosity and pore sizes on filling die and changes when pressed (Hg-porosimetry technique).

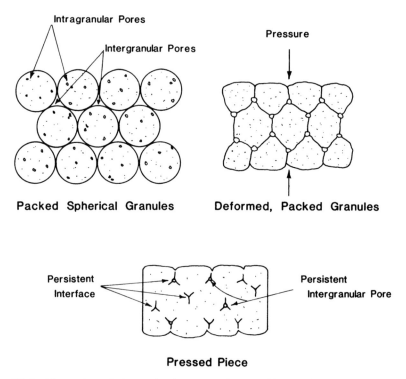

Fig. 22.9 Illustration of change in shape and change of bimodal pore size distribution during compaction.

than about 1 MPa when the binder system in the granules is soft and ductile. Dense granules with a higher binder content and/or a poorly plasticized binder system (Fig. 22.10) resist deformation, and a higher pressure is needed to compact these to an equivalent density. Qualitatively, the dependence P_Y of the granules on parameters of the binder and granules is described by Equation 15.19,

$$P_Y = \text{constant } (PF/(1 - PF))(V_b/V_p)S_0 \tag{22.2}$$

It is in Stage II that the major densification occurs. Granule deformation is the predominant mechanism, but granule densification occurs to a significant extent when the granules are of low density. In Stage II, the compact density may be approximated by the equation

$$D_{\text{compact}} = D_{\text{fill}} + m \log (P_a/P_Y) \tag{22.3}$$

where D_{compact} is the compact density at an applied pressure P_a and m is a compaction constant that depends on the deformability and densification of the granules. The slope m is defined as

$$m = (D_{10P} - D_P)/\log (10P/P) = D_{10P} - D_P \tag{22.4}$$

where 10P is a decade of applied pressures within Stage II, and indicates the compactability of the powder. It is observed that powders containing high-density granules may produce a high compact density with a relatively low

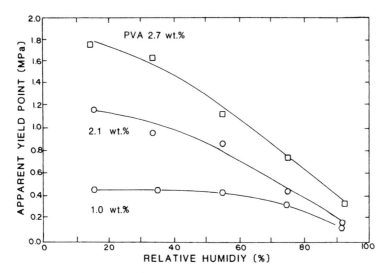

Fig. 22.10 Dependence of yield pressure P_Y of granules on both concentration of PVA binder and plasticizing produced by absorbed moisture.

value of *m*. But the elimination of all large intergranular pores is favored when powders contain granules of somewhat lower density which have an intermediate value of *m* of about 7–10%. For each powder there is an optimum granule density, fill density, and *m* value which produces the best compromise of compaction ratio and the elimination of large pores.

Stage III begins at a high pressure when densification occurs by the sliding and rearrangement of particles into a slightly denser packing configuration (Fig. 22.11). Ideally the larger pores between deformed granules have disappeared (monomodal pore size distribution) and interfaces between granules do not exist. The high applied pressure and concentrated stress at contacts causes fracture of some aggregates and/or anisometric particles blocking densification. In powders containing granules varying in yield strength or density, the transition between the stages does not occur uniformly throughout the compact. Deformation and densification of granules may occur simultaneously when the granule density is only moderately high.

Interfaces between smaller and softer granules begin to be eliminated in Stage II, and groups merge into a more homogeneous mass. It is observed that large interstices between large granules and large pores within large donut-shaped granules persist into Stage III. Large interstices are also observed between relatively hard granules which resist deformation. Both as-pressed external surfaces and fracture surfaces appear more rough and less uniform large pores and large granules persist, as shown in Fig. 22.12. These defects are not eliminated by sintering and reduce the mechanical reliability of the product

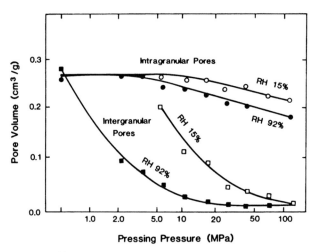

Fig. 22.11 Change of intergranular and intragranular porosity during compaction and dependence on storage relative humidity which controls adsorbed moisture and plasticizing. [From R. A. DiMilia and J. S. Reed, *Am. Ceram. Soc. Bull.*, **62**(4), 484–488 (1983).]

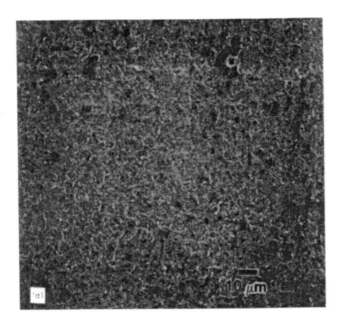

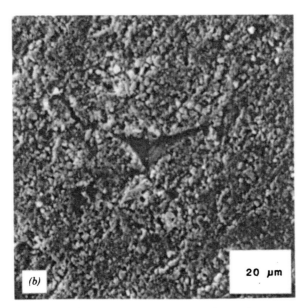

Fig. 22.12 (a) Large circular pore from undeformed donut-shaped granule on pressed surface and (b) fracture surface of compact showing large angular intergranular interstice between incompletely deformed granules.

(Fig. 22.13). Control of the size range of granules and their deformability is essential when processing ceramics for high-performance applications.

The compaction of a bulky powder containing particles coated with a deformable binder (or spray-dried granules reduced in size by hand grinding) and a powder containing granules with the binder removed proceeds somewhat differently than described above, as shown in Fig. 22.14. The fill density of a bulky powder may be quite similar to that of the granulated powder, but the pore size distribution is broad and monomodal. In the absence of granules, P_Y = 0. Pressing causes particle sliding and rearrangement, which decreases the volume of the larger interstices. Particle translation is predominantly axial. Eventually the compaction imitates Stage III compaction. As shown in Fig. 22.15, the compaction of the bulky powder initially leads that of the granulated powder. The intersection at a high pressure of the densification curves for the

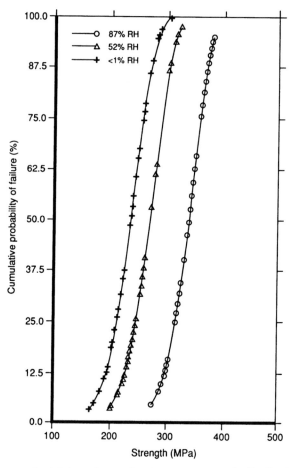

Fig. 22.13 Better plasticizing of granules improves strength distribution of the fired compacts (absorbed moisture is the plasticizer).

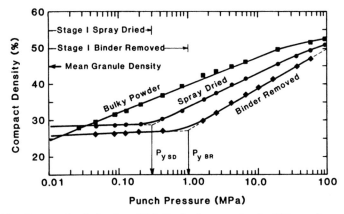

Fig. 22.14 Compaction behavior of MnZn ferrite granules in different form—as spray-dried granules comminuted into fine-size (bulky powder) using mortar and pestle, with binder removed thermally. [After S. J. Lukasiewicz and J. S. Reed, *Am. Ceram. Soc. Bull.*, **57**(9), 798–801 (1978).]

granulated powder and the bulky powder defines a joining pressure P_J and the achievement of uniformly small pores. Deformation-resistant granules will displace the joining pressure to a higher value. For a granulated powder with the binder removed by heating, the P_Y of the granules is relatively high and the granules are brittle. Stage II begins with granule fracture and densification occurs when fracture fragments enter large neighboring interstices. The densification with pressure is retarded and Stage III compaction may not become apparent.

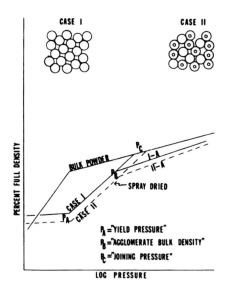

Fig. 22.15 Using poorly plasticized granules, the compact density is lower on entering Stage III and the pressure at which granulated and nongranulated powders achieve equal density P_J is increased.

22.4 EJECTION AND TRANSFER

Elastic compression begins in Stage II and increases in Stage III of compaction. Elastic energy stored in the compact produces an increase in the dimensions of the pressed part on ejection, called springback. Some differential springback between the compact and hardened steel punch is necessary to cause the compact to separate from the punch. A linear springback less than about 0.75% is considered desirable. Excessive springback may cause compact defects on ejection which are described below.

Springback is higher when the pressing pressure is higher. Generally springback is higher when the concentration of organics is higher and the pressing temperature is below the glass transition temperature T_g of the binder system (Table 22.6). The presence of an additive which increases the plasticity of the binder system, such as moisture, reduces the springback. In Fig. 22.16, for the low storage humidity, plasticizing from the greater amount of adsorbed moisture with hygroscopic gum arabic present predominates. But for the higher storage humidity plasticizing from adsorbed moisture is adequate and the greater

TABLE 22.6 Mean Diametral Springback on Ejecting Compacts of Barium Titanate Granules[a]

T_g (°C)	P_y (MPa)	Initial Springback Pressure (MPa)	Pressed at 400 MPa Springback (%)
< 10	0.3	22	0.31
22	0.5	—	0.65
42	1.0	7	1.6

[a] 2 wt% polyvinyl alcohol binder, 24°C.

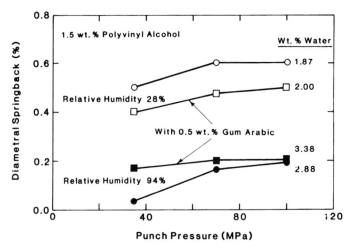

Fig. 22.16 Springback of compacts of an alumina powder pressed at 20°C depends on pressing pressure and plasticizing.

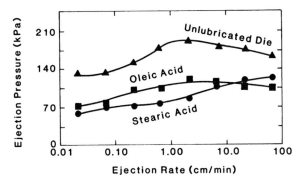

Fig. 22.17 Lubrication significantly reduces the ejection pressure for an alumina compact containing a polyvinyl binder at 23°C.

springback is apparently due to the springback contributed by the adsorbed molecules.

The force for ejection depends on the taper and surface condition of the die, elastic stress within the compact, lubrication of the die wall, and the ejection rate. The ejection pressure is smaller using a tapered die and for a die with a smooth surface. Lubrication provided by a plasticized binder may reduce the ejection pressure by 80–90%. The use of a die wall lubricant can reduce the ejection force still further (Fig. 22.17). Lubrication reduces the tendency for slip-stick sliding and die wear, as shown in Fig. 22.18.

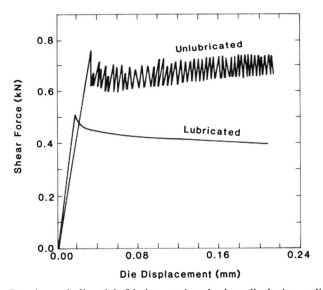

Fig. 22.18 Regular and slip-stick friction produced when displacing a die while the compact is compressed. [From R. A. DiMilia and J. S. Reed, *J. Am. Ceram. Soc.*, **66**(9), 667–672 (1983).]

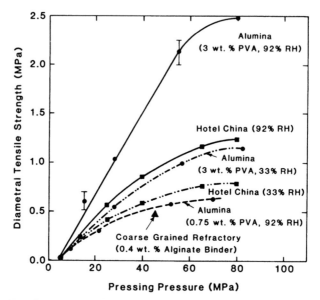

Fig. 22.19 Tensile strength of dried compacts is higher when granules were better plasticized and when pressed at a higher pressure.

Transfer of the ejected part to a conveyer is accomplished by pushing with the front edge of the filling shoe or other mechanical means, a vacuum lifting device, or gravity fall when using a horizontal pressing action. Pressed parts must absorb impact stress and have enough strength off the press to survive handling. The topic of green strength and its determination was discussed in Chapter 15. Microscale laminations between poorly joined granules and large pores reduces the tensile strength. Compacts pressed at a higher pressure using well-plasticized granules and more binder are mechanically stronger, as shown in Fig. 22.19. A diametral tensile strength above 0.5 MPa is needed to obtain compacts with sharp edges. Strength values needed for forming and finishing were discussed in Chapter 21.

22.5 DIE WALL EFFECTS AND PRESSURE TRANSMISSION

A portion of the applied load is transferred to the die wall during compaction. The resisting force produced by die wall friction produces pressure gradients and density gradients in the compact. For uniaxial compaction (Fig. 22.20), the mean shearing stress at the wall $\bar{\tau}_w$ is related to the mean axial pressure P by the equation

$$\bar{\tau}_w = fK_{h/v}\bar{P} + A_w \tag{22.5}$$

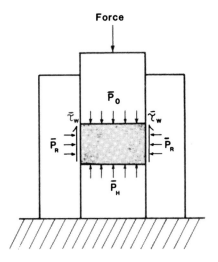

Fig. 22.20 Schematic diagram of mean stresses on a compact during unidirectional compaction.

where f is the wall friction, $K_{h/v}$ is the ratio of horizontal/vertical pressure (see Chapter 15), and A_w is the adhesion at the wall. The shear stress at the wall increases with pressing pressure, but is reduced by lubrication of the die (see Fig. 22.21). The parameter $K_{h/v}$ is lower when the yield strength of the granules is higher and when the pressing speed is higher. Lubricant added in excess of a fraction of a percent may not reduce f further but may increase $K_{h/v}$. A higher $\bar{\tau}_w$ reduces the pressure transmitted into the depth of the compact and the mean compact density.

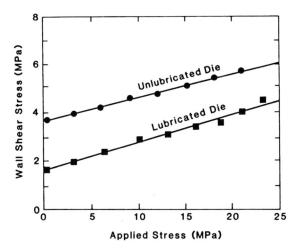

Fig. 22.21 Wall shear stress is higher for an unlubricated die and increases with applied pressure for an alumina compact containing a polyvinyl alcohol binder (see Fig. 22.20).

For one-directional pressing at an applied pressure P_a, the mean axial pressure transmitted P_H at a depth H in the compact is approximated by the equation

$$P_H/P_a = \exp -(fK_{h/v}A_{friction}/A_{pressing}) \qquad (22.6)$$

where $A_{friction}/A_{pressing}$ is the ratio of the friction area/pressing area of the compact. When pressing a cylindrical compact of diameter Dia and thickness H, the $A_{friction}/A_{pressing} = 4H/Dia$ and

$$\text{Single action } P_H/P_a = \exp -(4fK_{h/v}H/Dia) \qquad (22.7)$$

For double-action pressing, the $A_{friction}/A_{pressing}$ is one-half that for single-action pressing and the minimum pressure occurs at midheight. The corresponding equation for double-action pressing is

$$\text{Double action } P_{min}/P_a = \exp -(2fK_{h/v}H/Dia) \qquad (22.8)$$

The stress transmission in single action pressing for a granulated, micron-size alumina powder containing a well-plasticized polyvinyl alcohol binder is shown in Fig. 22.22 and is well described by Eq. 22.7. Near the end of Stage II of compaction, the parameter $K_{h/v}$ was determined to be approximately 0.4 for a compact of alumina powder containing a polyvinyl alcohol binder plasticized with adsorbed moisture. The low value indicates that plasticized binder adsorbed on particles had been extruded from contracts between particles. The

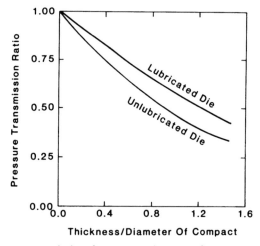

Fig. 22.22 Pressure transmission from top to bottom of compact is greater for compacts of smaller thickness/diameter and when die wall is lubricated. (Unidirectional pressing in polished steel die, alumina powder containing polyvinyl alcohol binder, stearic acid lubricant; note relative effect of lubricant is greater when compact has a greater thickness.)

friction factor f was determined to be approximately 0.30 for an unlubricated die and 0.24 for an lubricated die.*

Note in Fig. 22.22 for an unlubricated die when H/Dia of the part is 0.8, $P_{min}/P_a = P_H/P_a$ is only 0.55. Double-action pressing effectively halves the $A_{friction}/A_{pressing}$ and the H/Dia in Fig. 22.22, and P_{min}/P_a is about 0.75. Die wall lubrication is effective even when the $A_{friction}/A_{pressing}$ (H/Dia in Fig. 22.22) is not large. Granules may contain a lubricant added during granulation or are coated with a lubricant after granulation. Granule sliding will rub lubricant onto the die wall during pressing. Empirical results indicate that the admixed lubricant ($< 1\%$) should be added in proportion to the ratio of friction area/pressing area. Lubricants must be admixed homogeneously and used sparingly, as they can increase springback. During fast pressing, a dwell time in proportion to $A_{friction}/A_{pressing}$ may reduce pressure gradients.

Studies of the movement of bulk powder during pressing indicate that the relative axial displacement of particles is greater in the center than near the die wall, and the average movement decreases at a greater distance from the moving punch. Some radial translation is also observed, and this is a minimum at the center of the surface of the punch. The displacement vector is inclined relative to the pressing direction. Pressure profiles determined for the unidirectional pressing of bulky powders, using various microscale experimental techniques, indicate the general pressure gradients at the beginning and end of Stage II, as shown in Fig. 22.23. The diagonally directed pressure profile is sometimes called the Y-thrust. The maximum pressure occurs near the top corner of the compact and diminishes in depth toward the central axis. Near the end of Stage II, a region of lower pressure develops just below the center of the punch. Gradients in both the pressed density and the elastic springback are a concom-

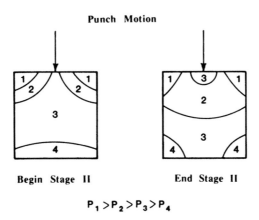

$$P_1 > P_2 > P_3 > P_4$$

Fig. 22.23 General pressure profiles at beginning and at end of Stage II for pressing only from top. Note low-pressure region near top surface below center of punch produced by friction. Mirror image of profile obtains when pressing from below.

*See reference 8.

itant result of these pressure gradients. Density gradients may lead to dimensional distortion due to shrinkage gradients when the compact is sintered.

22.6 CONTROL OF COMPACT DEFECTS

The compact must survive ejection and handling without failure and should be free of defects. The most common defects in dry-pressed compacts are laminations and cracks. Most laminations and cracks in pressed parts are caused by stresses produced by differential springback when the part is ejected. Differential springback within the compact or between the compact and the die occurs for the following reasons:

1. Pressure gradients within the compact produced by die wall friction.
2. Nonuniformities in elastic compression in the compact due to variable granules, nonuniform filling, or compressed air.
3. Frictional restraint at the die wall on ejection due to surface roughness and/or poor die wall lubrication.
4. The differential springback between the ejected portion and the unejected portion of the part constrained by the die.

In general, the tendency for laminations is decreased by lowering the pressing pressure to reduce the average springback, changing the composition of the additives to increase the compact strength and reduce the springback, lubricating the die to decrease pressure gradients, and using a die of sufficient stiffness with a smooth wall and an entry bevel. Steps to eliminate several commonly observed defects in pressed parts (see Fig. 22.24) are as follows:

1. *Laminations.* Laminations appear as periodic circumferential cracks on the frictional surface, and are oriented perpendicular to the pressing direction.

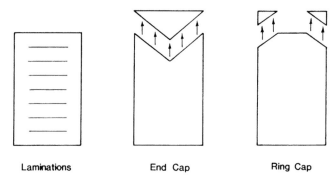

Laminations   End Cap   Ring Cap

Fig. 22.24 Common defects in pressed compacts. (Note angle of end cap is exaggerated.)

This defect is observed when die wall friction is high and slip-stick in nature, the springback of the part is high, and the compact strength is low. Much better die-wall lubrication, and reformulating the additives to plasticize the granules better, will often eliminate this defect.

2. *End Capping.* An end cap is a shallow wedge-shaped section which separates at an angle of $10°-20°$ from the end of the compact on ejection. This defect is observed when the springback is relatively high, the compact strength is low, and differential springback occurs within the part. Adhesion of the part to the punch surface may aggravate this defect. An end-capping index CI has been proposed to predict end capping,*

$$CI = P_w/S_t \tag{22.9}$$

where P_w is the extrapolated residual die wall pressure and S_t is the tensile strength of the compact. End capping was observed when $CI > 1$. The CI may be reduced by increasing the content of well-plasticized binder and using a stronger binder to increase S_t and altering parameters to reduce the springback. Adding more of a poorly plasticized binder will increase CI. In bodies containing anisometric particles, such as tile bodies, the springback is relatively high owing to particle bending during pressing. Control of the powder granules to obtain a sufficient pressed density at a moderate pressure may reduce the CI. Reducing the friction between the compact and the punch and the compact and the die wall reduces the differential springback. A high polish on the punch and die and a die tapered to partially accommodate the diametral springback may reduce the tendency for end capping. Adhesion of the compact to the punch surface may increase the end-capping tendency. A punch material which gives less adhesion or the elimination of an additive which causes adhesion may reduce end capping. Sometimes ejection of the compact with the punch in contact, creating a small compressive stress before the punch is separated will eliminate this defect. An effective lubricant must reduce die wall friction without increasing springback significantly.

3. *Ring Capping.* A ring cap is caused by relatively high differential springback at the corner of the compact. The escape of air on compression tends to draw granules into the gap between the die and punch. High springback near edges is commonly produced by the wedging of powder between the punch and the die wall. Tooling should be maintained to keep the gap between the punch and the die wall smaller than the granule size.

4. *Vertical Cracks in the Exterior Region.* These cracks are often caused by differential springback from compressed air and may be concentrated in the center of the compact. The tendency for this defect is high when compaction ratio and punch velocity are high, the thickness of the compact is relatively large, and the compact is of low gas permeability and of low strength. Changes in the powder preparation, such as granulation to increase the fill density and

*K. Sugimore, S. Mori, and Y. Kawashima, *Ad. Powder Technol.*, **1**(1), 25–37 (1990).

permeability and a longer pressing cycle, may eliminate this defect. Also, a greater fill height near the axis or other factor that increases the central spring-back can produce this defect.

5. *Shape Distortion on Sintering.* Nonuniform shrinkage on sintering may be caused by density gradients in the pressed part. Top and bottom punch displacements are usually controlled to produce the minimum in the transmitted stress at the centroid of the part. Compositions and pressing procedures which reduce die wall friction and improve stress transmission should reduce density gradients. A higher compaction rate may improve Stage II stress transmission in parts with an unfavorable H/D ratio.

6. *Extremely Large Pores in Compacts.* Large interstices may be a defect in advanced ceramics which are sintered without the presence of a glassy phase. This defect was described in Section 22.3. Steps to eliminate large pores are:

1. Reduce the maximum size of granules.
2. Use granulated feed that does not contain donut shapes.
3. Use granules of sufficient compressibility.

The maximum granule size is controlled by screening the granulated feed. Donut-shaped granules are eliminated by altering the additives in the formulation and altering spray drying, as was described in Chapter 20. Granules of sufficient compressibility are produced by formulating the slurry for spray drying to produce granules of moderately high density.

7. *Surface Defects.* The quality of the surface of the compact depends directly on the friction, the smoothness of the die and punches, the absence of adhesion of granules on the punch, the size and deformation of the granules, and the pressing pressure. Larger pores on the pressed surface are observed between relatively large granules which resist deformation and are poorly joined and within undeformed large donut-shaped granules (see Fig. 22.12)

22.7 ISOSTATIC COMPACTION

Products with one elongated dimension, a complex shape, or a large volume are not easily dry pressed and are produced using isostatic pressing, often referred to as isopressing. In the wet bag process, shown in Fig. 22.25, flexible molds are filled and sealed at a separate station. Molds may be filled while supported on a vibratory table to obtain a more uniform packing. Molds are often partially evacuated. The filled molds are then submerged in a liquid pressure chamber and pressed. After decompression, molds are removed, and the part is ejected. Dry bag isopressing (Fig. 22.26) imitates dry pressing except that pressure is applied radially by means of the pressurized liquid medium between a flexible mold and a rigid shell. Flexible tooling for isopressing is formed from synthetic rubber, polyurethane, or silicon rubber. Tool steel is used for end fixtures, mandrels, and perforated wet bag supports.

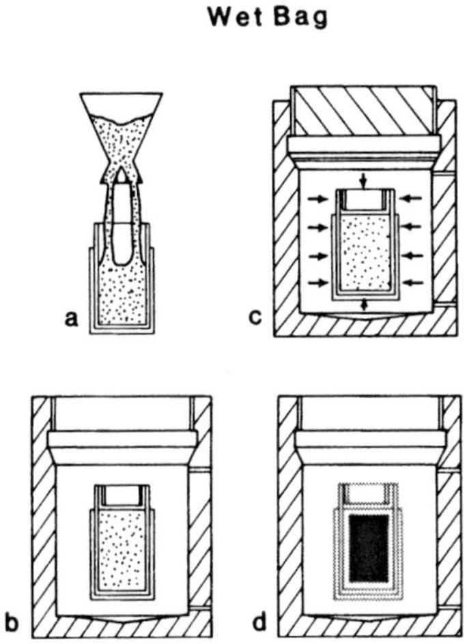

Wet Bag

Fig. 22.25 Stages in wet bag isopressing are (*a*) filling, (*b*) loading, (*c*) pressing, and (*d*) decompression prior to removing part. (Courtesy of Loomis Products Co., Levittown, PA.)

Wet bag isopressing is used for pressing complex shapes of various sizes and large refractories. Dry bag isopressing is commonly used for pressing small elongated parts such as tubes and spark plug insulators and small parts where a very high, uniform density is required such as grinding media. Operating pressures up to 200 MPa are common in isopressing, and wet bag presses operating to 500 MPa are used for pressing large blocks and billets. Pressing powders are very similar to those used for dry pressing except that granules are commonly plasticized to be more deformable. Shear stresses which deform granules are low in isopressing owing to the different stress path (see Chapter 15).

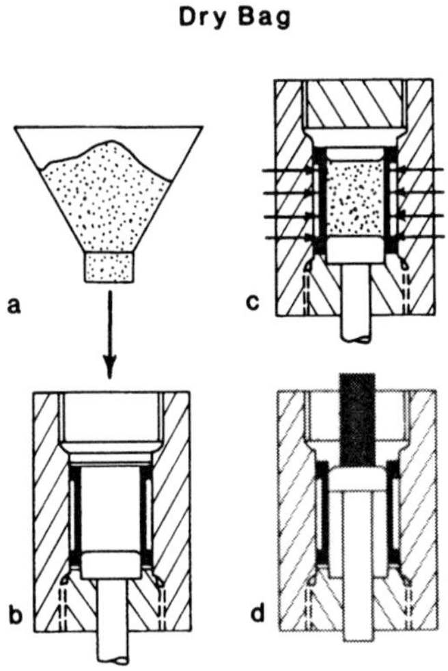

Dry Bag

Fig. 22.26 Steps in dry-bag isopressing are (*a* and *b*) loading of powder into die, (*c*) pressing in *x-y* plane, and (*d*) ejection of pressed part through top or bottom of die. (Courtesy of Loomis Products Co., Levittown, PA.)

Density gradients are minimized when isopressing thick parts and long parts, because the $A_{friction}/A_{pressing}$ is significantly reduced relative to that in dry pressing, and die wall friction is reduced. Sources of compact defects are similar to those described for dry pressing except that the nonrigidity and greater elastic expansion (springback) of the tooling must be considered. Slow decompression below 2 MPa is important to reduce differential springback within the part from air pressure and between the part and thick-walled tooling in contact with the part.

22.8 COMBINATION PRESSING

Large flatware items such as dinnerware having a thickness less than 0.5 cm and a diameter-to-thickness ratio ranging up to about 50 are pressed by a combination of dry-pressing and dry bag isopressing. The feed material for large flatware products is characteristically coarser (Fig. 22.4). In combination pressing, a relatively rigid steel or polyurethane plastic die component applies pressure to the simpler surface, and a flexible diaphragm pressurized by an internal liquid applies pressure to the surface with more complex relief (Fig. 22.27). Feed material on the bottom die set is contoured using a rotating tool prior to pressing. Filling, contouring, pressing, and extraction occur simultaneously on a multistation, rotating table. Production rates may range up to several hundred pieces per hour. Vertical fill and a horizontal pressing mode has also been used to increase the production rate. Advantages over plastic forming include a greater dimensional precision, the elimination of drying and the drying shrinkage in the piece, and energy savings in spray drying relative to conventional drying.

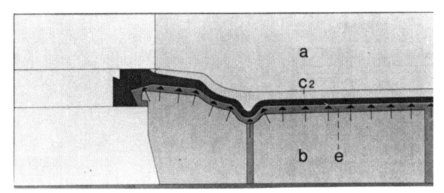

Fig. 22.27 In combination pressing of a flatware piece, mechanical pressure is produced by (*a*) hard punch surface and a complex lower punch consisting of (*b*) hard material and (*e*) pressurized hydraulic liquid pressing up against a flexible surface membrane in contact with (C$_2$) the flatware piece. (Courtesy of Dorst Maschinen and Anlagenbau, Kochel A. See, Germany.)

22.9 ROLL PRESSING

Granulated powders may be compacted continuously between hard rolls into a thick sheet or segmented tablets such as substrates and chip carriers. Compaction pressures are high relative to dry pressing, because the area of powder contacted by the rolls is very small. Production rates range up to several centimeters per second. Roll pressing is used to produce substrates up to 30 cm in width with a uniform thickness in the range of 0.5–1.5 mm and for the high-productivity forming of simple shapes in very large production lots.

SUMMARY

Dry pressing is a very important fabrication process used to produce products that are of relatively high density and are dimensionally precise. Pressed products vary widely in size, shape, and composition. Isostatic pressing is used to press relatively large products, products that must have the maximum pressed density, and products having a complex shape. Substrates and briquetted material are produced using roll compaction.

Properly formulated and well-controlled feed material is a key to successful pressing operations. Feed material for pressing is commonly a granulated powder or, for coarse-grained products, a semicohesive mass. The system of processing additives and characteristics of the granules must be carefully controlled to produce pressing granules with the necessary flow and compaction properties. Pressing should eliminate large pores and produce a product having a uniform density and adequate strength to survive ejection and handling. Die wall friction causes a gradient in pressure and a density gradient in the part. The density gradient is reduced by admixing a lubricant in the pressing powder and by using double-action pressing or isostatic pressing. Elastic energy stored in the pressed part produces a volume expansion called springback when the pressed part is ejected. Springback must be controlled to prevent laminations and cracks in pressed products.

SUGGESTED READING

1. B. D. Mosser, J. S. Reed, and J. R. Varner, "Strength and Weibull Modulus of Sintered Compacts of Spray Dried Granules," *Am. Ceram. Soc.*, **71**(1), 105–109 (1992).

2. P. Balakrishna, K. P. Chakraborthy, and P. S. A. Narayanan, "Cracking, End-Capping, and Other Defects in Pressed Ceramic Compacts," *Interceram.*, **41**(3), 152–156 (1992).

3. K. Uematsu, M. Miyashita, J. Y. Kim, Z. Kato, and N. Uchida, "Effect of Forming Pressure on the Internal Structure of Alumina Green Bodies Examined with Immersion Liquid Technique," *J. Am. Ceram. Soc.*, **74**(9), 2170–2174 (1991).

4. M. Takahashi and S. Suzuki, "Compaction Behavior and Mechanical Characteristics of Ceramic Powders," in *Ceramic Transactions*, Vol. 12, G. L. Messing, S. Hirano, and H. Hausner, Eds., American Ceramic Society, Westerville, OH, 1990, pp. 473–484.

5. R. L. K. Matsumoto, "Analysis of Powder Compaction Using a Compaction Rate Diagram," *J. Am. Ceram. Soc.*, **73**(2), 465–468 (1990).

6. R. G. Frey and J. W. Halloran, "Compaction Behavior of Spray Dried Alumina," *J. Am. Ceram. Soc.*, **67**(3), 199–203 (1984).

7. C. W. Nies and G. L. Messing, "The Role of Binder Glass Transition Temperature During Powder Compaction," *J. Am. Ceram. Soc.*, **67**(4), 301–304 (1984).

8. R. A. DiMilia and J. S. Reed, "Stress Transmission During the Compaction of a Spray-Dried Powder in a Steel Die," *J. Am. Ceram. Soc.*, **66**(9), 667–672 (1983).

9. R. A. DiMilia and J. S. Reed, "Dependence of Compaction on the Glass Transition Temperature of the Binder Phase," *Am. Ceram. Soc. Bull.*, **62**(4), 484–488 (1983).

10. H. Niffka, "Isostatic Dry Pressing of Flatware," *Am. Ceram. Soc. Bull.*, **59**(12), 1220 (1980).

11. P. Popper, K. Bloor, R. D. Brett, and D. E. Lloyd, "Isostatic Pressing with Particular Reference to Tooling," in *Proceedings of the 3rd CIMTEC International Meeting on Modern Ceramic Technologies*, Rimini, Italy, 1976, pp. 101–107.

12. E. L. J. Papen, H. D. B. Raes, and G. Deplace, "Development and Application of the Isostatic Process in the Refractories, Ceramic, and Electrical Porcelain Industries," *Interceram*, **3**, 204–211 (1974).

13. E. J. Motyl, "Spray Drying, Pressing Lubricants Upgrade Ferrite Production," *Ceram. Age*, **2**, 45–48 (1964).

PROBLEMS

22.1   What are the granule characteristics which produce a high fill density?

22.2   Why must both the maximum and minimum granule sizes be controlled? Give at least three reasons for each.

22.3   The fill density is 32% when the granule density $D_G = 55\%$. Estimate the PF of the granules in the fill.

22.4   Calculate the compaction ratio for the granulated tile composition in Fig. 22.6 and compare it to the guideline.

22.5   If the two-level part in Fig. 22.1 has heights of 0.5 and 1 cm, widths of 1.5 and 2.5 cm, and a hole diameter of 0.5 cm, sketch the dimensions of the fill cavity if CR = 2.

22.6   Estimate the fill density for alumina granules of uniform size. Assume that the PF within the granules is 0.52 and the PF of the granules is 0.58. Calculate the number of granules in 1 cm^3 of fill.

22.7   Estimate the PF of the alumina granules in Fig. 22.6. Assume that the granules pack with PF = 0.50 and the particle density is 3.98 Mg/m^3.

22.8   Estimate the compaction constant m for the data shown in Fig. 22.14.

22.9   Calculate the intergranular and intragranular porosity in the fill and after pressing from the curves in Fig. 22.8. What is PF of granules?

22.10 Design the taper of a die for a compact 3 cm in diameter if the spring-back is 0.7% and the length of the tapered section is 1.5 cm. Allow for 0.75% of the springback in your design.

22.11 Estimate the effect of lubrication on the ejection force when the ejection rate is 60 cm/min in Fig. 22.14.

22.12 Calculate $A_{friction}/A_{pressing}$ when pressing a rectangular compact $3 \times 2 \times 1$ cm for single-action, double-action, and dry bag isostatic pressing. When dry pressing, the punch presses on the 2×3 cm face.

22.13 For confined compression, $K_{h/v} = \nu/(1 - \nu)$. Estimate $K_{h/v}$ for a liquid ($\nu = 0.5$), an organic polymer ($\nu = 0.4$), and porous alumina ($\nu = 0.23$).

22.14 Calculate the minimum pressure for double-action pressing when pressing a compact with $H/Dia = 1.0$ and $P_a = 100$ MPa. Assume $fK_{h/v} = 0.15$.

22.15 Estimate P_{min} for the pressing situations in Problem 22.12 when $P_a = 84$ MPa and $fK_{h/v} = 0.12$.

22.16 For the results in Problem 22.12, estimate the range of density using the compaction diagram in Fig. 22.7.

22.17 Use the compaction results in Fig. 22.7 to explain why compacts containing granules of different deformation resistance will not be of uniform pore size.

22.18 Sketch the pressure profiles for double-action pressing using the profiles for single-action pressing in Fig. 22.23 as a reference.

22.19 Calculate the end-capping index for the unlubricated die in Fig. 22.21. Assume that $A_w = 0$, $fK_{h/v} = 0.15$, and tensile strength is 2.5 MPa.

22.20 A company is experiencing laminations and end capping in parts pressed from a granulated alumina powder containing 3 wt% polyvinyl alcohol binder. The parts pressed at 90 MPa are of variable strength and weaker compacts are of lower density. What is your analysis and recommended solution?

EXAMPLES

Example 22.1   The D_{Fill} of a spray-dried powder is 34% and the mean density of the granules D_G is determined to be 58%. In the bulk fill in the die, what is the distribution of the porosity within granules ϕ_w and in interstices ϕ_I?

Solution. In the granules, the volume fraction solids $f_s^v = 0.58$ and the pore fraction $\phi_G = 0.42$. The total pore fraction $\phi_T = \phi_I + \phi_W = \phi_I + PF_G(\phi_G)$. And in the fill, $PF_G = 1 - \phi_I$.

$$\phi_T = 1 - D_{Fill}/100 = 0.66 = \phi_I + PF_G(\phi_G)$$

$$0.66 = \phi_I + (1 - \phi_I)(0.42)$$

and

$$\phi_I = 0.41 \qquad \phi_W = \phi_T - \phi_I = 0.66 - 0.41 = 0.25$$

Of the 66% total porosity in the fill, 41% exists as interstices and 25% as porosity within the granules.

Example 22.2   A large CR indicates a relatively long punch stroke and relatively more air compressed during granule deformation. Is the CR of the spray-dried technical alumina granules in Fig. 20.6 within the guidelines for a good pressing powder?

Solution. CR = fill volume/pressed volume = pressed density/fill density. CR = 53%/27% = 1.96, which is within the guideline of CR < 2.

Example 22.3   Contrast the ratio of friction area A_f to pressing area A_p for the one-directional pressing of an electronic substrate $30 \times 30 \times 2$ mm and the same substrate with 100 holes, 0.5 mm in diameter.

Solution. For one-directional pressing of the rectangular substrate,

$$A_f/A_p = 4(30 \text{ mm} \times 2 \text{ mm})/(30 \text{ mm} \times 30 \text{ mm}) = 240/900 = 0.27$$

For the substrate with 100 holes,

$$A_f/A_p = [240 + 100(\pi \text{ Dia } H)]/[900 - 100(\pi (\text{Dia})^2/4]$$

$$A_f = \{240 \text{ mm}^2 + 100[\pi(0.5 \text{ mm})^2 \text{ mm}]\} = 240 + 314 = 554 \text{ mm}^2$$

$$A_p = \{900 \text{ mm}^2 - 100[\pi(0.5 \text{ mm})^2/4\} = 900 - 20 = 880 \text{ mm}^2$$

$$A_f/A_p = 554/880 = 0.63$$

The increase with the holes is $(0.63 - 0.27)\,100/0.27 = 133\%$.

Example 22.4   For the double-action pressing of a compact with $H/\text{Dia} = 0.8$, it is necessary to produce a pressure exceeding 40 MPa for good compaction of the granules. Assuming $K_{h/v} = 0.39$ for the granules and $f = 0.26$ for the die wall friction, what punch pressure P_a should be used?

Solution. The punch pressure P_a can be estimated from the equation

$$P_{min}/P_a = \exp - [fK_{h/v}(A_f/A_p)]$$

For two-directional pressing,

$$P_{min} = P_{middle} \quad \text{and} \quad A_f/A_p = (\pi \text{ Dia } H)/2[\pi(\text{Dia})^2/4] = 2 \text{ } H/\text{Dia}$$

Accordingly,

$$P_{middle}/P_a = \exp - [2fK_{h/v}(H/\text{Dia})]$$
$$P_{middle}/P_a = \exp -2(0.26)(0.39)(0.8) = 0.85$$
$$P_a = P_{middle}/0.85 = 40 \text{ MPa}/0.85 = 47 \text{ MPa required}$$

Example 22.5 A part with a triangular cross section $5 \times 5 \times 8$ cm and 8 cm height is to be pressed on end in a metal die (double action) or by isostatic pressing. Can a difference in the pressed density be expected when using the two methods?

Solution. For dry pressing on end,

$$A_f/A_p = \{[2(8 \times 5 \text{ cm}^2) + 8 \times 8 \text{ cm}^2]/2\}/[2(0.5)(8 \times 3) \text{ cm}^2]$$
$$= 144/24 = 6.0/1$$

But for isostatic pressing, $A_f/A_p < 1/6 = 0.17$. Because of the more favorable A_f/A_p and higher applied pressure in isostatic pressing, a higher pressed density and a significant improvement in pressure transmission and reduced density gradients can be anticipated when isostatic pressing.

Example 22.6 Powder granules for pressing have a mean size of 70 μm and their packing factor is $PF = 0.56$. Estimate the number of granules in a filled die.

Solution. The number of granules per 1 cm³ of bulk volume can be calculated using Equation 13.1, ie.

$$N_p = 6(PF)/\pi a^3 = 6(0.56)/\pi(70 \times 10^{-4} \text{ cm})^3 = 3 \text{ million/cm}^3$$

CHAPTER 23

EXTRUSION AND PLASTIC DEFORMATION FORMING

Extrusion is shaping by forcing a cohesive plastic material through the orifice of a rigid die. A lineated extrudate with a controlled cross section is formed which is then cut to length to form the product. The plastic consistency is produced using clay binder, a polymeric organic binder, or a mixture of the two types. Extrusion is a very productive forming technique that is used for the mass production of both large products ranging up to more than 1 ton and small products weighing only a few grams. Extruded compositions include both oxide and nonoxide ceramics such as carbides and nitrides. Traditional construction materials such as brick and tile, refractories such as thermocouple protection tubes, furnace tubes, silicon carbide heat exchanger tubes, and kiln furniture, porcelain electrical insulators, magnets and electronic substrates, "honeycomb" cellular catalyst supports, and transparent ceramic tubes for efficient lamps are produced by extrusion. Substrates may be extruded at a thickness < 1 mm. Warm extrusion is used when forming graphite electrodes, and extrusion is used for forming composites.

Extrudate may be reshaped in a second extrusion operation and by plastic pressing and molding. In jiggering, a section of deaired extrudate is compressed and sheared between the surfaces of a lubricated template; in roller tool forming it is shaped between a roller tool and the surface of a rotating mold. Jiggering and roller tool forming are now highly mechanized and are used for forming thin-walled porcelain items such as dinnerware and hotel china. Thicker-walled clay-plasticized products are produced by pressing a section of the extrudate between porous dies or between a rotating metal plunger and a porous mold. Shrinkage on drying separates the product from the mold, but pneumatic separation is also used. In this chapter we discuss the composition of plastic bodies, extrusion mechanics and technology, plastic transfer pressing technology, and causes of defects in plastically formed products.

23.1   EQUIPMENT AND MATERIAL VARIABLES IN EXTRUSION

Plastic feed material for extrusion is commonly prepared by directly batching and mixing the raw materials and additives in a high-shear mixer. A binder in powder form is commonly mixed with the ceramic materials and then the liquid is added, rather than admixing a very viscous binder solution. This procedure improves its macroscale mixedness and the dispersion and reduces the liquid diffusion required for wetting the binder. Measurement of the torque of mixing with mixing time may be used to determine the effects of procedures and additives and for control purposes, as shown in Fig. 23.1. To obtain a more homogeneous material, the batch may be mixed first as a slurry. Feed material for porcelain extrusion is often produced by filter pressing the slurry to form a plastic filter cake. When using an organic binder to plasticize the body, filter pressing is not feasible because of the low liquid permeability of the filter cake, but plastic material may be prepared by mixing slurry and spray-dried slurry in a high-shear mixer. Further dispersion and mixing may occur in a pug mill in which an auger with broken flights mixes the feed material and forces it through a shredder into a vacuum chamber (Fig. 23.2). Shredded material that is small in cross section is more uniformly deaired without surface drying. Material is also mixed more completely while flowing through the feeding section of a twin-screw extrusion device (Table 23.1).

A piston (ram), single auger, or twin-screw auger is then used to consolidate the material and extrude it through the die. Sectioned extrudate is used as the as-formed product and may be used as feed material in a second shaping

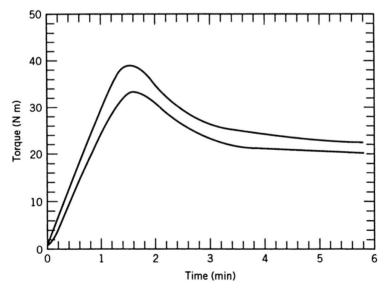

Fig. 23.1   Direct mixing of a prewetted extrusion batch containing a methylcellulose binder produces a torque which varies with mixing time as indicated. (Courtesy of Greg Dillon, Alfred University.)

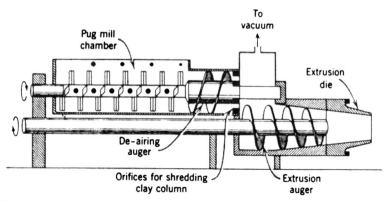

Fig. 23.2 Industrial pug mill with deairing chamber and extrusion auger. (From W. D. Kingery, *Introduction to Ceramics*, 1st ed. Wiley, New York, 1960.)

operation. Extruded billets or rods may be plastically deformed to shape on a lathe or dried and dry machined to produce the final shape (Fig. 23.3).

The stages in extrusion are (1) material feeding, (2) consolidation and flow of the feed material in the barrel, (3) flow through a tapered die or orifice, (4) flow through the die-land of constant or nearly constant cross section, and (5) ejection. The cross sections of piston and auger extruders are shown in Fig. 23.4. In piston extrusion, feed material is commonly a deaired billet from the pug mill, which is sometimes enclosed in a sleeve, but may be segmented material. After inserting the piston and deairing, the feed material is compressed and forced to flow down the barrel by the moving piston. An advantage of piston extrusion is that very high pressures can be generated; however, it is an intermittent, low-capacity batch process requiring incremental loading.

Pressure zones in feeding extrudate using an augur are shown in Fig. 23.5. Segmented feed is conveyed and densified when compacted. The material becomes continuous in the metering zone. Pressure and velocity gradients are redistributed on leaving the auger and entering the barrel in Zone 4.

For feeding using a single auger, the material must not slip on the wall of

TABLE 23.1   Comparison of Single and Twin Screw Extruders

Factor	Single Screw	Twin Screw
Energy loss	Viscous flow	Heat transfer
Transport	Wall friction dependent	Positive displacement
Throughput	Pressure dependent	Independent
Shear forces	Large	Small
Power/kg of product	Higher	Lower
Temperature gradients	Higher	Lower
Air evacuation	Simple	Difficult
Capital cost	Moderate	High

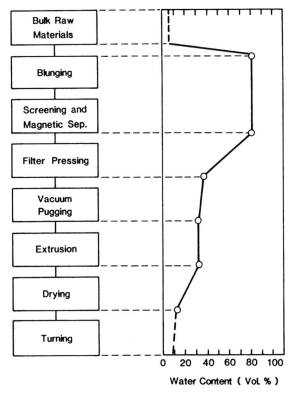

Fig. 23.3 Processing steps and change of liquid content of body during the processing of electrical porcelain.

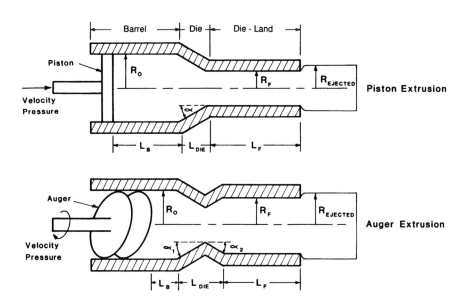

Fig. 23.4 Schematic diagram of (top) piston extruder and (bottom) auger extruder indicating design parameters.

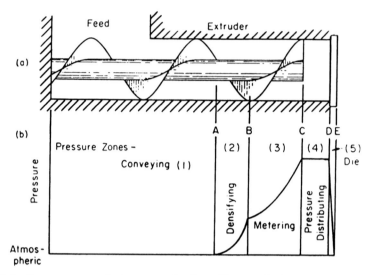

Fig. 23.5   Pressure zones during auger feeding. [Source: J. R. Parks and M. J. Hill, *J. Am. Ceram. Soc.*, **42**(1), 2 (1959).]

the barrel. Accordingly, the adhesion of the body on the wall and the area wall/area auger must be sufficiently high. Augers should be highly polished to facilitate slippage. Larger augers or tapered augers may be used to produce a higher extrusion pressure. The number of flights on the auger controls the number of feed columns displaced. A larger helix angle increases the potential delivery rate but reduces the compressive thrust on the material; a helix angle of approximately 20–25° is commonly used. The required ratio of the auger diameter to the product diameter increases as the yield strength of the material increases and when the friction area of the die land increases. Heating or cooling the extrudate may change its adhesion on the metal surface and its flow resistance. Higher pressure may be developed using a twin screw extruder, which provides a positive material displacement (Fig. 23.6).

In single auger extrusion, material is thrust forward but also continues to twist for some distance downstream from the auger; near-center material moves inward to fill the cavity in front of the hub of the auger. The axial velocity of material discharged by the auger is a minimum at the hub and at the wall of the barrel, and is a relative maximum near the surface of the flight of the auger (see Fig. 23.7). The differential axial velocity is changed as material flows down the barrel. Material nearer the center axis accelerates and becomes of highest velocity as the extrudate enters the die. The die alters the differential flow, reduces the cross section, and produces a particular cross-sectional geometry. Geometrical parameters of the complete die are the entrance angle α, the ratio of the area of the barrel/area of the die orifice called the reduction ratio, and the wall friction area/cross-sectional area of the die land.

Complex dies may contain small channels for injecting a die-wall lubricant.

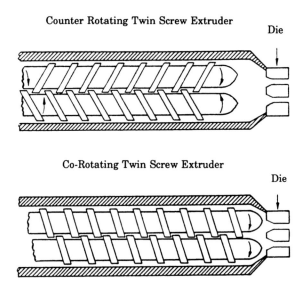

For the conventional extrusion of hollow items, a system of arms, called a spider, is attached to the die and supports the central core rod; flow is diverted around these arms and the flow streams must join downstream from the spider (Fig. 23.8). Flow through the die land improves the knitting of material separated by the spider. The complete extrusion die must generate an internal pressure and flow pattern that minimizes defects. Measurable elastic springback occurs on ejection of the extrudate.

Ceramic substrates with a "honeycomb" cross section with up to 200 channels per square centimeter and a material thickness ranging down to 0.1 mm are now extruded using a special patented* die. The die consists of a solid plate with entrance holes intersecting the center of perpendicular grooves which form the honeycomb shape. The body is of a relatively fine particle size, and the processing additives and rheology must be very precisely controlled.

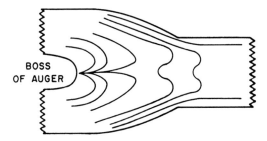

Fig. 23.7 Flow velocity profile downstream from auger in barrel. (From W. O. Williamson, *Ceram. Age* (3)29 (1966)).

*Corning Inc, Corning, NY.

Fig. 23.8 Spider in throat of die supports core rod needed to form hollow tube.

Industrial extrusion pressures range up to about 4 MPa for porcelain bodies and up to about 15 MPa for some organically plasticized materials. Capacities range widely, depending on the product size, but may approach 100 T/h for large products. Industrial extrusion rates in terms of velocity of ejected material also vary widely and are controlled in part by the rates of cutting and conveying the extruded material. An extrudate velocity of about 1 m/min is common for the extrusion of many large products.

Compositions of plastic bodies are listed in Table 23.2. Clay-based porcelain bodies commonly have a close-packing particle size distribution with 25–40 vol% finer than 1 μm. The particle size distribution of nonclay bodies may vary widely owing to constraints imposed by densification and grain sizes desired after sintering. However, a minimum amount of colloidal particles and binder material of about 20 vol% appears to be requisite for plastic forming.

Organic binders used as a body plasticizer are of medium to high viscosity grade. Binders and other additives used in aqueous extrusion are listed in Table 23.3. The liquid phase is commonly water, but liquid wax, petroleum oils,

TABLE 23.2 Compositions of Extrusion Bodies (vol%)

Silicon Carbide		High Alumina		Electrical Porcelain	
Silicon carbide	50	Alumina	46	Quartz	
Hydroxyethyl		Ball clay	4	powder	16
cellulose[a]	6	Methyl		Feldspar	16
Water	42	cellulose[a]	2	Kaolin	16
Polyethylene glycol[b]	2	Water	48	Ball clay	16
				Water	36
				CaCO$_3$[c]	<1

[a]Viscosity grade 4400.
[b]20,000 MW.
[c]Coagulant.

TABLE 23.3   Additives Used in Nonclay Extrusion Bodies

Binder	Coagulant	Lubricant
Methyl cellulose	$CaCO_3$	Sodium stearate
Hydroxyethyl cellulose	$MgSO_4$	Polyethylene glycol
Polyvinyl alcohol	Acid/base	Silicones, oleic acid
Polyethylene glycol		Colloidal talc/graphite
Polyethylene[a]		Mineral oil
Polypropylene[e]		Liquid wax

[a]Warm extrusion, nonaqueous.

ethylene glycol, alcohol, and so on are used in nonaqueous extrusion systems. Extrudate feed is typically coagulated using <1% of a suitable additive. A lubricant is used to reduce the extrusion pressure when the friction area/product cross section is high and also to improve surface quality. The concentration of processing additives is somewhat dependent on the particle packing and the proportion of colloidal particles in the body.

Extrusion bodies are elastic-plastic in nature, as was discussed in Chapter 15. Less flexible cellulosic binders may increase the elasticity of the extrudate and provide a relatively high wet strength which aids shape retention in handling. High-molecular-weight binders and flocculation increase the elastic compressibility and reduce the liquid migration in the extrudate. An adsorbed gel film of colloidal clay or binder molecules on the hard particles reduces the friction between particles, and the colloid–liquid matrix provides plastic flow. This distribution of binder and liquid reduces the effective stress between particles and the shear stress for flow, as discussed in Chapter 15. The adsorbed binder and coagulated structure provides cohesion and strength after the product is ejected.

23.2   EXTRUSION MECHANICS

During extrusion, the driving force produced by the auger or piston must exceed the resistive force of the plastic material and the die-wall friction. The pressure motivating flow is highest in the barrel and decreases along the axis of the extruder, as shown in Fig. 23.9. Flow during extrusion may occur by one of several mechanisms:

1. Laminar flow within the extrudate and slippage at the wall. This type of flow commonly occurs in the barrel and in the die-entry regions.
2. Plug flow of the extrudate and slippage at the wall. This type of flow commonly occurs in the die land.
3. Plug flow near the center and laminar flow near the wall. This type of flow may occur when the yield strength of the extrudate is very low.

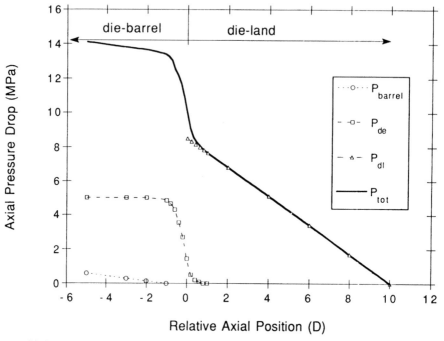

Fig. 23.9 Calculated mean axial pressure profile during extrusion and pressure drop in barrel, die entry, and die land. (Courtesy J. Zheng, Alfred University.)

These types of flow are illustrated in Fig. 23.10. On flowing into the die, material is accelerated and translates toward the central axis. The particular flow behavior depends on the geometry of the die and the flow properties of the material. Internal defects, surface smoothness, and particle orientation are dependant on the flow behavior during extrusion. The permeability constant of an extrusion body must be very low, $K_p < 10^{-18}$ m^2, to prevent the migration of the pressured liquid through the material during its residence in the extruder.

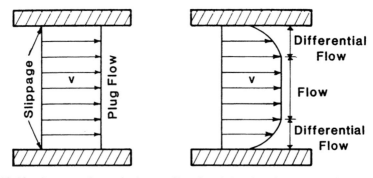

Fig. 23.10 Cross-section velocity profiles for (left) plug flow with slippage at wall and (right) differential laminar flow near wall.

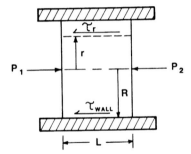

Fig. 23.11 Axial pressures and shear stress during steady-state extrusion in a rigid die land.

In the barrel of the extruder, the summation of forces for the cylindrical element undergoing steady-state laminar flow (Fig. 23.11) provides a relation between the shear stress τ and the radial position r:

$$\tau = r(P_1 - P_2)/2 L = r \, \Delta P/2L \qquad (23.1)$$

where $\Delta P/L$ is the differential axial pressure for the segment of length L. The maximum shear stress occurs at the wall when $r = R$,

$$\tau_{\text{wall}} = R \, \Delta P/2L \qquad (23.2)$$

The flow of colored slurries and pastes in capillary tubes was studied by Bingham* and later by Scott Blair and Crowthers.† They observed that the flow occurred first by slippage "en bloc" with a linear dependence of apparent velocity on pressure followed by a transition to laminar flow at a higher velocity. The tendency for plug flow was observed to increase as the content of solids increased. More recent studies of the flow of procelain‡ and organically plasticized extrudates have shown clearly that laminar flow within the extrudate occurs in the die-entry region and plug flow with slippage at the wall commonly occurs in the die land. Slippage occurs when the shear strength of the body exceeds the shear stress for slippage at the wall. Slippage occurs at the wall of a tapered die when the entrance angle is less than the critical angle.

A plastic extrusion body may be considered to be a compressible, cohesive solid in which the shear strength initially increases but then becomes constant when the body is compressed (Mohr–Coulomb behavior). In regions where DPS $= 1$, the properties are constant with position and time, and the flow behavior may be approximated by the Herschel–Bulkley equation (Eq. 16.7) for a shear thinning material having a yield stress τ_Y, that is,

$$\tau = \tau_Y + K\dot{\gamma}^n \qquad (23.4)$$

*E. C. Bingham, *Bull. Bur. Stand.*, **13**, 309–353 (1916–1917).
†G. W. Scott Blair and E. M. Crowthers, *J. Phys. Chem.*, **33**(3), 321–330 (1929).
‡D. Price and J. S. Reed, *Am. Ceram. Soc. Bull.*, **62**(12), 1346–1350 (1983).

For an extrudate velocity v, the pressure required (ΔP) for steady-state laminar shear flow of a Herschel–Bulkley material through a square entry die and plug flow with slippage at the wall in the die land is described by the modified Benbow equation;* $\Delta P = \Delta P_{\text{die-entry}} + \Delta P_{\text{die-land}}$, and

$$\Delta P = \ln (A_{\text{barrel}}/A_{\text{land}})[\tau_b + k_b v^n] + (A_{\text{friction}}/A_{\text{land}})[\tau_f + k_f v^m] \quad (23.5)$$

where τ_b is the internal yield strength of the plastic body; k_b and n are the velocity factor and shear thinning index, respectively, for the body; $A_{\text{friction}}/A_{\text{land}}$ is the die land surface/cross-section area; τ_f is yield strength of the slippage film on the wall of the die land; and k_f and m are the velocity factor and shear thinning exponent, respectively, for the slippage film. For extrusion of a circular cross section, the equation takes the form†

$$\Delta P = 2 \ln (D_0/D)[\tau_b + k_b'(v/D)^n] + (4L/D)[\tau_f + k_f v^m] \quad (23.6)$$

where D_0 is the barrel diameter, D is the die-land diameter, and k_b' is a thickness independent velocity factor for the body.

Zheng et al.† have shown that the Herschel–Bulkley parameters may be calculated from the flow parameters in Eq. 23.5. Extrusion through the square entry die and die land serves as an extrusion rheometer for the experimental determination of the flow properties during actual extrusion. Flow properties determined for a porcelain body and for a cordierite body without and with an added lubricant are listed in Table 23.4. The value $n = 0.3$ indicates that these industrial bodies are very shear thinning. The higher value of τ_b for the cordierite body occurs because of the bonding provided by the organic binder and the lower liquid content. The yield strength of the die land slippage film is smaller than the yield strength of the body by more than one order of magnitude

*J. J. Benbow, E. W. Oxley, and J. Bridgewater, *Chem. Eng. Sci.* **42**(9), 2151–2162 (1987).
†J. Zheng, W. B. Carlson, and J. S. Reed, *J. Am. Ceram. Soc.*, **74**(11), 3011–3016 (1992).

TABLE 23.4 Flow Properties Determined Using Extrusion Rheometer

Property	Electrical Porcelain[a]	Cordierite	Cordierite + Lube[a]
Die-entry			
τ_b (kPa)	160	850	520
k_b kPa/(mm/min)^n	20	30	13
Shear thin n	0.3	0.3	0.4
Die-land			
τ_f (kPa)	6	57	6
k_f kPa/(mm/min)^m	0.7	1.4	0.1
Shear thin m	0.5	0.5	0.8

[a]Porcelain body contains 50% Clay; cordierite body contains 3 wt% methylcellulose binder and 1 wt% sodium stearate lubricant.

and the shear thinning index is higher for the slippage film on the die land. These differences indicate that the film has a higher liquid content and a different composition. Changes in flow properties on adding sodium stearate into the cordierite body indicate that it acts both as an internal lubricant in the body and as an external lubricant on the wall of the die land.

The modified Benbow equation and measured extrudate properties provide important insights into extrusion. The pressure for flow in the die-entry region is larger when the die orifice is small in cross section and when the extrudate has a greater yield strength τ_b and is weakly shear thinning (n is large). The dependence on the reduction ratio varies only as a log function. The pressure drop in the die land is relatively larger when the friction area/cross-section area, the yield strength of the slippage film, and the extrudate velocity are high. An effective lubricant can reduce the yield strength and velocity sensitivity factor of the slippage film. These effects are illustrated in Fig. 23.12.

Laminar flow may occur in the die land when the shear resistance of the body $\tau_b < \tau_{wall}$. This occurs when the yield strength of the body is low. As is shown in Fig. 23.13, the shear stress developed in the die land is zero at the center axis and increases radially. At a position r_c, the shear stress may become equal to the yield stress. When the shear stress exceeds the yield stress, laminar shear will occur in the region $r_c < r < R$. Flow without shear occurs in the

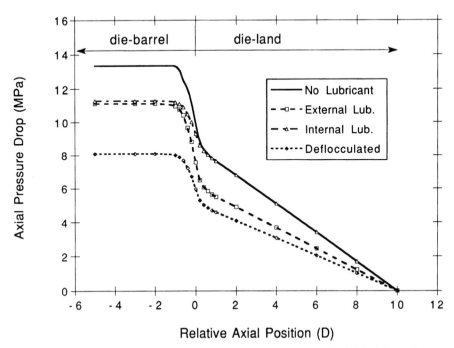

Fig. 23.12 Effect of internal lubricant, external lubricant, and added deflocculant on axial pressure profile.

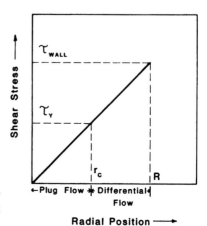

Fig. 23.13 Variation of shear stress with radial position in die land. Note complete plug flow occurs when $\tau_Y > \tau_{wall}$.

center region of radius $r \leqslant r_c$. At higher velocities, the extrusion pressure increases nearly linearly with flow rate. Complete plug flow ($r_c = R$) with a slippage film occurs in the die land when $\tau_{wall} < \tau_b$.

Material flowing through a tapered die is subjected to compressive and shearing stresses at the wall, as shown in Fig. 23.14. The shear stress/compressive stress increases as the entrance angle α of the die decreases. When the die angle is small, slippage occurs on the wall of the tapered die. On increasing the die angle, the shear stress at the wall decreases, and at some critical angle α_c, shear occurs in the material. Above the critical angle, a region of static material (dead zone) is present in the periphery of the die. The flow of a plastic material in a conical die involves both internal shear and slippage at the die wall when $\alpha < \alpha_c$. Static material may be a source of defects. For industrial extrusion the die entrance angle must be less than the critical angle. For an angle smaller than the critical angle, modeling studies indicate that the pressure P for extrusion in the conical die varies as $\cot \alpha / D^n$. This indicates that the extrusion pressure increases for a decreasing die angle and increasing reduction;

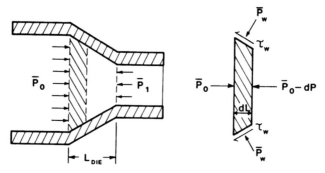

Fig. 23.14 Mean pressures and stresses for flow into conical die-entry region.

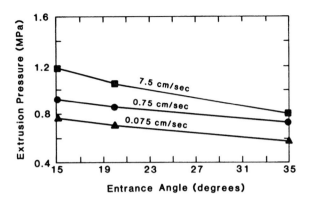

Fig. 23.15 Dependence of extrusion pressure on entrance angle for an electrical porcelain body.

this dependence is observed (Fig. 23.15) and is expected because of the increasing conical die surface.

The flow velocity in the tapered die is a function of both the radial position and axial location, and material nearer the central axis experiences more acceleration. Experiments with striated feed of different color indicate that the differential flow velocity between the center and the wall increases as the entrance angle increases, as indicated in Fig. 23.16.

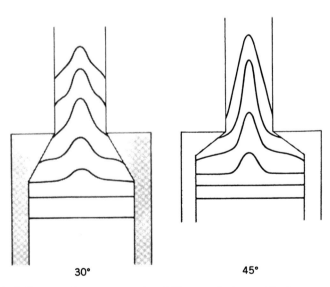

Fig. 23.16 Influence of entrance angle of die on velocity profile in extrusion. (From G. C. Robinson, in *Ceramic Processing Before Firing*, by L. L. Hench and G. Y. Onoda Jr., Eds., Wiley-Interscience, New York, 1978, p. 392.)

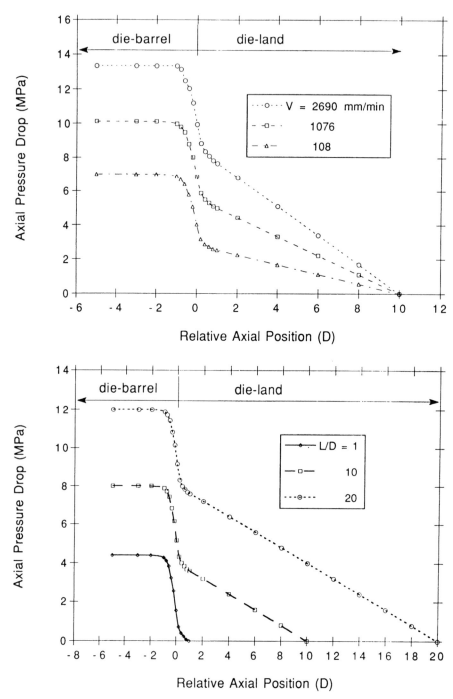

Fig. 23.17 Variation of extrusion pressure on (top) extrudate velocity and (bottom) land ratio. Note shear thinning exponent $m > n$ and pressure drop in die-land increases more rapidly with velocity. (Courtesy of Jingmin Zheng, Alfred University).

When extruding through a die and die land, the dependence of the pressure for extrusion on the flow velocity and land ratio has the form shown in Fig. 23.17. The initial change of slope is a function of the velocity factor and the shear thinning parameter of the body and the lubricating boundary film. A lubricating film of water and colloidal clay minerals has been distinctly observed on the surface of an extruded porcelain body (Fig. 23.18). It is this surface film which facilitates slippage flow in the die land. For organically plasticized materials, a film of binder, liquid inorganic colloids, and a lubricant, if added, can provide lubrication for slippage flow. Computer modeling studies of square-entry extrusion indicate a slight reduction of pressure near the wall on entering the die land, and this pressure drop may motivate the microscopic separation of the liquid and other components to the surface. A more gradual formation of a slippage film may be expected to occur on the surface of a tapered die. An increase in temperature may change the flow properties of both the body and the slippage film.

For dies having a square cross section or dies with several holes, the pressure for extrusion is described by Eq. 23.5. The first term is dominated by the ratio of cross-sectional flow areas and the second by the area over which the wall

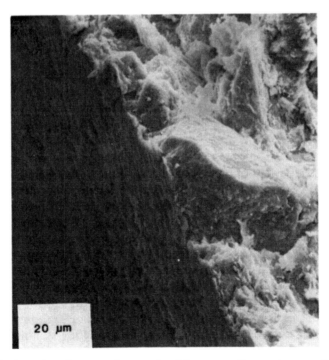

Fig. 23.18 Scanning electron micrograph of fracure surface and external surface of extrudate of an electrical porcelain body. Note smooth surface "skin" composed of oriented colloidal particles, and large angular particles and fine clay in extrudate below surface film.

shear stress is applied. For the square entry extrusion through a die with N circular holes, the extrusion pressure ΔP is given by*

$$\Delta P = 2 \ln (D_0/n^{0.5}D)[\tau_b + k_b v^n] + (4\ L/D)[\tau_f + k_f(4Q/N\pi D^2)^m]   (23.7)$$

where Q is the volume flow rate of extrudate.

Some elastic expansion can be expected on the ejection of the extrudate from the die land. An axial springback exceeding 5% and a radial springback of 0% were measured for an industrial porcelain body. The observed anisotropy of the springback is caused by the flow-induced orientation of the anisometric clay particles which are the major component in the body. Clay faces are observed to align tangentially to the velocity profile across the section. A nearly isotropic springback is expected when extruding powders which have an isometric particle shape. The diametral springback of an extruded alumina body plasticized with 4 vol% methyl cellulose was observed to be of the order of only 1% and approximately isotropic.

23.3   CONTROL OF EXTRUSION DEFECTS

The ejected extrudate must have adequate strength to survive handling without slumping or deformation. Large pores, cracks, and laminations are usually undesirable. Causes of common defects are as follows:

1. *Insufficient Strength and Stiffness*. The cohesive strength of the ejected extrudate must be high enough to resist deformation during handling for control of the product shape. Cohesive strength is increased by increasing the content of a high molecular weight binder and colloidal particles, coagulating the body, reducing the liquid content, and using a binder that gels during extrusion.

2. *Cracks and Laminations*. Differential shrinkage produces a tensile stress and differential drying shrinkage is the cause of most cracks and laminations in an extruded product. The shape of the crack near an inhomogeneity having a different shrinkage depends on the relative shrinkage of the inhomogeneity, as shown in Fig. 23.19. The mean drying shrinkage of extruded products is commonly in the range of 1–5%. The common origin of a "bird foot crack" (Fig. 23.20) is a small hard inclusion of low drying shrinkage; the greater shrinkage of surrounding material creates a circumferential tensile stress and radial fracture cracks. Hard inclusions may be agglomerates of the starting material that were poorly wetted during mixing and extraneous large particles or aggregates in incoming materials or from the wear of equipment. The entrance angle of the die should be small enough to ensure that no extrudate is retained which may become compacted and later break away into the extrudate stream. A moon-shaped crack may be produced by an inhomogeneity exhibiting greater shrinkage, such as a poorly mixed region of higher liquid content.

*Reference 4

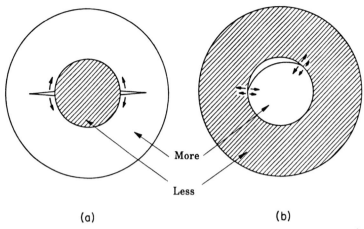

(a)   (b)

Fig. 23.19 Crack shape produced by inclusion having (*a*) less shrinkage and (*b*) more shrinkage than the surrounding material.

Another cause of differential shrinkage is liquid migration. The liquid permeability is reduced by increasing the content of colloidal particles and the amount and molecular weight of the binder in the body. Also, increasing particle coagulation, the liquid content of the body, and lubrication may reduce the extrusion pressure which is the driving force for liquid migration.

A group of oriented plate-like particles exhibits anisotropy in the drying shrinkage in proportion to the mean orientation and aspect ratio of the particles. Mixing may orient groups of clay particles or fibers. Differential shrinkage between regions having a different orientation may cause short cracks called checks.

It is observed that the long dimension of anisometric particles aligns tangentially to the velocity profile during extrusion. The anisotropy of the drying shrinkage may cause characteristic laminations in bodies containing a high concentration of anisometric particles. A die design that produces a more gradual gradient in the final velocity profile (Fig. 23.7) should reduce the maximum stress due to shrinkage anisotropy. For auger extrusion the use of a choke in

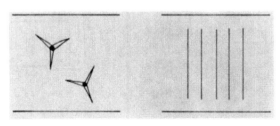

Crow Foot Cracks   Surface Laminations

Fig. 23.20 Bird-foot crack and periodic laminations on surface of extrudate.

the die to alter the velocity profile is common. The use of particles of low aspect ratio and an increase in the concentration of fine isometric filler particles aids in reducing the shrinkage anisotropy. Velocity gradients produced on extruding around a spider or a core rod, as shown in Fig. 23.21, may also produce an undesirable discontinuity in the particle orientation. Control of the mean drying shrinkage and the shrinkage anisotropy is required to avoid cracks and laminations. A higher green strength is also desirable.

3. *Surface Craters and Blisters.* Air in the body may dissolve in water at a pressure below 1 MPa. On decompression during ejection, exsolved air migrates through small channels which join to form a larger pore and a crater or blister surface defect that is visible to the eye. Evacuation of air from the feed material is common practice. The extruder must be well sealed to prevent air infiltration under a vacuum. Feed material should be shredded to less than about 3 mm to reduce the evacuation distance.

4. *Periodic Surface Laminations* (Fig. 23.20). Slip-stick wall friction and a high elastic springback can cause periodic surface cracks perpendicular to the flow direction. This defect is observed when the die is poorly lubricated and the extrusion pressure is high. Improving surface lubrication, increasing the liquid content, coagulating the body, and increasing the extrusion velocity may eliminate this defect.

5. *Curling of the Extrudate.* The die is normally centered on the auger system so that the pressure distribution is symmetrical (see Fig. 23.22). To produce uniform flow through an unsymmetrical die, the die must not be positioned too close to the termination of the auger and a die design providing compensating die wall friction may be needed. When extruding hollow items, the friction produced by the bridges and core rods must be balanced to prevent curling of the extrudate.

6. *Laminations from Unjoined Flow Streams.* Material leaving an auger is separated at the hub. Flow in front of the hub and radial twisting downstream creates a characteristic S-shaped interface which may persist as an internal S

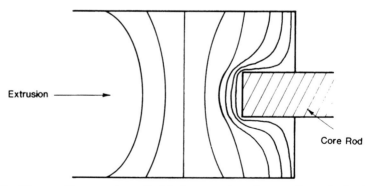

Extrusion →

Core Rod

Fig. 23.21 Flow profile around a cylindrical core rod produces a discontinuity in the orientation of anisometric particles near the corner of the rod.

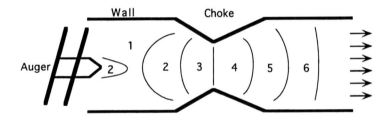

Pressure: 1 > 2 > 3 > 4 > 5 > 6

Fig. 23.22 Pressure distribution in an auger extruder. (Courtesy of Jingmin Zheng, Alfred University.)

crack on drying (Fig. 23.23). Divided flow is also produced by the spider used to form a hollow product. A weak interface may be produced where the flow lines join when the pressure is insufficient to produce bonding. Sufficient evacuation of air from the body is essential because the pressure of exsolved air may separate material at a weak interface.

7. *Poor Skin.* An insufficiently developed slippage film on the body may result in a product with small regions having a rougher surface texture. This problem may be corrected by dispersing colloidal particles more completely and increasing the content of fines, liquid, and/or lubricant. Polishing the surface of the die land may improve the skin quality of the extrudate.

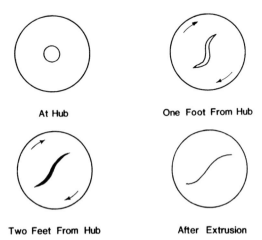

Fig. 23.23 Development of an S crack during auger extrusion. (After R. Lester, *Technological Innovations in Whitewares*, Alfred University Press, Alfred, NY, 1982, p. 151.

8. *Gradients in Extrudate Stiffness.* Liquid migration down the column during extrusion, called bleeding, will cause the material that is last out to be stiffer. Bleeding is reduced by decreasing the permeability of the body. Feed material should be of uniform composition.

9. *Tearing During Cutting.* Tearing during cutting is commonly produced by frictional drag. A lubricated cutting tool with a small friction area is desirable. The cutting action should be fast and the extrudate must have sufficient cohesive strength and elasticity.

10. *Curling on Drying.* Dimensional distortion due to differential drying conditions is discussed in Chapter 27.

23.4 JIGGERING, ROLLER TOOL FORMING, AND PLASTIC TRANSFER PRESSING

In ancient times, bricks and pottery were formed by pressing a plastic material in a mold by hand. Later, hollow ware was formed by pressing one's hands against material supported on a rotating wheel, which is called jiggering. A later modification was the replacement of a template with a roller tool. These processes are now highly mechanized and are used for forming a wide variety of refractory and electrical porcelain pieces with surface relief, institutional porcelain, cooking ware, and fine china. Highly automated roller tool systems are used for the industrial fabrication of hollow ware, as shown in Fig. 23.24.

Fig. 23.24 Industrial automatic roller-tool forming machine. (Photo courtesy of A. J. Wahl Inc., Brocton, NY.)

Feed material is commonly a section of deaired extrudate. Plastic feed may be pressed in a metal die or between porous molds of a relatively hard gypsum or other porous material. In jiggering, a porous mold or "bat" supports the body. A template or roller tool is used to deform the plastic body and contour the outer surface. Small electrical porcelain insulators are commonly formed by forcing a rotating, heated metal tool into a section of material supported on a porous mold. Lubrication is produced by heating metal tools to produce steam or by spraying a lubricant onto a high molecular weight polymer tool, to minimize sticking. Differential drying shrinkage separates the piece from the mold, but pneumatic separation is also used. Excess material is commonly eliminated by mechanical trimming. Mechanized fettling is used to smooth surfaces.

Parameters in plastic pressing and jiggering include the thickness and diameter of the feed, the thickness reduction, the material flow rate, surface features, the motion of the contouring tool and its lubrication, imposed vacuum, pressure, temperature, and the flow properties of the body. Fabrication begins as the feed material is cut and transferred to the cavity of the mold. On lowering the forming tool, the body makes contact with the mold and flows laterally in a plane between the tool and the mold (Fig. 23.25). In this stage, the shear strength of the body is only about 20–250 kPa. Material contacting the surface of the permeable mold becomes static, but may slide across polished metal. Air may be forced into the permeable mold to provide pneumatic release of the product.

Material flow is observed to be laminar except in protrusions such as the foot of a piece of dinnerware. In plastic pressing, the closure rate is relatively rapid until the die cavity is filled. The velocity of laminar flow is higher near the center of the piece. A roller tool produces a high degree of particle orientation because of the high shear near the surface. When pressing between permeable molds, consolidation and dewatering of the piece may occur if the permeability and dwell time under pressure are sufficiently high; this increases the yield strength for handling and reduces the drying shrinkage (Fig. 23.26).

Defects in jiggered and pressed plastic pieces include those observed in extruded material. A differential liquid content between the surface and center of a piece pressed between permeable molds may sometimes cause "case hardening" and shrinkage cracks in the center. During jiggering, "winding" near the base of the piece occurs if the applied vacuum is insufficient or when the

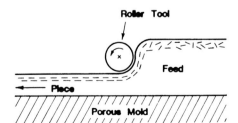

Fig. 23.25 Material flow and particle orientation near surface produced using a roller tool.

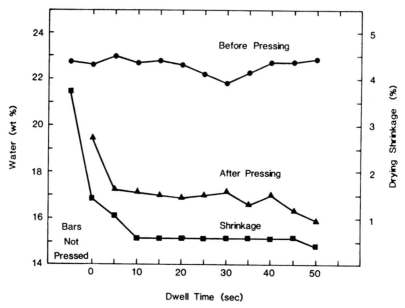

Fig. 23.26   Reduction in liquid content and drying shrinkage on pressing a hotel china body between permeable dies.

diameter of the ''bat'' of material is too large for the shape being formed. Splitting of a piece may occur when the rotation of the mold is too fast or when the surface of the roller tool is insufficiently lubricated.

Plastic transfer pressing is commonly used for forming large shapes with detailed surface relief and relatively deep nonsymmetrical shapes. Jiggering and roller tool forming are widely used for forming circular and elliptical shapes with a relatively thin wall, less than 5 mm in thickness, and for deeper circular shapes.

SUMMARY

Extrusion is a common plastic forming technique for both porcelain bodies and more advanced ceramics which are plasticized using molecular binders. Important variables in extrusion forming are the type and amount of binder, the particle size distribution and content of colloidal particles, liquid content, particle coagulation, temperature, deairing, lubrication, and die design.

During extrusion, material flows from a region of higher pressure to a region of lower pressure. Mechanisms of flow are commonly laminar flow of the body in the barrel and die-entry regions and plug flow with slippage in the die-land. The shear resistance of plastic extrusion bodies varies directly with the yield strength and is very velocity dependent. A lubricating film of different composition and liquid content is developed on the surface and this film facilitates slippage and may provide a smoother surface. The particular flow behavior is controlled by the die design and the shear resistance of the body. The cohesive strength must support the extruded product and resist deformation during handling. Extrusion is used to produce a wide variety of solid and hollow products with a uniform cross section. Variations in the drying shrinkage and deformation during handling reduce dimensional precision. Causes of defects in products are improperly dispersed components in the body, differential drying shrinkage, entrapped air, and the incomplete bonding of material separated and rejoined during flow.

Porcelain bodies similar to those used in extrusion are industrially fabricated into products using the methods of plastic transfer pressing, roller tool forming, and jiggering.

SUGGESTED READING

1. J. Zheng, W. B. Carlson, and J. S. Reed, "Flow Mechanics on Extrusion through a Square-Entry Die," *J. Am. Ceram. Soc.*, **75**(11), 3011–3016 (1993).

2. Irving Ruppel, "Extrusion," in *Engineered Materials Handbook*, Vol. 4, ASM International 1991, pp. 166–172

3. J. J. Benbow, T. A. Lawson, E. W Oxley, and J. Bridgewater, "Prediction of Paste Extrusion Pressure," *Am. Ceram. Soc. Bull.*, **68**(10), 1821–1824 (1989).

4. J. J. Benbow, E. W. Oxley, and J. Bridgewater, "The Extrusion Mechanics of Pastes—The Influence of Paste Formulations on the Extrusion Parameters," *Chem. Engr. Sci.*, **42**(9), 2151–2162 (1987).

5. J. J. Benbow and J. Bridgewater, "The Role of Frictional Forces in Extrusion," in *Tribology in Particulate Technology*, B. J. Briscoe and M. J. Adams, Eds. Adam Hilger, Bristol, UK, 1987, pp. 80–90.

6. J. E. Schuetz, "Methylcellulose Polymers as Binders for Extrusion of Ceramics," *Am. Ceram. Soc. Bull.*, **65**(12), 1556–1559 (1986).

7. G. A. Ackley and J. S. Reed, "Body Parameters Affecting Extrusion," in *Advances in Ceramics*, Vol. 9, J. A. Mangels and G. L. Messing, Eds., The American Ceramic Society, Westerville, OH, 1984, pp. 193–200.

8. D. B. Price and J. S. Reed, "Boundary Conditions in Electrical Porcelain Extrusion," *Am. Ceram. Soc. Bull.*, **62**(12), 1348–1350 (1983).

9. G. C. Robinson, in *Ceramic Processing Before Firing*, G. Y. Onoda and L. L. Hench, Eds., Wiley-Interscience, New York, 1978. pp. 391–407.

10. J. F. White and A. L. Clavel, "Extrusion Properties of Non-Clay Oxides," *Am. Ceram. Soc. Bull.*, **42**(11), 698–702 (1963).

11. F. Moore, "The Physics of Extrusion," *Claycraft*, **36**(2), 50–57 (1962).

12. Eugene C. Bingham, "An Investigation of the Laws of Plastic Flow," *Bull. Bur. Stand.*, **13**, 309–353 (1916–1917).

PROBLEMS

23.1   For extrusion, the liquid content of the body must exceed some critical amount. Explain why.

23.2   Explain why $\tau_{wall} > \tau_b$ is required for auger feeding.

23.3   Calculate and compare τ_{wall} and τ_b for extrusion of the lubricated cordierite body in Table 23.4. ΔP = 2.4 MPa, Dia = 6.47 mm, and L = 126.8 mm for the die land.

23.4   What is the volume of binder relative to the volume of particles in Example 23.1?

23.5   Using the flow properties for the porcelain body in Table 23.4, calculate the effect on the extrusion pressure of doubling the reduction ratio and doubling L/D of the die land from 3 to 6 at $v = 1$ m/min.

23.6   Calculate the apparent viscosity for the porcelain body and the slippage film when extruding at a velocity of 1800 mm/min. Flow properties are given in Table 23.4.

23.7   Using the properties for the porcelain body in Table 23.4, calculate and graph the variation of shear resistance with velocity for both the die entry and die land film.

23.8   Graph the results in Problem 23.7 as a function of shear rate and comment on the behavior.

23.9   Illustrate the orientation of anisometric particles for the flow profile in Fig. 23.16.

23.10   Explain why the shape of a crack around an inhomogeneity depends on the moisture content of the inhomogeneity.

23.11   Show that for the extrusion of a rectangular cross section, the $A_{friction}/A_{land}$ = $2L(a + b)/ab$ where a and b are the width and breadth of the die.

23.12   A two-stage deformation rate program is commonly used for plastic transfer pressing and for jiggering. Should the initial rate be higher or lower? Explain.

23.13   When pressing a plastic body between gypsum dies, does the time required to obtain a uniform moisture content depend on the permeability and the thickness of the piece? Why does the drying shrinkage of the pressed body remain at about 0.5%?

EXAMPLES

Example 23.1   An extrusion body is prepared using 74.1 wt% powder, 23.0 wt% water, and 2.9 wt% hydroxyethyl cellulose of viscosity grade 4400. Estimate the viscosity of the solution (consult Fig. 11.9).

Solution. The concentration of the binder dissolved in water is

$$2.9 \text{ g}(100)/(23.0 + 2.9)\text{g} = 11.2 \text{ wt}\%$$

From Fig. 11.6, the viscosity is estimated to be $>1,000,000$ mPa·s. Some binder and water molecules adsorb directly on the particles. The very viscous binder solution of high molecular weight distributed between particles facilitates particle sliding and resists migration during extrusion.

Example 23.2   Compare the ratio $100V_{\text{binder}}/V_{\text{particles}} = 100V_b/V_p$ for the extrusion body in Example 23.1 with the guidelines in Chapter 21? The powder is silicon nitride Si_3N_4 with a particle density of 3.20 Mg/m^3 and the binder has a density of 1.3 Mg/m^3.

Solution. The ratio $100V_b/V_p$ is calculated as

$$100V_b/V_p = 100(2.9 \text{ g}/1.3 \text{ g/cm}^3)/(74.1 \text{ g}/3.20 \text{ g/cm}^3) = 9.6.$$

Yes, this falls within the guideline, but is on the low side.

Example 23.3   For extrusion of rods with Dia $= 1$ cm at 1800 mm/min, estimate the shear rate in the die entry and in the die land. Assume that the thickness e of the slippage film is 3 μm.

Solution. From the modified Benbow equation, the shear rate $\dot{\gamma}$ in the die entry is $\dot{\gamma} = v/(\text{Dia}/2)$.

$$\dot{\gamma} = v/(\text{Dia}/2) = 1800 \text{ mm/min } (1 \text{ min}/60 \text{ s})/(10 \text{ mm}/2) = 6 \text{ s}^{-1}$$

In the die land, shear occurs in the slippage film of thickness e.

$$\dot{\gamma} = v/e = 1800 \text{ mm/min } (1 \text{ min}/60 \text{ s})/0.003 \text{ mm} = 10,000 \text{ s}^{-1}$$

The shear rate is much higher in the microthin slippage film. Note that the shear rate in the die is higher when extruding a thin cross section at a fast speed.

Example 23.4   Do the properties for the cordierite body in Table 23.4 confirm the presence of a slippage film? For extrusion $\Delta P = 4.5$ MPa, Dia $= 6.47$ mm, and $L = 126.8$ mm.

Solution. The wall shear stress τ_{wall} is

$$\tau_{\text{wall}} = R \, \Delta P/2L = (6.47 \text{ mm}/2)(4.5 \text{ MPa})/2(126.8 \text{ mm}) = 57 \text{ kPa}$$

The shear resistance of the body is $\tau_Y = 850$ kPa. Since $\tau_{\text{wall}} \ll \tau_Y$, slippage at the wall of the die is expected.

Example 23.5 What is the origin of shrinkage anisotropy for a region containing oriented anisometric particles?

Solution. Shrinkage occurs when liquid is removed from between particles and the surfaces move closer together. Consider the oriented particles shown, which are 1.0×0.2 μm plates. If on drying, the surfaces move 0.01 μm closer together, the linear shrinkage will be:

Horizontal shrinkage $\quad \Delta L\%/L_0 = (0.01 \ \mu\text{m})100/1.0 \ \mu\text{m} = 1.0\%$

Vertical shrinkage $\quad \Delta L\%/L_0 = (0.01 \ \mu\text{m})100/0.2 \ \mu\text{m} = 5.0\%$

Note that the aspect ratio of the particles is 5 and the shrinkage ratio is 5 for the perfectly aligned particles. The maximum linear shrinkage and the anisotropy of the shrinkage are lower when the particles are less well oriented.

reduced shrinkage

Example 23.5 Particle orientation produces anisotropic shrinkage.

CHAPTER 24

INJECTION MOLDING

Polymer products varying widely in size and shape are produced with high productivity using the technique of injection molding. Typical compositions are thermoplastic materials such as polyethylene, polystyrene, and polypropylene. The injection molding of thermoplastic polymers involves premixing the polymer with pigments and opacifiers, feeding and compressing material by means of a plunger or screw, heating to form a consolidated viscous material, and then forcing the viscous material into the cavity of a cool mold. The thermoplastic polymer strengthens on cooling and is separated from the mold.

A recent adaption of this process has been to premix a relatively high concentration of inorganic powder with the polymer, injection mold a product using the polymer to provide flow in shaping and strength for handling, and then remove the organic binder from the shaped part before sintering. Powder injection molding is now used for the industrial production of small ceramic parts of complex shape and high dimensional precision. In this chapter we examine the process of powder injection molding as a forming technique for ceramics.

24.1 EQUIPMENT AND MATERIAL VARIABLES IN INJECTION MOLDING

The steps involved in the powder injection molding process are shown in Fig. 24.1. The components of a batch for injection molding are the ceramic powder, polymer binder, secondary binder, solvent/plasticizer, and lubricant. The goal in mixing is to disperse powder agglomerates, coat the particles with the organic system, and produce a small scale of inhomogeneity in the mixture. Mixing is

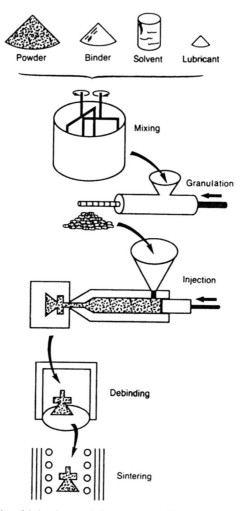

Fig. 24.1 Schematic of injection molding process. [From R. M. German and K. F. Hens, *Am. Ceram. Soc. Bull.*, **70**(8), 1294 (1991).]

conducted in the absence of air to avoid oxidative degradation of the organic components and at a temperature below which the organic components would be degraded. Commonly the highest melting organic components are mixed and the temperature is adjusted down before adding the lower-temperature components and then the powder. The mixing torque rises significantly when the powder is added into the binder system and a high-shear mixer with temperature control must be used. Fig. 24.2 shows the change of mixing torque with mixing time for a powder that is wetted by the binder (A) and cases where heat causes the binder to decompose (B) or harden (C). Curve D occurs when the powder concentration is too large. Mixers used include the double plane-

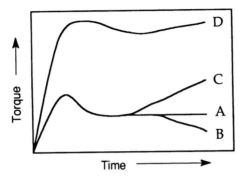

Fig. 24.2 Change of torque during mixing for cases of (A) wetting binder, (B) binder decomposition, (C) binder hardening, and (D) excessive powder.

tary, sigma, single screw, and twin screw designs. Extremely high torque and prolonged mixing significantly wear metal mixing elements, introducing contamination, and are undesirable. The shear stress $\tau = \eta \times \dot{\gamma}$ and a high shear stress needed to disperse agglomerates is developed by reducing the temperature to increase the viscosity or mixing at a faster rate. Mixing is conducted at a temperature where the shearing stress is sufficient to disperse agglomerates but at which the mixture has no yield stress. After the torque has subsided and the powder is mostly wetted, mixing at a high shear rate in a mixer with narrow clearance between moving and static elements will aid in reducing the maximum agglomerate size in the batch. The apparent viscosity is now between 10^3 and 10^6 mPa·s.

The mixed feed stock is commonly cooled and mixed with recycled material before further processing. Recycling is important because for small parts the actual part may be only a minority of the injected material. Material is cut into a finer size, screened, and then fed into a heated pelletizing extruder. Cooled extrudate of about 4 mm diameter is cut on exiting the extruder to form a pelletized feed material. Air bubbles in the pellets must be avoided.

As shown in Table 24.1, binder selection and proportions of binder, solvent/plasticizer, and lubricant may vary considerably. The more common additive system contains a backbone binder that is polyethylene, polypropylene, or polystyrene of low-medium molecular weight to produce a satisfactory flow viscosity when mixed with a high concentration of powder but yet avoid degradation in mixing and molecular orientation during molding. Wax may serve as a liquid at the molding temperature and as a binder when cooled below its softening temperature. Vegetable or mineral oil serves as a liquid solvent/plasticizer and the stearate is a lubricant. Other minor additives may include a plasticizer of low vapor pressure such as a phthalate. Because of the high concentration of organics in the body, commonly about 40 vol%, removal of the organics without loss of integrity of the product must be considered in the design of the organic system.

Epoxy resin that crosslinks on heating may be used for a thermosetting

Table 24.1   Binder Systems for Injection Molding (wt %)

Type	Binder #1/Binder #2	Solvent/Plasticizer	Lubricant
Wax	Paraffin wax (70%) Microcr. wax (20%)	Methyl ethyl ketone (10%)	—
Resin	Polypropylene (67%) Microcr. wax (22%)	(Molten wax)	Stearic acid (11%)
Resin	Polystyrene (45%) Polyethylene (5%)	Vegetable oil (45%)	Stearic acid (5%)
Epoxy	Epoxy resin (65%) Paraffin wax (25%)	(Molten wax)	Butyl stearate (10%)
Aqueous	Methylcellulose (4%)	Water (96%)	

Source: R. M. German, *Powder Injection Molding*, Metal Powder Industries Federation, Princeton, NJ, 1990.

system. Other thermosetting binders used are phenolformaldehyde and phenolfurfural.

Low-pressure injection molding is accomplished using a wax or an aqueous system (see Table 24.1). Systems containing only a wax binder may be molded at a relatively lower temperature and at a low pressure. Aqueous injection molding is accomplished by using water-soluble binders which gel on heating to $>60°C$, or by freezing the molded part and removing the ice by sublimation.

The binder content in the batch 100 V_b/V_p may range from about 0.15 to 0.50 depending on the type of additive system used and the powder concentration. The powder content in the body should be as high as possible to obtain a relatively high packing density before sintering. But the viscosity should be $<10^5$ mPa·s for satisfactory flow on molding. The structure of the molded material is that of a slurry containing a high-viscosity liquid. Shape distortion, particle flow, and particle/binder segregation are more of a problem when the binder is removed from a part with an excess of organics. Powder size distributions which have a higher PF_{max} should permit a higher powder loading, but the wide size distribution required may not be desirable for a fine-grained sintered ceramic. Angular shapes and surface roughness also contribute to the resistance to particle flow. An organic system containing several components which can be extracted or eliminated in steps is more suitable because pores formed on removal of the first component provide channels facilitating removal of remaining components. The backbone binder remains to bond particles and resist stresses produced. It may be removed with relatively little stress when vaporized through open pores.

An injection molding machine is shown in Fig. 24.3. A plunger or screw is used to consolidate the feed and convey it to the injection chamber where it is heated. Note that the material cross section thins to provide more uniform heating. A plunger pressure in the range 30–100 MPa is required to force material to flow into the mold and consolidate into a monolithic product. Material heated in the range 125–160°C exists from the nozzle through a sprue

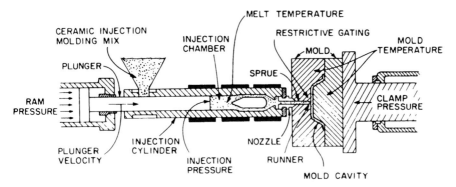

Fig. 24.3 Schematic diagram of a plunger type injection molding machine (Note: dark regions are heaters used to control temperature of material just upstream from nozzle). (From J. A. Mangels and W. Trela, in *Advances in Ceramics*, Vol. 9, American Ceramic Society, Westerville, OH, 1984.)

where it may divide into runners when feeding multiple mold cavities. Dies for complex shapes may contain removable pins and shaped inserts. Machine and material variables are summarized in Table 24.2.

It is important to note that a small stream of heated material feeds into the mold cavity. Temperature control along the flow path into the mold cavity must be carefully controlled. As shown in Fig. 24.4, the pattern of flow into the mold cavity may take many forms. Filling should occur as uniformly as possible so that flow units meld together into a cohesive product and air is displaced to mold vents and not entrapped in the material. The lubricant reduces the pressure for flow and mold filling, and facilitates separation of the molded product from the mold surface. Tooling is commonly a hardened tool steel or stainless steel. Steel with a nitrided surface or cemented carbides are used in regions where a higher wear resistance is required.

Screw or plunger motions and die motions must be coordinated during a fabrication cycle. Feed displacement produced by the screw or plunger displacement is coupled with the heating of the feedstock to the required temper-

Table 24.2 Machine and Material Variables in Powder Injection Molding

Machine Variables	Material Variables
Feeding mechanism	Viscosity at flow shear rate
Barrel temperature	Viscosity change on cooling
Sprue, runner, and gate geometry	Solidification temperature
Plunger velocity	Volumetric shrinkage
Mold temperature and geometry	Thermal diffusivity
Injection pressure–time program	Mechanical strength
Post injection pressure–time program	Elastic modulus

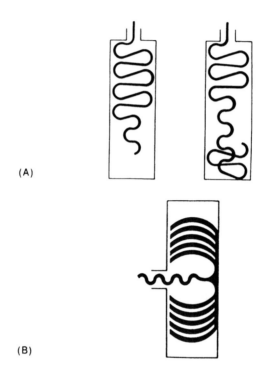

Fig. 24.4 Mode of filling and air displacement depends on locations of entrance to mold and design of mold cavity and more uniform filling (B) is preferred. (After J. A. Mangels and W. Trela, in *Advances in Ceramics*, Vol, 9, American Ceramic Society, Westerville, OH, 1984, p. 228.)

ature. The dwell time in the heater must be sufficient to heat the material uniformly. Molding begins with a rapid rise of pressure produced by a forward motion of the screw or plunger. When using a screw, a check ring seals against the screw and the screw assembly acts as a plunger. Pressure is held until the mold is filled and material in the gate freezes. After the binder has cooled sufficiently for the compact to hold its shape, the die is opened and the push of ejector pins ejects the part. The time for a molding cycle may be less than 2 min.

24.2 MECHANICS OF FLOW

The microstructure of an injection molding material is that of a concentrated slurry (DPS > 1) with a high concentration of viscous binder. Bonding between the nonpolar molecules is low and at the molding temperature the material is considered not to have a yield strength. The flow viscosity is commonly determined by using a heated capillary tube rheometer. No slippage at the wall is assumed. A power law shear thinning (pseudoplastic) model as described in Chapter 15, that is,

$$\eta_a = K\dot{\gamma}^{n-1} \tag{24.1}$$

approximates the flow. Flow properties are determined using a capillary rheometer of radius R and length L, by measuring the pressure ΔP developed for flow at a velocity $\bar{v}$. The mechanics of flow is described by the equations

$$\eta = [n/(3n + 1)](\Delta P R^2/2L\bar{v}) \tag{24.2}$$

and

$$\dot{\gamma}_{wall} = [(3n + 1)/n]\bar{v}/R \tag{24.3}$$

As shown in Fig. 24.5, the variation of the log (apparent viscosity) with log (shear rate) commonly exhibits a range of linearity in the range of shear rate above about 10 s^{-1} which is the primary range of shear rates during flow on molding. The shear thinning exponent n is calculated from the slope of the curve. Properties at the molding temperature are $\eta_a < 10^6$ mPa·s and an n of about 0.5 at a shear rate of 100 s^{-1}.

The flow properties of the ceramic molding material are shear thinning as is the binder system, and are influenced by the shape, size distribution, and packing of the powder. Properties at the forming temperature for an alumina powder plasticized with a polypropylene–wax–stearic acid additive system (MW

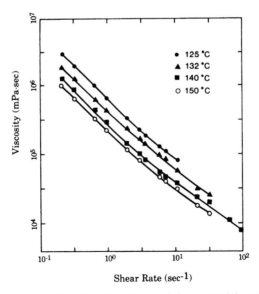

Fig. 24.5 Apparent viscosity of an injection molding material varies with shear rate and temperature. (After J. Mangels and W. Trela, in *Advances in Ceramics*, Vol. 9, American Ceramic Society, Westerville, OH, 1984, p. 225.)

polypropylene = 32,000; 6/2/1 weight proportions, respectively) are reported in Table 24.3. The behavior of shear stress with shear rate was approximated well using the Casson equation and the variation of high shear rate viscosity with volume fraction of powder could be approximated using the equation of Chong et al. (see Chapter 16). These results indicate shear thinning behavior and an increase in the viscosity, but an essentially constant shear thinning index, as the powder volume fraction increases and the polymer matrix thins between particles.

Cooling, which occurs as material flows from the nozzle, causes a rapid increase in viscosity, as indicated in Fig. 24.5. For a thermoplastic binder system, the viscosity over the temperature range is approximated by the exponential expression

$$\eta_a = \eta_0 \exp [(Q/R)(1/T - 1/T_0)] \tag{24.4}$$

where Q is the activation energy for flow. A lower Q indicates a less-drastic viscosity increase on cooling, and less sensitivity of the flow to temperature variations.

Flow into the cool mold causes both a drop in shear rate and cooling, which increases the viscosity several orders of magnitude. The ceramic powder in the body reduces the specific heat and increases the thermal diffusivity. Volumetric changes on cooling include contraction due to crystallization of the polymer and its reversible thermal contraction. Relative to 100% polymer systems, ceramic molding materials have a much lower compressibility and much less "die swell" when flowing from the gate into the die cavity and very little elastic recovery. These effects increase the tendency for jetting. Cooling speeds are faster and the ceramic molding material exhibits a larger pressure drop in the mold, which may lead to an inferior fusion of the weld sections. Gate design is more critical. Pressure transmission in the die cavity can be prolonged when the sprue, runners, and gate are relatively large.

During solidification, a solid layer forms adjacent to the mold surface immediately after injection. Shrinkage occurs as the solidified layer increases in

Table 24.3 Viscosity Calculated for an Alumina Body With a Polypropylene–Wax–Stearic Acid Binder System ($\dot{\gamma} = 108\ \text{s}^{-1}$; $n = 0.5$)

Alumina (vol%)	η_r
0	1
48	5.42
54	9.00
58	10.80
60	12.60
64	24.30

Source: M. J. Edirisinghe, H. M. Shaw, and K. L. Tomkins, Ceram. Intl., **18**, 193–200 (1992).

thickness. The static pressure hold compensates for this shrinkage when it produces continued filling until the center freezes. The relatively high thermal contraction of the polymer phase produces stresses where the material contacts the die and first cools. A low solidification temperature and a minimum of binder crystallization may reduce the stresses due to differential cooling during molding. Adequate mechanical strength is needed to resist the formation of cracks due to thermal stress.

24.3 DEFECTS AND THEIR CONTROL

Defects in powder injection molded parts may have their origin at several stages in the process, as is indicated in Table 24.4. Examples of different defects observed are shown in Fig. 24.6.

Steps to reduce the size of agglomerates and segregated binder in mixing have been discussed above. Causes of defects which originate during molding are numerous. A mold-filling design which produces a more uniform flow and filling (see Fig. 24.4) can eliminate both voids due to incomplete filling and

Table 24.4 Defects in Injection Molded Products

Stage of Origin	Defect
Mixing	Agglomerates
	Segregated binder
	Wear product contamination
Molding	Weld lines
	Short shots
	Voids from water adsorbed on materials
	Voids due to differential shrinkage
	Cracks due to differential shrinkage
	Mold ejection defects
	Blistering on mold release; poor skin
	Cracks from stress relaxation
	Mold surface wear contamination
Binder removal	
$T < T_{soften}$	Deformation on relaxation of residual stress
	Cracks
$T = T_{soften}$	Bloating
	Slumping
$T \gg T_{soften}$	Cracks on decomposition of binder
	Cracks from differential binder removal
	Cracks from improper support of product
	Delamination of surface skin
	Binder ash residue contamination

Source: M. J. Edirisinghe, *Am. Ceram. Soc. Bull.*, **70**(5) 824–28 (1991); R. M. German and K. F. Hens, *ibid.*, **70**(8) 1294–1302 (1991).

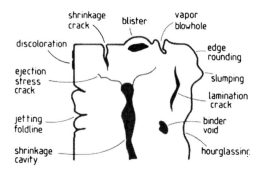

Fig. 24.6 Illustration of possible defects in an injection molded material after debinding. (Source: R. M. German, *Powder Injection Molding*, Metal Powder Industries Federation, Princeton, NJ, 1990.)

knit lines where contacting flow streams do not fuse together well. Solidification of material in the gate must not precede complete filling of the die cavity and the volume associated with bulk shrinkage on solidification. Improved temperature control of the runners and sprue may be required to eliminate filling defects. Porosity in a molding may be caused by air and moisture in the feedstock and by air entrapped when filling the mold. Flow during filling should displace air to vents in the die. Knit lines due to the improper fusion of flow streams are often visible to the eye. One cause of knit lines is random flow due to the "jetting" of a thin stream of material into the mold. Another is when a stream contacts the wall and cools before joining with another stream. Control of material flow and the pressure and temperature distribution in the die is essential. In the sintered part, knit lines will appear as large laminations or cracks.

The removal of the organics from the injection molded part must be carried out very carefully because of the high concentration of organic material which separates the particles and fills the interstices among the particles. For large sections, the elimination of all organics by thermal vaporization may require weeks to accomplish. Particles are dispersed in the organic additive system in the molded part and DPS > 1. Binder elimination causes the PF of the particles to increase and DPS < 1. Elimination of the organics by thermal means creates gas and causes differential shrinkage, producing stress, and the part is weakened when the binder is eliminated.

Other techniques for eliminating the binder may be desirable. The binder may be removed by one or more of the following mechanisms:

1. Liquid flow
2. Solvent extraction
3. Vaporization or sublimation
4. Thermal or oxidative decomposition

The advantage of the first two mechanisms is removal without forming large volumes of gas. Capillary flow of organic liquid into a porous support may be used to remove low-viscosity liquid such as liquid wax. Capillary flow is motivated by gravity and the capillary suction

$$\Delta P = 2\gamma_{LV}(\cos \theta_1/R_1 - \cos \theta_2/R_2) \tag{24.5}$$

where γ_{LV} is the surface tension of the organic liquid, θ is the wetting angle, R is the pore radius, and the numerical subscripts refer to the porous support and the body, respectively. The rate of flow varies inversely with the liquid viscosity. Accordingly, this technique is limited to low molecular weight organic components.

Solvent extraction may be used when plasticizer, lubricant, and/or the secondary binder but not the backbone binder is soluble in a solvent. The solvent is then removed by drying. Swelling of the part commonly occurs on infiltration of the solvent and contraction on drying, and a backbone binder is needed to maintain product integrity. A benefit of these techniques is that pore channels are opened to facilitate the removal of remaining components.

The organic binder is most commonly removed by pyrolysis. Chemical factors include the gaseous species produced and residual solid matter such as carbon; physical aspects are the mass and thermal transfer and local and bulk changes in the particle packing. The rate of production of gas must be low and controlled and the gas must be able to diffuse to the surface without causing a defect. Low molecular weight species such as oils and plasticizers may be lost by vaporization. A pressurized inert gas may be used to avoid boiling. The thermal degradation of the binder may produce a range of gaseous species. Oxidative degradation may occur on the surface of oxide particles. The presence of gaseous oxygen in the atmosphere alters the gaseous species produced and the rate of thermal decomposition. Because gaseous elimination is diffusion controlled, the time for removal depends on the product size and shape.

The allowable initial heating rate below the softening point of the binder varies inversely with the product size. Above the softening temperature during which binder removal occurs, heating must occur at a very slow rate in the range of 2–5°C/h. Defects produced during binder removal are cracks from the sources indicated in Table 24.4. Support of the product on a molded powder bed is common to prevent slumping or a distortion of the product. Bloating or blisters may form when air or gas moves through the liquid binder.

24.4 COMPRESSION MOLDING AND ROLL FORMING

A blank of prepared molding material may be forged into a particular shape by pressing the viscous material in a die at a temperature that causes the material to stiffen when heat is transferred. This technique is especially adaptable for shaping thermosetting materials in warm dies.

In roll forming, a moldable feedstock is pulled, compressed, and effectively extruded between steel rolls rotating in opposite directions. Either a thermoplastic or a thermosetting material may be used. The reduction ratio in one pass is limited and a thin sheet with a well-controlled thickness is produced using multiple passes through a series of rolls having a diminishing gap size. Resin and rubber-bonded abrasive products have traditionally been fabricated using warm roll forming.

SUMMARY

Powder injection molding is used to manufacture ceramics that are of complex shape and small in cross section. The thermoplastic molding material has the structure of a slurry. The powder is dispersed in an organic matrix consisting of a backbone polymer, a solvent/plasticizer, a lubricant, and other additives such as a surfactant (coupling agent) to improving wetting. The very viscous slurry is heated above the softening temperature of the thermoplastic polymer system and injected into the cavity of a cool mold. Solidification on cooling provides strength for removal and handling. The large proportion of organic component is removed slowly by drying, capillary flow extraction into a porous support, and pyrolysis. Factors limiting the wider application of injection molding are problems associated with dispersing powder agglomerates while mixing the viscous mass, the removal of the large volume of organic components prior to sintering, and the high initial tooling costs. Systems containing a thermosetting binder system are used to a limited extent for some products. These molding systems are also formed into shapes using plastic transfer pressing and roll forming.

SUGGESTED READING

1. J. Edirisinghe, H. M. Shaw, and K. L. Tomkins, "Flow Behaviour of Ceramic Injection Moulding Suspensions," *Ceram. Int.*, **18**, 193–200 (1992).
2. R. M. German and K. F. Hens, "Key Issues in Powder Injection Molding," *Am. Ceram. Soc. Bull.*, **70**(8), 1294–1302 (1991).
3. Mohan J. Edirisinghe, "Fabrication of Engineering Ceramics by Injection Molding," *Am. Ceram. Soc. Bull.*, **70**(5), 824–828 (1991).
4. J. R. G. Evans and M. J. Edirisinghe, "Interfacial Factors Affecting the Incidence of Defects in Ceramic Mouldings," *J. Mater. Sci.*, **26**, 2081–2088 (1991).
5. B. C. Mutsuddy, "Injection Molding," in *Engineered Materials Handbook*, Vol. 4, ASM International, 1991, pp. 173–180.
6. Randall M. German, *Powder Injection Molding*, Metal Powder Industries Federation, Princeton, NJ, 1990.
7. M. J. Edirisinghe and J. R. G. Evans, "Review: Fabrication of Engineering Ceramics by Injection Molding. I. Materials Selection," *Int. J. High Technol. Ceram.*, **2**, 1–31 (1986); "II. Techniques," *ibid*, **2**, 249–278 (1986).

8. R. M. German, "Theory of Thermal Debinding," *Int. J. Powder Metall.*, **23**, 237–245 (1987).

9. J. A. Mangels and W. Trela, "Ceramic Components by Injection Molding," in *Advances in Ceramics*, Vol. 9, J. A. Mangels and G. L. Messing, Eds., American Ceramic Society, Westerville, OH, 1984, pp. 220–233.

PROBLEMS

24.1 For a thermoplastic injection molding material, graph the DPS and PF of the particles beginning with feedstock and again when the material is heated, molded, cooled, and the binder is removed.

24.2 Sketch two particles separated by the backbone polymer and the secondary binder of lower molecular weight.

24.3 Would you select angular particles with some surface roughness or smooth spherical particles for an injection molding system? Assume particle shape does not significantly influence sintering?

24.4 Explain why temperature control is important when mixing the thermoplastic batch.

24.5 Is the maximum size of the molded part limited by molding factors or debinding factors, or both? Explain.

24.6 Compare the relative change in viscosity on heating from 125 to 150°C at a shear rate of 10 s^{-1} and on increasing the shear rate from 1 to 100 s^{-1} at 150°C, for the material in Fig. 24.5.

24.7 Compare the thermomechanical properties of polyethylene and polypropylene for injection molding.

24.8 For an injection molding mixture, the apparent viscosity determined using a capillary viscometer is 100,000 mPa·s when the shear rate at the wall is 100 s^{-1}. Calculate the consistency K and graph viscosity versus shear rate. (Assume $n = 0.5$).

24.9 Graph the results in Table 24.3 for viscosity and relative viscosity versus powder loading. Then determine the f_{cr}'' and k parameters, assuming the Chong equation (Eq. 16.33) is a good approximation.

24.10 Sketch the solidification rate for a part with both thin and thick sections, assuming heat transfer into the die is not rate limiting.

EXAMPLES

Example 24.1 Explain why it is difficult to achieve a small scale of homogeneity throughout the bulk material when mixing a powder and a liquid to form a mixture of very high viscosity.

Solution. The diagram of viscosity versus powder content is very helpful to understanding this problem. At the beginning of mixing, regions deficient in powder are supersaturated (DPS >> 1) and of lower viscosity and regions rich in powder have DPS << 1 and are significantly higher in viscosity. Homogeneity in mixing is produced by transferring material between these dissimilar regions to produce an intermediate composition and DPS, and a higher average viscosity, as is indicated by the arrows on the diagram. A high shear stress is required to disperse agglomerates and highly viscous regions, to reduce the scale of segregation, but considerable shear flow is needed to achieve uniformity in the bulk. The shear stress in mixing $\tau = \eta\dot{\gamma}$. After bulk mixing, mixing at a higher shear rate or at a lower temperature (in injection molding) to increase η may produce a higher shear stress to further break down resistant agglomerates. Subsequent mixing during feeding in the forming operation may also reduce the scale of inhomogeneity.

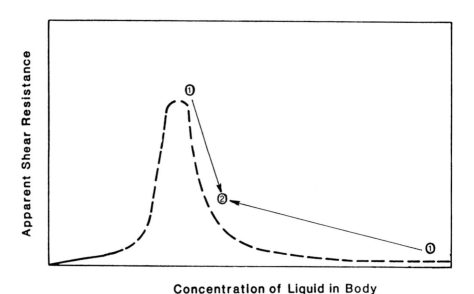

Concentration of Liquid in Body

Example 24.1   Mixing path diagram when preparing an injection molding material.

Example 24.2   Trace a viscosity path during injection molding using a viscosity vs. shear rate diagram as an aid.

Solution. Material enters the mold at a relatively high shear rate and temperature, as indicated by Point 1. As the mold becomes filled and contact between warm material and the cooled mold increases, the temperature and shear rate decrease and the apparent viscosity increases to Point 3, as indicated on the diagram.

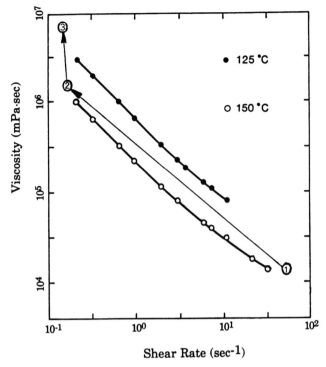

Viscosity (mPa·sec)

Shear Rate (sec⁻¹)

Example 24.2   Change of viscosity on cooling during injection molding.

Example 24.3   What is the volume change that may be expected during cooling of the thermoplastic material in the die?

Solution. For a polypropylene–powder system, contraction occurs on the partial crystallization of the polymer and from the thermal contraction of the polymer and particles on cooling. The volume change of the polymer on crystallization is of the order of 1.75 vol%. The thermal expansion coefficient of the polymer exceeds that of the ceramic powder and the polymer forms a continuum. For a volume coefficient of expansion of 300 ppm/K, the volume contraction on cooling from 150° to 25°C would be 37,500 ppm (3.75 vol%). The sum is $1.75 + 3.75 = 5$ vol%.

Example 24.4.   What parameters favor the more complete removal of organics from an injection molding body by capillary flow?

Solution. From Chapter 2, the volume flow rate $Q = \pi \,\Delta P R_c^4 / 8\eta L$. Flow is enhanced when the binder system contains a minimum of the backbone binder and more of a low-viscosity binder phase, and when the pore size is relatively larger, as when using larger particles. Polymer adsorbed on particles reduces the capillary radius R_c^4.

CHAPTER 25

DRAIN CASTING AND SOLID CASTING

Casting processes begin by filling a mold with a ceramic slurry having a pourable consistency. The cast is produced when a physical, chemical, and/or thermal change causes the slurry to develop a yield strength. In drain casting, the cast forms adjacent to the mold surface, and after the wall has grown to the desired thickness, excess slurry is drained from the mold. Solid casting produces a solid cast having the shape of the cavity of the mold. The cast may or may not be dried before it is removed from the mold.

An aqueous slurry containing fine clay is called a slip. Slip casting is the conventional casting of a slip or a slurry in a porous gypsum mold. Capillary suction of water into the mold concentrates and coagulates the particles in the slip near the mold surface, forming the cast. Pressure applied to the slurry (pressure casting), a vacuum applied to the mold (vacuum-assisted casting), or centrifuging (centrifugal casting) may be used to increase the casting rate. Slurries containing a reactive bonding additive such as a hydraulic cement or a binder that gels may be solid cast in molds or cast "in place," such as in a furnace or a dental cavity. In a process called "gunning," a castable batch containing a sticky binder is mixed with water and blown through a pipe to patch wear areas on the surface of a structure or to apply the material as a thick coating.

Drain casting and solid casting are used to produce a variety of traditional porcelain products such as fine china and dinnerware and industrial porcelain components. Slip casting processes are used to make dense refractories that are of a complex shape, thin-walled products such as crucibles and closed-end tubes, and structural refractories that are large in cross section. Coarse-grained structural refractories and concrete products are widely produced by solid casting. Vacuum casting is used for forming porous refractory insulation products of various sizes and shapes. Newer processes of slurry pressing and gel casting are being developed for the production of more advanced ceramics.

The advantages of casting processes are the more complete powder dispersion in a relatively low-viscosity liquid, the complexity of product shape permissible, and the relatively low capital cost. The disadvantages of casting are the lower production rates and lower dimensional precision commonly obtained. Here we will examine the very old applied art and the young science of casting processes. Knowledge of many principles discussed in prior chapters is essential to understanding this very interesting topic.

25.1 SLIP CASTING PRACTICE AND MECHANICS

The formation of a cast by drain casting and solid casting is shown in Fig. 25.1. Steps involved in the slip casting process are the preparation of the slurry, beneficiation of the slurry by screening or some other form of classification and evacuation of air, mold filling, casting, draining, partial drying while in the mold, separation from the mold, and sometimes trimming and surface finishing before the final drying.

Figure 25.2 presents a processing flow diagram for a well-controlled slip casting process. Ideally the behavior of the casting slurry should not be sensitive

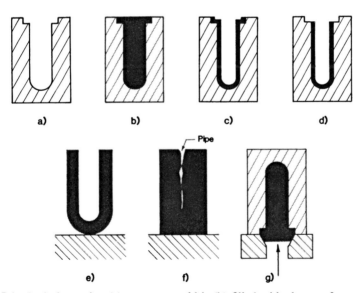

Fig. 25.1 In drain casting (a) a porous mold is (b) filled with slurry; after a period of casting (c) excess slurry is drained from the mold leaving a hollow cast which (d) may be trimmed in place if of sufficient strength and toughness. A longer casting time will produce a cast having (e) a thicker wall and when continued (f) a nearly solid cast is formed. In (g) pressure casting, the slurry is pressurized and pumped into the mold.

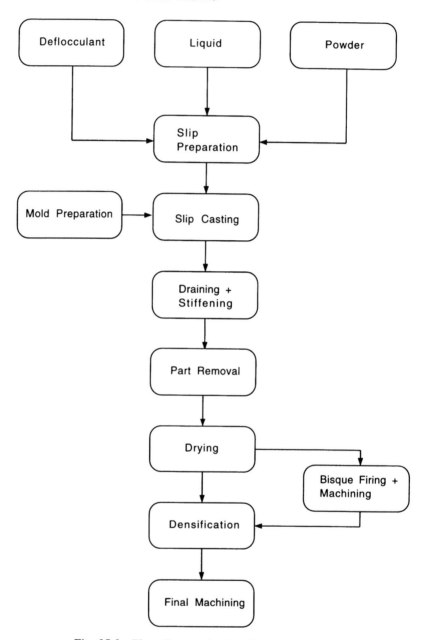

Fig. 25.2 Flow diagram for the slip casting process.

to slight variations in chemical composition or particle size distribution. The slurry should have a reasonable storage life and aging may be required to obtain a time-stable slurry. Control parameters are the proportions and dispersion of components in the slurry, slurry rheology during mold filling, the casting rate, the density and yield strength of the cast, flow rheology on draining, shrinkage and release of the cast from the mold, the strength and mechanical toughness of the product removed from the mold, and trimming and surface finishing.

Slurries containing liquids, solids, and processing additives (see Table 25.1) may be prepared in a variety of mixers and in ball mills, planetary ball mills, and vibratory mills. Batch size may range from a fraction of a liter for a special ceramic to several thousand gallons when casting porcelain products. Intensive mixing and milling disperse agglomerates more completely and shorten the mixing time; however, low-intensity mixing for a longer time is used when solid materials can be comminuted in mixing and the proportion of submicron sizes must be relatively low. Additives that can degrade during mixing, such as binders of medium to high molecular weight, are often added near the end of the mixing cycle, and their potential degradation may influence the choice of mixing operation. Casting slips containing up to 50% clay are often blunged for 8–48 h to produce an "aged," time-stable slip with a reproducible casting behavior. The mixing time is reduced when the clay is received as an aged slurry. Postmixing processes include classification, separation of magnetic material, temperature control, and deairing. A mixed slip may be conveyed in the mixing container or pumped into portable holding tanks for distribution.

Casting slurries are typically formulated to be shear thinnings and have an apparent viscosity below 2000 mPa·s at a mold-filling shear rate of $1-10$ s^{-1}, as shown in Fig. 25.3. This viscosity behavior is convenient for mixing and pumping, and for flow when the mold is filled. A lower viscosity below about 1000 mPa·s is required when the slurry is screened and air bubbles must be

Table 25.1 Compositions for Casting

Material	Concentration (vol%)	
	Porcelain	Alumina
Alumina		40–50
Silica	10–15	
Feldspar	10–15	
Clay	15–25	
Water	45–60	50–60
	100 vol%	100 vol%
Processing additives (wt%[a])		
Deflocculant		
Na silicate	<0.5	
NH$_4$ polyacrylate		0.5–2
Na citrate		0.0–0.5
Coagulant		
CaCO$_3$	<0.1	
BaCO$_3$	<0.1	
Binder		
Na carboxymethylcellulose		0.0–0.5

[a]Based on weight of solids in slurry.

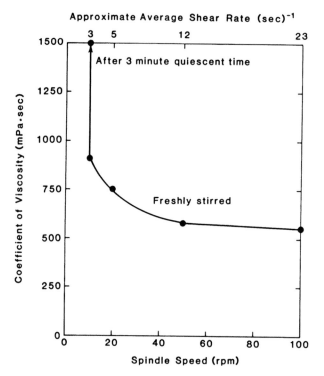

Fig. 25.3 Porcelain slip for casting is shear thinning; gelation when at rest after pouring causes the apparent viscosity to increase (53 vol % solids in slip).

quickly eliminated from the slurry. The density and apparent viscosity or yield stress of the slurry should be high enough to minimize particle settling during casting. For particles larger than a few microns in diameter, a yield stress τ_Y sufficient to suspend a particle is

$$\tau_Y > (2/3)\, a(D_p - D_{slurry})g \qquad (25.1)$$

where D_{slurry} is the density of the slurry and a and D_p are the diameter and apparent density of the suspended particle. When the slurry does not have a yield stress, a high viscosity at the low shear rate during casting (some vibration commonly occurs) will retard settling. The density of the slurry is higher when the solids loading in the slurry is higher, and a working concentration for the state of deflocculation may be obtained from the graph of viscosity (Fig. 25.4). The maximum proportion of solids which still gives a suitable viscosity is higher when the PF_{max} of the particle distribution and the deflocculation are higher, as explained in Chapter 16.

Permeable molds for casting are commonly gypsum with 40–50% porosity and an effective pore size of 1–5 μm. The casting time may range from a few minutes for a thin cast from a partially coagulated porcelain slurry to about

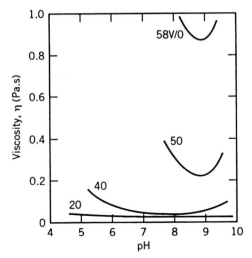

Fig. 25.4 Viscosity at 9 s^{-1} versus pH for deflocculated alumina slurries of 20–58% solids content. [Polymethacrylic acid deflocculant. From J. Cesarano, III and I. A. Aksay, *J. Am. Ceram. Soc.*, **71**(12), 1064 (1988).]

1 h for a thick porcelain cast or a thin, dense cast from a well-deflocculated slurry of a powder containing submicron particles. The casting time for a dense solid cast refractory of fine particle size may range up to several weeks for thicknesses of 30 cm.

In slip casting, the consolidated layer of particles formed on the mold surface occurs by the process of filtration. For uniaxial filtration, the cast thickness L as a function of casting time t is

$$L = [(2J\Delta Pt/\eta R_c) + (R'_m/R_c)^2]^{0.5} - (R'_m/R_c) \qquad (25.2)$$

where J = volume of cast/volume of liquid removed, R_c is the resistivity to liquid transport in the cast, ΔP is the apparent mold suction, η is the viscosity of the liquid transported, and R'_m is the liquid transport resistance of the mold. It is assumed in deriving Eq. 25.2 that the parameters other than time on the right-hand side of the equation are constant during casting. When the slip "ages" during casting, changes in the interparticle forces may cause a change in the particle packing and, therewith, J and R_c.

Commonly the ratio R'_m/R_c is small, and the simplified equation

$$L = [2J\Delta Pt/\eta R_c]^{0.5} \qquad (25.3)$$

approximates the behavior, as shown in Fig. 25.5. The parameter $R_c = 1/K_p$, where K_p is the liquid permeability of the cast (K_p was described in Chapter 19). Equation 25.3 describes the casting when the liquid transport through the cast is rate limiting, which is the case when the porosity × (pore size)^2 is

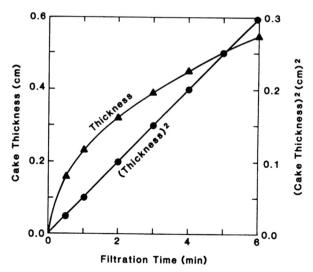

Fig. 25.5 Parabolic behavior of cast thickness with time and linear dependence of cast (thickness)^2 with time when slip casting a porcelain slip.

smaller for the cast than for the mold. The transport of liquid in the mold due to capillary suction depends on the residual liquid in the mold and the depth of liquid penetration. The mold may become rate limiting after it has absorbed liquid from several casts, and the mold is said to be "waterlogged."

Parameter adjustments to increase the casting rate are indicated by Eq. 25.3. The parameter J is higher when the difference in liquid concentration between the slurry and the cast is small. A higher solids loading in the slurry will increase J. Heating the slurry may reduce the viscosity of the liquid, but heating may also change the state of deflocculation which would change R_c. A partially coagulated slip, which produces a relatively high cast porosity, will increase J and lower R_c and speed casting. This is acceptable for porcelain products which are densified by the large amount of glassy phase produced during sintering. When producing dense sintered crystalline ceramics, a dense cast from a well-deflocculated slurry is commonly required and J/R_c and the casting rate are lower. The effect of parameter adjustments on the casting kinetics are shown in Fig. 25.6. For purposes of quantitative comparison, the time-independent casting rate is conveniently expressed as L^2/t. Table 25.2 presents casting rate results for both porcelain and engineering ceramic formulations. Rates for a pressure of 140 kPa are for ordinary casting in a gypsum mold. Note that the casting rate when 0.5% binder is included in the slurry is about 0.1 time that for the slurry prepared without an organic binder. From the form of the curves in Fig. 25.6, it is apparent that product items with thin walls are manufactured more productively. Casting times may be reduced significantly by applying an external pressure on the slurry. Pressure casting is discussed in Section 25.2.

In drain casting, the excess slurry must drain readily from the surface of the

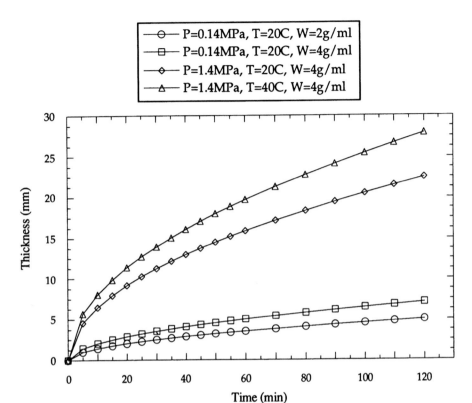

Fig. 25.6 Effect of combinations of pressure, solids in slurry, and temperature on thickness of cast with time (calculated for slurry of alumina powder deflocculated with polyacrylate deflocculant).

TABLE 25.2   Casting Rate L^2/t for Slip Casting and Pressure Casting

Pressure (kPa)	Porcelain[a]	Casting Rate (mm²/min)		Zirconia[d]
		Alumina[b]	Alumina[c]	
140	2.0	2.4	0.3	
280	3.1	4.9		
560	5.2	10		0.1
1400		25		0.3
2800		50		0.8

[a]Semivitreous porcelain body.
[b]Alumina powder, no binder.
[c]Alumina powder, with 0.5 wt % PVA binder.
[d]Particle size <50 nm.

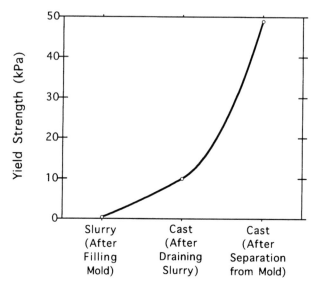

Fig. 25.7 Yield strength of porcelain composition increases at each step in casting process.

cast but the cast must have a yield strength sufficient to resist flow. An important factor is the differential (and yield strength) between the slip and the cast, as shown in Fig. 25.7. In some systems, and particularly when using partially coagulated clay slurries, coagulation forces in the cast contribute significantly to the strength. This coagulation is commonly called gelation when casting porcelain. For engineering ceramics, gelation may occur from the flocculation of an organic binder in the cast. The content of a molecular binder in common slip casting must be <1 vol% for reasons mentioned above. A binder which is adsorbed or of high molecular weight should resist migration with the liquid. The anionic polymer binder sodium carboxymethylcellulose has been shown to act beneficially both as a deflocculant and a binder in aqueous slip casting.*

Figure 25.7 shows the yield strength of a cast during casting, after draining, and at separation after partial drying. The strength varies directly with the PF in the cast, the strength of interparticle coagulation forces, and bonding produced by a binder. The coagulation between clay colloids contributes to the strength of a porcelain cast. However, excessive coagulation will produce a cast having a low PF and a low yield strength, and the cast may flow slowly when draining the slip. The strength imparted by a binder increases with its flocculation and desolvation on drying.

Separation of the piece from the mold normally occurs from the shrinkage of the cast on drying. Air may be blown into the back of the mold for pneumatic release. When casting porcelain, molds are stripped 20–60 min after draining. The partially dried pieces must be strong enough to resist fracture and plastic

*A. J. Ruys and C. C. Sorrell, *Am. Ceram. Soc. Bull.*, **69**(5), 828–832 (1990).

deformation during handling. Casts that are plastic have toughness and may be trimmed and surface-finished while damp. Thin-walled brittle casts are easily fractured. Casts are commonly loaded onto conveyors or racks for the final drying. Brittle casts are commonly surface-finished after drying, called fettling, or after partial or final sintering.

Examples of compositions for porcelain and refractory alumina casting slurries are listed in Table 25.1. In the porcelain slip, the clay content of 15–20 vol% is lower than in an extrusion body because of the need for liquid permeability. Casting clays used contain relatively less colloidal material and are relatively free of bentonite (see Table 5.5) so that the total submicron content of the body is less than about 30% when fully dispersed. The partial coagulation of the clay slip agglomerates colloidal particles, and the working viscosity is controlled above the minimum that is possible for the solids loading (Fig. 25.8). The type and amount of additives used to control the state of partial coagulation depend on the type and amount of colloids present, soluble impurities in the raw materials, and impurities in the water.

Slurries for casting oxides, carbides, nitrides, and so on range more widely in composition. A fraction of a percent of an organic binder increases the pseudoplasticity of the slurry and the strength of the cast, but reduces the permeability of the cast and the casting rate (Fig. 25.9). A slurry containing colloidal particles which is well deflocculated will produce a dense cast having adequate green strength to survive very careful handling. Compositions containing a small amount of colloidal clay and/or an organic binder may exhibit limited toughness and marginal tolerance to stresses in finishing and handling.

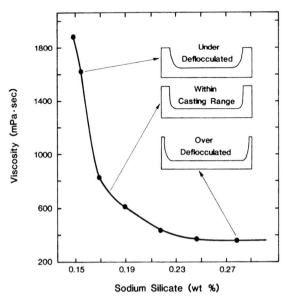

Fig. 25.8 Effect of deflocculant concentration on the viscosity and casting behavior for a whiteware slip (viscosity at $\dot{\gamma} = 3 \text{ s}^{-1}$ just before filling).

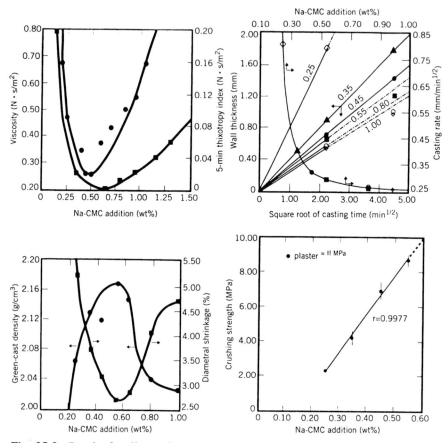

Fig. 25.9   Results for slip casting an alumina slurry using different concentrations of Na carboxymethyl cellulose deflocculant/binder: (top left) slurry viscosity and thixotropy, (top right) casting rate, (bottom left) density and shrinkage of cast, and (bottom right) strength of dried cast. [From A. J. Ruys and C. C. Sorrel, *Am. Ceram. Soc. Bull.*, **69**(5) 830–831 (1990).]

Ceramic materials such as magnesium oxide and calcium oxide, which hydrate quickly in water, may be slip cast by using a nonaqueous liquid such as an alcohol, trichlorethylene, or methyl ethyl ketone and an appropriate deflocculant. Ceramics such as silicon carbide and silicon nitride are also slip cast from nonaqueous slurries. However, carbide and nitride powders handled in air commonly have an oxide surface layer, and particle charging may often be used to produce stable aqueous slurries. An exception is AlN powder which reacts with water,

$$AlN + 2H_2O = AlOOH + NH_{3(gas)} \tag{25.4}$$

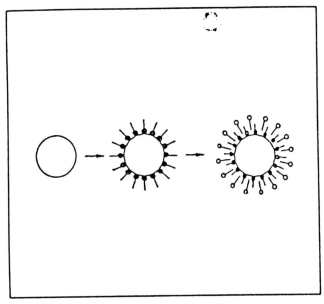

Fig. 25.10 Depiction of double coating of surfactant on AlN particle used to resist hydration but provide dispersion.

The hydrolysis of AlN powder may be significantly retarded by an adsorbed surfactant which produces a hydrophobic coating. Adsorption of a second coating of surfactant (Fig. 25.10) enables the powder to be dispersed as an aqueous slurry for casting and also as a precursor for spray drying.

25.2 PRESSURE CASTING, SLURRY PRESSING, VACUUM CASTING, AND CENTRIFUGAL CASTING

The parameter that directly impacts the slip casting rate is the pressure ΔP motivating liquid migration, as indicated by Eq. 25.3. The capillary suction of a gypsum casting mold is less than about 200 kPa, and for a used mold is commonly less than 150 kPa. An external pressure or suction can increase the driving force and reduce the casting time.

Pressure Casting

Pressure casting has been investigated as a forming technique for porcelain and complex refractory shapes, and mechanized, relatively automatic systems are now used in some industries. In pressure casting, the mold serves as a shaped filtration support, and the casting time is controlled by the slip pressure. The casting time for a particular thickness can be reduced significantly, as indicated

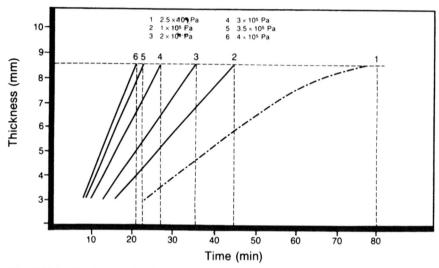

Fig. 25.11 Casting results for traditional battery casting and pressure casting. (Courtesy of Netzsch Inc., Exton, PA.)

in Fig. 25.11. In studies of the pressure casting of porcelain products,* is was found that a slurry pressure of up to 1.5 MPa increased the cast thickness by a factor of 2–3. Pressure casting reduced the water content of the cast and the drying shrinkage from about 3–3.5% to 1–1.5%, but increased the shrinkage anisotropy. The casting time for a wall approximately 1 cm thick cast from a heated slip containing very little deflocculant was reduced from 1 to 2 h to less than 20 min. Compressed air was also used for drying and to release the piece from the mold.

Casting rate results for both porcelain and engineering ceramic compositions are listed in Table 25.2. Increasing the casting pressure reduces the cycle time for a cast, but the higher pressure increases the cost in terms of mold materials and equipment requirements. High-pressure slip casting at up to 4 MPa is used for the small volume production of small shapes and is especially advantageous when floor space is very limited. Special polymer molds are used, which are described below. Molds are not dried. High-pressure casting occurs in a manner similar to injection molding, except that filtering rather than a thermal change is used to set the material. Slip is supplied to the die by a diaphragm pump. After casting, the die is opened and the part is removed by means of a vacuum pickup. The part is fettled at the filling point and at the mold seam, and then removed for drying. Dies may be single cavity or multicavity in nature. Medium pressure casting is used for higher production levels and for larger product items. Either gypsum or polymer molds are used. Single molds or a series of molds are mounted on a bench-type frame (Fig. 25.12). Slip is pumped simul-

*T. M. Gainer and J. A. Carter, *Am. Ceram. Soc. Bull.*, **43**(11), 9–11 (1964).

Fig. 25.12 Medium pressure casting machine with a series of polymer molds mounted on a bench-type frame. (Courtesy of Netzsch Inc., Exton, PA.)

taneously into each mold at a pressure of about 350 kPa. After casting is completed, casts are demolded individually and removed for drying. Molds are immediately ready for reuse. Production advantages are the lower inventory of molds, hydraulic control of the casting rate, shorter drying times because of the lower liquid content and shrinkage of the cast, lower drying stresses between thin and thick sections in a part, and improved working conditions. Disadvantages include adapting the forming of porous molds from nontraditional materials and the higher capital cost. Quality advantages include a higher shape precision, green strength, and density.

Pressure casting produces a larger differential pressure across the cast. When using a very high pressure, a large gradient in the effective stress $d\bar{\sigma}/dx$ will occur from the difference between the imposed pressure P and the liquid pressure u within the cast:

$$d\bar{\sigma}/dx = dP/dx - du/dx \qquad (25.5)$$

The gradient in effective stress (Fig. 25.13) will produce a density gradient in a cast which has significant mechanical compressibility, such as a cast of colloidal particles or one from a partially coagulated slurry. The density gradient may cause differential shrinkage and shape distortion or a crack on drying.

Slurry Pressing (Pressure Filtration)

A variation of pressure casting is used to form hard ferrite magnets which have a highly oriented microstructure. The ferrite particles are submicron tabular crystallites that have a preferential magnetization in a direction normal to the faces. A die set with pistons which are magnetically permeable and also drilled to permit the flow of liquid is charged with a well-dispersed slurry. A magnetic

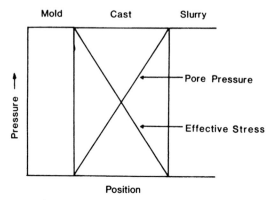

Fig. 25.13 Nominal pressure profile and effective stress in cast during casting.

field is applied to orient the particles and the pistons are forced into the die to form a dense cast with oriented particles.

A similar process has been investigated for forming advanced ceramics from slurries of colloidal particles. Processing in slurry form facilitates particles dispersion. The high slurry pressure speeds up casting and the mechanical pressing when the pistons contact the cast reduces the density gradient. Cycle times are long for colloidal particles because the R_c of the cast is high. However, porous refractory insulation, which has a very low R_c, is conveniently cast into thick sheets by low-pressure slurry pressing.

Vacuum Casting

Vacuum casting is widely used for forming very porous refractory insulation having a complex shape. The slurry typically contains partially deflocculated ceramic powder and chopped refractory fiber. The fiber increases the viscosity and liquid requirement and increases the porosity of the cast. An expendable, permeable preform is coupled to a vacuum line and then submerged in the slurry. After the desired wall thickness has been cast, the preform is withdrawn and removed for drying and a chemical set.

In slip casting, the external surface of a porous mold may also be subjected to a vacuum to effectively increase the casting rate.

Centrifugal Casting

Centrifugal casting has been investigated for forming advanced ceramics having a complex shape where a very high cast density is needed. Centrifuging increases the driving force for the settling of the particles and liquid is displaced to the top of the centrifuge cell. For a relatively dilute suspension, the settling rate is described by Stokes' law (Eq. 7.3), and particles having a greater size or density settle at a faster rate and a structural gradient is produced.

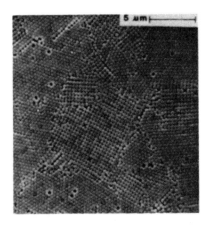

Fig. 25.14 Surface of consolidated pressure-filtered monosize silica spheres. (From E. Barringer et al., in *Ultrastructure Processing: Ceramics, Glasses, and Composites*, Wiley-Interscience, New York, 1984, p. 321.)

Suspensions of spherical monosize colloidal particles have been centrifugally cast into dense packings consisting of ordered domains of close-packed spheres separated by interdomain regions of higher porosity (Fig. 25.14). In very concentrated slurries of the type used for slip casting, hindered settling occurs and segregation by size or density is significantly retarded. The casting rate increases with centrifugal speed and the flow rate of the liquid away from the deposition surface. Casts of complex shape having a uniform high density, and sedimentation rates of 0.6 mm/min have been reported.

25.3 POROUS MOLD MATERIALS

The most commonly used porous mold material for slip casting is gypsum ($CaSO_4 \cdot 2H_2O$) formed from the reaction between plaster of Paris ($CaSO_4 \cdot 0.5H_2O$) and water:

$$CaSO_4 \cdot 0.5H_2O + 1.5H_2O \rightarrow CaSO_4 \cdot 2H_2O \qquad (25.6)$$

This technology is used because of the ability to fabricate molds with good surface smoothness and detail, high ultimate porosity and micron-size pores, the short setting time, the small dimensional expansion (about 0.17%) on setting which aids release from models, and its relatively low cost.

From Eq. 25.6, the weight ratio of water to plaster required for hydration forming gypsum is 18.6/100. A range of 60/100 to 80/100 water/plaster weight ratio is used in slurries for production molds. Heated water is commonly used to obtain a more consistent mold structure. During setting, an interconnected network of needle and platelike arrangements of gypsum crystals is formed, which gives the mold strength. Molds with a maximum pore size of about 5 μm and an apparent porosity of 40–50% are formed after drying, as shown in Fig. 25.15. The high water/plaster ratio used for industrial casting molds pro-

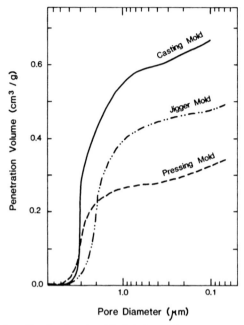

Fig. 25.15 Pore size distribution in dried industrial gypsum molds prepared using water/plaster ratios of 80, 65, and 40/100. (From L. Hollem, MS Thesis, Alfred University, Alfred, NY, 1969.)

duces the highest porosity and a slightly larger pore size, which increases the water absorption but lowers the strength (Fig. 25.16). Stronger molds for mechanical pressing are prepared at a water/plaster ratio of about 40/100; air must be blown through the mold after the initial set to maintain continuous pore channels. The setting time and pore structure of the gypsum mold depends on the temperature and electrolytes in the water and the mixing intensity and time. These variables must be controlled to obtain molds with reproducible behavior.

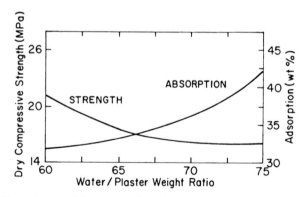

Fig. 25.16 Dependence of the water adsorption and compressive strength (dry) on the water/plaster ratio used in fabricating the molds.

TABLE 25.3 **Properties of Conventional Gypsum Molds**

Aqueous solubility at 25°C (g/L)[a]	2.6
Tensile strength (dry, MPa)	3
Compressive strength (MPa)	
Dry	14
Wet	7
Thermal expansion (mm/mm/K)	155
Desiccation	Dehydrates in dry air

Source: C. M. Lambe, in *Ceramic Fabrication Processes*. W. D. Kingery, Ed., MIT Press, Cambridge, MA, 1958.
[a]Increases with temperature and sodium sulfate in water.

New molds are brought to a low water content on drying in a mold drying room or a dryer. Water evaporation from the molds occurs rapidly. Overheating above about 40°C may cause dehydration and should be avoided. The mold is conditioned by wetting the mold with less than 15% water; this will prevent very strong water adsorption when slip contacts the surface on the first cast. With each casting cycle, the water content of the mold increases and the casting rate will be reduced for the next cast. After a certain number of casts, the casting rate becomes too slow and the "waterlogged" mold must be partially dried.

Limitations of gypsum molds are their low compressive strength when partially saturated with water, erosion in use due to their low abrasion resistance and the significant solubility of gypsum in water, their relatively low thermal shock resistance, and the potential desiccation of the gypsum when heated above 40°C during drying, as indicated in Table 25.3. The life of gypsum molds is lower when using acidic aqueous slurries or an alcohol medium. Mold coatings such as alginate, talc, and paper pulp are sometimes applied on the mold surface to facilitate separation of a cast part during drying.

Improved porous polymer molds which combine higher porosity, relatively high mechanical strength, and good elasticity have been developed for pressure casting applications. The formulation may be cast into a complex shape having sections of different thickness. Elimination of the solvent produces the porous structure. In a multipiece mold, components should fit well and the mating surfaces seal better when flat. Less finishing of the cast product is needed when the mold parts fit and seal well. A degree of elasticity in the mold is needed to compensate for small mating differences and will reduce the clamping force required to obtain a good seal.

25.4 SLIP CASTING PROBLEMS AND CONTROLS

When casting porcelain products, the casting rate of the partially coagulated slip is relatively high because J/R_c is relatively high. The differential liquid content between the slip and the cast is small. A liquid concentration gradient in the cast is observed. In contrast, the casting rate of a well-deflocculated slurry is relatively low because J/R_c is low. The differential liquid content between the slurry and cast is much larger, and the liquid concentration across the section is more uniform. These differences are shown in Fig. 25.17.

A porcelain slip has a small yield stress, is shear thinning, and is thixotropic. At rest after agitation, the viscosity increases as is indicated in Fig. 25.3. The slightly lower water content and particle coagulation (gelation of colloids) increase the apparent yield strength and viscosity of the cast. Experimental determination of viscosity and gel strength were discussed in Chapter 16. The partially coagulated slip contains porous agglomerates called flocs which must be of a controlled size and content. Gel formation in the cast is critical, and the gel strength varies with the proportion of clay minerals, colloid content, particle shape, coagulation caused by chemical additives, and casting time. A cast with a balance between firmness, density, and plasticity is requisite to prevent fracture or shape distortion during handling, trimming, drying, and surface finishing. A more uniform distribution of both the liquid and particle sizes throughout the cast reduces stress and distortion on drying and firing. Gelation minimizes mold staining and improves mold release by reducing the mobility of colloids through the cast.

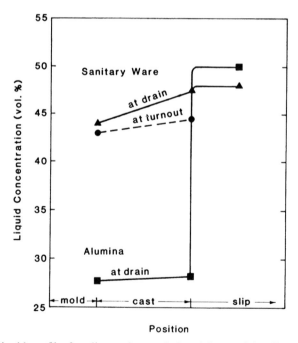

Fig. 25.17 Liquid profile for slip casting an industrial porcelain slip and a calcined alumina slurry. (After W. Brodie, *Technological Innovations in Whitewares*, Alfred University Press, Alfred, NY, 1982.)

When the viscosity of the slip is used as a control parameter, it is commonly observed that slips within a narrow range of viscosity and gelation produce casts with the best compromise of casting rate, cast density, and handling properties (see Table 25.4). A dense, brittle cast is produced from a slip that is nearly completely deflocculated, and a relatively porous, soft cast is produced from a significantly coagulated slip.

In industrial casting, a satisfactory slip must be tolerant to small fluctuations in the particle size distribution and soluble impurities in materials and chemical additives. The behavior of relatively tolerant (wide casting range) and intolerant slips is shown in Fig. 25.18. Control of the microstructure of the slip, gelation behavior, and the microstructure of the cast when using industrial minerals and unrefined water is not an easy task. Water used for molds and casting slurries is now filtered and deionized in many industrial operations. Some clays are received in tank cars as an aged slurry. Ball clays contain both complex organic (lignite) and inorganic colloids which are altered during mixing. The proper balance of the type and amount of colloids and soluble electrolytes from the clay materials is critical to gelation and microstructure development in the cast. Coagulating ions from materials such as Ca^{2+} and Mg^{2+} are precipitated as silicates on the addition of a Na_2SiO_3, or chelated by sodium polymethacrylate deflocculant, and sulfate ions on the addition of $BaCO_3$. The amount of added electrolytes depends on the purity of the water and solubility of impurities and fine particles in the raw materials.

The casting rate (L^2/t) is constant with time when the parameters in Eq. 25.3 and the mold suction and resistance are invariant with time. The time to achieve a particular wall thickness is proportional to $(\eta_L R_c / J\Delta P)$, as described above. Water in a gypsum mold reduces the absorption of the mold and the casting rate owing to the presence of water in finer pores. During its lifetime, the gradual enlargement of pores occurs as gypsum is dissolved (Fig. 25.19). The

TABLE 25.4   Porcelain Slip Casting Problems

Flow of Slip	Observation
Viscosity too low	Long casting time
	Wreathing
	Cracks
Viscosity too high	Pin holes
	Poorly drained surface
Gelation (thixotropy) too low	Long casting time
	Brittle cast
	Poor fettling
	Wreathing
Gelation (thixotropy) too high	Flabby cast
	Poorly drained surface
	Long drying time

Source: W. E. Worral, *Clays and Ceramic Raw Materials*, Halsted Press, New York, 1975.

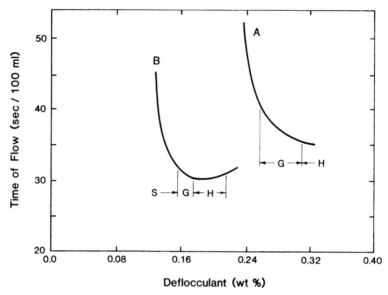

Fig. 25.18   Variation of slip flow time (nominal viscosity) with content of deflocculant for desirable case A with wide range of good casting and unfavorable case B. H = hard, G = good, S = soft cast. (After G. W. Phelps and J. Van Wunnik, *United Clay Mines Manual*, Cyprus Industrial Minerals, Sandersville, GA.)

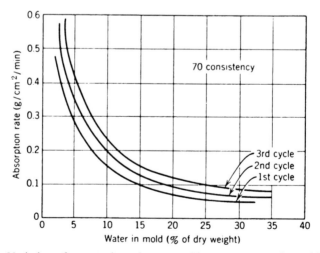

Fig. 25.19   Variation of water adsorption rate with water content in mold for several casting cycles. (From C. M. Lambe, in *Ceramic Fabrication Processes*, MIT Press, Cambridge, MA, 1958, p. 38.)

"staining" of the mold over time blocks surface pores and increases its resistance to liquid migration.

Slurries for casting refractories and fine-grained technical ceramics are commonly nearly completely deflocculated (Fig. 25.9) in order to obtain a dense cast. The density of the cast also depends on the PF_{max} of the particles. Systems containing a significant amount of colloidal clay minerals may be handled much like standard whiteware compositions. Slurries that do not contain clay are cast with more difficulty. Control of the pH and deflocculation using one or more deflocculants such as ammonium or sodium polymethacrylate and sodium citrate is required to develop the proper particle charging and rheology (Fig. 25.20). A very small amount of binder such as sodium carboxymethyl cellulose or sodium or ammonium alginate of moderate to high molecular weight may be added. The adsorbed ionic binder which coats the particles and attracts water plays a secondary role in deflocculation, retards settling in the slurry, and causes a gelling mechanism in the cast. Binder migration is undesirable and its presence retards liquid migration during casting and drying. The amount must be very limited to obtain a compromise of slurry properties, casting rate, cast properties, and drying behavior. A sufficient quantity of colloidal inorganic particles and their deflocculation is essential to achieving a satisfactory slip casting process.

Heating an aqueous slip can increase the deflocculation and widen the deflocculation range. Increased deflocculation will increase the PF in the cast and reduce the casting rate, L^2/t, but the reduction of the liquid viscosity (η_L) is greater, and L^2/t may be increased by as much as 25%. Control of the mold temperature and the slip temperature is required to control both the firmness of the cast and the draining behavior. Adjustments in slip temperature and deflocculation are commonly based on the mold temperature.

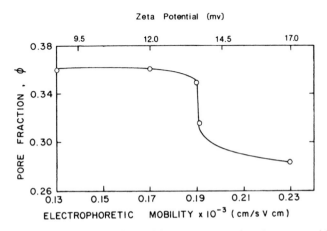

Fig. 25.20 Pore fraction in casts formed by pressure casting decreases with improved deflocculation indicated by zeta potential and particle mobility. (Alumina slurry deflocculated using an ammonium polymethacrylate defloculant.)

25.5 GEL CASTING

Casting slurries containing a binder which may be polymerized or gelled or which reacts with the liquid or particles to form a bond after filling the mold, may be solid cast in a nonporous mold when the volume change is small. Metal or polymer molds may be used and the shape may be quite complex. The process is often called gel casting when a binder that polymerizes or gels is used.

   This technology has been used for many years to produce a variety of both dense and porous refractory products and ceramic molds for metal casting, to obtain dental impressions and to fill cavities, and to repair the surface of structural components. For dense materials, the mold or cavity is usually filled with a very concentrated slurry containing a minimal liquid content. Mechanical vibration is sometimes used to induce flow and eliminate air pockets. The strength depends on the bond type, the $V_{bond}/V_{particles}$, and the PF of the particles in the material.

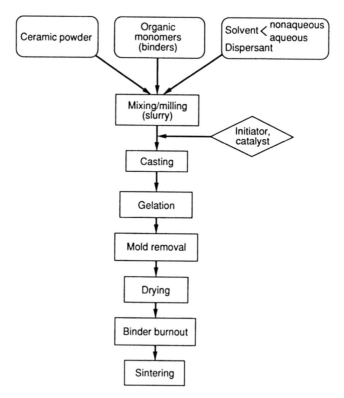

Fig. 25.21 Processing flow diagram for gel casting. [From A. C. Young, O. O. Omatete, M. A. Janney, and P. A. Menchhofer, *J. Am. Ceram. Soc.*, **74**(3), 612 (1991).]

The yield strength and viscosity of the cast are increased by the initial "set." The set eliminates particle settling, and the strength increases as the polymerization reaction continues. A process flow chart for the aqueous gel casting* of fine-grained engineering ceramics is shown in Fig. 25.21. The reactive organic monomers are first dissolved in deionized water. A polyacrylate deflocculant and ceramic powder are dispersed in the solution to form a slurry. The slurry is evacuated to remove air and then an initiator is added. The polymerization reaction is accelerated by the catalyst or by heating. Cross linking of the polymer produces an elastic hydrogel which serves as a binder.

Molds of wax, glass, and a variety of metals were found to be acceptable. The gelation time was in the range 5–60 min and could be controlled by the process conditions. After gelation, the part was demolded and moved into a controlled humidity drying chamber for controlled drying. The linear drying shrinkage was in the range 0.5–1.5%. The volume of binding phase/volume of particles ($100V_b/V_p$) was approximately 16, which is significantly lower than for a common injection molding system. Gel casting has also been accomplished using molecular binders which may be induced chemically to polymerize, such as sodium silicate, or thermally, such as methylcellulose on heating and agar and alginate on cooling. A distinct advantage of gel casting is the use of nonporous molds. Gel casting has been used satisfactorily to produce parts of complex shape having a uniform sintered microstructure.

25.6   CASTING CONCRETE

Aqueous slurries containing rather coarse particles and hydraulic cement have been used for centuries to cast engineering structures of all types. Concrete containing Portland cement is cast in a wide variety of structures such as the foundations of buildings and other engineering structures and highways. Concrete containing Portland cement and an organic binder are now used in some applications to produce macrodefect-free structures and coatings.

Refractory concretes commonly contain a high-alumina cement, a reactive bond, or a gelling bond (bonding materials were described in Chapter 11). Refractory concretes may be cast in molds to form specific products or cast "in place," as in forming a lining in a kiln or furnace. Dense concretes are prepared using discrete particle sizes as indicated in Table 25.5. Large particles range up to several centimeters in size. Mechanical vibration is often used to assist the flow of these concretes or to enable mold filling with a lower liquid content in the slurry. In a process called "gunning," a slurry similar to one used for casting, but containing a small amount of a secondary binder such as bentonite, which makes the material sticky, is mixed with water and pumped through a nozzle to apply the material as a coating or to patch the floor or the wall of a structure.

*See Reference 2

TABLE 25.5 Castable Refractory Batch

Material	Composition (wt %)
Refractory aggregate (μm)	
4760–2000	25
2000–595	25
<210	40
High alumina cement	10
	100%
Water	As required for forming

Bond contents relative to the amount of ceramic materials depend on the strength required. For castable refractories, 10–30 vol % phosphate bond or 10–35 vol % hydraulic cement is commonly used. The water requirement for slurrying exceeds that needed for the bond reaction. A mixture slurried and cast with a higher particle PF and less water will commonly produce a stronger material.

Foamed concretes are used to produce structures of low bulk density, to increase the resistance to disintegration during freezing and thawing conditions, and to reduce the bulk thermal conductivity. Foaming is produced by adding a foam agent that stabilizes air bubbles during mixing or by the introduction of an additive that reacts with one of the components to produce gas bubbles just before the material "sets."

25.7 OTHER CASTING TECHNIQUES

Soluble Mold and Model Techniques

These two techniques are designed to take advantage of a binder which is insoluble in one type of solvent but soluble in another. In the soluble mold technique, the mold is insoluble in the solvent used for the slurry and casting may proceed in a normal fashion. The mold is then immersed in a solvent which will dissolve the binder of the mold, and the mold is slowly removed to expose the cast. This technique may be used when casting a very complex shape such as the vanes of a gas turbine (Fig. 25.22), which would otherwise require a complex multicomponent mold with small sections that can be removed separately. Examples of binder systems are water-soluble waxes, non-aqueous-soluble waxes, and polyethylene glycol.

Differential dissolving may also be used to advantage when preparing a mold of a complex model. After the mold has been cast around the model and set, the model is dissolved in a solvent which does not dissolve the mold.

Fig. 25.22   Cast experimental gas turbine shape (green body). (Courtesy Allied Signal Aerospace, Torrance, CA.)

Electrophoretic Casting

The electrophoretic deposition of particles onto the surface of a conducting mold or substrate has been investigated as a means for forming thin tubular shapes and coatings on metals. Electrically charged particles are attracted to a surface of opposite polarity. Cross sections are limited in thickness because the rate is relatively slow.

Sponge Technique

A variation of gel casting used to form reticulate (open-cell) porous ceramics is the sponge technique.* An open-celled polymeric sponge is immersed in a ceramic slurry. After absorbing the slurry, the sponge is compressed or passed between rollers to remove 25–75% of the slurry. Subsequent steps are drying to remove the low-viscosity liquid and firing to eliminate the polymer and to sinter the ceramic powder into a reticulate structure.

*J. Saggio-Woyansky and C. E. Scott, *Am. Ceram. Soc. Bull.*, **71**(11) 1674–1682 (1992).

Foamed Castables

Slurries varying widely in particle size distribution and organic additive system may be formed into either a closed- or an open-cell structure. Ceramic constituents, liquid, a polymer, and reactants which create gas bubbles are mixed and cast. The foamed cast is then dried and fired to produce a porous ceramic product. The systems and methods developed have been summarized by Saggio-Woyansky and Scott.*

25.8 CONTROL OF DEFECTS IN CASTS

The cast must have strength adequate for removal from the mold without forming a stress crack, for handling without shape deformation, and to resist cracks on drying. Large- and small-scale defects observed in casts are summarized in Table 25.6. The causes of the defects are described below:

1. *Inadequate Control of Wall Thickness.* Variable wall thickness in a slip cast product is commonly caused by either a nonreproducible suction or liquid transport in the mold or a variable casting rate of the slip. The suction of a mold varies inversely with its pore size, which is influenced by the water/plaster ratio, temperature and electrolytes in the water, and the mixing intensity and time before the mold is cast. The suction of the mold also depends on its water content. The pore size of a mold enlarges with time in use because of the slow dissolving of the gypsum and its transport to the surface during drying (effloresence). Changes in the mold over time may not occur uniformly. A variable casting rate is commonly caused by insufficient control of the batch proportions, the particle size distribution, the deflocculation conditions, and aging of the slurry before casting. Aging of the slurry before casting is com-

---

*J. Saggio-Woyansky and C. E. Scott, *Am. Ceram. Soc. Bull.*, **71**(11) 1674–1682 (1992).

TABLE 25.6   Defects in Casts

Macroscopic Defects	Microscopic Defects
Variable wall thickness	Large pores
Differential wall thickness	Particle size segregation
Shape distortion	Particle density segregation
Macroscopic cracks	Small cracks, laminations
Voids within the part	Packing fraction gradients
Voids on mold/cast surface	
Bubbles in cast	
Pin holes	
Wavelike drained surface	
Uneven drained surface	

monly desirable so that the state of dispersion in the slurry does not change during the period of casting.

Differential thickness is commonly caused by uneven suction of portions of the mold due to differential liquid content in the mold and/or differential pore enlargement during the life of the mold. The blockage of fine pores on the surface of the mold may be caused by particles adsorbed from the slip (staining) or by soluble salts in the water, such as magnesium, which migrate when drying the mold and precipitate on the surface (hard spot). Differential thickness is also caused by particles settling in a slip with an insufficient viscosity; settling will cause horizontal sections to be thicker than side walls.

2. *Shape Distortion.* Shape distortion is caused by mechanical stresses in handling or from the weight of the product, which cause the cast to plastically deform, or from differential drying shrinkage.

Each cast must have a yield strength sufficient to support its weight and any external loads introduced in handling. The cast strength varies directly with the packing factor of the particles, volume of binder/volume of particles, bonding strength of binder, and particle coagulation, as discussed in Chapter 15.

Warping is caused by differential shrinkage on drying. Differential shrinkage is caused by gradients in PF or particle size and the orientation of anisometric particles; this has been discussed at length for extruded products (see Section 23.3). In slip casting, a PF gradient and concomitant liquid concentration gradient occur because of the gradient in pressure across the thickness. This effect is enhanced when the pressure imposed on the slurry is higher and when the cast is more compressible. Warping is minimized by reducing the average shrinkage, the liquid gradient, and the segregation of particles by size or shape. A linear drying shrinkage of 2–4% is rather typical for cast products; pressure casting may reduce the average drying shrinkage to < 1%. In casts of colloidal particles, the average drying shrinkage may exceed 5% and warpage is more of the problem.

3. *Macroscopic Cracks.* Cracks are caused when the product in the mold is restrained from shrinking owing to the mold geometry, from adhesion on some region of the mold, and from friction where heavy products are supported. Shrinkage of a part around a rigid insert may increase the difficulty of removing the insert and the tensile stress may cause a crack in the cast parallel to the axis of the insert. Molds with several removable sections and inserts may be required for complex shapes.

Large cracks are caused by differential volumetric shrinkage between regions differing in liquid content (a differential PF) or particle size and at junctures where the anisotropy of shrinkage changes. Shrinkage occurs when the separation between particles decreases. The shrinkage is lower when the particle dimension is larger because of the fewer particle surfaces; the shrinkage is also lower when the liquid content is lower and the mean separation of particles is smaller. Causes of differential shrinkage include (1) differential coagulation and settling within a very large cast with a long casting time, (2) a differential mean liquid content between thick and thin sections caused by differential suction over time because of the greater depth of liquid penetration into the mold from the larger portion (improper mold design), or (3) time-dependent coagulation within the thicker portion, the settling of coarser particles, and evaporation from an exposed surface during casting. Particle orientation effects are enhanced by sudden changes in the orientation of the mold surface and at corners. Differential particle orientation also occurs near the joints of multipiece molds. In general, any controls that reduce the mean drying shrinkage and improve the uniformity of the shrinkage will reduce the tendency for large cracks.

4. *Voids.* Mold filling should occur by laminar flow in a manner which displaces air above the surface of the rising slip or into the mold. Air may be entrapped where the slip suddenly increases in viscosity on contact with the mold and the mold should not be completely dry. Sometimes molds are tilted, subjected to a shearing motion, or vibrated to assist the flow of slurry into a difficult-to-fill region.

When slip casting a thick section, a "pipe" is formed in the center of the cast. Slip is added as the casting continues. A portion of the pipe may close off prematurely, producing a lineated void (Fig. 25.1).

5. *Bubbles and Pin Holes.* Bubbles within the cast are caused by air bubbles in the slurry or by gas produced during casting. Air bubbles are eliminated by adding a surfactant to improve the wetting of particles or an antifoam additive to break a foam, and by evacuating air from the slurry.

Pin holes observed on the mold contact surface may be caused by air bubbles in the slip or by air displaced into the slip at the surface from the uneven suction of liquid into the mold (which displaces air). Uneven suction may be caused by a bubble in the gypsum just below the surface. Pin holes on the drained surface are common when the draining slip contains bubbles and is of high viscosity.

6. *Surface Irregularities.* Surface irregularities on the molded surface are caused by uneven wear of the mold, uneven suction of the mold, or uneven filling. A case mold is used for casting the working molds. Soaping the case mold with a mold soap seals the surface and enables the working mold to release and be removed. Water adsorption into a region of the surface of the case mold due to improper soaping can result in a dense spot on the surface of the working mold.

A wave-like feature on the drained surface of a porcelain cast is called wreathing and is caused by variations in casting rate which depend in part on particle orientation created when the mold is whirled when it is filled. Whirling can influence the orientation of particles on the mold surface and the thickness of "seams" of different particle orientation near the joints of multipiece molds.

Surface irregularities are also produced when the draining slip is very viscous or the suction of the mold is very high and some of the draining slip remains on the cast.

7. *Microscopic Defects.* In addition to the defects described above, micro-scopic defects are produced by incomplete dispersion of material during mixing, wear product contamination, poor deflocculation, and the migration of colloidal particles and unadsorbed additives during casting.

SUMMARY

Casting processes have long been used to produce traditional ceramics and are used more for forming advanced ceramics. Casting modes include drain casting and solid casting. In conventional slip casting, the cast is formed by filtration on the surface of a porous mold which provides suction for the removal of the liquid. The packing factor of particles is higher in the cast and bonding is provided by particle coagulation and from the flocculation of a binder. The slip must be properly formulated and deflocculated to produce a balance between the filtration rate and the coagulation rate, to produce an adequate casting rate, cast density, and strength. Reproducibility in casting depends on consistency of slip and mold properties. Pressure casting and vacuum casting reduce the casting time and drying shrinkage of the cast. Organic binders used in casting nonclay ceramics improve the mechanical properties of the cast but reduce the casting rate. Casting slurries containing a binder system which gels or develops a chemical set are used to produce both dense and porous refractory structures, dental materials, and other structural materials. Defects in casts originate from insufficient strength for handling, improper dispersion of materials, air bubbles, and differential shrinkage on drying.

SUGGESTED READING

1. B. Kostic and M. Gasic, "Influence of Temperature and Solid Content on Fused Silica Slip-Casting Kinetics," *Ceram. Int.*, **18**, 65–68 (1992).

2. A. C. Young, O. O. Omatete, M. A. Janney, and P. A. Menchhofer, "Gelcasting of Alumina," *J. Am. Ceram. Soc.*, **74**(3), 612–618 (1991).

3. A. J. Ruys and C. C. Sorrell, "Slip Casting of High Purity Alumina Using Sodium Carboxymethylcellulose as a Deflocculant/Binder," *Am. Ceram. Soc. Bull.*, **69**(5), 828–832 (1990).

4. F. F. Lang, "Powder Processing Science and Technology for Increased Reliability," *J. Am. Ceram. Soc.*, **72**(1), 3–15 (1989).

5. J. Cesarano, III and I. A. Aksay, "Processing of Highly Concentrated Al_2O_3 Suspensions with Polyelectrolytes," *J. Am. Ceram. Soc.*, **71**(4), 1062–1067 (1988).

6. W. A. Sanders, J. D. Kiser, and M. R. Freedman, "Slurry-Pressing Consolidation of Silicon Nitride," *Am. Ceram. Soc. Bull.*, **68**(10), 1836–1841 (1989).

7. R. Moreno, J. Requena, and J. S. Moya, "Slip Casting of Yttria-Stabilized Tetragonal Zirconia Polycrystals," *J. Am. Ceram. Soc.*, **71**(12), 1036–1045 (1988).

8. E. G. Blanchard, "Pressure Casting Improves Productivity," *Am. Ceram. Soc. Bull.*, **67**(10), 1680–1683 (1988).

9. F. M. Tiller and C. Tsai, "Theory of Filtration of Ceramics: I. Slip Casting," *J. Am. Ceram. Soc.*, **69**(12), 882–887 (1986).

10. T. J. Fennelly and J. S. Reed, "Mechanics of Pressure Slip Casting," *J. Am. Ceram. Soc.*, **55**(5), 264–268 (1972).

11. R. P. Heilich, G. Maczura, and F. J. Rohr, "Precision Cast 92–97% Alumina Ceramics Bonded with Calcium Aluminate Cement," *Am. Ceram. Soc. Bull.*, **50**(6), 548–554 (1971).

12. R. R. Rowlands, "A Review of the Slip Casting Process," *Am. Ceram. Soc. Bull.*, **45**(1), 16–19 (1966).

13. P. E. Rempes, B. C. Weber, and M. A. Schwartz, "Slip Casting of Metals, Ceramics, and Cements," *Am. Ceram. Soc. Bull.*, **37**(7), 334–338 (1958).

14. D. S. Adcock and I. C. McDowall, "The Mechanism of Filter Pressing and Slip Casting," *J. Am. Ceram. Soc.*, **40**(10), 355–360 (1957).

PROBLEMS

25.1 A porcelain casting slip has a bulk density of 1.82 Mg/m^3. Estimate the volume of solids in the slip. Assume a mean particle density of 2.60 Mg/m^3.

25.2 What is the minimum yield stress needed to suspend 44-μm ($<$325 mesh) particles in Problem 22.1.

25.3 Estimate the porosity and aqueous suction of the casting mold in Fig. 25.15. The density of gypsum is 2.32 Mg/m^3. (25°C and cos $\theta = 1$).

25.4 Calculate the parameter J = cast volume/filtrate volume from the results for the porcelain and alumina data in Fig. 25.17.

25.5 The slip casting time for a large refractory block is 2 wk when cast conventionally. Estimate the casting time when 1.6 MPa pressure is applied to the slurry and other factors are constant.

25.6 Why does the deflocculation of a slip depend on both the content of submicron particles and the concentration of deflocculant?

25.7 Illustrate the different microstructure in a cast when using a slurry with completely dispersed particles and a slurry that is partially coagulated.

25.8 Calculate the J parameter for an aqueous slurry (25°C) containing 300 g alumina powder and 100 mL water when the PF of the cast is 0.60.

25.9 For the slurry conditions and results in Problem 25.8, calculate and graph the cast thickness as a function of casting time for the following conditions: $R_c = 10^{17}$ m^{-2}; $\Delta P = 140$ kPa; $R'_m \ll R_c$.

25.10 Calculate and graph the cast thickness with time for the conditions in Problem 25.9, but with a slurry pressure of 700 kPa.

25.11 Pressure filtration results for casting two powders differing in particle size distribution give $R_c = 10^{18}$ m^{-2} and 10^{16} m^{-2}. For each cast the PF = 0.58. Estimate the effective pore diameter in each cast (consult Chapter 19).

25.12 Explain why the effective stress changes through the thickness of the cast during pressure casting. What is the benefit of mechanical pressing of the cast?

25.13 The following data was obtained for the uniaxial pressure filtration of a casting slurry: aqueous, 25°C; Dia = 7.8 cm; ΔP = 170 kPa; $V_{filtrate}^2/t = 5 \times 10^{-4}$ m^6/s; J = 2.53. Calculate R_c of the cast assuming an incompressible cast and that parameters are constant with time.

EXAMPLES

Example 25.1 Estimate the diameter a of the largest particle of zirconia (D_p = 6.0 Mg/m^3) that will be suspended in a concentrated slurry (D_{slurry} = 3.5 Mg/m^3) having a yield stress of 0.7 Pa.

Solution. The size is calculated using Eq. 25.1:

$$a < \tfrac{3}{2} \tau_y/[(D_p - D_{slurry})g]$$
$$a < 1.5 \, (0.7 \text{ N/m}^2)/(6000 \text{ kg/m}^3 - 3500 \text{ kg/m}^3) \, 9.8 \text{ m/s}^2$$
$$a < 43 \; \mu\text{m}$$

Example 25.2 Explain the dependence of cast density on the zeta potential in the slurry of oxide particles shown in Fig. 25.20.

Solution. When the particles have a high zeta potential, the forces of repulsion are high and they are well dispersed in the slurry. On casting, the free particles approach closely before the coagulation force causes them to agglomerate. The PF of the cast is relatively high. For a lower zeta potential, some agglomerates exist in the slurry and the particles and agglomerates coagulate in the cast before a close packing is achieved. The PF of the cast is lower.

Example 25.3 Explain why the slip casting rate depends on the PF of the cast as well as the solids content of the slurry.

Solution. The parameter J = volume of cast/volume of filtrate is a function of both parameters. From Chapter 19, $J = W/(D_S - \phi D_S - W\phi)$ where W = mass of solids/volume of liquid in the slurry. D_S = solids density, ϕ =

void fraction in the cast. For $W = 3.0$ g/mL and $D_S = 3.98$ g/cm^3, the value of J is

$$\phi = 0.4 \quad J = 3.0/[3.98 - 0.4(3.98) - 3.0(0.4)] = 2.5$$

but for

$$\phi = 0.5 \quad J = 3.0/[3.98 - 0.5(3.98) - 3.0(0.5)] = 6.1$$

When $\phi = 0.5$, less water is removed to form an equal volume of cast.

Example 25.4   Calculate and compare the slip cast thickness L after 30 min for the porcelain slurry and the alumina slurry with binder using the results for 140 kPa in Table 25.2

Solution. From Table 25.2, $L^2/t = 2.0$ mm^2/min for the porcelain and 0.3 mm^2/min for the alumina. The thickness L is calculated as $L = [(L^2/t)\ 30]^{0.5}$.

$$\text{Porcelain: } L = [(2.0 \text{ mm}^2/\text{min})(30 \text{ min})]^{0.5} = 7.7 \text{ mm}$$
$$\text{Alumina: } L = [(0.3 \text{ mm}^2/\text{min})(30 \text{ min})]^{0.5} = 3.0 \text{ mm}$$

For a thick cast, the time differential is large. For $L = 1$ cm, the casting time is 50 min for the porcelain but 5.6 h for the alumina slurry.

Example 25.5   How are anisometric particles oriented near a corner during slip casting? Why is a corner crack observed?

Solution. Dispersed plate-like particles and fibers will orient with their long dimension paralleled to the plane of the mold surface during casting, as shown in the figure. Drying shrinkage is anisotropic and greater perpendicular to the cast surface. Strain from differential drying shrinkage in the corner, where a sharp transition in particle orientation occurs, may produce enough tensile stress to produce a crack.

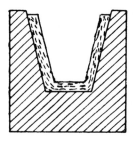

Example 25.5   Differential shrinkage occurs at corner of cast where platelike particles have a different orientation.

CHAPTER 26

TAPE CASTING

Paper-thin, flexible green sheets of various ceramic compositions are produced with high productivity using the continuous process of tape casting. First a concentrated slurry containing deflocculated powders mixed with a relatively high concentration of binder and plasticizers is prepared. The tape is formed when the slurry flows beneath a blade, forming a film on a moving carrier substrate, and is then dried. Thin sheets of ceramic may also be formed by pouring the slurry onto a flat surface and moving a blade over the surface to form the film. The dried tape is rubbery and flexible and has a very smooth surface. Green ceramic sheets are separated from the substrate and cut to size. Electronic conductors, resistors, and capacitive materials are printed as films on the sectioned tape (substrate) to produce electronic packages. Product items may be a single printed substrate or a multilayer laminated package. The substrate and electronic materials densify and bond together during firing to produce a monolithic electronic product. Ceramic tape is also used as a substrate for photovoltaic cells, electrical sensors, and solid electrolytes for batteries. Tape may be rolled into tubes and laminated to form multilayer tubular refractory products.

In Chapter 26, we will learn about tape casting technology, essential processing additives, slurry formulations, and processing controls required for modern tape casting.

26.1 TAPE CASTING PROCESS

Steps in the tape casting process are shown in Fig. 26.1. The first step is the preparation of a suitable slurry. Nonaqueous slurries are more commonly used

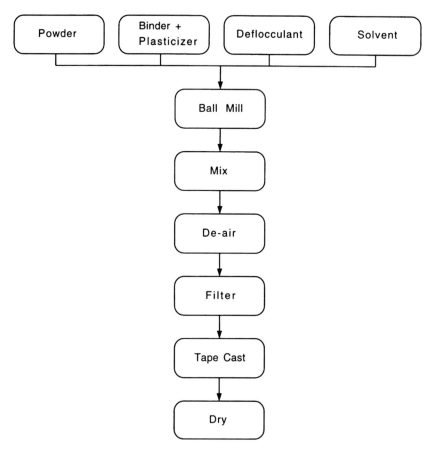

Fig. 26.1 Processing flow diagram for tape casting process.

because of the greater ease of powder dispersion and slurry drying, but aqueous tape casting is of growing interest for health and environmental reasons. Table 26.1 presents examples of batch formulations used industrially. Processing additives include the liquid/solvent, dispersant (deflocculant), binder, plasticizers, and sometimes a surfactant. Each component must be characterized and each step must be carried out under well-controlled conditions to produce tape that consistently meets specifications.

Slurry Composition

Inorganic powders used for tape casting commonly have a maximum particle size in the range of 1–5 μm and a specific surface area of about 2–5 m^2/g. The size distribution is commonly dictated by sintering considerations. Finer powders will produce a smoother surface. Dispersion of particle aggregates and powder agglomerates is accomplished by ball milling or vibratory milling with

TABLE 26.1   Compositions for Nonaqueous Tape Casting Slurries

Function	Composition	Vol%	Composition	Vol%
Powder	Alumina	27	Titanate	2.8
Solvent	Trichloroethylene	42	Methylethyl ketone	3.3
	Ethyl alcohol	16	Ethyl alcohol	16
Deflocculant	Menhaden oil	1.8	Menhaden oil	1.7
Binder	Polyvinyl butyral	4.4	Acrylic emulsion	6.7
Plasticizer	Polyethylene glycol	4.8	Polyethylene glycol	6.7
	Octyl phthalate	4.0	Butylbenzylphthalate	6.7
Wetting agent			Cyclohexanone	1.2

Source: Advances in Ceramics, Vol. 9, J. A. Mangels and G. L. Messing, Eds., American Ceramic Society, Westerville, OH, 1984.

the liquid and the dispersant. The dispersion step is essential because tape homogeneity must be very high to consistently obtain the high-quality surface required when printing narrow circuit patterns and the controlled deformation necessary to produce a flaw-free multilayer laminate. Powder dispersion is also essential to obtain uniform shrinkage without coarse grains when the package is sintered. Anisometric particles may preferentially align during casting and produce a texture which is beneficial for some applications, such as oriented ferroelectrics.

The liquid/solvent system must dissolve the additives but yet permit their adsorption on the particles, and a solvent blend is commonly used. Organic solvents have a low viscosity, low boiling point, low heat of vaporization, and high vapor pressure; these properties promote drying in a short time with less heating, which limits degradation of other additives (see Table 26.2). An azeotropic solution which does not change composition on vaporization may minimize the segregation of an additive during drying. Blends of an alcohol and another solvent such as ketone or tolulene may be used, but must be very carefully controlled because of health concerns in the workplace. Blends of alcohols are sometimes used. Relative to the organic solvents, water has a higher boiling point and heat of vaporization. A longer drying time and higher drying temperature are required when using an aqueous system.

TABLE 26.2   Boiling Point and Heat of Vaporization of Solvent Systems

Solvent	Boiling Point (°C)	Heat of Vaporization (J/g)
Water	100	2257
Ethanol	70	856
Trichloroethylene	86	240
Trichloroethylene/ethanol blend	71	—

Source: Andreas Roosen, in Ceramic Transactions, Vol. 1, G. L. Messing, E. R. Fuller, and H. Hausner, Eds., American Ceramic Society, Westerville, OH, 1988.

$$CH_3(CH_2)_7CH=CH(CH_2)_7COOCH_2$$
$$CH_3(CH_2)_7CH=CH(CH_2)_7COOCH$$
$$CH_3(CH_2)_7CH=CH(CH_2)_7COOCH_2$$

Glyceryl trioleate

$$CH_3(CH_2)_{18}COOCH_2$$
$$CH_3(CH_2)_7CH=CH(CH_2)_7COOCH$$
$$CH_3(CH_2CH=CH)_6(CH_2)_3COOCH_2$$

Fig. 26.2 Molecular structure of glyceryl tri-oleate and triglycerides in fish oil deflocculant.

Structures of triglycerides.

Powder dispersion and deflocculation in the nonpolar solvent is promoted by the steric repulsion faces of adsorbed dispersant molecules. (See Chapter 10 for a discussion of deflocculation.) Powder is charged into the dispersant–solvent solution and continuous mechanical agitation is used to disperse agglomerates and bring surfaces in contact with the dispersant. In aqueous systems, particle charging is also a dispersion mechanism. Menhaden fish oil which contains several molecular species and is a complex mixture of triglycerides (see Fig. 26.2) is widely used as a deflocculant in nonpolar solvent systems. Iodine reacts with the double bond in the molecule and an Iodine Number is used to characterize the extent of molecular unsaturation. The repulsion force depends on the surface coverage, thickness, and configuration of the adsorbed layer. The menhaden oil is significantly absorbed on the particles, as indicated in Fig. 26.3. An additive called a coupling agent, which provides a stronger covalent bonding of the dispersant on the surface, is sometimes used. Measurement of slurry viscosity combined with a sediment height test can be used to judge the relative effectiveness of the dispersant and the concentration required.

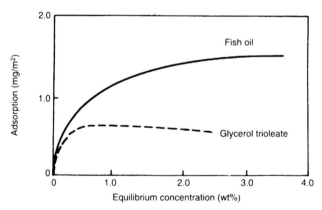

Fig. 26.3 The adsorption of menhaden fish oil and glycerol trioleate from tolulene onto an alumina powder. (From E. S. Tormey, R. L. Pober, and P. D. Calvert, *Advances in Ceramics*, Vol. 9, J. A. Mangels and G. L. Messing, Eds., American Ceramic Society, Westerville, OH, 1984, p. 143.)

Binder and plasticizer concentrations for tape casting are much higher ($100V_b/V_p = 15$ to 25) than for pressing, extrusion, and slip casting systems. More flexible molecules of the vinyl and acrylic type are popular (see Table 26.3). The binder is of high molecular weight and must be satisfactorily dispersable in the slurry. The long molecules orient during casting, providing more toughness and strength, and resist migration on drying. Binders that polymerize with the solvent have also been investigated. Separation from the carrier film and clean thermal elimination over a sufficient range of temperature to avoid a sudden evolution of gas is also a requirement for a candidate binder. The plasticizers of lower vapor pressure than the solvent are retained after drying. Plasticizers reduce the glass transition temperature of the binder and contribute to the flexibility and toughness of the tape in handling and laminating (see Figs. 15.20 and 26.4). But the plasticizer also reduces the strength of the binder and increases the V_b/V_p requirement in the tape. The combination of binder and plasticizer must be balanced to provide the required mechanical properties and still permit a high concentration of inorganic particles in the slurry. An excess of binder may increase the particle separation and reduce the PF of the particles in the tape (Fig. 26.5). Two plasticizers that decompose at different temperatures are commonly used to reduce the rate of gas evolution when the organics are eliminated.

Other additives sometimes used include a wetting agent to promote spreading of the slurry on the carrier substrate, a "homogenizer" which contributes to a better surface quality, and an antifoaming agent to prevent foaming during mixing. An antifoam is needed when using an emulsified binder system. A wetting agent and an antifoaming additive are especially required for an aqueous system.

Slurry Processing

Dispersing and mixing are commonly accomplished using a two-stage milling process. The liquid, dispersant, powder, and perhaps other additives of low molecular weight are mixed for 12–24 h at a relatively higher viscosity, at

TABLE 26.3   Binders and Plasticizers Used in Tape Casting Systems

	Binder	Plasticizer
Nonaqueous	Polyvinyl butyral	Dioctyl phthalate
	Polymethylmethacrylate	Dibutyl phthalate
	Polyvinyl alcohol	Benzyl butyl phthalate
	Polyethylene	Polyethylene glycol
Aqueous	Acrylics	Glycerine
	Methyl cellulose	Polyethylene glycol
	Polyvinyl alcohol	Dibutyl phthalate

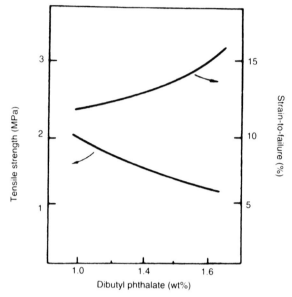

Fig. 26.4 Plasticizer reduces tensile strength of tape but increases strain to failure. (After A. Roosen, in *Ceramic Transactions*, Vol. 1, G. L. Messing, E. R. Fuller, and H. Hausner, Eds., American Ceramic Society, Westerville, OH, 1988.)

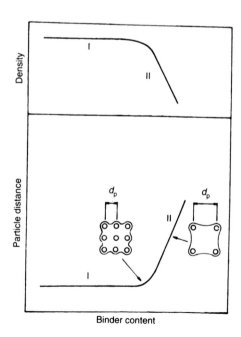

Fig. 26.5 Effect of binder content on particle separation in green tape. [Modified from R. A. Gardner and R. W. Nufer, *Solid State Technol.*, **5**, 40 (1974).]

which shear forces to disperse agglomerates are higher. When grinding to reduce the particle size is not required, smaller media and a reduced media charge may be used for mixing with less wear and product contamination. Next the binder and plasticizers are added and milling continues for another 2–24 h to complete the mixing. The two-stage milling process is used to reduce the scale of inhomogeneity in the slurry without degrading the high molecular weight binder molecules.

After milling, the slurry is screened to remove any coarse inhomogeneity. Agglomerates and undissolved organic material must be much finer than the thickness of the tape for proper forming and for ultimate thermal and mechanical properties. Sieves with an aperture as fine as 10 μm are sometimes used. Dissolved gas in the slurry is commonly removed in a closed container under vacuum. Viscosity is monitored to assure that the solvent loss is controlled.

Casting

A tape casting machine is shown in Fig. 26.6. The powder concentration in the slurry must be very reproducible and viscosity of the slurry feed must be well controlled. The casting slurry should be significantly shear thinning (see Fig. 26.7) and have a viscosity above about 4000 mPa·s at the shear rate of casting. Temperature has a significant effect on the viscosity and the temperature of the slurry should be controlled above the highest ambient value expected.

Slurry is cast on a clean, smooth, impervious, insoluble surface such as Mylar$^@$, Teflon$^@$, or cellulose acetate. The tape carrier may be coated with a surfactant to facilitate release of the tape and to extend cycles of reuse. The thickness of the tape varies directly with the height of the blade above the carrier, the speed of the carrier, and the drying shrinkage. A flow model (see Fig. 26.8) for tape casting, combining pressure flow under the blade and planar laminar flow produced by the moving surface, gives the dependence of the tape thickness H on the carrier velocity v, as[*]

$$H = AD_r h_0[1 + h_0^2 \Delta P/(6 \eta_S vL)] \tag{26.1}$$

where A is a constant that depends on the amount of side flow, D_r is the ratio of the density of the slurry/density of as-dried tape, h_0 is the cast thickness at the blade. ΔP is the pressure-motivating flow, η_S is the viscosity of the slurry on casting, and L is the length of the cast. During drying, significant shrinkage occurs in the thickness, but almost no lateral shrinkage occurs. The thickness of the tape depends on both the slurry viscosity and the velocity during casting. Equation 26.1 indicates that the tape thickness is nearly independent of variations in the speed v when h_0/η_S is relatively small. A high slurry viscosity and a velocity exceeding about 0.5 cm/s is desired for thickness uniformity, as indicated in Fig. 26.9.

[*]Y. T. Chou, Y. T. Ko, and M. F. Yan, *J. Am. Ceram. Soc.*, **70**(10), C280–C282 (1987).

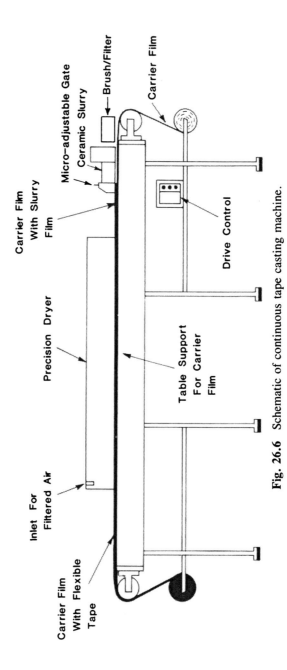

Carrier Film
With Slurry
Film

Micro–adjustable Gate

Ceramic Slurry

Brush/Filter

Carrier Film

Precision Dryer

Inlet For
Filtered Air

Carrier Film
With Flexible
Tape

Table Support
For Carrier
Film

Drive Control

Fig. 26.6 Schematic of continuous tape casting machine.

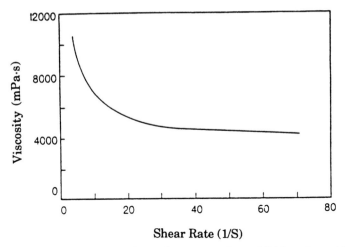

Fig. 26.7 Viscosity of alumina slurry for tape casting exhibiting shear thinning behavior and lower viscosity of about 4000 mPa · s.

Casting machines ranging up to 25 m in length, several meters in width, and a speed up to 1500 mm/min are common for industrial casting. Tapes thickness is in the range of 25–1250 μm.

Drying

A viscoelastic tape is formed as the slurry film moves through a drying tunnel and the solvent vaporizes. Adhesion to the carrier should be sufficient to prevent curling, but not eventual separation. The drying rate varies directly with the solvent concentration on the surface and the temperature and solvent content in the drying air. Initial vaporization is relatively slow because the counter-flowing air contains much solvent and is cooler. At this stage it is important that a skin not form which will trap bubbles and reduce the drying rate. The

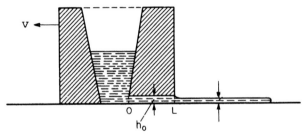

Fig. 26.8 Tape casting flow model used for deriving Eq. 26.1. [From Y. T. Chou, Y. T. Ko, and M. F. Yan, *J. Am. Ceram. Soc.*, **70**(10), C280 (1987).]

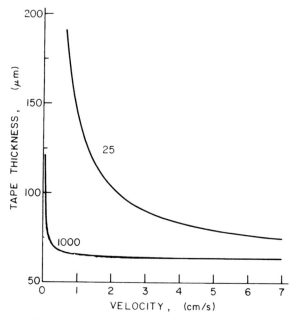

Fig. 26.9 Tape thickness is more uniform when cast at a velocity exceeding 1 cm/s using a slurry of high viscosity (units of viscosity are mPa · s).

temperature must be controlled and below the boiling point of the solvent system.

Capillary forces transport the liquid to the drying surface. Shrinkage occurs as solvent is lost and the particles move closer together. Shrinkage occurs in the thickness and the tape thickness is commonly about one-half the blade height. Segregation of particles or organic phases should not occur during drying. Some migration of the plasticizer to the drying surface occurs, but the migration of high molecular weight binder and gelled binder is very minimal when binder molecules are adsorbed and a fine powder is used. The liquid surface is interrupted and drying by vapor transport begins when DPS < 1. The viscoelastic tape becomes more elastic as drying continues. The PF of the particles in the dried tape is about 55–60%, and the remaining volume is about 35 vol% organics and 15 vol% porosity. Dried tape may be used directly or stored on a take-up roll or reel.

26.2 SHAPING AND LAMINATING

Shaping

Steps involved in producing a laminated array are shown in Fig. 26.10. The surface of the tape in contact with the carrier is smoother and metallization lines or other printed material are commonly printed on this surface. Subsequent

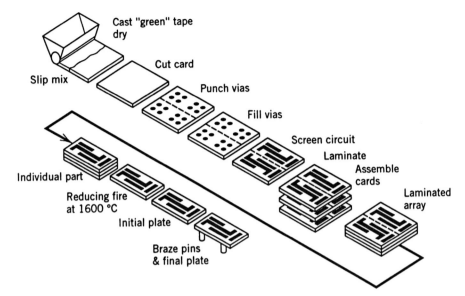

Fig. 26.10 Schematic of cofired multilayer manufacturing process for alumina electronic packaging using tape.

processing is conducted in a clean room. The leather-hard tape is cut or blanked to size and holes are punched (Fig. 26.11). Blanks are cut using a blanking press or using wheel knives. Holes larger than about 0.5 mm are often punched during blanking. Smaller holes are punched in a subsequent operation for each sheet. Small holes may be produced by laser drilling, and a series of these holes is called laser scribing. Variations on the size and separation of holes due to mechanical wear or stretching of the tape must be very carefully controlled, especially for the via holes, which are filled automatically by injecting a metal paste.

Laminating

Tape properties of importance are the tensile stress–strain behavior, compressibility, laminate bond strength, and gas permeability. The elastic modulus, strength, and toughness may be determined from a simple tensile test (see Chapter 15). The ambient temperature and atmosphere during the test should be well controlled. After drying, the relative volume of organics V_0/V_p is about 0.7 and $V_{porosity}/V_{particles}$ is about 0.25. The plasticized binder and porous microstructure permit deformation and some compaction of tape around printed material on the adjoining substrate and the joining of layers into a monolithic product. The strength varies with the relative volume of binder V_b/V_p in the tape (see Fig. 15.20). In general, for some binder content, a larger plasticizer content or a higher temperature will reduce the elastic modulus and strength,

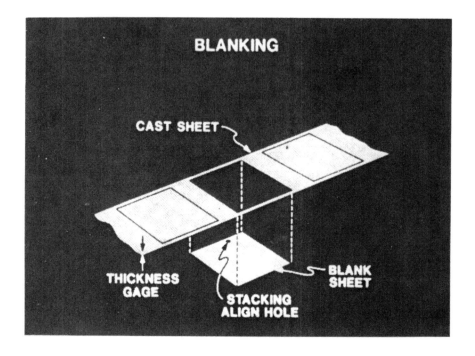

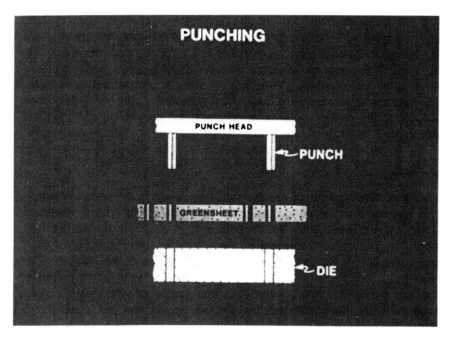

Fig. 26.11 The banking and punching of ceramic tape. [From R. A. Rinne and D. R. Barbour, *Electrocomponent Sci. Technol.*, **10**, 40 (1982).]

but increase the toughness. Compressibility, which is important for satisfactory laminating, depends on both the toughness and the porosity, which permits some densification. Finished sheets heated to 50–80°C are stacked and laminated at a pressure in the range 3–30 MPa. The bond strength between layers is important for maintaining product integrity during binder burnout and subsequent processes such as pinning and brazing. The bond strength S_b, determined in tension, is approximately by the equation*

$$S_b = K_1 + K_2 \log (P \cdot t \cdot T/T_g) \tag{26.2}$$

where T_g is the glass transition temperature of the binder system, P, t, and T are the pressure, time, and temperature during lamination, and K_1 and K_2 are constants which depend on the PF of particles in the tape, the binder type and amount, and the plasticizing of the binder system. Laminated tape is commonly 2–8% higher in density.

Binder Burnout

The gas permeability K_p of the green laminated product must not be too low because gas must be vented during binder thermolysis. Porosity in the tape increases K_p. Binder thermolysis is discussed in Chapter 29.

26.3 DEFECTS IN TAPE AND LAMINATES

Defects in sintered tape include cracks, camber, regions of low density, and surface defects consisting of unacceptable surface roughness and large pores. Cracks are caused by a differential shrinkage. A low-density region may be produced by retarded sintering in powder agglomerates in the tape. Surface roughness is reduced when a finer powder is used and the sintered grain size is smaller. The average surface roughness measure by profilometry may be less than 2 μm after sintering. Surface pores may be caused by bubbles in the slurry and by poor wetting of the carrier film by the slurry. Pores are especially undesirable because they may interrupt thin printed metallization lines. Pores are also caused by segregated organic material and pullouts where the tape adheres to the carrier. Potential defects in tape cast products are summarized in Table 26.4.

Defects in the laminated assembly include those for individual tape which includes bubbles and a poor surface. The surface adjacent to the carrier surface is smoother when differential adhesion of particles is not a problem. Additional defects are those associated with incomplete deformation of the tape around the metallization and incomplete bonding between two layers of tape (Fig. 26.12). The lamination process produces stress gradients around the printed

*R. A. Gardner and R. W. Nufer, *Solid State Technol.*, 5 38–43 (1974).

TABLE 26.4   Potential Defects in Tape Cast Products

In tape	In laminate	Postthermolysis
Bubbles	Incomplete deformation	Delamination
Agglomerates	Incomplete adjoining	Blisters
Surface craters		Print break
Cracks		
Tears		
Warping		
Print break		

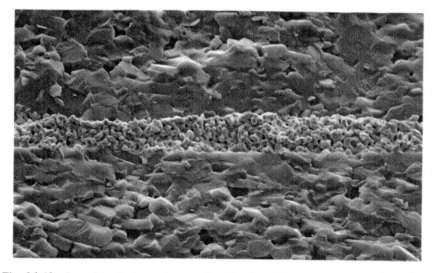

Fig. 26.12   Scanning electron micrograph of fired multilayer electronic package showing metallization between ceramic layers. (The thickness of the metal layer is 10 μm.)

material, which may cause a differential density and differential shrinkage on sintering, and cracks. Control of the geometry of the printed material, the lamination stress, and the deformation properties of the tape is essential. Poorly joined regions are local areas of weakness which may separate from the gas pressure produced during binder thermolysis, and are not healed during sintering. Joining requires uniformity of the deformation of the tape and adherence of the surfaces. This type of defect may lead to premature failure and reduces the reliability of the product.

SUMMARY

Thin ceramic sheets having a very smooth surface are mass produced by tape casting and sheet casting. The tape is produced when a controlled, viscous slurry film on a carrier surface is dried to a viscoelastic film which is then separated from the carrier. Precise control of incoming materials, the slurry rheology, and drying are essential. Slurries are formulated with a relatively high binder and plasticizer content to produce a tape with mechanical properties satisfactory for both mechanical handling, deformation, and bonding during laminating. Strict process controls are required to obtain a high yield of tape and laminated product.

SUGGESTED READING

1. Rodrigo Moreno, "The Role of Slip Additives in Tape Casting Technology: Part I—Solvents and Dispersants," *Am. Ceram. Soc. Bull.*, **71**(10), 1521–1531 (1992). "Part II—Binders and Plasticizers," *ibid*, **71**(11), 1647–1657 (1992).

2. R. E. Mistler, "Tape Casting," in *Engineered Materials Handbook*, Vol. 4, ASM International., Materials Park, OH, 1991, pp. 161–165.

3. Andreas Roosen, "Basic Requirements for Tape Casting of Ceramic Products," in *Ceramic Transactions, Vol. 1*, G. L. Messing, E. R. Fuller, and H. Hausner, Eds., American Ceramic Society, Westerville, OH, 1988, pp. 675–692.

4. A. Karas, T. Kumagai, and W. R. Cannon, "Casting Behavior and Tensile Strength of Cast $BaTiO_3$ Tape," *Ad. Ceram. Mater.*, **3**(4), 374–377 (1988).

5. Y. T. Chou, Y. T. Ko, and M. F. Yan, "Fluid Flow Model for Ceramic Tape Casting," *J. Am. Ceram. Soc.*, **70**(10), C280–C282 (1987).

6. Edmond P. Hyatt, "Making Thin, Flat Ceramics—A Review," *Am. Ceram. Soc. Bull.*, **65**(4), 637–638 (1986).

7. P. A. Calvert, E. S. Tormey, and R. L. Pober, "Fish Oil and Triglycerides as Dispersants for Alumina," *Am. Ceram. Soc. Bull.*, **65**(4), 669–672 (1986).

8. D. J. Shanefield, "Competing Adsorptions in Tape Casting," in *Advances in Ceramics*, Vol. 19, J. B. Blum and W. R. Cannon, Eds., American Ceramic Society, Westerville, OH, 1986, pp. 155–160.

9. L. Braun, J. R. Morris, Jr., and W. R. Cannon, "Viscosity of Tape Cast Slips," *Am. Ceram. Soc. Bull.*, **64**(5), 727–729 (1985).

10. R. Mistler, D. Shanefield, and R. Runk, "Tape Casting of Ceramics," in *Ceramic Processing Before Firing*, G. Onoda and L. Hench, Eds., Wiley-Interscience, New York, 1978, pp. 411–488.

11. J. C. Williams, "Doctor Blade Process," in *Treatise on Materials Science and Technology*, Vol. 9, F. Y. Y. Wang, Ed., Academic Press, New York, 1976, pp. 173–198.

12. R. A. Gardner and R. W. Nufer, "Properties of Multilayer Ceramic Green Sheets" *Solid State Technol.*, **5**, 38–43 (1974).

PROBLEMS

26.1 If the final tape is 35 μm thick, and during sintering the largest grains grow 25% in diameter, what is the maximum particle size in the slurry to be assured of > 10 grains across the thickness of the final tape?

26.2   Calculate $100V_b/V_p$ for the slurries in Table 26.1.

26.3   Explain why a shear thinning slurry is desired for tape casting.

26.4   Estimate the shear rate during tape casting when the substrate velocity is 3 cm/s and the blade height is 50 μm. Correlate your answer with the information in Fig. 16.5.

26.5   Sketch fired tape surfaces which show why the roughness varies with the sintered grain size.

26.6   Illustrate why a large surface pore will likely interrupt a printed metal line on the surface.

26.7   Explain the lamination process using a sketch of two pieces of bread with slices of hard peperroni between them for your model.

26.8   What are similarities and differences between the joining of granules in compaction, flow streams in injection molding, and tape in the lamination process?

26.9   Explain the mechanism of one-dimensional drying shrinkage.

26.10 Alumina tape after drying contains 12 wt% organic material. The mean bulk density is determined to be 2.55 Mg/m^3 and the apparent mean density is 2.95 Mg/m^3. Calculate the open porosity.

EXAMPLES

Example 26.1   Tape is cast at a velocity of 1200 mm/min and the blade height is 0.4 mm. Estimate the shear rate (s^{-1}) and compare the value to that reported in Table 21.4.

Solution. The velocity of 1200 mm/min corresponds to 20 mm/s. The shear rate is calculated as dv/dy and for a planar geometry is the velocity/separation between the moving and static surfaces:

Shear rate $= v$ (mm/s)/h_0 (mm) $=$ (20 mm/s)/(0.4 mm) $= 50$ s^{-1}

This shear rate is not high as is seen when comparing 50 s^{-1} with the range 10–2000 s^{-1} in Table 21.4. The shear rate would be higher for a thinner tape cast at this velocity.

Example 26.2   Explain how the particle packing in the tape depends on the binder content.

Solution. The density and particle separation in the tape as a function of binder content is shown in Fig. 26.5. Ceramic particles are in contact in the tape with a small amount of binder and the porosity depends largely on the particle size distribution. On increasing the concentration of binder, voids are filled, but eventually the particles becomes separated and the separation increases rapidly with binder content. Compressibility and gas permeability depend on the porosity in the tape.

Example 26.3 Why does tape handling depend on the thickness of the tape as well as its strength?

Solution. The tensile force causing elongation is the product of the yield strength (σ_Y) and the cross-sectional area. For tape that is 150 mm in width and 0.4 mm thick having a yield strength of 560 kPa, the force producing yield (F_Y) is

$$F_Y = \sigma_Y(hw) = 560 \text{ kPa } (0.0004 \text{ m})(0.15 \text{ m}) = 34 \text{ N}$$

But for tape of 0.1 mm thickness and equal strength, $F_Y = 8.5$ N, which is equal to about 2 lb of force.

The strength of the tape depends on the binder content, plasticizing of the binder, and temperature relative to T_g (see Fig. 15.20).

PART IX

POSTFORMING PROCESSES

Postforming processes include drying, film printing and coating, firing, and surface finishing and machining processes. Products formed by plastic forming and casting and many types of coatings must be dried in a controlled process prior to further processing and/or firing. Chapter 27 describes drying processes, mechanisms of drying, and drying shrinkage and causes of defects. The wide variety of processes for film printing on surfaces and for coating ceramics are described and discussed in Chapter 28. Firing transforms the relatively fragile, porous green object into what we recognize as a ceramic product. In Chapter 29 we consider firing systems, the thermal decomposition of organic and inorganic constituents on heating, and the very large topic of sintering. Chapter 30 introduces principles and describes techniques for the machining of porous or dense ceramics and for smoothing surfaces.

CHAPTER 27

DRYING

Drying is the removal of liquid from a porous material by means of its transport and evaporation into a surrounding unsaturated gas or, in some cases, a desiccating liquid. It is an important operation prior to firing in processing bulk raw materials, products shaped by plastic forming and casting, and decorations and coatings on surfaces.

The evaporation of processing liquids is relatively energy-intensive, and drying efficiency is always an important consideration. Drying must be carefully controlled, because stresses produced by differential shrinkage or gas pressure may cause defects in the product. In this chapter we consider drying practices, mechanisms of drying, and causes of defects.

27.1 DRYING SYSTEMS

Drying costs are a significant factor in the selling price of industrial minerals, and the natural drying action of the sun and wind is considered in mining and storing these materials. Wet-processed raw materials are commonly dried in large rotary dryers in which the material is tumbled, on belts in continuous tunnel dryers, by spray-drying, or supported on trays in a chamber dryer. The flow of a warm dry gas through a permeable material, called fluidized bed drying, is used when a granular material is very temperature-sensitive. Material that has been partially dewatered mechanically and is in the form of a ribbon, pellets, granules, and so on is usually dried more uniformly.

Freshly cast gypsum molds and very large wet-processed shapes are commonly dried slowly by "open-air drying." Shaped ceramic products requiring drying and working molds are usually dried in a controlled manner in fabricated metal dryers. Chamber dryers are used for drying large, free-standing shapes and smaller products supported on shelves or suspended from the structural framework on dryer cars. Continuous dryers may convey the ware up and down through a baffled chamber by means of shelves supported on continuous chains in a mangle dryer, or on rack dryer cars in a tunnel dryer (Fig. 27.1). Air circulation is maintained and controlled by means of fans. Heat sources include direct fired air heaters, steam coils, waste warm air from kilns and furnaces, and infrared or microwave radiation.

Damp material may be heated by the mechanisms of convection, conduction, and radiation (Fig. 3.10). When using convective heating, which is most common, the product temperature during drying is lower than the temperature of the hot circulating gas. Conductive heating in which heat passes to the product from a heated surface supporting the product is used in drying some thin substrates and slurries in belt and rotary drum dryers. Infrared radiation that is of a long wavelength does not penetrate deeply into wet ceramics but may be adsorbed by the liquid and transported into the interior by conduction. It is more often used for drying thin substrates, films, and coatings. Heating is produced by the coupling between the infrared radiation of 4–8 μm and the molecular vibration of O—H groups. Radiation of very long wavelength in the microwave range penetrates deeply into most ceramics; energy dissipation from the polarization of the water molecules absorbs the radiation and heats the

Fig. 27.1 Products supported on gypsum molds entering a continuous dryer. (Photo courtesy of A. J. Wahl Inc., Brocton, NY.)

liquid. Dielectric and microwave drying are used for the drying of liquid saturated products where the drying must be relatively rapid, the maximum temperature of solids must be relatively low, or the product integrity is sensitive to liquid concentration gradients and capillary stresses. Large cast and extruded products containing water have also been partially dried by inserting electrodes and passing a high dc current at low voltage through the piece.

The design and capacities of industrial dryers vary considerably and are a function of the size, shape, and drying behavior of the product; the dryer loading; setting configuration; the production rate; the temperature, humidity, and velocity of the drying air; and the mode of heating. The temperature and humidity of inlet air and the circulation of air within the dryer must be monitored to control the performance of an industrial dryer.

27.2 MECHANISMS IN DRYING

Drying involves the transport of energy into the product; liquid is transported through pores to the meniscus, where evaporation occurs, and by vapor transport through pores. In a drying system, heat energy must be brought to the surface of the product, and vapors must be carried away. The mechanisms of evaporation and mass and thermal transport must be considered before discussing the drying process.

Liquid placed in a closed container will evaporate to establish a pressure of vapor above the liquid. This pressure will increase with the temperature of the liquid and is dependent on the radius of curvature of the meniscus (Eq. 2.7). The constant vapor pressure at a particular temperature is referred to as the saturation vapor pressure. Because evaporation involves the loss of molecules having the highest kinetic energy, evaporation is a cooling process, and heat must be supplied to maintain an isothermal condition. The difference between the temperature of thermometers with dry and wet bulbs in flowing air indicates the cooling due to evaporation. Heats of vaporization for several processing liquids are listed in Table 27.1. The boiling point of a liquid is the temperature at which its vapor pressure becomes equal to the external pressure acting on the surface of the liquid, which for water is 100°C at 760 mm Hg, or 1 atm.

TABLE 27.1 Latent Heat of Vaporization of Several Processing Liquids

Temperature (°C)	Heat of Vaporization (MJ/kg)		
	H_2O	CH_3OH	C_2H_5OH
20	2.45	1.17	0.91
40	2.40	1.14	0.90
60	2.36	1.11	0.88
80	2.31	1.06	0.85
100	2.26	1.01	0.81

The critical temperature and pressure of a liquid are the conditions for which the physical properties of the liquid and vapor become identical (no meniscus); for water these are 374°C and 220 atm.

The evaporation rate of water into air below its boiling point varies directly with the temperature and surface area of the liquid and inversely with the concentration of that liquid in the air. Above the boiling point, the evaporation rate is dependent on the rate at which heat is supplied and independent of the concentration of liquid vapor in the air. The humidity is defined as the mass of liquid/mass of air, and the relative humidity as the humidity/maximum possible humidity at that temperature. The dew point is the temperature at which air of a certain humidity becomes saturated and precipitates as liquid droplets; it is lower when the content of liquid in the air is lower. During evaporation into moving air, a static boundary layer of air and vapor exist between the mobile air and the stationary product. Diffusion through this boundary layer controls the liquid transport between the air and the product. The apparent evaporation rate at a particular temperature varies indirectly with the salt content of the liquid and the forces bonding liquid on the surfaces of solids.

Thermal transport to the product may occur by convection, conduction, and radiation. Convective heating of the product is limited by the heat transfer coefficient (h_b) for the static boundary layer between the moving air and the static liquid on the surface. Mobile air of higher velocity produced by forced convection reduces the thickness of the boundary layer and increases thermal transport and the apparent evaporation rate. If this heating is the only source of energy for evaporation, the rate of evaporation R_E is

$$R_E = \frac{h_b(T_A - T_L)}{L_E} \tag{27.1}$$

where $T_A - T_L$ is the difference in temperature across the boundary layer and L_E is the latent heat of evaporation. In contrast, radiation transport to the product is relatively independent of the static boundary layer and the airflow conditions.

Within the product, thermal transport occurs by conduction and radiation. Conduction through the solid particles occurs by phonon propagation, and in the liquid between particles and within pores by molecular vibrations. The thermal conductivity of the ceramic particles is significantly higher than the conductivity of water. Infrared radiation does not commonly penetrate deeply into the product, because the waves are scattered by the particle system; microwaves of much longer wavelengths may penetrate deeply into an electrically insulating ceramic material but are adsorbed by water.

Liquid in a porous solid may be chemically or physically adsorbed on solid surfaces, exist as bulk liquid in pores of microscopic size, and be distributed as an external surface film. The migration of liquid to the surface may occur by means of capillary flow, chemical diffusion, and thermal diffusion. Capillary migration is motivated by capillary forces and, as discussed in Chapter 2, is very dependent on the radius of the pores and the surface tension and viscosity of the liquid. Liquid and vapor may diffuse along a gradient in concentration. Thermal diffusion is the migration of liquid or vapor along a thermal gradient and is important when using conduction heating and in dielectric and microwave drying.

27.3 THE DRYING PROCESS

Drying is often regarded as occurring in three stages corresponding to the ranges of liquid content for which the drying rate is increasing, constant, and decreasing, as shown in Fig. 27.2.

In a saturated material, liquid is initially removed by evaporation from the wet external surface (Fig. 27.3A). The drying rate, expressed as a weight loss per unit time, increases on heating when the relative humidity is less than 100%. The drying rate is strictly constant when the evaporation rate and evaporation surface area are constant (B in Fig. 27.2). In this stage, the product temperature is normally equal to the wet-bulb temperature of the environment. The mass of water evaporating per unit (area · time) R_E is

$$R_E = K_E(P_w - P_o) \tag{27.2}$$

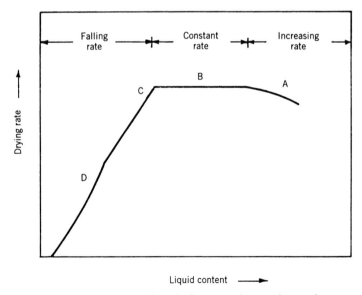

Fig. 27.2 Change in drying rate when drying a product under moderate conditions. The letters on the diagram correspond to those in Fig. 27.3 and discussion in the text. Point C corresponds to the end of shrinkage in all regions only when the liquid is uniform throughout the volume of material.

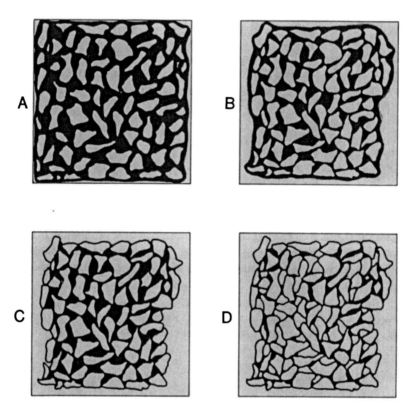

Fig. 27.3 Distribution of liquid among particles during slow drying: (A) as cast, (B) just preceding the end of the constant rate period, (C) entering the falling rate period, and (D) near the end of the falling rate period.

where K_E is the evaporation constant that is dependent on air flow conditions, P_w is the vapor pressure of the liquid at the evaporation temperature, and P_o is the partial pressure of liquid in the surrounding atmosphere. Evaporated liquid may be replenished by interparticle liquid transported to the external surface by fast diffusion and capillary flow. The loss of interparticle liquid coupled with external capillary stress may cause dimensional shrinkage (Fig. 27.3B) and an increase in the plastic shear strength. Adsorbed binders, binders of higher molecular weight, and gelled binder will resist migration with the liquid, but dissolved salts and dispersed colloidal particles or molecules may migrate to the surface. Shrinkage and chemical migration may cause a perceptible lowering of the drying rate. For the constant rate (CR) period, the practical drying rate R_{CR} is given by the equation

$$R_{CR} = \frac{W_1 - W_2}{At} \qquad (27.3)$$

where W_1 and W_2 are the liquid contents before and after drying for a time t, and A is the surface area for evaporation. Dynamic changes in the temperature of the product or absorptive capacity of the atmosphere can reduce or preclude the appearance of a constant rate period.

Evaporation from the menisci of liquid in pores (Fig. 27.3C) begins to occur when the rate of internal liquid transport is lower than the evaporation rate and when the pores in the body becomes unsaturated; the concomitant decrease in the drying rate is called the falling rate period (C in Fig. 27.2). The temperature of the surface of the product may rise rapidly to the dry-bulb temperature in those areas where the evaporation surface recedes into the pores. Thermal diffusion into the body is retarded by heat consumed in supplying the latent heat of vaporization. Transport by vapor diffusion through pores increases when the liquid menisci recede into the body. The termination of capillary migration may cause a noticeable change in slope in the falling rate period. Large pores and large accessible interstices are emptied first (Fig. 27.3D) in an isothermal body; an increasing amount of energy is needed to remove water from smaller interstices and interparticle contacts where the radius of curvature of the liquid meniscus is relatively small. Dissolved salts in residual liquid may also reduce the vapor pressure slightly.

A larger capillary stress is produced as the menisci recede. The shape of the drying curve during the falling rate period is strongly influenced by the driving force for liquid evaporation and the pore structure of the product which moderates liquid and vapor transport. Spatially nonuniform drying may occur in bodies with a nonuniform microstructure.

Adsorbed liquid remains after the bulk liquid has been removed (Fig. 27.3D). Heating above the boiling point of the liquid or long exposure to a desiccating environment can remove physically adsorbed liquid and may sometimes remove chemically combined liquids such as the water of crystallization in some inorganic salts and gypsum molds.

During drying, heating increases both the vapor pressure of the liquid and the adsorptive capacity of the drying air. The forced convection of hot air of lower relative humidity maintains the product temperature and flushes away air of higher humidity. Impinging air thins the boundary layer and increases the evaporation rate, particularly during the constant rate period. Setting patterns and the circulation of air into cavities in products must be considered to improve the uniformity of drying as well as the efficiency. Heating becomes more important than airflow during the falling rate period.

27.4 DRYING SHRINKAGE AND DEFECTS

Shrinkage occurs during drying as the liquid between the particles is removed and the interparticle separation decreases. When the shrinkage is isotropic; the volume shrinkage ($\Delta V/V_0$) is related to the linear shrinkage ($\Delta L/L_0$) by the equation

$$\frac{\Delta V}{V_0} = 1 - \left(1 - \frac{\Delta L}{L_0}\right)^3 \tag{27.4}$$

The linear shrinkage is proportional to the mean reduction in interparticle spacing Δl and the mean number of interparticle liquid films per unit length N_1 when particle sliding and rearrangement do not contribute significantly to the shrinkage:

$$\frac{\Delta L}{L_0} = \overline{N}_1 \Delta \overline{l} \tag{27.5}$$

A variation of $\overline{N}_1$ or $\Delta \overline{l}$ with direction due to particle orientation or liquid gradients may cause the linear shrinkage to be anisotropic; variations with position produce differential shrinkage.

Shrinkage during drying can be reduced by forming the product at a lower liquid content to reduce $\Delta \overline{l}$ and increasing the mean particle size to decrease N_1. Bonds that develop a chemical set may reduce or eliminate $\Delta \overline{l}$ and the shrinkage on drying.

For common extruded and cast products, the linear drying shrinkage is usually in the range of 1.5–4%. The dependence of the volume of the piece on the liquid content for homogeneous drying where the liquid content is spatially uniform is shown in Fig. 27.4; shrinkage ceases at a particular liquid content, which is called the leatherhard liquid content.

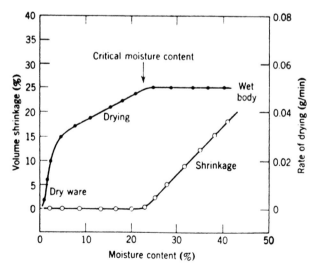

Fig. 27.4 Change in bulk volume when uniformly drying a ceramic body. (From W. D. Kingery, Introduction to Ceramics 1st Edition, Wiley-Interscience, 1960.)

During drying, a liquid concentration gradient at the beginning of the falling rate period may cause differential shrinkage and a tensile stress in the surface. The transition from a plastic body to an elastic body occurs with a loss of liquid content of 5–15 vol % (Fig. 27.5). Unfired ceramics are normally quite weak and can sustain stresses of only a few 100 kPa without deformation or local fracture. Surface cracks may form when material near the surface becomes brittle and the differential shrinkage produces a stress that exceeds the tensile strength. Regions of low strength such as laminations are especially susceptible. Stresses produced during drying may be increased by the pressure of vapor in pores. Small cracks called checks are produced by differential shrinkage in a small region. Parameters that increase the dry strength, discussed in Chapter 15, may increase the stress required to form cracks. Case hardening occurs when shrinkage of surface material has ended, but internal material with a higher liquid content continues to shrink; internal tensile stress may produce an internal crack. Differential shrinkage may be produced by gradients in particle size and at junctures between differentially oriented particles of high aspect ratio, as discussed in Chapter 25.

Differential shrinkage due to a differential liquid content when the product was formed or a differential drying rate across the surface of the product may also produce warping (Fig. 27.6). Warping is caused by stresses accompanying nonsymmetrical shrinkage, which produces plastic elongation in regions with a lower shrinkage rate. Warping is reduced by increasing the uniformity of drying and reducing the average drying shrinkage of the body. The tendency for warping may be increased by nonuniform external films or coatings, particle

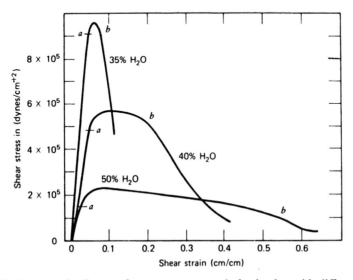

Fig. 27.5 Stress-strain diagram for an uncompressed plastic clay with different liquid contents. (Liquid concentration in wt%; after F. H. Norton, Elements of Ceramics, Addison-Wesley Press, Cambridge, MA, 1952.)

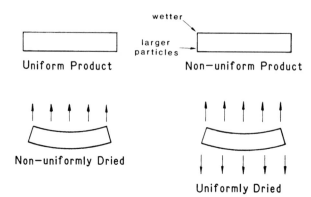

Fig. 27.6 Shape distortion is produced by nonuniform drying shrinkage.

orientation, or binder migration, which produce a nonuniform surface permeability, and by nonuniformities in the circulation or temperature of drying air due to setting patterns and supporting hardware. Ideally, drying should occur symmetrically in an isotropic material, and all shrinkage should end before entering the falling rate period. An insufficient rate of liquid transport to the surface relative to the evaporation rate will reduce the constant rate period and increase the transition range during which the deleterious differential shrinkage can occur.

The mechanical restraint of shrinkage may also produce stress and cracks. Contact friction between a rigid support and the shrinking product may cause a crack, especially when the product is heavy and the surface is rough. Restrained shrinkage is also produced between sections drying at different rates, as in a nonuniform cross section, and when material adheres to a porous mold. When drying shapes with a deep cavity, the forced convection of drying air into the cavity may be required to prevent the formation of a crack due to differential drying rates or the restraint of material adhering to the mold.

Other sources of defects are the migration of colloids to the surface producing a skin having different properties and internal gas pressure produced by too rapid volatilization of liquids and insufficient permeability of the pores.

27.5 MODES OF DRYING

Conduction and convection drying are commonly used for drying ceramic products, as discussed in Section 27.1. Floors or shelves supporting the product may be heated by waste heat or steam. Convection drying is used to heat the product and remove vapors, and the air circulation may be designed to facilitate the drying of a particular size, setting, or shape. Infrared drying can be used to reduce the drying time and for the drying of coatings such as decorations and glazes. Vacuum-assisted drying reduces the partial pressure of vapors. Other modes of drying deserving special comment are discussed below.

Controlled Humidity Drying

For every product with a finite drying shrinkage, there is some critical drying rate that will cause defects in the product. When drying thick products or a product of very low K_P, drying must be conducted in such a way that the liquid concentration gradients are not large. The isothermal capillary flow rate to the drying surface of a slab is proportional to K_P/η_L. The safe drying rate may be increased by reformulating the body to increase K_P or by using controlled humidity drying. Heating the product in humid air arrests the surface evaporation and reduces the viscosity of the liquid prior to drying. On reducing the humidity of the drying air, drying can occur at a greater rate without an increase in the liquid concentration gradient.

Microwave Drying

Microwaves are generally reflected by electrical conductors, transmitted by electrical insulators, and adsorbed by dielectrics. Liquid water behaves like a dielectric, because the molecule is polar and the direction of polarization cycles when subjected to a microwave field. Microwave adsorption causes heating in proportion to the field strength and the product of the frequency and dielectric loss factor. At a particular field strength, penetration of the dielectric varies inversely with power adsorption. Microwave energy may be used to heat and evaporate liquid in large cross sections relatively rapidly, independently of the thermal conductivity of the solid. When drying ceramic insulators, the microwaves are preferentially adsorbed by the water, and the product temperature during drying may never exceed 50°C; that is, the high surface temperatures in conventional drying are avoided. The apparent penetration of microwaves increases as water is vaporized and diffuses as a gas to the surface. Potential uses for microwave drying are in the processing of temperature-sensitive products, more rapid drying, drying products of large cross sections and large gypsum molds, and drying products containing colloidal materials such as gels, pigments, and clay of extremely low liquid permeability.

Slurry Drying

Slurried raw materials supported on a metal belt and tape-cast films are dried continuously and relatively rapidly. In drying a mineral slurry on a metal belt, heat is supplied by conduction through the belt and by the hot drying air, which may exceed 400°C. Turbulence produced by boiling may retard the formation of a low-permeability surface layer, as is shown in Fig. 27.7; the drying rate is dependent on the solid's concentration of the slurry and the viscosity and permeability of the partially dried material.

In tape casting, the thin slurry film cannot be dried too rapidly, because the drying is one-directional and the liquid permeability of the tape is very low. The concentration of liquid in the air at the beginning of drying is relatively high, as in controlled humidity drying. The maximum liquid temperature must

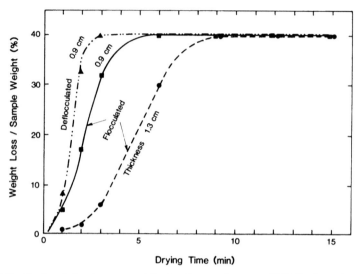

Fig. 27.7 Drying behavior on rapidly heating a slurry to 500°C; flocculated slurry develops a skin of low permeability.

be well below the boiling point of the liquid during the constant rate period and is commonly less than 50°C for nonaqueous systems. Filtered air enters the exit end and is solvent-loaded on exiting above the entering tape. The airflow and temperature are carefully controlled to maximize the constant rate period and minimize liquid concentration gradients until shrinkage has ceased (Fig. 27.8). Non aqueous solvent vapors are collected and reclaimed.

Supercritical Drying and Freeze Drying

Supercritical drying may be used to minimize effects of the surface tension of the liquid during drying. In supercritical drying, the product is heated in an autoclave until the liquid becomes a supercritical fluid. Supercritical drying occurs during isothermal depressurization. Freeze drying (discussed in Chapter 4) is used for drying products where a liquid formation and product heating must be avoided.

Spray Drying

Spray drying is relatively efficient compared to the convection drying of formed products, because the material is well dispersed in the drying medium, the diffusion path is shorter, and the high specific surface area contributes to a higher rate of evaporation per unit mass of product. Spray drying was discussed in Chapter 20.

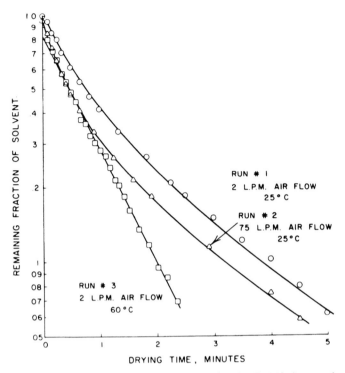

Fig. 27.8 Behavior on drying tape cast slurry. (After R. E. Mistler, et al., p. 432 in Ceramic Processing Before Firing, G. Onoda and L. Hench, Eds., Wiley-Interscience, New York, 1978.)

SUMMARY

Drying is an important process in producing ceramic raw materials and shaped products ready for firing. During drying, heat is transported to the liquid in the body, and evaporated liquid is transported into the surrounding atmosphere. The drying rate depends on the temperature of the liquid in the body and the temperature, humidity, and flow rate of the drying air. Radiation may be used to augment conduction and convective heating or as the primary heating source. After initial heating, and product dries at a constant rate during which shrinkage commonly occurs. Transition to a decreasing drying rate occurs when the external surface of the product is incompletely covered with liquid. When the drying rate is very fast, or nonuniform, the constant rate period is relatively short, and the differential shrinkage can cause cracks. Warping is produced by nonuniform drying when the body is shrinking and can deform plastically. Dried products are commonly hygroscopic and may readsorb moisture in proportion to the relative humidity of the atmosphere.

SUGGESTED READING

1. R. C. Chiu and M. J. Cima, "Drying of Granular Ceramic Films: II, Drying Stress and Saturation Uniformity," *J. Am. Ceram. Soc.*, **76**(11), 2769–2777 (1993).
2. George W. Scherer, "Theory of Drying," *J. Am. Ceram. Soc.*, **73**(1), 3–14 (1990).
3. R. B. Keey, *Introduction to Industrial Drying Operations*, Pergamon Press, Elmsford, NY, 1978.
4. F. H. Norton, *Fine Ceramics*, Robert E. Krieger, Malabar, FL, 1978.
5. A. R. Cooper, "Quantitative Theory of Cracking and Warping During the Drying of Clay Bodies," in *Ceramic Processing Before Firing*, George Y. Onoda and Larry L. Hench, Eds., Wiley-Interscience, New York, 1978.
6. W. E. Brownell, *Structural Clay Products*, Springer-Verlag, New York, 1976.
7. J. E. Funk, "Simultaneous Weight Loss and Shrinkage of Clays," *Am. Ceram. Soc. Bull.*, **53**, 450–452 (1974).
8. Rex W. Grimshaw, *The Chemistry and Physics of Clays*, Wiley-Interscience, New York, 1971.
9. W. O. Williamson, "Dimensional Changes and Microstructures of Unfired Clay Products," *Interceram.*, **31**, 199–204 (1968).
10. R. W. Ford, *Institute of Ceramics Textbook Series*, Vol. 3, Maclaren and Sons, London, 1964.

PROBLEMS

27.1 Sketch the specimen weight versus liquid content for drying corresponding to Fig. 27.2.

27.2 Estimate the mean shrinkage between particles $\Delta \bar{l}$ when the mean particle diameter is 1 μm and the measured drying shrinkage is 4%. Assume simple cubic packing of particles.

27.3 Contrast the drying shrinkage for particles of 1.0 and 0.01 μm in simple cubic packing when the interparticle binder + liquid contracts from 5 to 1 nm on drying.

27.4 A film of 4 water molecules in thickness is about 1 nm. What is your answer in Problem 27.3 in terms of molecular layers of water?

27.5 The diametral shrinkage is measured as 3.6%. Calculate the volume shrinkage assuming the shrinkage is isotropic.

27.6 Derive a general relationship for the dependence of volume shrinkage on linear shrinkage for the case of anisotropic shrinkage.

27.7 The diametral shrinkage is 1.6% and the shrinkage in thickness is 2.4%. What is the volume shrinkage?

27.8 Contrast the energy required to heat water from 20 to 100°C to the energy required to evaporate it at 100°C.

27.9 Explain why a crack may form at the interface of piece/foot when drying a product resting on a support.

27.10 At what point in drying does the sheen disappear from the surface?

27.11 When initially heating an unsaturated product in the dryer, liquid may be driven into the center of the product. What drying conditions will minimize this?

27.12 An extruded product has an interstitial porosity of 36% which is 95% saturated when the product enters the dryer. What is the relative volume of liquid removed when shrinkage stops. The linear shrinkage is 1.4% and isotropic.

27.13 A cylindrical capillary (pore) extends from the surface to the interior and is nearly filled with liquid. In which direction would liquid flow when (1) the product is initially heated rapidly in a humid atmosphere, and (2) the product has been uniformly heated at 100% RH and the humidity is rapidly reduced.

27.14 Does the stress causing cracks on drying depend on the pore size and the wetting angle?

27.15 Compare the tensile stress produced when the drying shrinkage is 4.0% and 1.5%. Assume the Young's modulus = 1.4 GPa for the material.

27.16 Does the drying rate during the falling rate period depend on the thickness of the product? Explain.

27.17 A flatware shape cast in a gypsum mold is dried using forced convection while remaining in the mold. When the center of the exposed surface dries more rapidly than the rim, how will the shape be distorted when completely dry?

EXAMPLES

Example 27.1 What are the different states of water in an as-formed, nearly saturated product?

Solution. The water may be considered to exist as (1) bulk water, (2) physically adsorbed water, (3) chemically adsorbed water, and (4) lattice water included in the composition of a mineral phase, such as in gypsum and kaolinite. Bulk water is present as a liquid and as a vapor in interstices and between particles. Water loss producing shrinkage on drying occurs in the bulk water between particles and the adsorbed water on surfaces. The loss of lattice water by thermal decomposition or desiccation may cause the densification of phases and shrinkage. On heating from room temperature, water will be lost in the sequence 1–4. The majority of drying shrinkage occurs on removal of the bulk water between particles early in the drying process.

Example 27.2   At what point in drying is heating rate control most crucial.

Solution. Control of heating is requisite to minimize differential shrinkage in the piece during drying and to minimize internal gas pressure from liquid vapor. Differential shrinkage is reduced by maintaining a small gradient in liquid content near the critical point C. Just below the critical point, stress is caused by the capillary force of the receding liquid menisci and by discontinuous liquid of high surface tension in pores. The tensile stress in the surface is in proportion to the differential shrinkage between the surface and the interior. The product is becoming leather hard (or even brittle when the surface is very dry) and a tensile stress in the surface can cause plastic deformation and/or a crack. As seen in the figure, fast heating will cause the surface to reach its critical point C_2 ahead of the interior C_1, which will increase the surface stress. Liquid transport to the surface is limited by the permeability K_P. After all shrinkage has ceased, heating may be increased when K_P is sufficient to vent pore vapors.

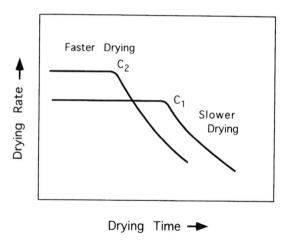

Example 27.2   The constant rate period is shorter for faster drying.

Example 27.3   The drying shrinkage of a cast product having a mean particle size of 3 μm is $\Delta L/L_0 = 0.9\%$. Estimate the shrinkage if the particle size is reduced to 0.7 μm and the packing density after drying is the same.

Solution. For a chain of particles of diameter a separated by a distance $\Delta \bar{l}$, the linear shrinkage is $\Delta L/L_0 = \Delta \bar{l}/(a + \Delta \bar{l})$. For $a = 3$ μm,

$$\Delta L/L_0 = 0.9\%/100 = \Delta \bar{l}/(3 \ \mu m + \Delta \bar{l}) \quad \text{and} \quad \Delta \bar{l} = 0.027 \ \mu m$$

For $a = 0.7$ μm,

$$\Delta L/L_0 = 0.027 \ \mu m/(0.7 \ \mu m + 0.027 \ \mu m) = 3.7\%$$

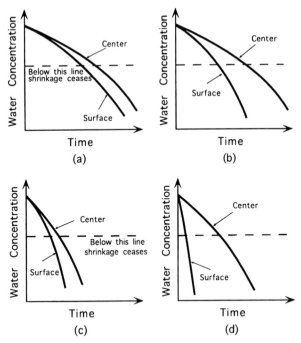

Example 27.4   Water content at surface and in center for different drying conditions.

The shrinkage is principally a function of the number of liquid films per length, which is controlled by the particle size. This estimate assumes that the shrinkage Δl between particle pairs is independent of the particle size.

Example 27.4   What are the effects of heating rate and moisture content of the drying air on the water concentration at the surface and in the center of a product.

Solution. The change of water content for surface material and material near the center, for drying into the falling rate stage, is shown in the diagram. This figure also indicates the differential expected: graph (a) indicates the change with time for slow heating in a humid atmosphere. Drying occurs relatively slowly, but the differential is small. Slow heating in a less humid atmosphere, graph (b), will increase surface evaporation and increase the differential. Fast heating in a humid atmosphere produces moderate to slow drying, depending on the relative humidity, and the differential is not large because evaporation is only moderate and the liquid mobility in the heated product is high. This is indicated in graph (c). But fast heating in a relatively dry atmosphere, indicated by graph (d), will rapidly evaporate moisture from the surface and produce a large differential. It is important to keep the differential liquid content small until shrinkage ceases. In controlled humidity drying, the conditions of case (c) are initially produced until shrinkage ends, and then the conditions are changed to those more like case (d).

CHAPTER 28

VAPOR DEPOSITION, PRINTING, AND COATING PROCESSES

Films and coatings having special electrical, magnetic, and optical properties and coatings providing special mechanical or chemical resistance are often applied on one or more surfaces of a ceramic product. Physical and chemical vapor deposition are used to apply thin films and film patterns. Screen printing processes are used widely for printing thick film conductors, resistors, and dielectrics on electronic substrates and decorative films on tile, institutional ware, and a variety of household ceramic products. Pad transfer printing and transfer decals are used for printing on flat or contoured surfaces. Other coating processes include sol-gel, thermal and plasma-sprayed, and electrophoretic coatings. Vitreous glaze coatings are widely applied on ceramic products to provide surface protection and to add beauty to a product. In this chapter we discuss the important topics of film application and coating processes.

28.1 THIN FILM PRINTING AND COATING BY VAPOR DEPOSITION

Vapor deposition is the condensation of an element or a compound from the vapor state, forming a thin film on a substrate. In physical vapor deposition, the film has the same composition as the vapor. Examples of physical vapor deposition are the deposition of a metal onto a masked substrate by vacuum deposition or by sputtering in a rarefied gas, as shown in Fig. 28.1. Applications include thin metal or alloy circuits on substrates and resistive films for electronic devices (microelectronics). The metallization must adhere well to the ceramic without degrading electrical properties and provide solderability. Vacuum deposition is the most popular thin film deposition technique. Vapors

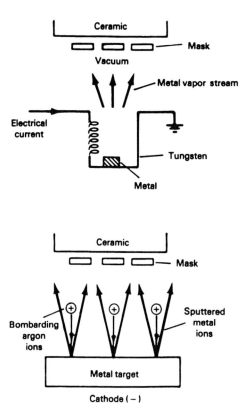

Fig. 28.1 Schematic showing (top) vacuum evaporation and (bottom) sputtering. (From J. Steinberg, in *Engineered Materials Handbook*, Vol. 4, ASM International, Materials Park, OH, 1991, p. 542.)

from the heated source material condense on the cooler substrate. Sputtering is used when elements in metal alloys may exhibit differential vaporization. In sputtering, particles from the surface of a target material are stripped off by the bombardment of positive ions. These ions are commonly an inert gas such as argon. Adhesion of the sputtered film is commonly very good. Sputtering is amenable to a wide variety of source materials.

In chemical vapor deposition CVD, a compound of the source material is vaporized and thermally decomposed or reacted with other gases or vapors to produce a nonvolatile reaction product which deposits onto the substrate as a thin coating (Fig. 28.2). Chemical vapor deposition is a molecular process in that the deposit is built up molecule by molecule. It has the capability of forming films of high density and uniform thickness on surfaces of complex shape. Chemical vapor deposition is used to form coatings of borides, carbides, nitrides, oxides, silicides and now polycrystalline diamond on the surfaces of metals and ceramics.

Components of the CVD reactor are the reactant supply, the CVD furnace, and an effluent gas handling system. Processes involved in CVD are summa-

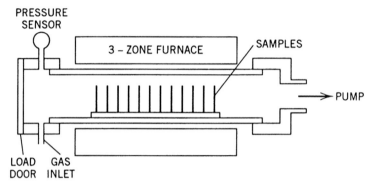

Fig. 28.2   Schematic of a reduced-pressure CVD reactor.

rized in Table 28.1. A common CVD process for the deposition of oxides is the hydrogen reduction of a metal chloride by water formed at a high temperature. An example is the formation of titanium diboride at about 900°C:

$$TiCl_{4(gas)} + 2BCl_{3(gas)} + 5H_{2(gas)} \rightarrow TiB_{2(solid)} + 10HCl_{(gas)} \quad (28.1)$$

Silicon carbide may be deposited below 1000°C from an organosilane compound such as CH_3SiCl_3 in the presence of hydrogen; the overall reaction is

$$CH_3SiCl_{3(gas)} \xrightarrow{H_2} SiC_{(solid)} + 3HCl_{(gas)} \quad (28.2)$$

but the decomposition of the organosilane in the hydrogen atmosphere precedes the formation of the silicon carbide.

Metal films may be deposited using CVD technology and precursors such as metal carbonyls and organometallics which are thermally decomposed. The decomposition of nickel carbonyl is expressed as

$$Ni(CO)_{4(gas)} \rightarrow Ni_{(solid)} + 4CO_{(gas)} \quad (28.3)$$

TABLE 28.1   Processes Involved in CVD

Gas phase reactions
Transport of reactants to surface
Adsorption of reactants onto surface
Surface reactions
Desorption of products from surface
Transport of products from surface

The purity of the deposited phase and the decomposition rate are functions of the initial compositions and the pressure and temperature of the reaction, as shown in Fig. 28.3 for the deposition of silicon from silane. Deposition rates range up to a few microns per minute. Chemical vapor deposition is used for forming very dense printed film patterns and coatings. Applications include SiO_2 and Si_3N_4 insulators and W, Si, Al, Cu, and TiN conductors for microelectronics, SiC and Si_3N_4 coatings for mechanical resistance, and optical coatings. Chemical vapor deposition is also used for coating fibers for composites. Coating requirements include adhesion, conformality, density, thickness, and smoothness.

A very significant scientific development in CVD processing has been the low-pressure synthesis of diamond films, which has stemmed from research by Soviet scientists in the late 1970s. A key to diamond synthesis is the careful control of the gas pressure and high concentration of hydrogen in the gas phase. Diamond is formed from the production of a supersaturation of atomic hydrogen with a supersaturation of carbonic species and a substrate temperature in the range of 975–1275 K. Dense polycrystalline films with a grain size of about 1 μm are formed (Fig. 28.4). The CVD diamond thin film technology is now used to produce protective coatings on products having electronic, optical, and mechanical functions.

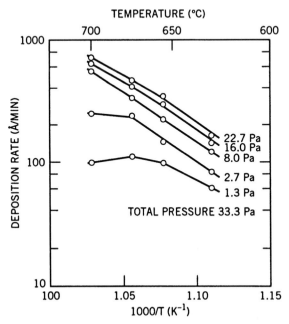

Fig. 28.3 Deposition rate of silicon increases with temperature and partial pressure of silane.

Fig. 28.4 Scanning electron micrograph of CVD polycrystalline diamond coating. (Photo courtesy of Robert Davis, North Carolina State University, Raleigh, NC.)

28.2 THICK FILM PRINTING

Film patterns of metals and a wide variety of ceramic materials are applied using techniques known as "thick film processes." The processes of screen printing, pad transfer printing, and decalcomania are discussed here. Additional coating processes are discussed in Section 28.3.

Screen Printing

In screen printing, an open pattern in a stencil screen defines the printed pattern. A thick paste is forced through a stencil screen onto the surface below, using a squeegee or a traversing paste reservoir and nozzle assembly. Process variables include mechanical indexing, screen variables, squeegee variables, the composition and rheology of the paste, and the surface roughness of the substrate.

Equipment and paste compositions for screen printing are produced commercially. Characteristics of common screen printing are print thickness in the range of 2–25 μm, print patterns ranging up to about 15 × 15 cm, line widths as small as 0.25 mm, squeegee speeds up to about 25 cm/s, and a cycle time of about 2 s. Special techniques are required for filling vias and achieving line widths finer than 0.25 mm. Paste compositions include the primary metal or ceramic powder, as indicated in Table 28.2, a powder sintering aid which typically vitrifies, such as a lead borosilicate fit, an organic liquid of low volatility, a high-molecular-weight binder, and other additives such as a lubricant. Specific compositions are highly proprietary. Powders are typically finer than 10 μm; mixtures must be well milled and mixed to produce a homogeneous dispersion. Soluble impurities must be carefully monitored and controlled.

TABLE 28.2   General Compositions of Powders for Screen Printing

Conductor	Resistor	Dielectric	Colorant
Au, Au/Pt	Pd/Ag	$BaTiO_3$	Transition metal oxides
Ag, Ag/Pd	RuO_2	Glass	ZrO_2: doped with V, Pr
Ni, Cu, Mo	$Bi_2Ru_2O_7$	Glass–ceramic	$ZrSiO_4$: doped

To print a specific image, printing paste must be allowed to pass only through specific areas of the screen. A popular means for producing the stencil pattern is to utilize a photosensitive emulsion. The liquid emulsion is applied in sequential layers to the stretched screen until a smooth, uniform coat has been achieved. After drying, the emulsion is exposed to the photographic positive of the printing image. Areas of the emulsion exposed to light become hardened, owing to cross-linking of the polymer, but unexposed areas remain soft and are washed out of the screen. The exposure intensity and time must be carefully controlled to produce an accurate stencil.

Screen frames must be dimensionally stable. The fabric of the screen must be of a regular weave, elastic, and abrasion-resistant to register a shape image with dimension reproducibility. Popular fabrics are monofilament nylon and polyester, stainless steel, and metalized polyester filament. For precise registration, polyester is used because it is less elastic and adsorbs very little moisture. Stainless steel is used for printing very abrasive or thermoplastic pastes on flat surfaces. Metalized monofilament polyester is used to improve the abrasion resistance.

The mesh and thread diameter of the screen define the width, height, the spacing of the rectangles of paste contacting the substrate. As these rectangles flow and merge into a continuous pattern, the thickness becomes one-fourth to one-third of the filament diameter (Fig. 28.5). The choice of the filament thickness and mesh size depends on the detail of the image, rheology of the paste, shape and roughness of the surface, and print thickness. The mesh opening must be several times larger than the particle size of the paste composition, but small enough to prevent bleeding of the printed images. Printed paste that is 60–80 μm thick on contact is 20–30 μm thick after merging and drying.

The action of the squeegee is shown in Fig. 28.6. For a uniform print, the squeegee pressure must be uniform. This is achieved using a squeegee that

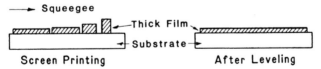

Fig. 28.5   Leveling and merging of screen-printed paste.

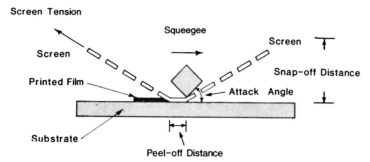

Fig. 28.6 Parameters in the screen-printing process using a squeegee.

does not excessively overlap the stencil opening and an adequate clearance between the end of the squeegee and the screen frame. A soft squeegee will conform more readily to an irregular surface or a previously printed surface and produce a more uniform print. The print thickness is more uniform when the surface is flat and the stroke of the squeegee is parallel.

Important properties of the paste for printing are its coefficient of viscosity, yield strength, and rate of drying. The shear rate during flow through the screen is in the range of $100-1000$ s^{-1}. The coefficient of viscosity of the paste for this shear rate is usually in the range of $70,000-150,000$ mPa·s and ideally independent of temperature between 20 and 30°C. The paste must be shear thinning and thixotropic so that under pressure of the squeegee it will flow readily through the apertures of the screen and the print blocks will merge into a monolithic film. Ethyl cellulose or an acrylic resin of very high molecular weight dispersed in terpineal or butyl carbitol is commonly used as a binder-liquid system. The viscosity must increase several hundred percent after leveling to prevent bleeding. For a particular screen, the viscosity of the binder and the solids loading can be coordinated to provide some flexibility in the fired thickness. The solids loading and binder viscosity can be adjusted to produce a particular paste viscosity. However, the use of a lower-viscosity binder and a higher powder loading will produce a thicker film after firing. Multiple printings are used to produce the desired combination of electrical components.

Defects include voids and pinholes in prints, improper print thickness, variation in print thickness, and poor line resolution. Voids and pinholes are caused by entrapped air, clogged screen apertures, and poor peel. Clogging is minimized by good mixing, control of the vapor pressure of the liquid solvent and airflow effecting drying, and the use of a larger mesh opening. Peeling behavior is dependent on the shear resistance between the paste and filament, screen tension, and the snap-off distance which controls the force normal to the print surface. The screen tension must be high enough for peeling, but not too high to cause nonuniform squeegee pressure. The snap-off distance should be about 5% of the screen size. The peeling speed should match the squeegee speed. Control of the screen mesh and filament diameter, emulsion thickness, and squeegee parameters is required for control of the print thickness. The uniformity of the print thickness is higher when the stroke of the squeegee is parallel to the surface and the hardness and pressure on the squeegee are carefully controlled. Carefully controlled peel and the use of a screen mesh having a finer filament diameter and concomitantly larger openings are required when printing fine lines. A surface with some roughness aids in paste transfer and adhesion, but the coating thickness becomes unacceptable above a particular roughness (Fig. 28.7).

Pad Transfer Printing

Pad transfer printings is used for decorating surfaces that may vary widely in shape and surface texture and has the ability to print multicolors without intermediate drying. In this process, thin paste called ink, in the recesses of an engraved plate, is transferred to a surface by means of a rubber pad that first contacts the plate and then the surface (see Fig. 28.8). Photoengraved plates are commonly made by coating a lapped tool steel plate with a thin layer of photosensitive etching resist, exposing the plate using a film positive, and then etching it to a depth of 30–60 μm. Most pads are of silicone rubber. The inks used in pad printing are less viscous and more slowly drying than paste used in screen printing.

Fig. 28.7 Scanning electron micrograph of the surface of ceramic tape before printing.

Fig. 28.8   Silicone pad printing machine. (Photo courtesy of Malkin Limited, Stoke-on-Trent, England.)

Pad transfer printing depends on the ink having some affinity for the rubber which is necessary for peel up, but a greater affinity for the surface, which is necessary for transfer. The plate is first flooded with ink using a squeegee. The descending pad contacts the plate for a specific time and then rises. During the transition of the pad to the printing surface, the plate is again flooded with ink. The pad descends and contacts the indexed part for a specific time, during which most or all of the ink is transferred. The pad then ascends, and the process is repeated. Cycle times may be as short as about 2 s. Image sizes may range up to about 15 × 15 cm.

Because the pad is flexible, the surface of the part need not be flat, and printing on surfaces with moderate relief and imprecise shapes may sometimes be done routinely. Multiple printing without drying is accomplished when the ink has a much greater affinity for the dry surface than for the pad, and the pad does not deform the previously printed image.

Decalcomania Printing

An indirect process for printing a design is the decal process. Decalcomanias or printed transfers, commonly called decals, are designs printed on specially prepared paper from which the design may be transferred (Fig. 28.9). Ceramic decals are printed using a prepared formulation similar to that used for screen printing. A simplex decal consists of the design printed on a coating of binder adhesive supported on a porous paper. After the simplex decal has been saturated with warm water or suitable solvent, the design is slid off onto the surface and smoothed with a sponge or cloth. Most ceramic decals are of the duplex type and consist of a tissue-like paper supported on a heavy paper backing, coated with the binder adhesive. The design is printed on the adhesive and then

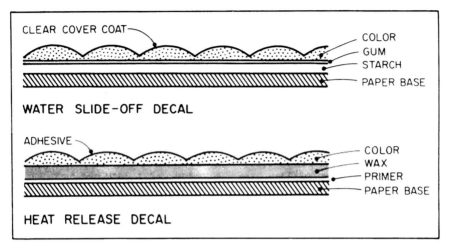

Fig. 28.9 Cross-sectional views of (top) solvent-release and (bottom) heat-release decals. [From J. J. Svec, *Ceram. Ind. Mag.*, **12**, 49, (1980).]

dried. Using a solvent-mount decal, the tissue and design are stripped from the paper backing and soaked for a specific time in a suitable solvent. The softened decal is squeezed onto the surface, and the tissue is removed using a damp sponge. In the varnish mount process, the surface of the product is coated with a varnish or similar adhesive or sizing. The decal on tissue paper is stripped from the backing and applied printed side down on the varnished surface, using a damp sponge, roller, or brush, and the tissue is removed. The varnish mount process has been adapted for application on surfaces of unfired whiteware products.

Heat-release decals consists of a paper base, a special primer coat, a layer of wax on which the ceramic color is printed, and a heat-sensitive top coat. The application of the decals can be highly automated. Heat from the product warms and activates the adhesive and softens the wax, enabling transfer of the design and removal of the paper backing.

28.3 COATING PROCESSES

Several different processes may be used to apply polycrystalline metal or ceramic coatings and vitreous glazes on ceramics. Sol-gel coatings, glaze coatings, electrophoretic coatings, and thermal and plasma spray coatings are discussed here.

Sol-gel Coating Process

Sol-gel reactions were introduced in Chapter 4 as a means for producing novel powders. But sol-gel technology may also be used for coating ceramics. Precursors, which have received the most interest, are the alkoxides ($-OC_nH_{2n+1}$) commonly represented as $-OR$. When $n = 2, 3$, and 4, the specific compounds are ethoxide, isopropoxide, and butoxide, respectively. Partial hydrolysis of an alkoxide such as $Ti(OR)_4$ on mixing in water (solvent) occurs by the reaction

$$Ti(OR)_4 + 4H_2O \rightarrow (RO)_3Ti-OH + ROH \qquad (28.4)$$

Condensation produces a polymerized oxide aquagel of much higher viscosity, having a yield point

$$(RO)_3Ti-OH + HO-Ti(OR)_3 \rightarrow (RO)_3-Ti-O-Ti(OR)_3 + H_2O$$

$$(28.5)$$

The hydrolysis of alkoxides has been used to prepare gel coatings of several metals oxides including aluminum and zirconium.

When a multicomponent alkoxide solution or single alkoxide and a soluble inorganic salt are mixed in solution, a multicomponent alkoxide or complex alkoxide may form. The reactions of a partially hydrolyzed alkoxide $ROM-OH$ with an alkoxide specie $M'(OR')_4$ may form a double alkoxide:

$$(RO)_3M-OH + R'O-M'(OR')_3 \rightarrow (RO)_3M-O-M'(OR')_3 + R'-OH$$

$$(28.6)$$

The overall chemistry forming TiO_2 is depicted by

$$Ti(OR)_4 + 2H_2O \rightarrow TiO_2 + 4ROH \qquad (28.7)$$

which indicates that water is consumed. The reaction indicated by Eq. 28.4 is rate-limited by the availability of water, and the reaction indicated by Eq. 28.5 is rate-limited by the availability of products from the first reaction. In the sol-gel preparation of SiO_2, the presence of a low water content, and an acid catalyst which speeds hydrolysis, long polymer chains form to produce the gel. But in a base-catalyzed sol, three-dimensional polymerization forms particles that gel by agglomeration.

Another system which has been used successfully for forming coatings is a nonaqueous solution of an organic acid, such as acrylic acid $R-COOH$, and a dissolved metal salt. Polymerization on drying may produce a tenacious film that is several hundred nanometers in thickness. Other sol-gel systems are described in Chapter 4.

The techniques used for applying sol-gel coatings are dipping, draining, spinning, and spraying. Dipping has been used to produce uniform fired coatings about 100 nm in thickness on nonporous surfaces. The coating thickness H is described by the equation*

$$H = K(v\eta_{sol}/D_{sol})^{0.5} \qquad (28.8)$$

*L. D. Landau and V. G. Levich, *Acta Phys. Chem. URSS*, **17**, 42 (1942).

where K is a constant which depends on the angle of the surface, η_{sol} and D_{sol} are the viscosity and density of the sol, and v is the withdrawal velocity. Note that the coating thickness on the nonporous substrate varies directly with the pulling rate. The process is commonly a batch process and several sides may be coated simultaneously. Drain coating is a variation in which the substrate is stationary and the vessel containing the sol is raised and lowered. Repeated dippings may be used to build up a thicker coating. Spin coating (Fig. 28.10) may be used to coat one side of a substrate. In spin coating, the solution drips onto the center of a substrate rotated at about 3–4 Hz. The solution may also be sprayed onto the surface of the rotating substrate.

Thin, properly formulated gel coatings with a flexible, cross-linked polymer structure are observed to shrink only in their thickness when dried and do not exhibit cracks. The dried coating is porous and firing is required to produce a dense adherent coating. Sol-gel processing may be used to coat fibers for composites. A very fine pore size is formed when the gel is dried into a xerogel. The xerogel commonly has a porosity in the range of 50–70%, extremely fine pores in the range 1–50 nm, and a specific surface area often exceeding 100 m²/g.

Glaze Coating

Many ceramic products are coated with a slurry which forms a vitreous or partially vitreous coating when fired. An example of a spray glaze slurry formulation is presented in Table 28.3. Frit is glass powder which on firing speeds the reaction and spreading of the vitrifying coating. Other inorganic additives serve to develop the proper softening temperature, reaction with the substrate, and thermal expansion coefficient of the fired glaze. The remaining processing additives provide dispersion in mixing, the proper rheology, and green strength.

Spraying is the controlled atomization of a slurry and the directed flow of

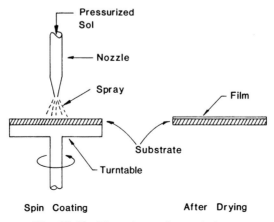

Fig. 28.10 The spin-coating technique.

TABLE 28.3   Example of a Spray Glaze Batch
(− 325 mesh)

Component	Weight %
Glass powder (frit)	46
Quartz powder	20
Feldspar	15
Whiting	14
Clay	5
Bentonite	0.25
Deflocculant	0.03
Methylcellulose	0.40
Polyethylene glycol	0.22

the droplets onto the surface of the product. On impact, the droplets deform and coalesce into a thick film. The cross section of the spray is controlled by the design of the nozzle and the surrounding spreader, the air pressure and rheology of the slurry, and the working distance. Using a needle-type spray nozzle, the airflow and nozzle opening are varied simultaneously. The spreader determines the shape of the spray pattern, which is usually elliptical and of a particular aspect ratio. Substrates commonly translate normal to the direction of the spray sweep. Hollow-ware shapes are commonly rotated during the spray application. The thickness of the sprayed film is dependent on the spray geometry, solids content of the slurry, working distance, spraying time or sequence, rebound loss, and film flow. The reproducibility of the thickness of sprayed coatings is lower than in screen or pad printing. Film patterns may be controlled by using templates or masks or by coating the surface with a film such as wax that prevents wetting.

The orifice of the nozzle must be much larger than the largest particles in the slurry, and slurries are commonly screened before being pumped to the nozzle. A smaller nozzle tends to produce a narrower distribution of droplet sizes and a more controlled spray. The slurry should be shear thinning (Fig. 28.11) to permit flow through the nozzle, but resist running on the product due to gravity, air currents, or mechanical vibrations. The yield strength and viscosity of the film are increased by the adsorption of liquid, gelation, and drying. When using a slurry containing a thermally gelling binder, as methyl cellulose, the slurry may be sprayed onto a heated product. Liquids used in nonaqueous slurries must not be too volatile. Spraying machines automatically spray continuously moving and rotating items. Disk and sonic atomization eliminates problems associated with compressed air and can produce coatings equal to those produced by spraying. A typical film thickness for a dried glaze is in the range of 100–200 μm.

Other methods for glaze application are listed in Table 28.4. Porous ceramics can adsorb a thin, relatively uniform film of slurry. Translation of the product under a waterfall (Fig. 28.12) or above a fountain may be used to glaze one

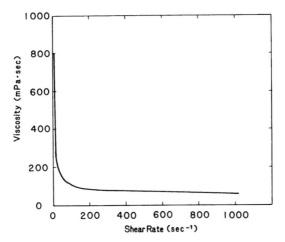

Fig. 28.11 Shear thinning rheological behavior of a glaze for spraying.

surface. Advantages over spraying include better thickness uniformity and smoothness and significantly lower collector losses. Dipping submerges the product, and a continuous coating is formed. The thickness H of the glaze film is

$$H = K_1(v\eta_S/D_S)^{0.5} + K_2 t^{0.5} \tag{28.9}$$

where η_S and D_S are the viscosity and density of the slurry, v is the withdrawal speed, t is the submersion time, K_1 is a constant that depends on the surface tension and angle of inclination, and K_2 is a constant that varies directly with the solids content of the slurry and the rate of adsorption of liquid into the product. Slinger, splatter, and flash/mist applications are used to produce coarse, fine, and irregular surface texture or color effects. Dry applications are relatively unrefined, but may be used to produce particular aesthetic effects. Lines and bands are often applied as the piece rotates against a brush, sponge, or other porous material filled with slurry.

TABLE 28.4   Glaze Application Processes

Wet Process	Dry Process
Spraying	Screening
Waterfall	Press
Fountain	
Dipping	
Slinger	
Splatter	
Flash/mist	

GLAZE WATERFALL CURTAIN

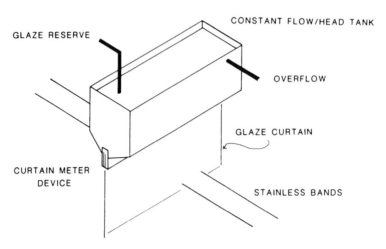

Fig. 28.12 Waterfall curtain for applying glaze. (From R. F. Jaeger, in *Technological Innovations in Whitewares*, Alfred University Press, Alfred, NY, 1982.)

Electrophoretic and Electrostatic Coating

Conductive substrates may be coated using electrophoretic or highly automated electrostatic techniques. Advantages include a high degree of thickness uniformity and smoothness and good coverage of shape edges and surfaces of drilled holes or small cavities. In electrophoretic deposition, a slurry containing about 25 vol% powder is well deflocculated to provide maximum particle mobility. The dispersed powder must be extremely fine to minimize settling. For an aqueous slurry, the substrate is normally the anode. Electrostatic application is now widely used for applying paints and enamels on metal. In the dry process, powder suspended in air is pumped through a powder gun where ion bombardment charges the particles. Charged particles accelerate to the surface of the grounded substrate forming a coating. In electrostatic spraying, a potential of about 100 kV is applied to the tip of the spray nozzle; charged droplets are formed which are attracted to the grounded substrate.

Thermal and Plasma Spraying

In thermal spraying, a fine powder is melted in a flame and deposited in a molten or semimolten state onto a prepared substrate (Fig. 28.13). In plasma spraying, a plasma created at a cathode tip by thermoionic emission in an inert gas passes through an anode nozzle near which the powder is injected. Particles become molten and on impact freeze almost instantly and adhere to the prepared surface. By cooling the surface, the substrate surface may be of almost any material which has a melting temperature above about 30°C. A nonoxidizing

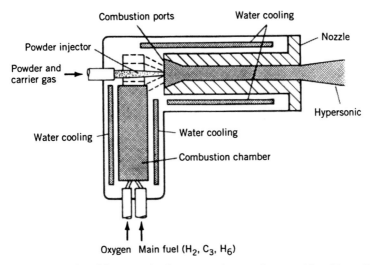

Oxygen Main fuel (H$_2$, C$_3$, H$_6$)

Fig. 28.13 Schematic of hypersonic flame-spraying torch assembly. (From P. Fauchais, in *Engineered Materials Handbook*, Vol. 4, ASM International, Materials Park, OH, 1991, p. 202.)

gas such as Ar—H$_2$ or Ar—He is used when depositing nonoxides and metals. Applications for flame and plasma spraying include refractory thermal barrier coatings on metals, metal coatings such as electrodes on ceramics, and wear-resistant coatings of borides and carbides. Surface preparation of the substrate commonly involves pneumatically abrading the surface with a hard grit. Chemical corrosion may be used when mechanical abrasion is not feasible.

When forming the coating, heating in the torch produces a temperature several times the melting temperature of the particles. Flight times are on the order of 1 ms. On impact with the surface, the partially or fully molten particles flatten and form thin splats on the surface of the substrate. Cooled splats freeze very rapidly and adhere to irregularities on the prepared surface and to each other. Impact and solidification of subsequent splats form a coating which has a lamellar structure, as shown in Fig. 28.14. The average roughness of the coating exceeds and is in proportion to the size of the particulates in the feed. The porosity of the coating is in the range of 5–20% and is commonly higher when the splat is only partially molten. The translation of the torch and/or substrate surface must be controlled well to produce a coating with a uniform thickness.

Temperature gradients in the coating are severe. The surface temperature is limited by the cooling of air or gas blown onto the surface. Stress in the quenched coating is tensile and is relieved by plastic yielding or by the formation of microcracks. Laminated bond coats may be used when the thermal expansion mismatch between the substrate and coating is high. Coating thicknesses may vary, depending on the application. Low-pressure grinding may be used to improve the surface of the coating.

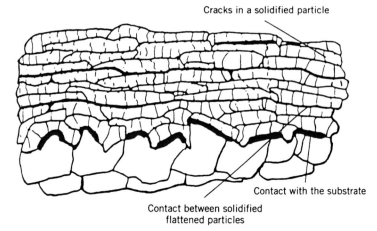

Cracks in a solidified particle

Contact with the substrate

Contact between solidified
flattened particles

Fig. 28.14 Simulated cross section of lamellar structure of coating produced by molten particle deposition and bonding to substrate. (From P. Fauchais, *Engineered Materials Handbook*, Vol. 4, ASM International, Materials Park, OH, 1991.)

28.4 DEFECTS IN THICK FILMS AND COATINGS

Common defects in slurry coatings are listed in Table 28.5. Voids and pinholes are caused by trapped air or gas formed from a spot contaminant in the film and insufficient flow during leveling. Pigment coagulation may cause color spots. Vertical runs occur when the adsorption of solvent or thermal gelling is slow and glaze flows down the vertical surface. Thickness variations are caused

TABLE 28.5 Common Defects in Coatings Applied as a Slurry

Defect Type	Cause
Voids and pinholes	Trapped air or gas in viscous coating
Color spots	Coagulation of pigment particles
Vertical runs	Gravity flow
Waves	Air currents or shear
Craters	Contamination such as dust, gel particle, and oil on the surface before applying glaze
Benard cells	Circulatory flow within coating caused by differential surface tension, temperature, or density when solvent evaporates
Orange peel	Localized variations in surface tension across coating
Picture framing	Surface tension gradient due to faster evaporation of solvent and particle migration near edge
Nonleveling	Surface tension gradients caused by differing evaporation rates of components
Crawling	Surface tension of coating is too high
Dewetting	Contaminated surface creates high interfacial tension

by improper application and leveling, a nonuniform adsorption into the porous surface, and gravity flow on inclined surfaces. Waviness is produced by cyclic air movement or vibrations during flow on inclined surfaces while the glaze is thickening.

Surface tension gradients may produce several types of defects, some of which are shown in Fig. 28.15. The surface tension may rise or fall as the solvent evaporates. Solvent evaporation and adsorption may be greater at the perimeter of contact of a drop on the surface. When the remaining perimeter material has a higher surface tension, polymer and other additives may migrate toward the perimeter if viscous drag can be overcome. Cratering (pinholing) is caused by a contaminant spot of low surface tension on the substrate and the flow of polymer and other additives away from the contaminant (Fig. 28.15). The tendency for Bernard cells can be reduced when the dependence of surface tension on temperature or concentration is minimal, the viscosity is increased, and the coating thickness is reduced. Orange peeling, which appears as slight mounds and dimples across the surface of the coating, is reduced by improving surface wetting, reducing variations in surface tension within the coating, and increasing the coating viscosity. Picture framing appears as a buildup of the coating at the edges. Phases higher in surface tension than the solvent will

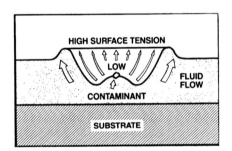
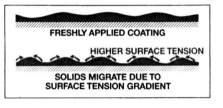

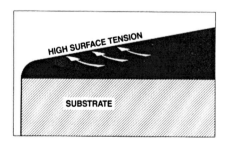
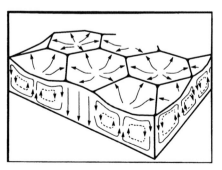

Fig. 28.15 Pictorial showing cause of defects in coatings applied as a slurry. Cratering (top left), nonleveling (top right), picture framing (bottom left), and Benard cells (bottom right). (Copyright 3M, all rights reserved. Reproduced with permission.)

migrate to the edge, creating a fat edge. Leveling of irregular thicknesses in the coating should be better when the surface tension of the coating is high, creating a strong driving force to reduce the surface area. However, surface tension gradients are exaggerated by the high value of the surface tension and may cause the flow of solids to peaks in the coating and retard leveling (Fig. 28.15).

Defects in thermal-spray coatings may include pores, microcracks, regions of poor adhesion, incompletely melted phases, and uneven thickness. Impact and flattening of individual droplets produces a coating with residual pores and a lamellar crystal structure. Metastable phases are sometimes observed. Flash cooling produces tensile stress which is relieved by plastic yielding and in ceramic coatings by microcracking. Bond coats about 80–180 μm thick are used when the thermal expansion mismatch between the coating and the substrate is high. Careful preparation of the substrate is needed to obtain a continuous adhesion of the coating. Control of the torch sweep relative to the substrate, overlap, the torch temperature, and the consistency of the powder feed are required to obtain a uniform thickness.

SUMMARY

Films, film patterns, and coatings of metals and ceramics are used as conductors, resistors, dielectrics, and decorations and for abrasion and corrosion protection. Thin films and film patterns are applied by physical and chemical vapor deposition processes. Thick film patterns are commonly applied using screen printing, pad transfer printing, and decals. The paste used for printing must be time-stable and carefully formulated to develop the pseudoplastic–thixotropic flow behavior for printing, without bleeding, and adhesion for attachment. Sol-gel technology is now well developed for applying thin coatings of a variety of materials. Glaze coatings which vitrify on firing are widely used to provide a nonporous protective surface on ceramics and for aesthetics reasons. Processing additives in the slurry are critical for developing the requisite flow for application, adhesion strength, and dry impact strength. Thermal and plasma spray coating processes offer a means for depositing thick, dense films of metals and ceramics on cooled surfaces without subsequent heating.

SUGGESTED READING

1. R. F. Davis, "Silicon Carbide and Diamond Semiconductor Thin Films: Growth, Defect Analysis, and Device Development," *Am. Ceram. Soc. Bull.*, **72**(7), 99–106 161 (1993).

2. D. P. Stinton, T. M. Besmann, and R. A. Lowden, "Vapor Deposition," in *Engineered Materials Handbook*, Vol. 4, ASM International, Materials Park, OH, 1991, pp. 215–222.

3. P. Fauchais, "Molten Particle Deposition," in *Engineered Materials Handbook*, Vol. 4, ASM International, Material Park, OH, 1991, pp. 202–208.

4. R. E. Johnson and J. Donohoe, "Decorating," in *Engineered Materials Handbook*, Vol. 4, ASM International, Material Park, OH, 1991, pp. 471–476.

5. D. P. Stinton, T. M. Besmann, and R. A. Lowden, "Advanced Ceramics by Chemical Vapor Deposition Techniques," *Am. Ceram. Soc. Bull.*, **67**(2), 350–356 (1988).

6. R. W. Vest, "Materials Science of Thick Film Technology," *Am. Ceram. Soc. Bull.*, **65**(4), 631–636 (1986).

7. S. Sakka, "Sol-Gel Synthesis of Glasses: Present and Future," *Am. Ceram. Soc. Bull.*, **64**(11), 1463–1466 (1985).

8. *Ultrastructure Processing of Ceramics, Glasses, and Composites*, L. L. Hench and D. R. Ulrich, Eds., Wiley-Interscience, New York, 1984.

9. *Better Ceramics Through Chemistry*, C. J. Brinker et al., Eds., North-Holland, New York, 1984.

10. R. J. Bacher, "Thick Film Processing for Higher Yield, 1: Screen Printing," *Insulations/Circuits*, **28**(8), 21–25 (1982).

11. D. A. Karlyn, "Pad Transfer Decorating," *Am. Ceram. Soc. Bull.*, **59**(2), 247 (1980).

12. Albert Kosloff, *Ceramic Screen Printing*, Signs of the Times Publishing, Cincinnati, 1977.

PROBLEMS

28.1 Write a chemical reaction for the chemical vapor deposition of zirconia.

28.2 Write a chemical reaction for the chemical vapor deposition of Si_3N_4.

28.3 Describe a procedure for preparing a mixed SiO_2 + TiO_2 gel. Illustrate its structure.

28.4 Explain why a sol-gel dip coating is thicker when the withdrawal velocity is faster.

28.5 Explain why the ink in pad printing must be of a lower viscosity than ink for screen printing.

28.6 What is the relation between snap-off distance and the force for peeling in screen printing?

28.7 When spraying a glaze slurry onto an inclined surface, where should the initial coating begin?

28.8 Describe the functions of the bentonite, methylcellulose, and polyethylene glycol in the glaze batch in Table 28.3.

28.9 For a waterfall slurry coating process for tile, what is the dependence of the glaze thickness H on the product velocity, the porosity of the tile, and the density and viscosity of the slurry?

28.10   How does gelation of the binder alter the yield strength, viscosity, and pseudoplastic behavior of a paste for printing?

28.11   Why is the reaction layer extremely minimal between a ceramic substrate and a metal thermal spray coating?

28.12   Explain the difference between the cause of cratering and ''orange peel'' of a viscous coating.

EXAMPLES

Example 28.1   What are the factors that control adhesion of a film or coating?

Solution. Adhesion can be produced by mechanical interlocking of the film or coating, by direct chemical bonding to the substrate material, and by the adhesion of one phase in a two or more phase coating. A roughened surface which has some porosity increases mechanical adhesion from interlocking. Mechanical abrasion and/or chemical etching may be used to increase the surface roughness. Interlocking of the coating with crystals on the surface or by penetration of the coating into pores increases the mechanical adhesion. Atom to atom bonding between the surface and the film produces chemical adhesion. This type of bonding is commonly produced when a film is applied by vapor deposition. Prior chemical cleaning of the surface may be requisite. For a film of a noble metal such as gold on 99% alumina, an intermediate metal film which bonds well to both the alumina and the gold may be needed. For a two-phase coating such as a frit-bonded Ag—Pd alloy or a glassy coating, both chemical and mechanical adhesion of the glassy phase may produce a tenacious coating. Surface roughening may aid both the chemical and mechanical adhesion. Particle sizes and the firing must be carefully controlled to produce one or more phases having the proper microstructure.

Example 28.2   How is a metallization technique selected?

Solution. Metallization techniques may be classified as bulk, particulate, and atomistic. Bulk techniques include cladding with a metal foil and those techniques mentioned for glazes and sol-gel coatings; in these a large amount of metal is quickly applied over a relatively large area. Subsequent drying and firing are required. Printing is used to apply the metallization over a smaller, precisely defined area. Flame spraying produces a coating directly over an area which may be controlled with a mask but less rapidly than for the techniques mentioned above; however, the coating may not require any additional heat treatment. Atomistic processes produce a film directly by nucleation and growth from a vapor and are used to form thin coatings and fine printed patterns. Annealing may be required to reduce residual stress and structural imperfections in the coating.

CHAPTER 29

FIRING

Products that have been dried and surface finished, traditionally called "green products," are heat treated in a kiln or furnace to develop the desired microstructure and properties. This process, called firing, proceeds in three stages: (1) reactions preliminary to sintering, which include organic burnout and the elimination of gaseous products of decomposition and oxidation; (2) sintering; and (3) cooling, which may include thermal and chemical annealing.

"Sintering" is the term used to describe the consolidation of the product during firing. Consolidation implies that within the product, particles have joined together into an aggregate that has strength. The term sintering is often interpreted to imply that shrinkage and densification have occurred; although this commonly happens, densification does not always occur. Highly porous, refractory insulation products may actually be less dense after they have been sintered.

In this chapter we consider general aspects of firing and principles involved in sintering simple powders and relatively complex, multiphase materials.

29.1 FIRING SYSTEMS

Ceramic materials and products are fired in a variety of kilns and furnaces* that are designed to operate either intermittently or continuously. Intermittent kilns for industrial firing, commonly called periodic kilns, are usually of the

---

*"Furnace" is a general term for an enclosed chamber in which heat is produced. The more specific term "kiln" is commonly used in the field of ceramics. A product is heated in contact with the hot combustion gases in a conventional kiln but is insulated from the combustion gases in a muffle kiln. The firing atmosphere may be varied more widely in a muffle kiln. Electrical heating is used in an electric kiln.

shuttle or elevator type. When loading a shuttle kiln, individual product items, product items supported on refractory setters and supporting members (kiln furniture), or product contained in refractory saggers are set on refractory shelves that are supported on a thermally insulated kiln car. The car mounted on a rail is pushed into the kiln for firing and withdrawn for unloading (Fig. 29.1a). An elevator kiln lined with relatively low mass thermal insulation is raised for setting the product and then lowered for the firing cycle; a variation is a car with a hearth that may be elevated into or lowered from the kiln.

Continuous tunnel kilns of the car, sled, roller hearth, and continuous belt type are commonly used for firing high-volume production items. When firing in a car or sled kiln, the product is set in a manner similar to that for a shuttle

Fig. 29.1 Photographs of (*a*) a shuttle kiln and (*b*) a three-tier roller hearth kiln. (Courtesy of Eisenman Corp. and Siti Kilns, respectively.)

kiln, but the cars or sleds move through the thermal gradient in the kiln. In a roller hearth kiln (Fig. 29.1*b*), large flat products such as ceramic tile or smaller products supported on refractory setter tile are conveyed on refractory rollers through the heated tunnel. A continuous refractory fiber carpet or mesh belt is also used to support and convey low-mass items through a kiln. A rotary kiln is an inclined, rotating or oscillating refractory cylinder used to continuously fire granular materials.

When firing products with a large mass or setting surface, the product is supported on mobile refractory supports, a granular bed, or partially fired "shrink plate" that can accommodate the differential shrinkage between the product and the setter. Products that have a tendency to deform during firing (slump) are supported on contoured refractory setters. In some setting configurations, the product is arranged so that shrinkage gradients and dimensional distortion during firing can compensate for density gradients or oversize dimensions in the green product. Elongated products are sometimes hung so that their weight acts to maintain linearity on firing.

Periodic kilns are used when a wide variety in product firing schedules or a relatively small or intermittent production volume requires flexibility in firing. Products from a continuous kiln that have had a minor body or glaze defect repaired are also often refired in a periodic kiln.

The car-type tunnel kiln (Fig. 29.2), introduced in 1916, is used widely for firing large products and heavy loads when the firing cycle exceeds about 6 h. Sled and roller hearth kilns provide very good temperature uniformity for single-tier settings and cycle flexibility. A sled kiln can be used for firing to a higher temperature than when using a roller kiln but requires a greater investment in refractories and a sled transfer system. Sled, roller hearth, and small periodic kilns are commonly used when the firing cycle is as short as 0.5 h.

Product heating is commonly produced by the combustion of natural gas or fuel oil or by electric heating. Radiant heating improves the temperature uniformity throughout the setting. When firing in a gas-fired muffle kiln or an electric kiln, the setting pattern should facilitate radiation transport to the center of the configuration, as shown in Fig. 29.2. Direct-fire combustion kilns also utilize convection heat transfer. In a car-type kiln, the setting pattern and hollow refractory supports are configured to facilitate the convection of hot gas throughout the car. High-velocity burners provide a higher velocity of convection and a better heat transfer and temperature uniformity.

Combustion heating is commonly produced from the ignition of a mixture of natural gas or fuel oil and air. The combustion of methane, the primary constituent in natural gas, in air is

$$CH_4 + 2O_2 \rightarrow 2H_2O + CO_2 \qquad (29.1)$$

Excess air is needed for an oxidizing atmosphere. Oxygen may be introduced at the burners to reduce the volume of air, to achieve a higher temperature, and to increase the oxygen partial pressure. A furnace temperature exceeding

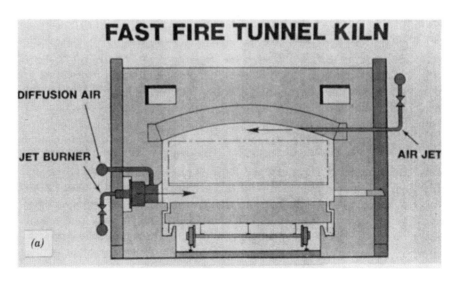

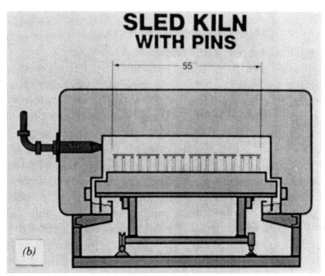

Fig. 29.2 Cross-section of (*a*) car-type and (*b*) sled-type tunnel kilns. Plan views of product settings for (*c*) axial convection heat transfer and (*d*) a gas-fired kiln with a muffle wall and a setting for radiant heat transfer. (*a* and *b* courtesy of Bickley Furnaces Inc.; *c* and *d* courtesy of Swindell Dressler International, Subsidiary of Rust International Co.)

1700°C may be achieved using oxygen and natural gas or another gas such as propane of higher caloric value.

Refractory metal alloy heating elements are used for heating to about 1150°C, silicon carbide to about 1550°C, and molybdenum silicide rods for heating to 1700°C in an oxidizing atmosphere. Molybdenum, tungsten, and graphite may

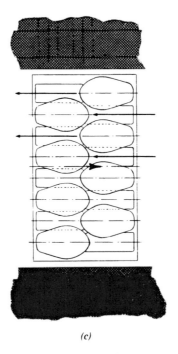

(c)

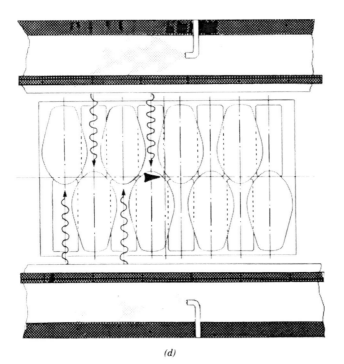

(d)

Fig. 29.2   *(Continued)*

be used for heating above 1700°C in an inert or reducing atmosphere. Induction, microwave, and plasma heating have been investigated. Significant improvements in the temperature uniformity and thermal efficiency have been achieved by using a greater amount of high-performance refractory insulation of much lower thermal mass and thermal conductivity, and higher-strength kiln furniture of thinner construction; these innovations enable the volume of product/furniture to increase in a setting and a faster firing when the kiln furniture is rate limiting.

29.2 PRESINTERING PROCESSES

Sintering does not commonly begin until the temperature in the product exceeds one-half to two-thirds the melting temperature, which is sufficient to cause significant atomic diffusion for solid-state sintering or significant diffusion and viscous flow when a liquid phase is present or produced by a chemical reaction. Material changes on heating prior to sintering may include drying, the decomposition of organic additions, the vaporization of chemically combined water from the surfaces of particles and from within inorganic phases containing water of crystallization, the pyrolysis of particulate organic materials introduced with the raw materials or as contamination during processing, changes in the oxidation states of some transition metal and rare-earth ions, and the decomposition or carbonates, sulfates, and so on, introduced as additives or as a constituent of the raw materials.

In this stage, stresses from the pressure of gas evolved or from differential thermal expansion of phases must not cause cracks or fracture of the fragile product. When firing a laminated structure or product with a surface coating, evolved gas must be eliminated while the surface permeability is high.

Presintering reactions are commonly investigated using the thermal analysis techniques discussed in Chapter 6. Initial heating may remove any liquid remaining after forming and drying the product and any moisture adsorbed from the atmosphere during transporting and setting. The adsorption of moisture can be quite significant when a warm porous product (internal surface area) is removed from the dryer but cools in moist air before firing. The adsorbed moisture may persist in the product up to a temperature exceeding 200°C.

Thermolysis (Organic Burnout)

Thermolysis of the organic additives not removed by drying is an important step prior to the densification on sintering. Incomplete binder removal and uncontrolled thermolysis may introduce a new population of product defects. These defects may reduce yields or impair the performance of the product. With the proper choice of binder for the product and controlled heating in an appropriate atmosphere, the green product should survive thermolysis without deformation, distortion, the formation of cracks, or expanded pores.

TABLE 29.1   Chemical and Physical Changes During Thermolysis

Chemical reactions	Decomposition
	Depolymerization
	Carborization
	Oxidation
Physical changes	Melting/softening (solid–liquid)
	Sublimation (solid–gas)
	Evaporation (liquid–gas)
Reaction zone	External
	Internal
	Uniform displacement
	Nonuniform displacement
Transport	Gas diffusion
	Liquid flow

Binder burnout is very dependent on the composition of the binder material and the composition and flow of the gas surrounding the product and in the pores of the product. It is also very dependent on the microstructure of the organic, powder, and porosity phases and dynamic changes in the microstructure as the binder is eliminated. Chemical and physical changes during thermolysis are summarized in Table 29.1 and a schematic of a simplified model of the binder burnout process is presented in Fig. 29.3.

The thermochemistry of the binder and additives, binder concentration, product size, product placement configuration, heating rate, and furnace atmosphere may influence the thermolysis behavior. Products for which $100V_o/V_p$ is less than about 10 generally have open pore channels sufficient to allow the transport of vapors and gases between the reaction zone and the product surface. This behavior can be expected in pressed compacts, slip cast products, and some extruded products. The time for thermolysis is controlled by the diffusion length for vapor phase transport rather than the characteristic dimension of the binder phase. Decomposition and vaporization of the organics will cause an internal

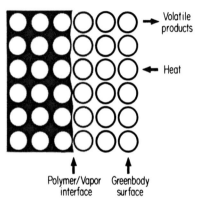

Fig. 29.3   A simple model of binder burnout process. [From M. J. Cima, J. A. Lewis, and A. D. Devoe, *J. Am. Ceram. Soc.*, **72**(7), 1193 (1989).]

gas pressure which will depend on the gas evolution rate, gas permeability K_p, and the compact size. The permeability varies with the binder loading and the pore structure and varies as $R_L^2 \phi$ of the pore phase. This dependence indicates that gas permeation rate is much lower in dense compacts of very fine particles. Exothermic and endothermic reactions on thermolysis will alter the internal temperature and the rate of reaction.

The burnout of polyvinyl alcohol from a compact of barium titanate powder in air is shown in Fig. 29.4. Initial endothermic behavior accompanies the loss of hydroxyl and hydrogen side groups leaving conjugated hydrocarbon. The hydrocarbon degrades in an exothermic manner at a higher temperature from about 150 to 500°C.

$$
\begin{array}{cccc}
\text{H} & \text{H} & \text{H} & \text{H} \\
| & | & | & | \\
\text{C} - \text{C} - \text{C} - \text{C} \\
| & | & | & | \\
\text{H} & \text{OH} & \text{H} & \text{OH}
\end{array}
\quad \xrightarrow{150\text{--}250°C} \quad
\begin{array}{cccc}
\text{H} & \text{H} & \text{H} & \text{H} \\
| & | & | & | \\
\text{C} - \text{C} = \text{C} - \text{C} + H_2O_{(gas)} \\
| & & & | \\
\text{H} & & & \text{OH}
\end{array}
\qquad (29.2)
$$

The volume of the gaseous products is several hundred times larger than the volume of the compact. Similar results are observed on heating vinyl binders with other common side groups and cellulose binders, except that the burnout range may be narrower.

Waxes and polyethylene glycols melt at a relatively low temperature and vaporize over a more narrow temperature range. Oxidation of the binder causes the temperature to increase more rapidly. The PEG binders, which contain oxygen in their backbone structure, are subject to exothermic oxidative degradation at about 180°C. For the thermolysis of polyethylene glycol in air, decomposition occurs by the mechanisms of chain scission and oxidative degradation. An added antioxidant which decomposes hyperoxides and quickly terminates chain reactions significantly reduces the decomposition rate and increases the oxidative degradation temperature, as indicated in Fig. 29.5.

Thermolysis of an organic binder in an inert atmosphere such as nitrogen, as is used when firing silicon nitride and aluminum nitride ceramics, proceeds differently. Some initial oxidation of the binder may occur from the adsorbed oxygen on particles or from oxygen in the binder molecules, such as in PEG. Vaporization is more predominant and in general the thermolysis is endothermic and is shifted to a higher temperature, as indicated in Figure 29.6. For PEG binders the onset temperature in a N_2 atmosphere is about 300°C. Some residual carbon may be expected in the absence of oxygen. The residual carbon commonly removes adsorbed oxygen from surfaces of particles and serves as a sintering aid in carbide and nitride ceramics.

In products containing a higher organic content, $100 V_o / V_p > 10$, such as in ceramic tape and injection molded products, the weight loss is much higher and the time for thermolysis without bubbling is relatively long. Diffusion in the condensed organic phase is rate limiting and binder removal from injection

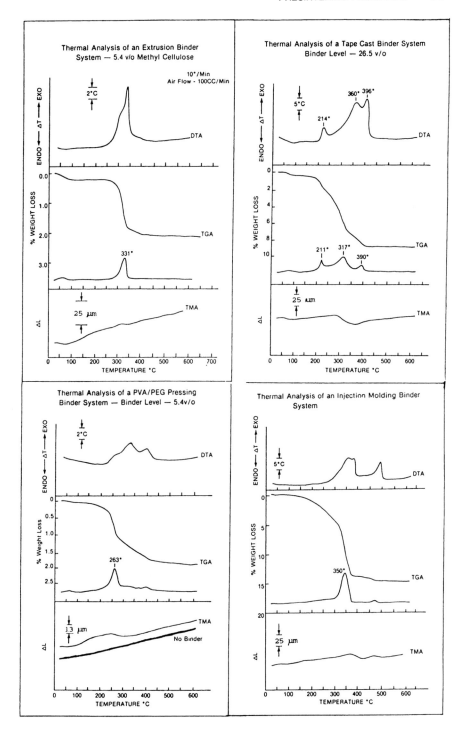

Fig. 29.4 Thermal analysis results for organic burnout from dried green material formed by extrusion, pressing, tape casting, and injection molding.

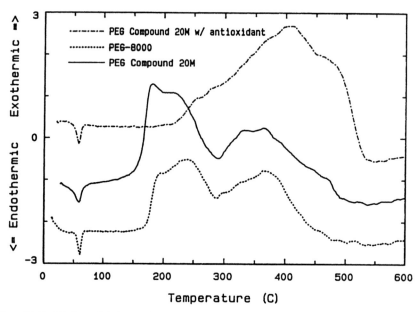

Fig. 29.5 Differential thermal analysis of alumina compacts containing polyethylene glycol binder indicates oxidative degradation is displaced to a higher temperature and rate is reduced when an antioxidant is present in the additive formulation. (Courtesy of W. J. Walker, Jr., Alfred University.)

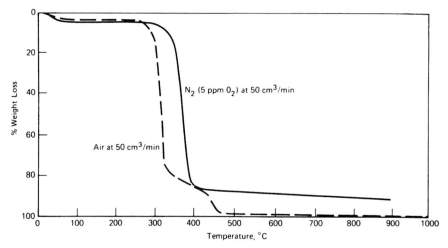

Fig. 29.6 Thermolysis of methylcellulose is displaced to a higher temperature and residue is higher when heated in nitrogen gas. [From N. Sarkar and G. Greminger, *Am. Ceram. Soc. Bull.*, **62**(11), 1284 (1983).]

molded parts more than a few millimeters in thickness can be expected to take several days. The development of penetrating porosity by capillary redistribution during the initial stage of binder elimination can reduce the path length to the order of the particle size, and significantly reduce the time requirement.* During the thermolysis of ceramic tape containing a polyvinyl butyral binder and a dibutyl phthalate plasticizer, capillary forces may cause the redistribution of the liquid nonvolatile component into the smaller pores. This redistribution causes the rate of thermolysis to be significantly higher.*

In injection molded compositions, it has been observed that low molecular weight vehicles, including wax, are lost by evaporation early, but high molecular weight polymers undergo thermal degradation followed by vaporization of the degradation products. Removal of the lower molecular weight liquid components by solvent extraction or liquid flow into the porous support for the product opens up penetrating pore channels, which significantly facilitates the subsequent removal of the higher molecular weight binder. The thermolysis of polyethylene occurs by thermal degradation involving random chain scission followed by oxidative degradation. In an oxidizing atmosphere, oxidative degradation is limited by oxygen diffusion and therefore product size.†

The burnout process may cause a slight volume expansion when the ceramic particles are in contact, as in a powder compact formed by dry pressing or slip casting, but a volume contraction when the particles are separated by the organic phase, as in a tape cast or injection molded material, as shown in Fig. 29.4. For the tape and injection molded material, the contraction varies directly with the binder loading. These linear changes may cause a stress, in addition to that produced by internal gas pressure, which may cause distortion or a crack in the product when the heating occurs too rapidly.

The ash content on binder thermolysis is relatively high for natural binders and binder containing metal ions. More refined binders such as polyvinyl alcohol and cellulose binders leave less ash but may contribute metallic contaminants of the order of 100–500 ppm. The residue after burnout of synthetic polymerized glycols and acrylic binders is relatively small, as indicated in Table 29.2. When firing nitride and carbide ceramics in an inert atmosphere, carbon residue is common, but may serve as a sintering aid.

Decomposition of Inorganic Components and Particulate Carbon

In bodies containing kaolin, the water of crystallization is eliminated between 450 and 700°C by the reaction producing metakaolin $Al_2Si_2O_6$,

$$Al_2Si_2O_5(OH)_4 \rightarrow Al_2Si_2O_6 + 2H_2O_{(gas)} \qquad (29.3)$$

The gas pressure may produce a volume expansion, and the heating rate must be carefully controlled in bodies containing a high clay content. Other dehy-

---

*M. J. Cima, J. A. Lewis, and A. D. Devoe, *J. Am. Ceram. Soc.*, **72**(7), 1192–1199 (1989).
†V. N. Shukla and D. C. Hill, *J. Am. Ceram. Soc.*, **72**(10), 1797–1803 (1989).

TABLE 29.2   Residue from Binder Thermolysis in Air

Binder Type	Residue (wt %)
Gums, waxes	> 1.5
Methyl cellulose	> 1.0
Polyvinyl alcohols	0.6–1.0
Hydroxyethyl cellulose	0.8
Polyvinyl butyral	0.4
Acrylic	0.14
Polyethylene glycols	0.05–0.07

dration reactions producing gas are the dehydration of aluminum hydrates in the range 320–560°C and talc at 900–1000°C. The decomposition of magnesium carbonate at 700°C and dolomite at 830–920°C produce CO_2. Magnesium sulfate decomposes as low as 970°C, but the decomposition of calcium sulfate occurs at a much higher temperature, in excess of 1050°C. The termination of decomposition depends on the particle size, the gas permeability of the product, and reactions between the salt and other phases present. Anion impurities in compacts of chemically prepared powders may persist into the sintering stage.

Fine organic matter in ball clay is commonly oxidized between 200 and 700°C, but coarse particulate carbon in the material may not be completely oxidized at 1000°C even when firing with excess air. Persistent carbon discolors the microstructure and causes the effect called "black coring" that may be observed on a fracture surface. Persistent carbon and a deficiency of oxygen produces carbon monoxide (CO) in the pores of the product. A chemical dopant such as a transition metal oxide in its higher oxidation state, such as Mn_2O_3, is added in some products as an internal oxidizing agent;

$$Mn_2O_3 + CO_{(gas)} \rightarrow 2MnO + CO_{2(gas)} \qquad (29.4)$$

Carbon monoxide in the product is not inert and may change the stoichiometry and properties of a transition metal oxide pigment or a magnetic ferrite:

$$CO_{(gas)} + Fe_2O_3 \rightarrow 2FeO + CO_{2(gas)} \qquad (29.5)$$

29.3   SOLID-STATE SINTERING

Sintered ceramic products represent a wide range of material systems that may vary widely in the number of components, particle characteristics, complexity of chemical reactions, and densification mechanisms during sintering. In this section we examine solid-state sintering, sintering of glass particles, sintering with reactions and dissolving, and the sintering of a glassy coating (glaze).

Examples of solid-state sintering include the sintering of a crystalline single phase such as α-Al_2O_3 and the sintering of a single phase containing a refractory dopant such as $Al_2O_3:0.5\%$ MgO, $ZrO_2:3\%$ Y_2O_3, and $SiC:2\%$ B_4C. An example of the densification behavior of a dense compact of a well-dispersed, nominally 1 μm magnesia-doped alumina powder during a constant rate of heating is shown in Fig. 29.7. The density increase of about 2% corresponding to the upturn of the curve is called the initial stage; the range of major densification, the intermediate stage; and the range of rapidly decreasing rate and cessation of densification, the final stage. Microstructural characteristics or changes observed in these stages in rather uniform packings of particles are listed in Table 29.3. In a very heterogeneous compact containing particles, agglomerates, and pores of widely different size and shape or chemical microinhomogeneities, the characteristic transitions between stages may proceed at different rates in different microscopic regions. Microstructural characteristics are determined using the techniques described in Chapter 6.

The driving force for sintering has its origin in the reduction in the total free energy ΔG_T of the system,

$$\Delta G_T = \Delta G_v + \Delta G_b + \Delta G_s \qquad (29.6)$$

where ΔG_v, ΔG_b, and ΔG_s represent the change in free energy associated with the volume, boundaries, and surfaces of the grains, respectively. The major driving force in conventional sintering is $\Delta G_s = \gamma_s \Delta A_s$ but the other terms may be significant in some stages for some material systems.

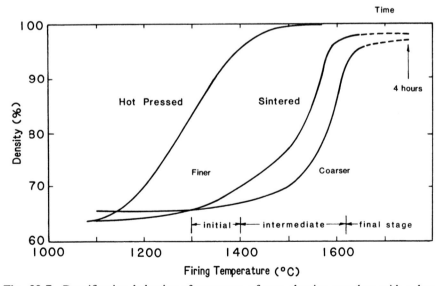

Fig. 29.7 Densification behavior of compacts of two alumina powders with a log-normal size distribution ($\bar{a}_{gM}$ = 1.3 and 0.8 μm) and the initial, intermediate and final stages of sintering for the coarser powder. Note that when hot pressing the compact of finer powder, densification begins and is completed at a lower temperature, and a higher maximum density is achieved.

TABLE 29.3 Microstructural Changes Observed in the Initial, Intermediate, and Final Stages of Solid-State Sintering (Powder Compact)

Stage	Observations
Initial	Surface smoothing of particles
	Grain boundaries form, neck growth
	Rounding of interconnected, open pores
	Diffusion of active, segregated dopants
	Porosity decreases $<12\%$
Intermediate	Shrinkage of open pores intersecting grain boundaries
	Mean porosity decreases significantly
	Slow grain growth
	(Differential pore shrinkage, grain growth in heterogeneous material)
Final (1)	Closed pores containing kiln gas form when density is $\approx 92\%$ ($>85\%$ in heterogeneous material)
	Closed pores intersect grain boundaries
	Pores shrink to a limited size or disappear
	Pores larger than grains shrink relatively slowly
Final (2)	Grains of much larger size appear rapidly
	Pores within larger grains shrink relatively slowly

Mechanisms for mass transport during sintering are listed in Table 29.4. Surface diffusion is a general transport mechanism that can produce surface smoothing, particle joining, and pore rounding, but it does not produce volume shrinkage. In materials where the vapor pressure is relatively high, sublimation and vapor transport to surfaces of lower vapor pressure also produce these effects. Diffusion along the grain boundaries and diffusion through the lattice of the grains produce both neck growth and volume shrinkage. The mechanisms of bulk viscous flow and plastic deformation may be effective when a wetting liquid is present and when a mechanical pressure is applied, respectively.

The shrinkage of an interstice in a uniform packing of uniform crystalline spheres (Fig. 29.8) has been considered by many investigators. Because of a difference in chemical potential, the concentration of vacancies beneath a concave surface is higher than beneath a flat or convex surface. The transport of

TABLE 29.4 Mass Transport Mechanisms in Sintering

Mechanism	Densification
Surface diffusion	No
Evaporation–condensation	No
Boundary diffusion	Yes
Lattice diffusion	Yes
Viscous flow	Yes
Plastic flow	Yes

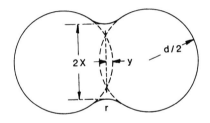

Fig. 29.8 A two-sphere sintering model for the initial stage.

vacancies from a concave surface can occur by the mechanisms of lattice and boundary diffusion, with a concomitant flow of atoms in the opposite direction. The effect is pore rounding and a decrease in the total surface free energy. Burke and Rosolowski* have pointed out that in the initial stage, pore rounding causes a decrease in ΔG_s in proportion to the reduction of the surface area, but the formation of a grain boundary causes an increase in ΔG_b. The angle of intersection of the pore at the pore–grain boundary juncture is

$$\cos\left(\frac{\phi}{2}\right) = \frac{\gamma_b}{2\gamma_s} \tag{29.7}$$

where γ_b is the interfacial tension of the boundary and γ_s is the surface tension. Neck growth should occur when $\gamma_b < \sqrt{3}\gamma_s$. For most materials, $\gamma_b < \gamma_s$, and the angle ϕ is large; however, in nonoxide materials, which sinter with difficulty, a dopant that concentrates in the boundary and reduces γ_b may be required for neck growth. Pore shrinkage occurs by the diffusion of vacancies to the grain boundaries and their annihilation there. The pore is a source of vacancies, and the grain boundary is a vacancy sink. Microcreep at the grain boundary eliminates vacancies and enables the centers of particles to approach, producing pore shrinkage.

The shrinkage $\Delta L/L_0$ with time in the initial stage for uniformly packed spheres has been modeled by numerous investigators† and is described by the equation

$$\frac{\Delta L}{L_0} = \left[\frac{KD_v\gamma_s V_v t}{k_B Td^n}\right]^m \tag{29.8}$$

where D_v is the apparent diffusivity of the vacancies of volume (V_v), d is the grain diameter, K is a constant dependent on the geometry, and m and n are constants that depend on the mechanisms of mass transport. The value of n is commonly observed to be approximately 3, which implies that surface diffusion

---

*J. E. Burke and J. H. Rosolowski, in *Treatise on Solid State Chemistry*, Vol. 4, N. B. Hannay, Ed., Plenum, New York, 1976, Chapter 10.
†R. L. Coble, *J. Am. Ceram. Soc.*, **56**(9), 461–466 (1973); R. W. Lay and R. E. Carter, *J. Am. Ceram. Soc.*, **52**(4), 189–191 (1969).

is the predominant mechanism, and m is commonly in the range 0.3–0.5. The shrinkage is very temperature-dependent in that the diffusivity (D_v) varies exponentially with temperature $[\exp(-Q/k_BT)]$. Equation 29.8 also indicates that smaller particles packed identically will produce an equivalent shrinkage in a shorter time.

In the intermediate stage, modeling of shrinkage is complicated by grain growth and a change in the pore geometry. More than one mass transport mechanism may be contributing significantly to the changes in microstructure. Sintering in this stage depends on the parameters in Eq. 29.8 but is also very dependent on the size, shape, and packing of the particles and chemical dopants that increase the vacancy flux of the slow-diffusing specie. Particle aggregates are a common source of a packing heterogeneity and inhomogeneous sintering (Fig. 29.9). As is seen in Fig. 29.10, the dispersion of particle aggregates by milling the powder before fabrication increased the compact bulk density only slightly; however, the particle packing and the intermediate sintering behavior were improved, and the transition to the final stage of sintering occurred at a significantly higher density in the more homogeneous compact.

In the intermediate and final stages, the densification behavior is very dependent on the association of pores with grain boundaries and the rate and mode of grain growth. Diffusion of atoms across the grain boundary, the disordered region between grains, causes the grain boundary to be displaced. Heating causes some grains to grow at the expense of others which shrink, and the net effect is an increase in the mean grain size and a reduction in the total grain boundary area. Because of the topological similarities between the grain boundaries in a solid and the cell walls in a foam, a model for the displacement of curved boundaries has been used as a classical description of grain growth in a solid. For grains with a nearly isotropic interfacial tension, it is observed in two dimensions that grains with fewer than six sides have convex boundaries and shrink and that those with more than six sides have concave boundaries and grow, as indicated in Fig. 29.11. Diffusion across the boundary will cause the boundary to be displaced toward its center of curvature at a velocity (v_b) given by the expression

$$v_b = M_b \gamma_b (1/r_1 - 1/r_2) \qquad (29.9)$$

where M_b is the mobility of the boundary, which varies as $[\exp(-Q_b/k_bT)]/T$, and r_1 and r_2 are the principal radii of curvature. When the mean grain size is directly proportional to the radius of curvature and M_b is a constant, the dependence of the mean grain size on time t is*

$$d_t^n - d_0^n = 2AM_b\gamma_b t \qquad (29.10)$$

*J. E. Burke and J. H. Rosolowski, in *Treatise on Solid State Chemistry*, Vol. 4, N. B. Hannay, Ed., Plenum, New York, 1976, Chapter 10.

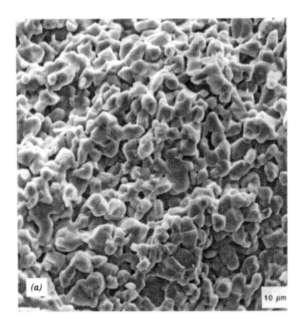

Fig. 29.9 Scanning electron micrograph showing (*a*) homogeneous sintering behavior in a compact of spinel powder and (*b*) inhomogeneous sintering in a compact of zirconia powder.

where d_0 is the original mean grain size and A is a constant that depends on the geometry. Much isothermal grain growth information is approximated by Eq. 29.10, and n is observed to have a value in the range of 2-3. A log-normal grain size distribution is commonly observed during regular grain growth. For a log-normal distribution of grain sizes, the mean grain size d may be calculated

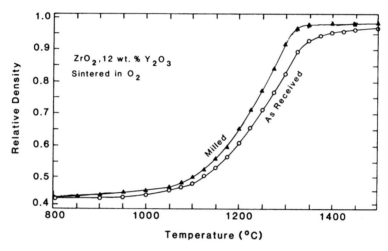

Fig. 29.10 Densification of compacts of submicron yttria-stabilized zirconia powder containing aggregates and with the aggregate size reduced by milling. (From C. Scott, PhD Thesis, Alfred University, 1977.)

from the mean intercept length $\bar{L}$ between boundaries using the relation $d = 1.56\bar{L}$, where $\bar{L}$ is determined using random test lines superimposed over a large number of grains in the plane of polish. An assumption in deriving Eq. 29.10 is that the form of the distribution of grain size is constant with time— that is, regular grain growth.

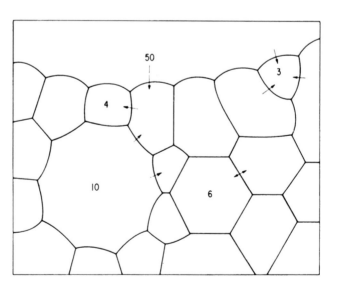

Fig. 29.11 Section through a sintered polycrystalline solid showing grains having a different number of sides and different boundary curvature. Arrows indicate direction of boundary displacement. (Reprinted with permission from J. E. Burke and J. H. Rosolowski, in *Treatise on Solid State Chemistry*, Vol. 4, N. B. Hannay, Ed., Plenum, New York, 1976, Chapter 10.)

Pores and solid inclusions smaller than the grains may intersect a grain boundary. When these inclusions disappear from the boundaries, grains exaggerated in size are commonly observed to appear in the microstructure, creating a bimodal grain size distribution (Fig. 29.12). This mode of grain coarsening is called exaggerated grain growth or discontinuous grain growth; it is usually undesirable unless a very coarse-grained microstructure is desired such as in a refractory material (where creep resistance is most important). An inclusion intersecting a boundary exerts a drag force on the boundary and reduces it effective mobility. The limiting grain size when immobile inclusions of radius (r_i) are dispersed along the boundary is*

$$d = r_i \left(\frac{8}{f_i^v} \right)^{1/2} \tag{29.11}$$

where f_i^v is the volume fraction of inclusions in the solid. In the intermediate stage and early part of the final stage, pores intersecting the grain boundary retard grain growth. Under the pull of the grain boundary, pores may migrate with a mobility M_p by means of mass transport across the pore. The condition for pore attachment depends on the ratio of M_p/M_b as well as f_p^v and the pore radius r_p. For pore migration by surface diffusion (surface diffusivity $= D_s$) the mobility has been expressed by Brook† as

$$M_p = \frac{KD_s}{Tr_p^4} \tag{29.12}$$

which indicates that M_p is larger for smaller pores. Pore growth can occur by either the coalescence of pores on the boundary or the diffusion of vacancies from smaller pores to larger pores rather than to the grain boundaries, which is called Ostwald ripening.

In the final stage of sintering, exaggerated grains separated from pores commonly appear in some regions of the microstructure when the processing is not carefully controlled. The tendency for exaggerated grains is higher when powder aggregates or dense, extremely coarse particles are present and the pore distribution is inhomogeneous. A high sintered density with minimal grain growth occurs when the material contains fine particles packed very densely and when an additive called a grain growth inhibitor is dispersed and homogeneously distributed. A classic grain growth inhibitor is MgO in alumina, which apparently inhibits exaggerated grain growth by increasing D_s and consequently M_p;‡ when this system is sintered in an appropriate atmosphere, the density may exceed 99.9%. Small second-phase solid inclusions on the bound-

---

*J. E. Burke and J. H. Rosolowski, in *Treatise on Solid State Chemistry*, Vol. 4, N. B. Hannay, Ed., Plenum, New York, 1976, Chapter 10.

†R. J. Brook, in *Treatise on Materials Science and Technology*, Vol. 9, F. F. Y. Wang, Ed., Academic Press, New York, 1976.

‡K. A. Berry and M. P. Harmer, *J. Am. Ceram. Soc.*, **69**(2), 143–149 (1986).

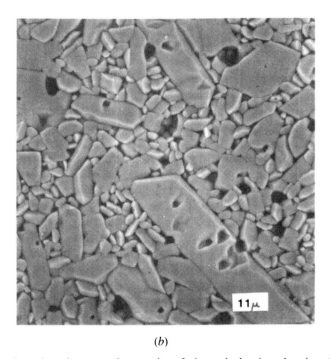

(b)

Fig. 29.12 Scanning electron micrographs of sintered alumina showing (a) regular grain growth and (b) exaggerated grain growth. (From D. Miller, MS Thesis, Alfred University, 1986.)

aries (Fig. 29.13) may also retard grain growth when the pores disappear, as has been demonstrated in yttria doped with La_2O_3 and ferrites and silicon nitride doped with Y_2O_3.* The general combinations of grain size and pore size for regular (pore-boundary attachment) and exaggerated (pore-boundary separation) grain growth and the effect of an inhibitor that reduces M_b are illustrated in Fig. 29.14.

A rapid heating schedule may produce densification with a smaller concomitant grain size. Surface diffusion, which predominates at low temperatures, can cause grain coarsening. When fast heating increases the vacancy diffusivity (D_v) and densification at a faster rate than diffusion causing grain coarsening,

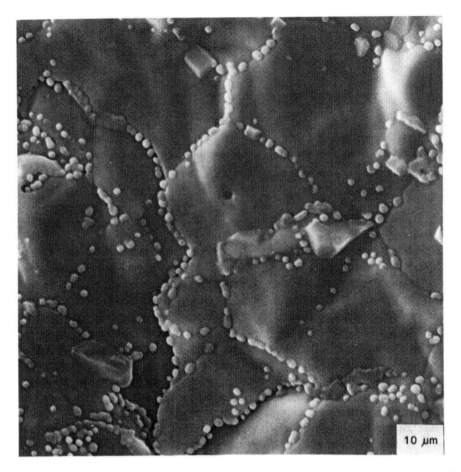

Fig. 29.13 Grain growth inhibiting inclusions segregated in the grain boundaries of alumina doped with strontium zirconate. (From C. Scott, MS Thesis, Alfred University, 1974.)

*M. F. Yan, in *Advances on Power Technology*, G. Y. Chin, Ed., American Society for Metals, Metals Park, OH, 1982, Chapter 6.

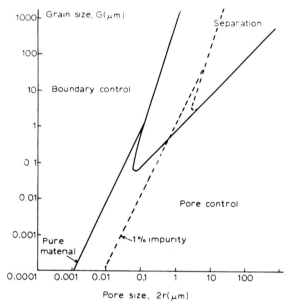

Fig. 29.14 Dependence of pore–boundary interaction and their separation on microstructural parameters in a system where the pores move by surface diffusion. [Reprinted with permission from R. J. Brook, *J. Am. Ceram. Soc.*, **52**(1), 57 (1969).]

fast firing with a minimum isothermal hold at the maximum temperature should be beneficial (see Fig. 29.15).

The atmosphere is also important in sintering. Gas trapped in closed pores will limit pore shrinkage unless the gas is soluble in the grain boundary and can diffuse from the pore. Alumina doped with MgO can be sintered to essentially zero porosity in an atmosphere of H_2 or O_2, which are soluble, but not in air, which contains insoluble nitrogen. Insoluble gas evolved into pores, such as SO_2 or Cl_2 from anion impurities in powders, may also limit pore shrinkage. The density of oxides sintered in air is commonly less than 98% and often only 92–96%. The sintering atmosphere is also important in that it may influence the sublimation or the stoichiometry of the principal particles or dopant. The oxygen pressure and partial pressure of PbO and ZnO must be controlled when sintering compounds such as lead titanates and zinc ferrites, because incongruent vaporization may produce PbO or ZnO vapor. The vaporization rate is lower when the oxygen pressure is higher and in a closed sagger saturated with the vapor. However, chromic oxide (Cr_2O_3) can be densified only on sintering at a low oxygen pressure (10^{-10} to 10^{-11} atm), because in an oxidizing atmosphere it vaporizes as CrO_2 or CrO_3.* In ceramic ferrites, the oxidation state of the transition metal ions and lattice vacancies depend on

*H. U. Anderson, *J. Am. Ceram. Soc.*, **57**(9), 34–37 (1974).

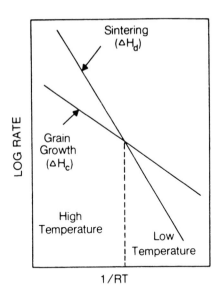

LOG RATE

Sintering
(ΔH_d)

Grain
Growth
(ΔH_c)

High
Temperature

Low
Temperature

1/RT

Fig. 29.15 Temperature dependence of grain growth and sintering rates for a material in which activation energy $\Delta H_{(densification)} > \Delta H_{(coarsening)}$. [From B. H. Fox, G. Dayton, P. Moses, and J. V. Biggers, *Am. Ceram. Soc. Bull.*, **64**(8), 1142 (1985); after R. J. Brook, *Proc. Brit. Ceram. Soc.*, **32**, 7–24(1982).]

the sintering atmosphere, and the partial pressure of oxygen is controlled to obtain a high density and the proper magnetic phases. When sintering nonoxide ceramics such as silicon carbide doped with B_4C at a temperature that may exceed 2000°C, a nitrogen atmosphere plays an important role by causing BN to form in the SiC lattice, which reduces the vaporization of the boron. The infiltration of a powder compact with a reactive gas may produce reaction bonding and a significant increase in density, as in producing reaction-bonded Si_3N_4 from silicon powder.

The sintering of ceramic materials depends importantly on the distribution of pore sizes and the homogeneity of the porosity in the compact.* On sintering an ideal homogeneous compact of uniform particles, the uniform interstices smaller than the grains should shrink at a relatively fast, uniform rate. However, large pores in a compact shrink relatively slowly (Fig. 29.16). Compacts pressed at a higher pressure that have a more narrow pore size range and a higher green density may achieve a higher sintered density when fired under identical conditions.

Inhomogeneous sintering and shrinkage may occur in materials containing relatively densely packed regions or regions containing finer particles that are separated by more porous regions containing larger pores. Regions containing smaller pores or particles that densify earlier may precede the more porous regions into the final stage of sintering, and the coarser pores in the boundary regions tend to persist. A similar effect may occur when a chemical densification aid is distributed inhomogeneously in the material.

*F. F. Lange, *J. Am. Ceram. Soc.*, **67**(2), 83–89 (1984).

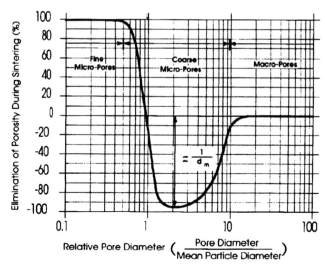

Fig. 29.16 Porosity elimination map during solid-state sintering for pores in different size ranges. The pore size has been normalized by dividing by the average particle size. Fine micropores are observed to shrink, coarse micropores grow, and very large macropores shrink at the average rate of the compact. [After J. Zheng and J. S. Reed, *Am. Ceram. Soc. Bull.*, **71**(9), 1414 (1992).]

26.4 SINTERING OF GLASS PARTICLES

A product composed only of glass particles may also be sintered to form a dense structure. In the initial stage, viscous flow produced by the driving force of surface tension causes neck growth. For two glass spheres of uniform diameter a, the initial shrinkage $\Delta L/L_0$ is given by the classic Frenkel equation*

$$\frac{\Delta L}{L_0} = \frac{3\gamma_s t}{2\eta a} \tag{29.13}$$

where t is the isothermal sintering time. The initial shrinkage behavior for a compact of a standard glass powder is of the form predicted by Eq. 29.13, as seen in Fig. 29.17.

In the intermediate stage, it can be assumed that gas diffuses rapidly through the interconnected pores and that the rate of pore shrinkage is proportional to γ_s/η, which is very dependent on temperature. Sintering occurs very rapidly when the glass is heated to its softening point of $10^{6.6}$ Pa s, but gravity-induced flow causes slumping at this temperature; the heating schedule must be controlled to obtain a moderate densification rate without slumping, which corre-

---

*J. Frenkel, *J. Physics (USSR)*, **9**(5), 305 (1945).

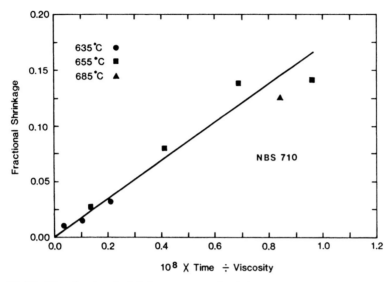

Fig. 29.17 The fractional shrinkage versus time/viscosity of compact of NBS 710 glass powder. The curve corresponds to the Frenkel equation with $\gamma = 300$ mN/m and $a = 10$ μm. (From T. Clark, PhD Thesis, Alfred University, 1984.)

sponds to an apparent viscosity of about 10^8 Pa·s. Gels may be sintered at or below T_g.

Closed pores form on entering the final stage, and these should be spherical when the medium is isotropic. The densification rate dD/dt of uniform spherical pores dispersed in an isotropic, incompressible viscous medium is described by the Mackenzie–Shuttleworth equation*

$$\frac{dD}{dt} = \frac{K\gamma_s n^{1/3}(1 - D)^{2/3}(D)^{1/3}}{\eta} \tag{29.14}$$

where D is the bulk density fraction of the porous glass containing n pores per unit volume of glass and K is a constant that depends on geometry.

When sintering a crushed-glass powder, the sintering rate is higher than predicted using Eq. 29.13, because the effective radius of curvature is smaller than the equivalent spherical radius. Water vapor in the sintering atmosphere may reduce η and increase the apparent densification rate. In the final stage, gas solubility and its diffusion in the glass and pore coalescence which decrease n, affect the final pore shrinkage. Finer pores have a larger driving force for shrinkage. An increase in the gas pressure due to an increase in temperature or from gas evolved from within the glass, or a reduction of the driving force due to pore coalescence, can cause pore enlargement and a volume expansion,

*J. Mackenzie and R. Shuttleworth; *Proc. Phys. Soc. (Lond.) B*, **62**, 833 (1949).

called bloating, which is usually undesirable. Some glasses may be crystallized after being densified by sintering, to produce a glass ceramic.

29.5 SINTERING IN THE PRESENCE OF A SMALL AMOUNT OF A WETTING LIQUID

A significant proportion of ceramic products used in low-temperature applications are predominantly crystalline materials containing a minor amount of a glassy phase distributed in the grain boundaries. Some examples are alumina substrates, grinding media, spark plug insulators containing a calcium magnesium aluminosilicate glass phase, titanates, and hard ferrites containing 2 vol% lead silicate glass. When the liquid coats each grain, the material may often be sintered to a higher density at a lower temperature (Fig. 29.18) with less of a tendency for exaggerated grain growth. Also, the glassy phase may be essential to improve the adherence of a printed thick film or a glaze. In systems where the glassy phase is inhomogeneously distributed, however, differential rates of densification and grain growth may produce an inhomogeneous microstructure and a lower average density.

Less than 1 vol% liquid phase is sufficient to coat the grains when the liquid is distributed uniformly in a material with a nominally 1-μm grain size, and at this concentration the viscosity of the liquid does not have to be high, to resist slumping during sintering. The liquid draws the particles together, and angular particles may rotate, enabling sliding and rearrangement into a denser configuration. Pores surrounded by liquid have a driving force for shrinkage that is opposed by solid particles in contact. Liquid-phase-assisted sintering continues

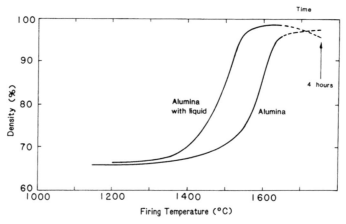

Fig. 29.18 Comparative densification behavior of a compact of a micron-size alumina powder and the powder containing 5 wt% of an alkaline earth aluminosilicate glass phase in the grain boundaries. Pore coalescence in the glass phase causes negative densification when the temperature is held at 1650°C, as indicated by the dashed line.

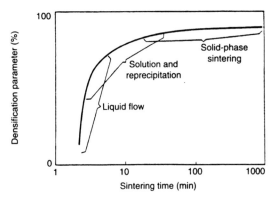

Fig. 29.19 Model for densification in the presence of a liquid phase indicating mechanisms. (After W. J. Huppmann in *Sintering and Catalysis*, Vol. 10, G. C. Kuczynski, Ed., Plenum, New York, 1975.)

when there is some solubility of the solid, and a higher diffusivity in the liquid may increase the rate of mass transport and shrinkage (Fig. 29.19). Sharp edges and small particles will preferentially dissolve, and diffusion through the liquid and crystallization in another region produce grain growth. A liquid phase that wets and dissolves the solid has been shown to rapidly penetrate between grains and disperse powder aggregates; this can reduce the tendency for exaggerated grain growth.* Also, viscous flow at boundaries may aid densification by providing a mechanism to accommodate strains between microscopic regions shrinking at slightly different rates. In some systems it may be desirable and possible to partially crystallize the glassy phase after densification. If the liquid becomes nonwetting or coalesces during sintering, a heterogeneous microstructure may be produced.

29.6   SINTERING OF WHITEWARE BODIES (VITRIFICATION)

The sintering of whitewares is complex, because densification occurs simultaneously with the reaction and dissolving of raw materials that produce new glass and crystalline phases. Fine clay coats the quartz and feldspar. The elimination of the water of crystallization in the clay minerals, the burnout of organic matter, and the decomposition of carbonate impurities occur prior to sintering. A characteristic shrinkage is observed when the metakaolin within the clay relics transforms into needle-shaped mullite crystals and silica glass in the range of 950–1000°C (Fig. 29.20).

Pure potassium and sodium feldspars melt at about 1150 and 1050°C, respectively. A liquid phase may form below 1000°C when potassium feldspar

*W. J. Huppmann, S. Pejovnik, and S. M. Han, in *Processing of Crystalline Ceramics*, Plenum, New York, 1978.

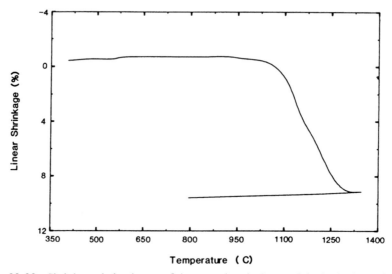

Fig. 29.20 Shrinkage behavior on firing an electrical porcelain body in a thermal dilatometer. The inflection at about 560°C corresponds to loss of H_2O from clay component, the shrinkage at 950°C to formation of mullite in metakaolin, and the inflection at about 1160°C to a change in the vitrification.

is in contact with silica and in a mixture of the feldspars when water vapor is present. The fluxing reaction of the feldspar with the kaolin above 1050°C produces a glass and needle-shaped (primary) mullite nearer the feldspar and platelike (secondary) mullite nearer the kaolin side. Shrinkage occurs by sintering in the presence of a liquid phase (Fig. 29.20).

Quartz in contact with the feldspar liquid dissolves slowly above 1250°C (Fig. 29.21). The rate of mullite formation and quartz dissolving is very dependent on the particle size of the materials and the type and concentration of impurities and secondary fluxing additives. Water absorbed from the kiln atmosphere reduces the viscosity of the glassy phase. Resistance to slumping is maintained by the low initial liquid content distributed somewhat heterogeneously and the increase in its effective viscosity due to the dispersed mullite crystals formed as the liquid content increases to above 50 vol%. At higher temperatures near the final stage of sintering, the mullite crystals become prismatic in shape, and some mullite may dissolve, but the dissolving of quartz increases the silica content and viscosity of the glass.

Coarse quartz may persist. The substitution of alumina for silica retards mullite formation but reduces the mullite dissolving at high temperatures. Because the reactions described above are controlled by chemical diffusion and viscous flow, the microstructure produced is dependent on the heating rate program and the maximum temperature. Sintering in the final stage after closed pores have formed is similar to that described in Section 29.5, except that in

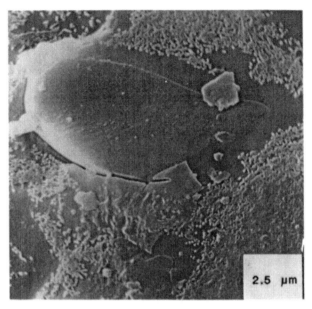

Fig. 29.21   Scanning electron micrograph of a fired electrical porcelain showing primary mullite and glass, secondary mullite, and a large quartz grain with a silica-rich glassy rim and a crack at the quartz–glass interface. (Photo courtesy of G. Steere, MS Thesis, Alfred University, 1977.)

the glassy phase fine mullite and coarser quartz are present and the glassy phases differ in composition and viscosity. Bloating occurs when gas is evolved on heating. A high green density and control of the heating program and maturing temperature are required to achieve a high sintered density without slumping and bloating. A small amount of a secondary flux as calcium carbonate or talc may reduce the firing temperature.

Steatite porcelains are produced from bodies containing clay and at least 60% talc. Fired bodies contain a high content of glass and enstatite ($MgSiO_3$) and cristobalite (SiO_2) crystals. Cordierite materials of very low thermal expansion are produced by firing a mixture of clay, talc, and alumina. Complex reactions between the raw materials produce transient crystalline phases such as cristobalite, mullite, and protoenstatite and a minor amount of a transient liquid phase. Cordierite forms slowly at 1250–1400°C. A long firing schedule is required to produce the cordierite product, and the maximum sintering temperature is quite close to its incongruent melting temperature of 1460°C. The quantitative microstructural analysis of these fired ceramics is a complex topic, as discussed by Kingery et al.*

---

*W. D. Kingery, H. K. Bowen, and D. R. Uhlmann, *Introduction to Ceramics*, Wiley-Interscience, New York, 1976.

29.7 SINTERING OF GLAZES AND GLASSY THICK FILMS

A dried glaze is a relatively thin coating commonly < 1 mm in thickness that contains raw materials that react to form a glassy phase, glass particles called frit, finely ground crystalline ceramic colors and opacifiers, and the processing additives. During firing, the thickness of the coating decreases as the particles fuse and the coating densifies. Reactions between the glaze and the near-surface substrate material and the glaze and the kiln gas have a significant effect on the bonding, appearance, and quality of the fired coating.

Presintering reactions described in Section 29.2 should be completed before the final stage of liquid phase sintering. Spreading of the vitreous phase is a function of the density and surface tension of the glaze, which commonly decrease slowly with temperature, and the viscosity, which decreases more rapidly. When the surface tension and contact angle are high, a break in the glaze coating due to a drying crack or the incomplete merging of screen-printed flow units may appear to widen—a process called crawling. The surface tension and contact angle may be reduced by lead oxide in the glaze; this is commonly added in the form of a lead frit in which the lead oxide is less soluble, and much less volatile on firing. Dispersed and agglomerated particles of the refractory colors and opacifiers increase the effective viscosity of the glaze and the temperature for its flow.

Air contained in closed pores must diffuse to the surface where it can be eliminated when bubbles break. Bubbles are produced rapidly on heating when gas dissolved in the frit or molten glaze or gas that is a decomposition product produces bubbles. In fast firing, the bubble growth may temporarily increase the thickness of the glaze. The maximum bubble size increases rapidly and may exceed several hundred microns (Fig. 29.22) before decreasing to below 100 μm, as the concentration of bubbles decreases. In a relatively thin glaze, the concentration of bubbles decreases more rapidly, and there is a practical upper thickness for a quality glaze. Smoothing of the surface depends on viscous flow after surface bubbles break.

Dense pigment particles may settle in a fluid glaze but at a slower rate in a higher-density lead glaze. The molten glaze commonly penetrates into pores and the glassy boundary phase of the substrate and may dissolve some crystalline phases preferentially. As the penetration-reaction layer is formed, the glaze coating thins. Some reaction of the glaze is desirable to increase the glaze-body interface area and to produce a stronger reaction bond. Dissolving of some constituents of the body, such as quartz or foreign particles during relatively slow firing, may decrease the solubility of gas in the glaze and produce a second growth of bubbles; a pinhole defect may appear when the glaze coating is thin. In a two-fire process, the firing of the glaze (glost fire) is programmed to facilitate its sintering and maturation on the sintered (bisque-fired) substrate. However, in a one-fire process, the sintering of the glaze must be synchronized with the sintering of the underlying body.

Glaze defects produced during the coating process are discussed in Chapter

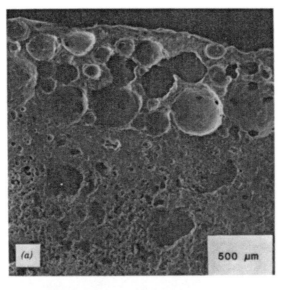

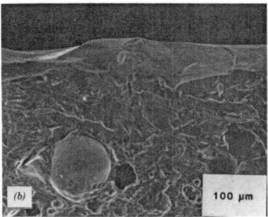

Fig. 29.22 Fracture surface of a glazed whiteware showing bubbles in glaze after (a) 10-min soaking and (b) 60-min soaking.

28. The appearances of several types of defects in a fired glaze is illustrated in Fig. 29.23 and described in Table 29.5. It is common to produce a glaze which on cooling develops a residual compressive stress in excess of 70 MPa. Crazing occurs when the stress in the glaze is tensile. It is eliminated by reducing the differential contraction between the glaze and the substrate. This may be accomplished by reducing the thermal expansion coefficient of the fired glaze and sometimes by reducing the glass transition temperature below which residual stress is produced during cooling. A body with a larger thermal contraction and/or greater volume stability in use may also eliminate crazing.

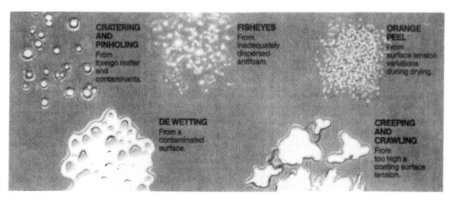

Fig. 29.23 Illustration of different types of defects in a fired glaze. (Copyright 3M, all rights reserved. Reproduced with permission)

Peeling (shivering) is corrected by reducing the compressive stress in the glaze. This may be accomplished by reducing the thermal contraction differential between the glaze and the body. Blisters may be caused by a wide variety of gas-producing sources such as dissolved water in the frit, particles that evolve gas or react to evolve gas near the maturing temperature of the glaze, and gas from the reaction of the glaze with the substrate. Resistance to blisters is increased by eliminating the sources of gas or adjusting the firing conditions to reduce the evolution of gas. Crawling during vitrification of the glaze is negative spreading in a region of the surface. The dissolving of agglomerated ceramic colors or clay in the glaze may sometimes increase the surface tension in that region and reduce wetting. Crawling is aggravated when the glaze

TABLE 29.5 Glaze Defects Observed After Firing

Defect	Observation
Blister	Raised surface from gas bubble produced near end of heating program in firing.
Crater/pinhole	Burst bubble incompletely healed by viscous flow.
Fisheyes	Agglomerated small bubbles under glaze.
Orange peel	Nonsmooth glaze surface (see Fig. 29.23).
Crawling/dewetting	Glaze withdraws from region of surface exposing body. (see Fig. 29.23).
Specking	Undesired inorganic particle in glaze.
Crazing	Network of fine cracks in glaze caused by tensile stress. Primary crazing occurs on cooling during firing; secondary crazing occurs when substrate expands in use.
Peeling/shivering	Glaze separates (peels) from body and sometimes flakes off body from excessive compressive stress in glaze (large shear stress is produced at interface).

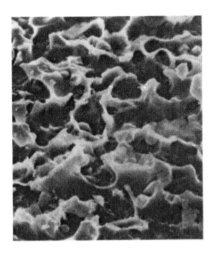

Fig. 29.24 Scanning electron micrograph of the interface of a frit-bonded Ag–Pd thick film fired on an alumina substrate (dispersed particles dissolved before photographing). [From R. W. Vest, *Am. Ceram. Soc. Bull.*, **65**(4), 631–636 (1986).]

coating is thicker and the specific surface area of refractory particles in the glaze is higher.

Thick film conductors, resistors, and dielectrics on electronic substrates and decorations contain glass frits that provide a vitreous fired bond. The frit must have a viscosity and surface tension dependence on temperature that is convenient for firing and a thermal contraction on cooling that is compatible with the substrate. For electronic applications, the frit should have a high resistivity and a low dielectric constant and loss. During sintering, the glassy phase must wet and bond the partially dispersed particles to the substrate so that shrinkage occurs only in the thickness. Some dissolving of the substrate commonly occurs, but the sintering time must be restricted. A discontinuous glass interface with particles in contact with the substrate decreases the tendency for a continuous crack separating the film from the substrate (Fig. 29.24).

29.8 COOLING

Fired ceramics that do not require either thermal or chemical annealing may be cooled quite rapidly, and the practical cooling rate is often controlled by the thermal shock resistance of the setters and the kiln furniture. Products with thicker cross sections and only average thermal stability are cooled more slowly. Some electrical ceramics are cooled in an atmosphere that is different from the sintering atmosphere to purposefully change the oxidation state of component ions, alter the stoichiometry, or alter the phase equilibria. In the industrial firing of manganese zinc ferrites (Fig. 29.25), a composition containing excess Fe_2O_3 is fired in an atmosphere with a relatively high partial pressure of oxygen to obtain a small grain size and to suppress the volatilization of zinc oxide.

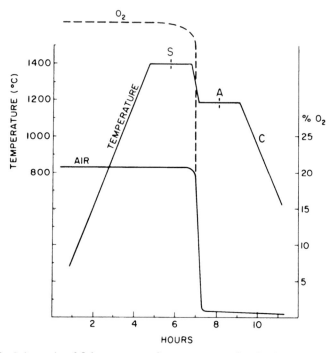

Fig. 29.25 Schematic of firing program for manganese zinc ferrite showing controlled atmospheres during presintering, sintering S, annealing A, and cooling cycle C. (Courtesy T. Reynolds, in *Treatise on Materials Science and Technology*, Vol. 9, F. F. Wang, Ed., Academic Press, New York, 1976, p. 211.)

During cooling, the oxygen partial pressure is reduced to establish the content of ferrous iron* and produce a single ferrite phase.

When cooling products containing a glassy matrix, it is common practice to cool slowly through the glass transition temperature (viscosity of 10^{12}–10^{13} Pa·s) to anneal stresses in the glass produced by thermal gradients. Slow cooling is also necessary when a crystalline phase undergoes a transformation causing a large change in volume, such as the thermal inversion of quartz at 573°C and cristobalite at 220–280°C. Stresses of sufficient magnitude caused by differential thermal contraction produce small cracks in the product or complete failure; in whiteware, this failure is called dunting. Tensile stress in a glaze produced by a differential contraction between the glaze and substrate may produce fine cracks in the glaze, a glaze defect called crazing; a high compressive stress may cause peeling.

*T. Reynolds, in *Treatise on Materials Science and Technology*, Vol. 9, F. F. Y. Wang, Ed., Academic Press, New York, 1976.

29.9 HOT PRESSING

Sintering during the application of an external pressure (P_e) is called hot pressing. A mechanical pressure can increase the driving pressure for densification (P_T) by acting against the internal pore pressure (P_i), without increasing the driving force for grain growth:*

$$P_T = P_e + \frac{2\gamma}{r_p} - P_i \qquad (29.15)$$

A practical consequence of hot pressing is that a ceramic may often be produced with a comparable density but a finer grain size or a comparable grain size but a higher density (see Fig. 29.7). In uniaxial hot pressing, the material is pressed and sintered in a refractory die manufactured from graphite, alumina, stabilized zirconia, or a refractory metal. The industrial use of uniaxial hot pressing is limited, because its productivity is relatively low and the die maintenance is expensive. It is used to produce some commercial materials when a sintering aid or grain growth inhibitor is unknown or unacceptable, and for producing specimens for research. The hot forging of ceramic containing a liquid phase has been investigated.

Hot isostatic pressing (called HIP, or hipping), in which a product previously sintered to the final stage of sintering is subjected to a hot pressurized gas, is used industrially to reduce the size of closed pores in high-performance structural materials. Applied pressures used are in the range of 10–200 MPa. Porous materials or products may be hipped if first enclosed in a refractory, compressible enclosure or when the surface is sealed.

SUMMARY

Ceramic products are manufactured in a wide variety of sizes, shapes, material compositions, and production rates, and many types of kilns and furnaces are used for industrial firing. Initial heating volatilizes organic processing additives and decomposable constituents in raw materials and may cause the diffusion and reaction of chemical dopants with particles of the primary materials. In the first stage of sintering, the formation of grain boundaries or a glassy matrix causes the particles to join together, and an aggregate having some strength is formed. A chemical dopant acting as a densification aid may reduce the temperature for the onset of densification and in some materials is requisite to complete neck growth by chemical diffusion.

Significant densification with a decrease in the sizes of pores and a small increase in the grain size occurs during the intermediate stage of sintering.

*D. S. Wilkinson and M. F. Ashby, in *Sintering and Catalysis*, G. C. Kuczynski, Ed., Plenum, New York, 1976, p. 473.

Grain growth reduces the boundary free energy but is retarded by pores that intersect the grain boundary. The transition from interconnected pores to closed pores marks the beginning of the final stage of sintering. The transition occurs rapidly in material containing pores, grains, and chemical dopants that are uniform and homogeneously distributed and more slowly when densification lags in some regions of the material.

Microstructural changes in the final stage of sintering may proceed in a desirable or an undesirable manner. To achieve a high final density and a small to moderate grain size, small pores should be associated with grain boundaries, and the gas in the pore should diffuse from the pore. The association of the pore and the boundary may be enhanced by doping the powder with an additive that either reduces the boundary mobility or increases the pore mobility, and using a faster firing schedule.

Powder aggregates are a common source of chemical and/or physical non-uniformity in the green material and exaggerated grain growth and heterogeneous sintering. A few percent of an additive that creates a reactive liquid in the grain boundaries may aid in dispersing aggregates and reduce the sintering temperature, and this type of additive is often used in producing nonrefractory products. Shaped products, glazes, and thick films containing a glassy matrix may be sintered into a monolithic product or coating. The sintering of ceramic whitewares to a high density without slumping depends on a complex progression of reactions which alters the crystalline phases and increases the proportion of the viscous glass in the microstructure.

The atmosphere and cooling schedule following sintering are important to develop the proper oxidation states and to anneal differential strain in products containing a glassy phase. Pressure sintering called hot pressing may be used to increase the density without increasing the sintering temperature.

SUGGESTED READING

1. Ahmet R. Selcuker and Michael A. Johnson, "Termination Sintering in Multilayer Ceramic Capacitors: Microstructural Interpretation," *Am. Ceram. Soc. Bull.*, **72**(11), 88–93 (1993).

2. Man F. Yan, "Solid-State Sintering," in *Engineered Materials Handbook*, Vol. 4, ASM International, Materials Park, OH, 1991, pp. 270–284.

3. Oh-Hun Kwan, "Liquid-Phase Sintering," in *Engineered Materials Handbook*, Vol. 4, ASM International, Materials Park, OH, 1991, pp. 238–290.

4. John S. Haggerty, "Reaction Sintering," in *Engineered Materials Handbook*, Vol. 4, ASM International, Materials Park, OH, 1991, pp. 291–295.

5. George E. Gazza, "Pressure Densification," in *Engineered Materials Handbook*, Vol. 4, ASM International, Materials Park, OH, 1991, pp. 296–303.

6. Jennifer A. Lewis and Michael J. Cima, "Diffusivities of Dialkyl Phthalates in Plasticized Poly(vinyl butyral): Impact on Binder Thermolysis," *J. Am. Ceram. Soc.*, **73**(9), 2702–2707 (1990).

7. Michael J. Cima, Mark Dudziak, and Jennifer A. Lewis, "Observation of Poly(Vinyl Butyral)—Dibutyl Phthalate Binder Capillary Migration," *J. Am. Ceram. Soc.*, **72**(6), 1087–1090 (1989).

8. Vishwa N. Shukla and David C. Hill, "Binder Evolution from Powder Compacts: Thermal Profile for Injection-Molded Articles," *J. Am. Ceram. Soc.*, **72**(10), 1797–1803 (1989).

9. Michael J. Cima, Jennifer A. Lewis, and Alan D. Devoe, "Binder Distribution in Ceramic Greenware During Thermolysis," *J. Am. Ceram. Soc.*, **72**(7), 1192–1199 (1989).

10. K. A. Berry and M. P. Harmer, "Effect of MgO Solute on Microstructure Development in Alumina," *J. Am. Ceram. Soc.*, **69**(2), 143–149 (1986).

11. T. M. Shaw, "Liquid Redistribution during Liquid Phase Sintering," *J. Am. Ceram. Soc.*, **69**(1), 27–34 (1986).

12. *Process Mineralogy of Ceramic Materials*, W. Baumgart, A. C. Dunham, and G. C. Amstutz, Eds., Elsevier, New York, 1984.

13. M. F. Yan, *Sintering of Ceramics and Metals*, Advances in Powder Technology, Gilbert Y. Chin, Ed., American Society for Metals, Metals Park, OH, 1982.

14. W. H. Rhodes, "Agglomerate and Particle Size Effects on Sintering Yttria-Stabilized Zirconia," *J. Am. Ceram. Soc.*, **64**(1), 19–22 (1981).

15. F. H. Norton, *Fine Ceramics*, Robert E. Krieger, Malabar, FL, 1978.

16. George W. Scherer, "Sintering of Low Density Glasses: I, II, and III," *J. Am. Ceram. Soc.*, **60**(5), 236–246 (1977).

17. W. D. Kingery, H. K. Bowen, and D. R. Uhlmann, *Introduction to Ceramics*, Wiley-Interscience, New York, 1976.

18. J. E. Burke and J. H. Rosolowski, "Sintering," in *Treatise on Solid State Chemistry*, Vol. 4, N. B. Hannay, Ed., Plenum, New York, 1976.

19. R. J. Brook, "Controlled Grain Growth," in *Treatise on Materials Science and Technology*, Vol. 9, Franklin F. Y. Wang, Ed., Academic Press, New York, 1975.

20. J. H. Rosolowski and C. Greskovich, "Theory of the Dependence of Densification on Grain Growth during Intermediate Stage Sintering," *J. Am. Ceram. Soc.*, **58**(5–6), 177–182 (1975).

21. F. Thummler and W. Thomma, "The Sintering Process," *Metal. Mater.*, **12**(115), 69–108 (1967).

22. G. C. Kuczynski, N. A. Hooten, and C. F. Gibson, Eds., *Sintering and Related Phenomena*, Gordon and Breach, New York, 1967.

23. G. C. Kuczynski, "Self Diffusion in Sintering of Metallic Particles," *Trans. Am. Inst. Min. Met. Eng.*, **185**, 169–178 (1949).

PROBLEMS

29.1   Estimate the ash residue after thermolysis in a fired ceramic when using wax, PVA, and PEG binders in an amount of 4 wt % in an alumina compact.

29.2 Explain why the temperature for complete decomposition of the binder is commonly higher when products of larger size or finer particle size are fired along with the regular product.

29.3 Explain the differences in the thermal expansion behavior for the compact, extruded body, tape, and injection molded material shown in the diagram of thermolysis of the binder.

29.4 What are causes of shrinkage gradients causing warping when firing ceramic products.

29.5 Explain how the intermediate and final stages of densification for a constant rate of heating would be altered relative to common behavior if (1) a dopant that increases the diffusion rate of vacancies is added, (2) the initial compact density is extremely uniform in pore size, and (3) some extremely large pores are present in the compact.

29.6 The mean intercept length determined for a polished and etched section of a sintered material exhibiting regular grain growth is 1.3 μm. What is the mean grain size?

29.7 Describe microstructural defects that might be expected after sintering compacts containing (1) powder aggregates that densify more rapidly than surrounding material, (2) powder aggregates that densify less rapidly, (3) clusters of pores much larger than the grain size.

29.8 On Figure 29.14 showing grain size versus pore size, beginning at a pore size of 5 μm and a grain size of 0.5 μm, draw an arrow showing the change expected during solid-state sintering. Then explain why no exaggerated grain growth is expected with 1% impurity retarding grain growth.

29.9 Using equations, explain why faster firing to a higher temperature may often produce a product with an equal density but a smaller grain size.

29.10 An unfired glaze coating is 0.4 mm thick and 45% porous. What is its thickness after firing when the fired glaze is 10% porous?

29.11 What would be the thickness of the fired glaze in Problem 29.10 if 20% of the fired glaze had penetrated into the body?

29.12 Explain using the equation $\Delta P = 2\gamma/r$ why pore coalescence in the matrix of a material sintered in the presence of a liquid phase may produce negative densification during isothermal sintering (bloating).

29.13 During sintering, a material forms closed pores when the bulk density is 92%. Estimate the grain size when the pore diameter is 0.3 μm. What are your assumptions?

29.14 On one diagram, sketch fractional grain size distributions for a sequence of times for regular grain growth and also on the appearance of exaggerated grains.

29.15 A resistive film printed on a fired alumina substrate is 0.02 mm thick and 40% porous. The fired density of the film is 90%. Estimate the fired thickness of the film.

29.16 Explain the difference between the mechanisms of cratering and crawling produced during the firing of a glaze.

29.17 During sintering, closed pores form at a pressure of 1 atm and the pore radius is 0.5 μm. What is the final radius of the pores when the driving force is $\gamma = 300$ mN/m in the viscous material?

EXAMPLES

Example 29.1   An alumina substrate made by dry pressing has a green density after dying (inorganic basis) of $D_g = 2.39$ Mg/m^3 and a fired density of $D_f = 3.96$ Mg/m^3. What is the percent volume and percent linear shrinkage assuming shrinkage is isotropic? What is the density and porosity of the fired material?

Solution. The volume shrinkage Sv is calculated from the green and fired density as

$$Sv = (1/D_g - 1/D_f)100/(1/D_g) = (1 - D_g/D_f)100$$

$$Sv = (1 - 2.39/3.96)100 = 39.6\%$$

The fired density is related to the inorganic green density and isotropic fractional linear shrinkage $\Delta L/L_0$ as

$$D_f = D_g/(1 - \Delta L/L_0)^3 \quad \text{or} \quad \Delta L/L_0 = 1 - (D_g/D_f)^{1/3}$$

$$\Delta L/L_0(\%) = [1 - (2.39/3.96)^{1/3}]100 = 15.5\%$$

The percent porosity removed is equal to the percent volume shrinkage of 39.6%. The residual porosity ϕ is

$$\phi = (1 - D_f/D_u)100 = (1 - 3.96/3.98)100 = 0.50\%$$

The fired density is 99.5% of the ultimate density.

Example 29.2   What are the desirable particle characteristics for solid-state sintering to a high density and small grain size?

Solution. Desirable characteristics are a fine particle size, absence of agglomerates, equiaxed particles, and a narrow particle size distribution. Herring's scaling law (see Eq. 29.8) indicates $t_2 = (d_2/d_1)^n t_1$ to achieve a comparable linear shrinkage. Since n is about 3, an increase in the mean size of a powder from 0.5 to 5 μm will increase the sintering time by about 1000.

However, the benefit of a finer powder is realized only when the particle aggregates and agglomerates are well dispersed and the ultimate particles are uniformly packed in the compact. Aggregates may densify more rapidly than the average of the compact, coarsen, and produce an inhomogeneous microstructure. Elongated particles, when randomly oriented, retard the maximum density achievable on sintering; elongated grains which grow during the final stage may be beneficial for some properties. A wide particle size distribution favors grain coarsening and increases the concentration of grain growth inhibitor required to prevent separation of pores from grain boundaries and exaggerated grain growth.

Example 29.3   How are dopants selected to aid solid-state sintering?

Solution. Chemical additives may aid sintering by (1) promoting neck growth, (2) increasing the diffusion rate of the slower diffusing specie, and (3) retarding grain growth. Additives that promote neck growth are especially important for the pressureless sintering of nitrides and carbides; examples include yttria added to AlN and yttria and silica added to Si_3N_4. Doping that increases the concentration of oxygen vacancies may promote the sintering of an oxide; a slight excess of the divalent metal oxide has been shown to aid the densification of ferrites. Grain growth is inhibited when the dopant exceeds the solubility limit and a well-dispersed second phase exists at the grain boundaries; a dopant that increases the pore migration velocity may also retard exaggerated grain growth. The effect of the sintering atmosphere on the stoichiometry of the powder and the segregation and wetting of the second phase may be important. Because of the different roles and chemical complexities, the literature for each powder type should be consulted.

Example 29.4   What is an estimate of the volume of gas evolved for the thermolysis of 2 wt% polyvinyl alcohol PVA $[-CH_2-CHOH-]_n$ from an alumina powder compact having PF = 0.60. Assume that the gas is evolved at 350°C and the combustion product gases are CO_2 and H_2O. The MW of a mer of PVA is 44 g/mol mers.

Solution. The mass M of alumina in 1 cm^3 of compact is

$$M_{alumina} = 1 \ cm^3(3.98 \ g/cm^3)(0.60) = 2.39 \ g$$

The mass of PVA is

$$M_{PVA} = 0.02(2.39 \ g) = 0.048 \ g$$

The reaction is

$$[-CH_2-CHOH-] + 5/2 \ O_2 = 2CO_{2gas} + 2H_2O_{gas}$$

$$moles \ PVA = 0.048 \ g/44 \ g/mol = 0.0011 \ mol$$

and

$$\text{moles of gas} = 4(0.0011) = 0.0044 \text{ mol}$$

From $V = nRT/P$,

$$V = 0.0044 \text{ mol } (0.082 \text{ L atm/K mol})(623 \text{ K})/1 \text{ atm} = 0.22 \text{ L gas}$$

When the 5/2 mol O_2 comes from infiltration of outside air,

$$V_{O_2} = (2.5/4) \ 0.22 \text{ L} = 0.14 \text{ L}$$
$$V_{air} = 0.14 \text{ L } (1 \text{ L air}/0.21 \text{ L } O_2) = 0.67 \text{ L air}$$

The volume of air and product gas is large and will be larger when the PVA binder content is higher.

Example 29.5   Why does pore coalescence in a glassy phase cause an increase in porosity and negative densification?

Solution.   Consider two pores of radius r_f, containing n moles of gas, which during isothermal coalescence become one pore of radius r_c. From a simple equality of pore volume, $2V_f = V_c$, and for spherical pores,

$$2(\tfrac{4}{3}) \pi r_f^3 = \tfrac{4}{3} \pi r_c^3 \quad \text{and} \quad r_c = 1.26 r_f$$

However, the coalescence reduces the driving force $2\gamma/r$ for pore shrinkage in the viscous glass, and from the gas law $PV = nRT$, $nRT = P_f(2V_f) = P_c V_c$, or

$$2(2\gamma/r_f)[(\tfrac{4}{3})\pi r_f^3] = (2\gamma/r_c')[(\tfrac{4}{3})\pi r_c'^3] \quad \text{and} \quad r_c' = 1.41 r_f$$

During coalescence, pore enlargement may occur by viscous flow, and because the driving force is reduced, r_c' is larger than r_c. Negative densification during sintering is seen in Fig. 29.18.

Example 29.6   What is the driving force for densification when hot isostatic pressing (HIP) at 35 MPa pressure relative to the driving force for conventional sintering? Assume a pore diameter of 5 μm and $\gamma_s = 700$ mN/m at the densification temperature.

Solution.   For conventional sintering, the driving force $= 2\gamma/r_p$. In HIP, the driving force for pore shrinkage is $P_e + 2\gamma/r_p$. This is opposed by the pore pressure P_i (see Eq. 29.15). In conventional sintering,

$$\text{Driving force} = 2(700 \times 10^{-3} \text{ N/m})/2.5 \times 10^{-6} \text{ m} = 0.56 \text{ MPa}$$

For HIP, driving force $= P_e + 2\gamma/r_p = 35$ MPa $+ 0.56$ MPa $= 35.6$ MPa $P_{HIP}/P_{sinter} = 35.6/0.56 = 64/1$. The driving force in conventional sintering is not large. Large pores not removed by sintering may be reduced in diameter or eliminated by HIP at a modest pressure.

Example 29.7   How is the thermal expansion matched between the coating and the substrate?

Solution. The coefficient of thermal expansion (CTE) of the coating is commonly either matched to the CTE of the substrate, or the CTE of the coating is designed so that a compressive stress develops in the coating on cooling to the use temperature. Glass is more than 10 times stronger in compression than in tension, and for a glassy glaze it is common to develop a residual compressive stress in excess of about 70 MPa. With a compression glaze, the ceramic product is commonly stronger and the coating exhibits better resistance to crazing and chemical corrosion. With a compressed coating, the interface is subjected to a shear stress in proportion to the CTE mismatch and the adhesion of the coating must be strong.

Residual stress in the coating is developed on cooling below its softening temperature (approximately T_g). The stress in the coating may be estimated from the equation

$$\text{Residual stress} = CE(\Delta CTE)(T_g - T)$$

where C is a shape factor, E is the Young's modulus of the coating, ΔCTE is the thermal contraction mismatch per degree of temperature change and T is the use temperature. Reaction between the coating and the substrate may alter the CTE and the T_g of the coating. Ductile coatings of thin metals such as Cu can be produced without significant residual stress when ductile flow of the metal relieves the stress produced by the differential CTE.

Example 29.8   Why may a relatively fine particle size and a high binder content contribute to a crawling defect in the fired glaze?

Solution. When drying the coating, detachment and cracks may occur in some regions because of its low permeability K_p. On vitrification during firing, surface tension may cause the detached region to recede and thicken (crawl).

CHAPTER 30

MACHINING AND FINISHING PROCESSES

Some ceramic products are initially formed to an approximate shape and then machined to the final shape in a subsequent operation. The workpiece for machining may be in the green state, partially sintered, or densely sintered material. Machining is used to produce undercut recesses, contour surfaces, and thread surfaces, to increase the radius of corners, and to remove material from a region that may be of different density or contain a defect. Machining of dried or very porous sintered material, called green machining, is used to produce contoured surfaces on spark plug insulators and large, high-tension electrical insulators, and to shape bioceramic implants, wear-resistant inserts, and threaded structural components. Highly densified parts are machined to a final precise size, shape, and surface condition in the process of hard machining.

Surface finishing operations are used to improve the surface finish of a green or densified product and to remove material such as "flash" produced by mold seams when forming the product. Precision cutting and slotting operations are very common when producing rods, tubes, and so on, which must be cut to a very precise length after drying or firing, or cut into smaller units of very precise length.

Machining and surface finishing processes may permit the use of an alternative forming technique which alone cannot produce the shape of interest. Smoothing and machining processes may contribute importantly to both the yields and the productivity. Here we will discuss principles of green and hard machining and finishing processes.

30.1 CUTTING, TRIMMING, AND SMOOTHING

The trimming and machining of an unfired or slightly sintered ceramic depends on its mechanical strength and toughness. When the liquid/plasticizer content exceeds that for the leatherhard state, as in a slip cast or extruded part before drying, the yield strength is low and the frictional resistance is relatively high, and the cutting or trimming tool causes distortion and tearing. In the leatherhard state, shrinkage liquid has been removed, DPS < 1, the strength is higher, and the adhesion of the body to the cutting tool is greatly diminished. A leatherhard material with sufficient binder will exhibit limited plastic behavior and sufficient toughness, enabling trimming using a knife or a contouring tool, much like the machining of a metal. Reducing the temperature of the body, or the plasticizer content which reduces the T_g of the molecular binder, produces the brittle glassy state which has low toughness. The mechanical properties of green ceramics are discussed in Chapter 15.

Trimming the material in the leatherhard state is a common operation when producing porcelain products. Damp sponging is used to remove high spots and the paste fills in low spots and smooths the surface. This may be mechanized to cause the product to translate under rotating sponges. Flash and irregularities on the edges of parts may be removed by dry fettling which may be mechanized using an abrasive pad or belt, or a carbide tool. The surface of a fettled clay product is commonly polished in an automated, damp sponging operation. Burnishing is smoothing of the surface when platelike particles at the surface are forced into alignment by a rotating polishing tool pressed against the surface.

Particles are directly but weakly bonded in a material which has been sintered to the Initial Stage and cooled. Relative to the green state, the elastic stiffness has increased, but the toughness of the material is commonly lower. A porous material having sufficient strength but low toughness may be cut and smoothed by the attrition of particles from the surface.

30.2 MACHINING FUNDAMENTALS

Terms used for machining processes are abrasive machining or hard machining for the grinding of a dense fired ceramic, and green machining (or dry grinding) for a dried but unfired product or a very porous, slightly sintered material. Rough and semifinish grinding involve the removal of excess material. Honing is fine grinding used to improve flatness and parallelism of surfaces. Finish grinding uses a finer grit size and is used to remove furrows, surface cracks, and material with residual strain.

The goal of a machining process is to enable production of a final product which will meet size and shape specifications and have a high-quality surface. Because an abrasive grinding operation for dense ceramics commonly employs diamond tooling and is $> 10 \times$ more costly than green machining, it is desirable to remove material from the part before firing if possible. Computer-aided machining technology can be applied to upgrade the machining processes. A machining step may provide manufacturing flexibility, enabling the production of many final shapes from a small number of as-formed blanks. Machining may also reduce the tooling costs for forming and the lead time for the fabrication of tooling. Machining may also improve product quality when regions of different density can be machined away and when sharp radii of curvature, which may act as stress risers, are removed by machining. When hard machining, finish grinding can increase the final strength and is especially important when the optical and magnetic quality of the surface is very important.

An abrasive machining operation is shown in Fig. 30.1. Efficiency of surface grinding may be defined as the volume of material removed per unit time and depends directly on the depth of cut, wheel width, and feed rate of the workpiece:

$$\text{Grinding efficiency} = (\text{depth})\,(\text{width})\,(\text{feed rate}) \qquad (30.1)$$

The product of the depth of cut and feed rate is the specific material removal rate:

$$\text{Specific removal rate} = \text{Volume removed}/(\text{width} \cdot \text{time}) \qquad (30.2)$$

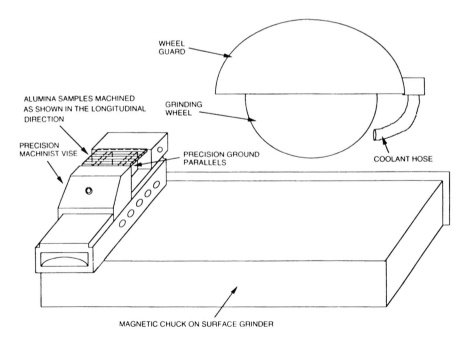

Fig. 30.1 Schematic of an abrasive machining operation. (Courtesy of Kristin Ploetz, Alfred University.)

The specific material removal rate is almost always achieved by using a small depth of cut and a high feed rate in conventional machining operations. In creep-feed grinding, the feed rate is low and the depth of cut is relatively high; this produces a relatively high contact area and the rigidity of the machine must be high to impart a large grinding force.

Machine parameters include the rigidity of the apparatus, vibration level, coolant if used, and the precision of tool movement. In addition to machine specifications, machining involves the cutting tool or abrasive specifications, workpiece material specifications, and operational parameters. Tool and abrasive specifications include the sharpness and angle of the working edge of the tool and the type, grit size and concentration, bond type, and geometry of the abrasive. Workpiece parameters include the geometry and consistency of the material, grain size and porosity, and its thermal and mechanical properties. Operational parameters include the dimensional precision and surface finish specifications. Considerable detail about machining operations for ceramics can be found in the first reference listed in the Suggested Reading.

30.3 GREEN MACHINING

Conventional grinding machines can be used for green machining. In rotary machining operations, the workpiece is fixed on a spindle and surface ground when rotating against a sharp, dense cutting tool or a contoured porous abrasive material (see Fig. 30.2). Grinding machines include vertical and horizontal lathes, mills, drill presses, and belt sanders. Power requirements for green machining are significantly less than for metal machining. Machining speeds vary depending on the strength and porosity of the green stock and the depth of cut. Dust collection is essential. Machinery and workers must be protected from the abrasive dust. Dust collection is also necessary to maintain tight tolerances (± 0.0125 mm). Closed systems are commonly used for high-speed machining.

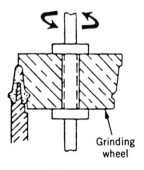

Grinding wheel

Contour grinding

Fig. 30.2 Contour grinding of unfired ceramic on a vertical lathe.

Tooling used may be single point when maximum setup flexibility and maximum material removal are needed. Sharpness of the working edge of the dense tool is a critical factor. An entire contour may be machined when a form cutting tool is used. The contour tool may be a shaped dense carbide tool or a porous abrasive tool which turns into the rotating green ceramic part. Spark plug insulators are commonly shaped using a contoured abrasive tool and large electrical insulators are contoured using a small sharp tool.

The tendency of binders in the green ceramic to build up in the pores of the abrasive is controlled by properly selecting the size of the grit and the porosity of the abrasive. The part must be held firmly in the fixture. Point contacts cause stress concentrations. Clamps must grip a wide area to reduce the contact stress. Hollow parts are commonly supported on a rotating mandrel. Pressure chucks and vacuum chucks may be used. The maximum machining speed is limited to that which produces an acceptable surface without cracks.

The most important parameter of the part being machined is its fracture strength. The workpiece must be strong enough to be resist the stresses in the grip and in machining without failure. The failure rate from fracture is high when the green strength of the workpiece is less than about 2 MPa. However, a high cutting resistance, short tool life, chatter, deflection, and catastrophic fracture may occur when the material is too strong. The workpiece strength depends on the strength of the binder and the volume of binder/volume of particles (V_b/V_p). For a partially sintered ceramic workpiece, the strength depends directly on its density and the strength of the sintered bond between particles (see Chapter 15). Green machining also depends on the particle sizes, the thermomechanical properties of the binder in the workpiece, and the temperature in the surface material during the machining. The binder type and amount influences the fracture toughness. Scrap rates increase when the strength and toughness are low because the part cannot withstand the forces imposed in the grinding process. The optimum strength will permit particle attrition without excessive force and catastrophic fracture. When using wax binders, their melting point and softness influences the grinding rate.

A broad particle size distribution (or granule size distribution in pressed compacts) containing some coarse particles may improve the grindability. Submicron particle sizes in the workpiece lead to problems due to loading of the wear product into the pores in the abrasive, and a dense cutting tool is commonly used. Particle loading in a porous tool increases with particle fineness, excessively fast removal rates, and high humidity. Flushing of the wheel with a liquid stream can sometimes be used when the liquid does not degrade the mechanical properties of the workpiece. Parameters such as microcracks, laminations, extremely large pores, and an inhomogeneous distribution of binder can impair the grinding behavior. Coarse particles in the material increase the grinding rate but also increase tool wear and reduce surface smoothness.

Parts containing a submicron average particle size are contoured using multiple-stage grinding or by turning on a lathe. Very large electrical porcelain insulators are contoured on a vertical lathe using a carbide or ceramic cutting tool. In bodies containing approximately 50% clay, the green strength is very dependent on the content of clay colloids and the moisture content in the material. Control of the moisture content below 3% and the microstructural homogeneity are essential for high-productivity contouring and a high-quality surface.

As-fired surface finishes achieved from the green machining operation are reported to range below a 2-μm finish using properly sharpened tools and sanding/sponging. A surface finish to below 0.5 μm is possible for dense fine-grained materials and well-controlled machining practice. Control of fired dimensions depends on maintaining control of the machined size and the firing shrinkage of the piece. Machined dimensions may be adjusted to compensate for small lot to lot variations in green density and firing shrinkage. Contact or noncontact probes are used to monitor dimensions.

30.4 HARD MACHINING

When grinding a metal, ductile behavior allows the abrasive particles to plow through the material and form well-defined furrows which are removed from the surface. Dense ceramics are brittle materials and material removal occurs by chipping and cracking produced by the action of the abrasive particles (Fig. 30.3). The abrasive interaction with the surface depends on both the grit shape and its bonding characteristics. A sharp, angular abrasive produces less structural damage. Diamond abrasive is commonly used for the initial cutting and grinding operations. Different modes of grinding using diamond grinding wheels are depicted in Fig. 30.4.

Cracks generated in the workpiece follow the path of least resistance, which is commonly along the grain boundaries. While an effective means of material removal, intergranular failure does not produce a smooth surface finish. Transgranular fracture or cleavage is also observed on machined surfaces. The trans-

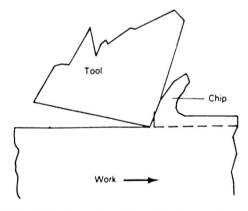

Fig. 30.3 Chip formation during abrasive cutting process.

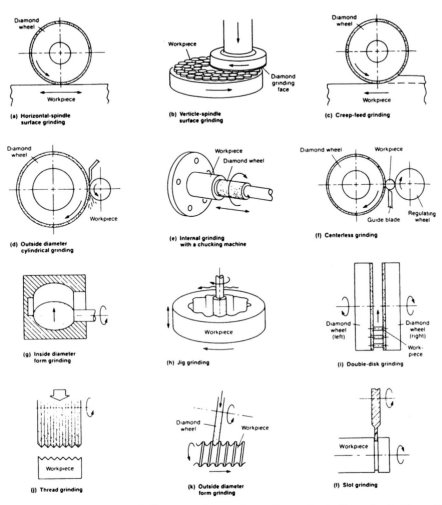

Fig. 30.4 Production grinding operations using a diamond grinding wheel. (From K. Subramanian and R. N. Kopp, *Engineered Materials Handbook*, Vol. 4, ASM International, Materials Park, OH, 1991.)

granular and intergranular fractures extend about two grain diameters below the surface in polycrystalline ceramics. Remnants of ductile flow are also observed. Deformed regions are on the order of a few microns. But isolated slip bands can sometimes extend >20 μm beneath the surface. The region of plastic flow is indicative of the extreme stress, temperature, and strain rate produced at the wheel–workpiece interface when chips are formed. Other factors that influence hard grinding are the hardness and crack resistance of the material, the distribution of microcracks and other defects, and the shape of the grains or particles being removed.

In hard machining, the structure and properties of the grinding wheel must be carefully controlled. Diamond wheel materials are well-standardized resinoid, metal, vitrified, and electroplated products. Resin-bonded wheels offer resilience which minimizes visually apparent, repeated grinding marks called grinding chatter. Metal-bonded wheels have a relatively long life because the diamond is held in a strong wear-resistant matrix. Vitrified wheels can be trued easily for the grinding of a complex form. The controlled porosity in vitrified wheels enables the coolant to be more effective. Electroplated wheels, which are a diamond-coated steel preform, are not widely used for ceramics.

Before hard machining, the wheel is balanced, trued, and dressed. The wheel is dynamically balanced on the machine to minimize vibrations. This will reduce chipping the workpiece. Trueing assures the geometric accuracy of the wheel face. Dressing, for example, cutting into silicon carbide stock, exposes the abrasive grit and aids the grinding action of the wheel.

Four types of surface interactions occur in the grinding zone between the workpiece and the grinding wheel, as shown in Fig. 30.5. The primary mechanism is the abrasion between the abrasive grain and the surface of the workpiece. Additionally, interactions at the chip–bond interface, chip–workpiece interface, and bond–workpiece interface occur. Maximizing the abrasive-workpiece interaction and reducing the three interactions between chips and the bond phase with the workpiece promotes efficient grinding.

The grinding direction has a significant influence on strength because stress concentrations produced when the product is loaded depend on the orientation of grinding artifacts. Strength is improved with an improved surface finish, but the effect is much larger when a sequence of ever-finer abrasive grits, which reduce the depth of surface damage, are used. The surface of a hard machined 99% alumina material is shown in Fig. 30.6.

Other machining techniques are ultrasonic machining, fluid jet machining, and laser machining. In ultrasonic machining, mechanical motion produced by

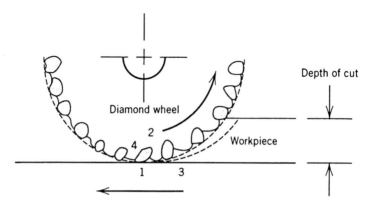

Fig. 30.5 Illustration of interactions between grinding wheel and workpiece: (1) abrasive/workpiece; (2) chip/bond; (3) chip/workpiece; (4) bond/workpiece. (From K. Subramanian, *Engineered Materials Handbook*, Vol. 4, ASM International, Materials Park, OH, 1991.)

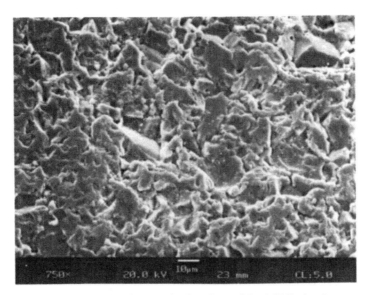

Fig. 30.6 Scanning electron micrograph of surface of fired 99% alumina ceramic after hard machining using 320 grit diamond wheel. (Photo courtesy of Kristin Ploetz, Alfred University.)

an electromechanical transducer causes abrasive grit in a slurry to abrade the surface at a high rate. The amplitude of the motion is very small, commonly <0.025 mm, but the frequency of 20 kHz produces rapid abrasion. The tooling is designed to resonate within the bandwidth of the transducer. Boron carbide grit, which has a relatively low phase density, is commonly used. Ultrasonic machining is used for the precise drilling of holes as small as 75 μm, holes of noncircular cross section, and for machining two- and three-dimensional shapes. An alternative to ultrasonic machining is abrasive jet machining in which a suspension or slurry of abrasive particles is forced through a nozzle to cut or machine the surface of a workpiece.

Laser scribing and laser beam machining are used to produce precision holes smaller than 0.3 mm in fired material. A pulsed laser melts the ceramic material and an orthogonal air jet blows the plume and molten debris away. A common scribe consists of 0.13 mm diameter holes one-half of the thickness spaced 0.15 mm apart. The part is subsequently snapped along the scribe line to produce two components. Laser machining may be used for precise contour machining, especially for small runs where using hard machine tools would be very expensive.

30.5 LAPPING

Final polishing is commonly conducted using an abrasive that produces a mechanochemical polishing action. Lapping is the polishing of a surface when

TABLE 30.1 Hardness of Abrasive Materials

Material	Mohs Hardness	Knoop Hardness
Zirconia	8	1200–1500
Calcined alumina	9	2100
Silicon carbide	9–10	2400
Boron carbide	9–10	2750
Boron nitride (cubic)	10	4500
Diamond	10	7000

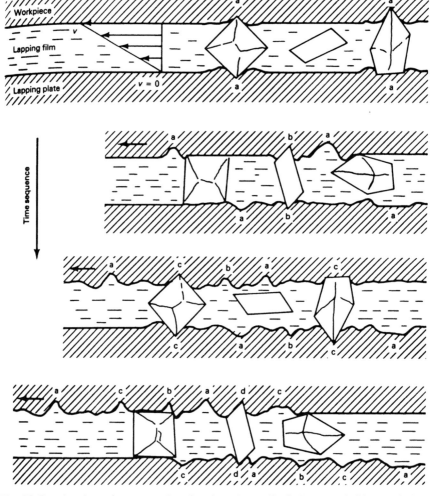

Fig. 30.7 Abrasive grit supported on lapping plate rolls, indents, and chips workpiece in lapping process. (From J. H. Indge, *Engineered Materials Handbook*, Vol. 4, ASM International, Materials Park, OH, 1991.)

TABLE 30.2   Surface Finishing Guidelines

Purpose	Abrasive		
	Size (μm)	Support	Bonding
Planar grinding			
Moderate rate, minimum damage	45–15	Hard surface	Free
Fast rate, more damage	125–40	Bonded disk	Fixed
Finish grinding	9–1	Medium soft platen	Semifixed
Polishing	0.25–0.10	Material-dependent	Semifixed

Source: After D. C. Zipperian, S. Chanat, and A. Trujillo, *Am. Ceram. Soc. Bull.*, **71**(7), 1077–1082, 1096–1098 (1992).

free abrasive particles entrained in a slurry supported by a lapping plate pass under the surface of the workpiece. The particles are of irregular shape and must have sharp edges and corners. The hardness of the abrasives used for lapping are listed in Table 30.1. These particles under low stress continuously indent and chip away microscopic fragments from the surface (Fig. 30.7). Surface finish improvement by lapping is accomplished by using a progressively finer grit size and carefully cleaning the surface after each treatment. Table 30.2 summarizes lapping guidelines. For flatness it is important that the abrasive be uniformly distributed across the surface of the lapping plate. In industrial lapping, the continuous flow of slurry must be applied evenly over the lapping area to maintain accuracy of flatness and finish.

SUMMARY

Unfired ceramic products that have sufficient mechanical strength and toughness may be trimmed and contoured while in the leatherhard state. Less tough material produced on eliminating liquid and plasticizer from the binder or by heating to the first stage of sintering is green machined using a hard, sharp cutting tool or a shaped, porous abrasive such as a grinding wheel. Green machining occurs by particle attrition and to be successful the strength of the interparticle bond relative to the bulk strength of the workpiece must be controlled within limits. Very dense ceramics are shaped in hard machining processes by the mechanism of chipping. Tooling containing wear-resistant diamond particles is commonly used for hard machining of ceramics. The grinding energy is relatively high. Ultrasonic machining and laser machining are used for producing precise slots and holes. Lapping is used to remove near-surface damaged material and to improve smoothness. Machining may aid manufacturing productivity by enabling the use of an alternative forming technique and by increasing the surface quality and/or the dimensional precision of the product.

SUGGESTED READING

1. D. C. Zipperian, S. Chanat, and A. Trujillo, "Ceramic Microstructural Analysis for Quality Control," *Am. Ceram. Soc. Bull.*, **71**(7), 1077–1098 (1992).
2. Dean Larson, "Green Machining," in *Engineered Materials Handbook*, Vol. 4, ASM International, Materials Park, OH, 1991, pp. 181–185.
3. K. Subramanian and S. Ramanath, "Principles of Abrasive Machining," in *Engineered Materials Handbook*, Vol. 4, ASM International, Materials Park, OH, 1991, pp. 315–328.
4. K. Subramanian and R. N. Napp, "Production Grinding Methods and Techniques," in *Engineered Materials in Handbook*, Vol. 4, ASM International, Materials Park, OH, 1991, pp. 336–350.
5. J. H. Indge, "Lapping, Honing, and Polishing," in *Engineered Materials Handbook*, Vol. 4, ASM International, Materials Park, OH, 1991, pp. 351–358.
6. *Intersociety Symposium on Machining of Advanced Ceramic Materials and Components*, American Society of Mechanical Engineering, New York, 1988.
7. J. G. Baldoni and S. T. Buljan, "Ceramics for Machining," *Am. Ceram. Soc. Bull.*, **67**(2), 381–387 (1988).
8. D. B. Quinn, R. E. Bedford, and F. L. Kennard, "Dry-Bag Isostatic Pressing and Contour Grinding of Technical Ceramics," in *Advances in Ceramics*, Vol. 9, J. A. Mangels, Ed., The American Ceramic Society, Westerville, OH, 1984, pp. 4–15.
9. *Machining Hard Materials*, D. R. Williams, Ed., The Society of Manufacturing Engineers, Dearborn, MI, 1982.

PROBLEMS

30.1 Explain why the trimming and green machining of a porcelain product depends on the concentration of water in the material.

30.2 Why does the green machining of an alumina ceramic containing a wax binder depend on the temperature created during machining?

30.3 In hard machining, how is the grinding force increased without increasing the frictional drag too much?

30.4 Explain why the grinding direction must be considered from the standpoint of the use of the product as well as convenience in surface machining.

30.5 Why is a moderate specific material removal rate common when hard machining ceramic products?

30.6 Why are diamond, aluminum oxide, and silicon carbide popular abrasives?

30.7 Why does the particular grinding action depend on the grit size, grit sharpness, and pressure of the abrasive against the workpiece?

30.8   Explain why the size of chip fragments depends directly on the fracture toughness of the workpiece material.

30.9   Why does an abrasive in a lapping compound have a relatively narrow particle size distribution?

30.10   What are mechanical differences between a grit particle that is a polycrystalline aggregate and a particle that is a single crystal?

EXAMPLES

Example 30.1   How does the grindability of ceramics and metals differ?

Solution. Grindability depends on the mechanical properties, thermal properties, and microstructure and composition of the workpiece material. Ceramics vary widely in strength. But ceramics have a higher Young's modulus and hardness than metals and deform less in grinding. Accordingly, ceramics may be ground to tighter tolerances in size, shape, and flatness. Ceramics are more stable chemically and on heating. The thermal conductivity of ceramics may range widely, as for metals and metal alloys. But ceramics are of much lower thermal shock resistance and fracture toughness, and their resistance to crack propagation is much lower.

Under the action of the abrasive, metals exhibit much plastic deformation and strain hardening before the formation of chips. When grinding most ceramics, microcracking with little plastic deformation precedes chip formation. The applied pressure and the size, cutting edge geometry, and hardness of the abrasive must be controlled to control the frequency of chipping for grinding efficiency and the size of chips for surface smoothness.

Example 30.2   Contrast grinding with lapping.

Solution. Grinding is a relatively high-energy process for stock material removal and is used to produce major changes in dimensions and shape. The grinding tool contacts only a portion of the working surface at one time and more structural damage occurs in the material below the abrasive. Lower pressures and a finer grit are commonly used for finish grinding.

Lapping involves a loose abrasive that is a sharp grit of irregular shape but of a carefully controlled size range. The grit may slide, roll, and inbed in the workpiece material during lapping. Sharp grit indents the surface, causing scratches, and produces microchips. In lapping, the entire working surface is polished. A finer grit size is used to produce a finer polish. Lapping is a relatively low-energy process used to produce minor alterations of shape and particularly to achieve high dimensional accuracy and surface finish.

APPENDIX 1

APERTURE SIZE OF U.S. STANDARD SIEVES

Sieve Number	Aperture (μm)	Sieve Number	Aperture (μm)
3.5[a]	5660	40	420
4	4760	45[a]	354
5[a]	4000	50	297
6	3360	60[a]	250
7[a]	2830	70	210
8	2380	80[a]	177
10[a]	2000	100[6]	149
12	1680	120[a]	125
14[a]	1410	140	105
16	1190	170[a]	88
18[a]	1000	200	74
20	841	230[a]	63
25[a]	707	270	53
30	595	325[a]	44
35[a]	500	400	37

[a]Recommended series.

APPENDIX 2

DENSITY OF CERAMIC MATERIALS

Material	Composition	Density (Mg/m^3)
Albite	$NaAlSi_3O_8$	2.61
Aluminum nitride	AlN	3.26
Anorthite	$CaAl_2Si_2O_8$	2.77
Andalusite	Al_2SiO_5	3.15
Anorthoclase	$KNaAl_2Si_6O_{16}$	2.58
Barium ferrite	$BaFe_{12}O_{19}$	5.31
Barium titanate	$BaTiO_3$	6.01
Beryllium oxide	BeO	3.0
Boron carbide	B_4C	2.5
Calcite	$CaCO_3$	2.71
Cordierite (beta)	$Mg_2Al_4Si_5O_{18}$	2.6
Corundum	Al_2O_3	3.98
Forsterite	Mg_2SiO_4	3.22
Galena	PbS	7.5
Graphite	C	2.25
Halloysite	$Al_2Si_2O_5(OH)_4 \cdot 2H_2O$	2.62
Hematite	Fe_2O_3	5.25
Ilmenite	$FeTiO_3$	4.7
Kaolinite	$Al_2Si_2O_5(OH)_4$	2.61
Kyanite	Al_2SiO_5	3.60
Lime	CaO	3.3
Magnesite	$MgCO_3$	2.96
Microcline	$KAlSi_3O_8$	2.58
Montmorillonite	$(Al_{1.67}Na_{0.33}Mg_{0.33})(Si_2O_5)_2(OH)_2$	2.50
Mullite	$Al_6Si_2O_{13}$	3.23

Material	Composition	Density (Mg/m^3)
Muscovite	$Al_2K(Si_{1.5}Al_{0.5}O_5)_2(OH)_2$	2.9
Orthoclase	$KAlSi_3O_8$	2.55
Periclase	MgO	3.58
Pyrite	FeS_2	5.02
Pyrophyllite	$Al_2(Si_2O_5)_2(OH)_2$	2.80
Quartz	SiO_2	2.65
Rutile	TiO_2	4.26
Silica, vitreous	SiO_2	2.20
Silicon carbide	SiC	3.21
Silicon nitride	Si_3N_4	3.2
Sillimanite	Al_2SiO_5	3.23
Spinel	$MgAl_2O_4$	3.60
Spodumene (beta)	$Li_2Al_2Si_4O_{12}$	2.35
Spodumene (alpha)		3.20
Talc	$Mg_3Si_4O_{10}(OH)_2$	2.75
Titania (see rutile)		
Tungsten carbide	WC	15.7
Zinc oxide	ZnO	5.68
Zircon	$SiZrO_4$	4.7
Zirconia, stabilized	$ZrO_2(8\%Y_2O_3)$	6.0

VISCOSITY AND DENSITY VALUES OF WATER AND AIR

Temperature (°C)	Viscosity of Water (mPa · s)	Density of Water (Mg/m^3)	Density of Air (Mg/m^3)
15	1.1404	0.999099	0.001226
16	1.1111	0.998943	0.001221
17	1.0828	0.998774	0.001217
18	1.0559	0.998595	0.001213
19	1.0299	0.998405	0.001209
20	1.0050	0.998203	0.001205
21	0.9810	0.997992	0.001201
22	0.9579	0.997770	0.001196
23	0.9358	0.997538	0.001192
24	0.9142	0.997296	0.001188
25	0.8937	0.997044	0.001184
26	0.8737	0.996783	0.001180
27	0.8545	0.996512	0.001176
28	0.8360	0.996232	0.001173
29	0.8180	0.995944	0.001169
30	0.8007	0.995646	0.001165

APPENDIX 4

CONVERSION FROM METRIC TO ENGLISH/AMERICAN UNITS

Parameter	From SI Units	Multiply by	To Get
Length	m	39.370	Inch
		3.2808	Foot
Area	m^2	1.5500×10^3	$Inch^2$
		10.764	$Foot^2$
Volume	m^3	6.1024×10^4	$Inch^3$
		35.315	$Foot^3$
		1.0567×10^3	Liquid quart (U.S.)
Mass	kg	2.2046	Pound
		1.1023×10^{-3}	Ton
Density	kg/m^3	3.6127×10^{-5}	$Pound/inch^3$
		6.2428×10^{-2}	$Pound/foot^3$
Speed	m/s	3.2808	Foot/second
Acceleration	m/s^2	3.2808	$Foot/second^2$
Force	N	0.22481	Pound force
		10^{5a}	Dyne
Pressure, stress	Pa or N/m^2	1.4504×10^{-4}	Psi
		10^a	$Dyne/cm^2$
		2.0885×10^{-2}	$Pound/foot^2$
		7.5006×10^{-3}	Torr
Energy	J	9.4845×10^{-4}	Btu
		0.23901	Calorie
		10^{7a}	Erg
		2.7778×10^{-7}	Kilowatt hour

Parameter	From SI Units	Multiply by	To Get
Power	W	9.4845×10^{-4}	Btu/second
		3.4144	Btu/hour
		0.23901	Calorie/second
		1.3410×10^{-3}	Horsepower
Heat capacity	$J/(kg \cdot K)$	2.3901×10^{-4}	$Cal/(g\,°C)$
		2.3901×10^{-4}	$Btu/(lb\ °F)$
Thermal conductivity	$W/(m \cdot K)$ or	6.9380	Btu in./(h ft^2 °F)
	$J/(s \cdot m \cdot K)$	2.3901×10^{-9}	$Cal/(s\ cm\ °C)$
		0.57816	Btu/(h ft °F)

[a]These factors are exact; others to five digits.

APPENDIX 5

ANSWERS TO SELECTED PROBLEMS

CHAPTER 1

1.1   Uniformity is sameness of characteristics within part; reproducibility is sameness within a set of parts and between sets

1.2   Performance is response of material/product to several/complex energy fields. Reliability is consistency of performance of material/product or production lot over time. Quality is a valuation that includes performance and reliability but also design, aesthetic, and cost factors

1.5   Better controls and improved technology

1.6   See Fig. 21.3 middle

1.7   Increased technical competence through education

1.10 Probably the ability to analyse microstructure and improve control of processing

CHAPTER 2

2.1 $r = 10.3$ nm

2.2   230 mN/m

2.4   90°

2.5   46 kPa

2.7   7.3 MPa

2.9   2 h

CHAPTER 3

3.1 crude-ind. inorg. chemical
   ind. mineral-ind. mineral
   ind. inorg. chemical-ind. mineral
   ind. inorg. chemical-ind. inorg. chemical
   ind. inorg. chemical-ind. inorg. chemical

3.4 differential oxidation pressure
   less volatilization in closed sagger
   maybe lower temperature in sagger

3.6 interdiffusion for reaction also produces partial sintering

3.7 $BaCO_3$- 846 g; TiO_2- 343 g

CHAPTER 4

4.1 conventional reaction T > 1250°C
   titanyl oxalate process T < 700°C

4.3 spray pyrolysis-solution droplets are dried and salt/compound decomposed; vapor phase reaction-mixed gases react and particulate product forms within product gas.

4.5 precipitation-control stoichiometry of salt and microhomogeneity; calcination-microhomogeneity may be lost when salt dissolves in its water of crystallization

4.9 8,930 L

CHAPTER 5

5.1 1% = 10,000 ppm

5.3 12.3 and 6.5

5.5 for capacitor grade 1500 ppm SO_3 and 900 ppm CO_2 undecomposed from raw materials

5.7 PCE and MOR are properties

5.9 some dye remains trapped within halloysite tubes

CHAPTER 6

6.1 purity = 99.976 wt%

6.2 Al_2O_3/ZnO = 0.997; 1.000

6.3 emission spectrometry (ICP)

6.7 chem analysis, XRD (lattice parameter), TGA

6.10 XRD

CHAPTER 7

7.3 nickel $\psi_A = 1.86$; $\psi_V = 0.072$; $\psi_A/\psi_V = 25.9$

7.5 50 μm

7.9 summation $\bar{a}_L = 53.1$, $\bar{a}_A = 57.5$, $\bar{a}_V = 61.3$, $\bar{a}_{V/A} = 69.7$ μm
equations $\bar{a}_L = 53.8$, $\bar{a}_A = 57.4$, $\bar{a}_V = 61.2$, $\bar{a}_{V/A} = 69.5$ μm

7.15 $\bar{a}_{gN} = 2.1$, $\bar{a}_{gM} = 15.8$, $\bar{a}_{V/A} = 8.7$ μm

7.16 gravity-7 day, 3 h, 36 min; centrifuge-6 min, 47 s

7.18 1 mm-1.87 mg, 0.27×10^{20} fu; 1 nm-1.87×10^{-21} g, 27 fu

7.21 2AR + 4

CHAPTER 8

8.1 total porosity = 25.0%; closed porosity = 5.0%

8.2 3.66 Mg/m^3

8.4 5.29 Mg/m^3

8.6 conventional 830 nm; chem prep 97 nm

8.7 barium titanate 0.47, zircon 1.48 m^2/g

8.10 kaolinite 9.2, montmorillonite 770 m^2/g

8.13 GA-P 10.4 wt%; GA-C 2.7 wt%

CHAPTER 9

9.2 TDS = 224 ppm

9.4 hard water with more impurity dissolves less CO_2 from air

9.8 0°; spreads

9.9 13.5 in water; 3.7 in oil

CHAPTER 10

10.1 halo surrounding particles

10.2 less hydrated when pH > PZC

10.5 n = 120; L = 37 nm

10.6 0.96 nm

10.9 a > 300 nm

10.10 zeta potential = 28.8 mv

CHAPTER 11

11.1   DS = 0.12
11.6   MW = 55,000, n = 156, L = 184 nm; MW = 295,000, n = 838, L = 989 nm
11.8   L = 372 nm
11.11   for 8 wt% solution, 0.42 wt%
11.14   wax 400, PE 100, alumina 6 ppm/K
11.18   58.8 wt% water

CHAPTER 12

12.1   midpoint of leathery region
12.4   0% glycerol, $-28°C$; 6.4°% glycerol, $-17°C$
12.5   3.0 nm vs. 2.0 nm
12.6   Zn stearate for ferrite; Na stearate for porcelain

CHAPTER 13

13.1   CN = 9.0 for PF = 0.65
13.2 N_p = about 1 billion/cm^3 for size of 10 μm
13.5   0.84
13.6 W_c = 67 wt%, W_f = 33 wt%
13.9   PF = 0.938; W_c = 63.7, W_m = 25.7, W_f = 10.6 wt%; intersticial porosity = 6.2%, total porosity = 14.1%
13.10   BT-PF(%) = 63.5; Z-PF(%) = 64.5

CHAPTER 14

14.1   0.44
14.4   2.5%
14.5   3.2%/1 MPa
14.9   alumina 100.00, PA 4.26, PVA 4.26, PEG 0.56, water 3.25 vol%
14.11 Si_3N_4 68.87, water 27.12, PA 2.87, PVA 1.14 wt%
14.13   AlN 84.4, PP 5.4, wax 8.6, stearate 1.6 wt%

CHAPTER 15

15.1   DPS > 1; u = P_a; effective stress = 0
15.3   6,703 MPa

15.5   L(surfactant) $\ll$ roughness
15.9   0.6
15.11 0.30
15.13 3.5 MPa
15.15 4.9 MPa
15.17 1.1%
15.19 0.86

CHAPTER 16

16.1   open packed—top, bottom, two midplanes
       closed packed—none unless volume dilation
       crowded—top, bottom, midplanes only with particle rotations
16.2   0.68 mPa s at 40°C
16.6   shear rate (b = 1.6 cm) at a is 6230 s^{-1} and at b 5473 s^{-1}
16.7   viscosity = 3×10^6 mPa s; shear rate at a = 173 s^{-1}
16.8   yield stress = 4.75 Pa; plastic viscosity = 6.4 mPa s
16.12 2.51 Pa
16.15 dynamic volume/static volume = 19 for larger prism
16.18 5.2; 11.8

CHAPTER 17

17.2   0.64/1.00/1.03/1.53/2.17/4.33
17.3   81/1
17.6   6.3/1
17.8   beads produce high N_c for attrition but low grinding force
17.10 temperature may alter wetting, deflocculation, viscosity, dissolving in
       slurry and dry adhesion
17.12 coarser alumina can act as fine media for grinding soft talc
17.13 dia = 12.5 cm

CHAPTER 18

18.1   25°
18.3   vibrations may increase PF (see Fig. 13.9) and f
18.5   belt-higher capacity, not enclosed, coarse-fines, abrasive particles okay
18.8   S = 2.08 and 1.71; 95% confidence intervals overlap. More samples
       are needed
18.10 $M_{max} = 1-(B/A)\phi_u$

CHAPTER 19

19.2 3.4 h
19.3 $55{,}000 \times 10^{-18}$ m^2
19.4 0.64 cm/s; $R_e = 0.0064$
19.5 110×10^{-16} to 160×10^{-18} m^2
19.8 47×10^{-16} m^2

CHAPTER 20

20.3 42.3% for 70 wt%
20.5 13.3%
20.6 10/1; 4/1
20.8 alumina 55% vs. 25%

CHAPTER 21

21.5 strength at low concentration, shear thinning rheology, low migration on drying
21.6 IM-flow but solidification on cooling without drying
21.8 to avoid flow-induced orientation and anisotropy
21.13 4.4%
21.14 26.0% and 8.2%
21.16 Sv = 0.412 and 0.381; L = 8.38 and 8.52 cm

CHAPTER 22

22.3 0.58
22.6 30%
22.9 in fill-intergranular porosity = 45.3%, granules are 48% porous
22.11 40% decrease
22.14 74 MPa
22.15 single-action = 69, double-action = 76, dry bag = 82 MPa
22.19 10; capping is expected

CHAPTER 23

23.3 stress-body = 520 kPa, stress-wall = 30 kPa
23.4 $100V_b/V_p = 12/1$

23.5 for diameter 1.6/1; for length 2.0/1

23.6 body viscosity $= 37 \times 10^6$ mPa s;
film viscosity $= 6 \times 10^3$ mPa s

23.12 higher for shear flow in bulk before drainage and increase in shear resistance

CHAPTER 24

24.4 to control shear resistance and dispersion of agglomerates which depends on viscosity of liquids, to avoid degradation of organics

24.5 both

24.6 temperature effect is 4×10^6 mPa s; shear rate effect is 2×10^6 mPa s

24.8 $K = 1 \times 10^3$ Pa $s^{0.5}$

24.9 $K_1 = 0.89$; $f_{cr}^v = 0.765$

CHAPTER 25

25.1 0.50

25.2 0.46 Pa

25.3 porosity $= 62\%$; suction $= 97$ kPa

25.5 20 h for pressure casting

25.11 4 and 44 nm

25.13 $R_c = 6.9 \times 10^{16}$ m^{-2}

CHAPTER 26

26.1 2.8 μm

26.2 $100V_b/V_p = 23.9$ for titanate

26.4 600 s^{-1}

26.10 13.6%

CHAPTER 27

27.2 42 nm

27.3 0.4% versus 26.7%

27.7 5.5%

27.10 near point C water film breaks and menisci recede below surface in some regions

27.12 12.1% volume change; DPS $= 0.87$

CHAPTER 28

28.1 $ZrCl_{4(gas)} + 2H_2O_{(gas)} = ZrO_{2(solid)} + 4\,HCl_{(gas)}$

28.4   drainage thins the coating; higher velocity reduces drainage

28.6   tension increases with snap off distance

28.10   gelation increases the yield strength, viscosity, and pseudoplasticity

28.12   cratering is caused by impurity under coating; orange peel is caused by incongruent vaporization and surface tension gradients

CHAPTER 29

29.1   wax-600 ppm; PEG-24 ppm

29.6   2.0 μm

29.10   0.24 mm

29.11   0.20 mm

29.13   1.5 μm

29.15   13 μm

29.17   0.14 μm

CHAPTER 30

30.1   moisture in contact region influences bond strength, Young's modulus, and mean strength

30.3   forcing sharp, angular grit into workpiece

30.4   direction of residual scratches may influence mechanical and optical performance of product

30.6   diamond—highest hardness; SiC and alumina—very high hardness and very high hardness/cost

30.10   polycrystalline grit may have a lower Young's modulus but may fracture along grain boundaries exposing second generation sharp, angular grit

INDEX